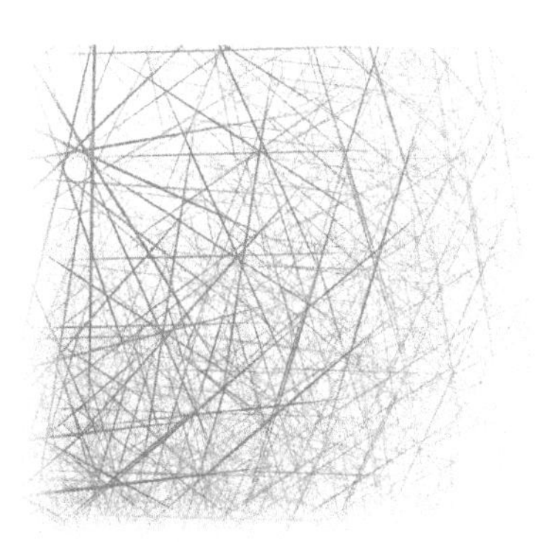

Stable Perturbations of Operators and Related Topics

Stable Perturbations of Operators and Related Topics

Yifeng Xue

East China Normal University, China

NEW JERSEY • LONDON • SINGAPORE • BEIJING • SHANGHAI • HONG KONG • TAIPEI • CHENNAI

Published by

World Scientific Publishing Co. Pte. Ltd.
5 Toh Tuck Link, Singapore 596224
USA office: 27 Warren Street, Suite 401-402, Hackensack, NJ 07601
UK office: 57 Shelton Street, Covent Garden, London WC2H 9HE

British Library Cataloguing-in-Publication Data
A catalogue record for this book is available from the British Library.

STABLE PERTURBATIONS OF OPERATORS AND RELATED TOPICS

ISBN-13 978-981-4383-59-2
ISBN-10 981-4383-59-7

Printed in Singapore.

To My Wife and My Son

Preface

The concept of generalized inverses was first introduced by Fredholm in his paper [Fredholm (1903)] in 1903. In order to solve some integral equations, he defined a pseudo–inverse of an integral operator. The generalized inverses of matrices was first introduced by Moore in his paper [Moore (1920)]. He defined a unique generalized inverse in terms of projectors of matrices. The concept of the generalized inverse of a unbounded operator on Hilbert spaces was first introduced by Tseng. In 1955, Penrose presented his new definition of Moore's generalized inverse in [Penrose (1955)]. This work promoted the study of generalized inverse of both matrices and operators on Hilbert spaces.

The Drazin inverse was introduced by Drazin in the context of semigroups and rings. It was later generalized to the sets of bounded (or unbounded) operators on Banach spaces and Banach algebras respectively.

The generalized inverse and the Drazin inverse of matrices or operators have been widely used in many research fields such as Statistics, Partial Differential Equations, Theory of Optimum Control and Approximate Theory etc.. This makes the theory of generalized inverses and Drazin inverses progress greatly. The development of this theory includes:

(1) The continuity of the Moore–Penrose inverse and Drazin inverse for both matrices and operators has been characterized and

(2) The error–bounds of the perturbation analysis of the Moore–Penrose inverse and the Drazin inverse for matrices and bounded operators have been given.

This Book is based on Guoliang Chen and the author's work on the perturbation analysis for the generalized inverse, Moore–Penrose inverse and Drazin inverse of operators under the stable perturbation. The book consists of six chapters. In Chapter 1, we introduce some basic and important results from Linear Functional Analysis, Banach Algebras and C^*–Algebras. We also present the relationships among the densely–defined operator with closed range, the reduced minimum modulus of the densely–defined operator and the δ–function of ranges (resp. kernels) of such two operators. In Chapter 2, we present the stable perturbation theory of densely–defined operators on Banach spaces or elements in a Banach algebra with with generalized inverses. In Chapter 3, we introduce the Moore–Penrose inverse of a densely–defined operator with closed range on Hilbert spaces and give the reverse order law of Moore–Penrose inverses and then present error–bounds of perturbation of the Moore–Penrose inverse and the least square problem under stable perturbation. We also discuss the Moore–Penrose inverse of elements in a C^*–algebra in this chapter. Chapter 4 is devoted to the group inverse, Drazin inverse and generalized Drazin inverse of operators on a Banach space and in a Banach algebra and their perturbation analyses under the stable perturbation. Chapter 5 presents applications of the stable perturbation theory of generalized inverses in the perturbation of generalized Bott–Duffin inverses and in the estimations of the condition numbers as well as in the analyses of certain holomorphic mappings and C^1–mappings. We present perturbation analysis of the reduced minimum modulus and a theory of Fredholm elements and their indices and discuss the approximately polar decomposition property of a C^*–algebra in last chapter.

The author would like to express his gratitude to Professor Guoliang Chen for his kindly help and sincere cooperation for long. He would like to thank Professor Yimin Wei and Professor Qingxiang Xu for their helpful discussions with the author. The author would like to thank Associate Professor Fapeng Du who read the manuscript and made various comments. The author is indebted to his wife Ying Lu for her typesetting half of the manuscript. Finally, the author would like to acknowledge that this book was written while he was partially supported by Natural Science Foundation of China (no.10771069) and Shanghai Leading Academic Discipline Project (no.B407).

Yifeng Xue

Contents

Chapter 1

Basics In Functional Analysis

In this chapter, we will overview some important results in linear functional analysis, Banach algebras as well as C^*–algebras and will give a complete description of the reduced minimum modulus of a densely–defined operator and the gap function of linear subspaces. For convenience, we may assume that X, Y, Z are Banach spaces over complex filed $\mathbb{C}$ and H, K, L are Hilbert spaces over complex field $\mathbb{C}$ throughout the chapter.

1.1 Banach spaces and Hilbert spaces

Let $B(X,Y)$ denote the linear space of all bounded linear operators from X to Y. $B(X,Y)$ becomes a Banach space under the norm

$$\|T\| = \sup\{\|Tx\| \mid \|x\| \le 1,\, x \in X\}, \quad \forall T \in B(X,Y).$$

When $Y = \mathbb{C}$, $B(X,Y)$ is called to be the dual space of X, denoted by X^*. When $X = Y$, we set $B(X) = B(X,X)$.

Set $S(X) = \{x \in X \mid \|x\| = 1\}$ and $D(X) = \{x \in X \mid \|x\| \le 1\}$. Let $x_0 \in X$, $\delta > 0$ and set

$$O(x_0,\delta) = \{x \in X \mid \|x - x_0\| < \delta\}, \quad D(x_0,\delta) = \{x \in X \mid \|x - x_0\| \le \delta\}.$$

For a subset A in X, we write $\overline{A}$, A° to denote the closure and the internal of A respectively.

Proposition 1.1.1. *Let A be a subset of X. Put*

$$f_A(x) = \operatorname{dist}(x,A) \triangleq \inf\{\|x - z\| \mid z \in A\}, \quad \forall x \in X.$$

(1) *$x \in \overline{A}$ iff $f_A(x) = 0$.*
(2) *$f_A(x) = f_{\overline{A}}(x)$, $\forall x \in X$.*
(3) *If A is a subspace of X, then f_A is a semi–norm on X.*

(4) *If A is a convex set, then f_A is a convex function.*
(5) $|f_A(x) - f_A(y)| \leq \|x - y\|, \forall\, x, y \in X$.

Proof. (1) If $x \in \overline{A}$, then there is $\{x_n\} \subset A$ such that $\|x - x_n\| \to 0$ as $n \to \infty$. Thus, $f_A(x) \leq \operatorname{dist}(x, x_n) \to 0$ $(n \to \infty)$, i.e., $f_A(x) = 0$. Conversely, when $f_A(x) = 0$, we can find $\{y_n\} \subset A$ such that

$$0 = \operatorname{dist}(x, A) > \|x - y_n\| - \frac{1}{n}, \quad n = 1, 2, \cdots.$$

So $\|x - y_n\| \to 0$ $(n \to \infty)$ and hence $x \in \overline{A}$.

(2) Clearly, $f_{\overline{A}}(x) \leq f_A(x), \forall\, x \in X$. Now for any $\epsilon > 0$, there is $a \in \overline{A}$ such that $f_{\overline{A}}(x) > \|x - a\| - \epsilon$. Choose $b \in A$ such that $\|a - b\| < \epsilon$. Then

$$f_{\overline{A}}(x) > \|x - b - (a - b)\| - \epsilon > \|x - b\| - 2\epsilon \geq f_A(x) - 2\epsilon$$

and hence $f_{\overline{A}}(x) \geq f_A(x)$ by letting $\epsilon \to 0$.

(3) Since $0 \in A$, $f_A(0\, x) = 0 = 0\, f_A(x), \forall\, x \in X$. When $\alpha \neq 0$, we have

$$f_A(\alpha x) = \inf\{\|\alpha x - z\| \,|\, z \in A\} = |\alpha| \inf\{\|x - \alpha^{-1} z\| \,|\, z \in A\} = |\alpha| f_A(x)$$

since A is a linear subspace. Let $x, y \in X$. Then for any $z_1, z_2 \in A$,

$$f_A(x + y) \leq \|x + y - (z_1 + z_2)\| \leq \|x - z_1\| + \|y - z_2\|.$$

This means that $f_A(x + y) \leq f_A(x) + f_A(y)$. So f_A is a semi–norm on X.

(4) Let $x, y \in X$ and $\alpha \in [0, 1]$. Then for any $z_1, z_2 \in A$,

$$\begin{aligned} f_A(\alpha x + (1 - \alpha) y) &\leq \|\alpha x + (1 - \alpha) y - (\alpha z_1 + (1 - \alpha) z_2)\| \\ &\leq \alpha \|x - z_1\| + (1 - \alpha)\|y - z_2\| \end{aligned}$$

and consequently, $f_A(\alpha x + (1 - \alpha) y) \leq \alpha f_A(x) + (1 - \alpha) f_A(y)$.

(5) For any $z \in A$, we have

$$f_A(x) = \operatorname{dist}(x, A) \leq \|x - z\| \leq \|x - y\| + \|y - z\|.$$

It follows that $f_A(x) \leq \|x - y\| + f_A(y)$. Similarly, $f_A(y) \leq \|x - y\| + f_A(x)$. Then the assertion follows. □

Definition 1.1.2. Let W be a linear space over field $\mathbb{R}$ (or $\mathbb{C}$) and let $p\colon W \to \mathbb{R}_+$ be a function, where $\mathbb{R}_+ = \{x \in \mathbb{R} \,|\, x \geq 0\}$. p is called to be a sub–functional if p satisfies following conditions:

(1) $p(\alpha x) = \alpha p(x), \forall\, x \in W$ and $\alpha > 0$.
(2) $p(x + y) \leq p(x) + p(y), \forall\, x, y \in W$.

p is called to be a semi–norm if p satisfies (2) and
(1') $p(\alpha x) = |\alpha|\, p(x), \forall\, x \in X$ and $\alpha \in \mathbb{R}$ (or $\mathbb{C}$).

Note that from (1) of Definition 1.1.2, we can deduce that $p(0) = 0$.

Theorem 1.1.3 (Hahn–Banach Theorem). *Let W be a linear space over $\mathbb{R}$ or $\mathbb{C}$ and V be a subspace of W.*

(1) *Assume that W is a real linear space and p is a sub-functional on W. Let f be linear functional on V such that $f(x) \leq p(x)$, $\forall\, x \in V$. Then there is a linear functional F on W such that $F(x) = f(x)$, $\forall\, x \in V$ and $F(x) \leq p(x)$, $\forall\, x \in W$.*
(2) *Assume that W is a complex linear space and p is a semi-norm on W. Let f be linear functional on V such that $|f(x)| \leq p(x)$, $\forall\, x \in V$. Then there is a linear functional F on W such that $F(x) = f(x)$, $\forall\, x \in V$ and $|F(x)| \leq p(x)$, $\forall\, x \in W$.*

Applying Theorem 1.1.3 (2) to the set of Banach spaces, we have

Proposition 1.1.4. *Let V be a closed subspace of X and f be a linear functional on V. There there exists $F \in X^*$ such that $F(x) = f(x)$, $\forall\, x \in V$ and $\|F\| = \|f\|_V$, where $\|f\|_V = \sup\{|f(x)|\,|\, x \in D(V)\}$.*

By using Proposition 1.1.4, we can obtain following corollary, which is widely used.

Corollary 1.1.5. *Let $x_0 \in X$ and $x_0 \neq 0$.*

(1) *There is $f \in X^*$ such that $f(x_0) = \|x_0\|$ and $\|f\| = 1$.*
(2) *Let V be a closed subspace of X and $x_0 \in X\backslash V$. Then there is $f \in X^*$ such that*

$$f(x) = 0,\ \forall\, x \in V,\ f(x_0) = 1 \text{ and } \|f\| = \frac{1}{\operatorname{dist}(x_0, V)}.$$

By means of Theorem 1.1.3 (1), we can deduce one of Mazur's theorems as follows.

Theorem 1.1.6. *Let V be a closed convex subset in X, which contains 0. Let $x_0 \in X\backslash V$. Then there is $f \in X^*$ such that $\operatorname{Re}(f(x_0)) > 1$ and $\operatorname{Re}(f(v)) \leq 1$, $\forall\, v \in V$.*

Proof. By Proposition 1.1.1 (1), $d = \operatorname{dist}(x_0, V) > 0$. Put

$$U = \{x \in X\,|\, f_V(x) < d/2\}.$$

Then $0 \in V \subset U$ and U is open and convex by Proposition 1.1.1. Furthermore, we can find a $\delta > 0$ such that $O(0, \delta) \subset U$. Thus, for any $x \in X\backslash\{0\}$,

$\dfrac{\delta}{2\|x\|}x \in U$. Set $P_U(x) = \inf\{\lambda |\, \lambda > 0,\ \lambda^{-1}x \in U\}$. Obviously,

$$P_U(v) \leq 1,\ \forall\, v \in V,\ P_U(0) = 0 \text{ and } P_U(x) \leq \frac{2}{\delta}\|x\|,\ \forall\, x \in X\backslash\{0\}.$$

Moreover, we also have $P_U(x_0) > 1$. In fact, if $P_U(x_0) \leq 1$, then for any $\epsilon > 0$, we can find $\lambda_0 > 0$ such that $\lambda_0^{-1}x_0 \in U$ and $1 \geq P_U(x_0) > \lambda_0 - \epsilon$. Since U is convex and $0 \in U$, we have

$$(1+\epsilon)^{-1}x_0 = \frac{\lambda_0}{1+\epsilon}(\lambda_0^{-1}x_0) + \Big(1 - \frac{\lambda_0}{1+\epsilon}\Big)0 \in U.$$

Thus, $f_V((1+\epsilon)^{-1}x_0) < d/2$ and so that $f_V(x_0) \leq d/2$ by letting $\epsilon \to 0$. But this is impossible because $f_V(x_0) = d$.

We now show that P_U is a sub–functional on X. Let $\alpha > 0$ and $x,\ y \in X$. Then

$$\begin{aligned} P_U(\alpha x) &= \inf\{\lambda |\, \lambda > 0,\ \lambda^{-1}(\alpha x) \in U\} \\ &= \alpha \inf\{\alpha^{-1}\lambda |\, \lambda > 0,\ (\alpha^{-1}\lambda)^{-1}x \in U\} \\ &= \alpha P_U(x). \end{aligned}$$

Let $\lambda,\ \mu > 0$ such that $\lambda^{-1}x,\ \mu^{-1}y \in U$. Noting that

$$(\lambda+\mu)^{-1}(x+y) = \frac{\lambda}{\lambda+\mu}(\lambda^{-1}x) + \frac{\mu}{\lambda+\mu}(\mu^{-1}y) \in U,$$

we have $P_U(x+y) \leq \lambda + \mu$. Thus, $P_U(x+y) \leq P_U(x) + P_U(y)$.

Put $L = \{\lambda x_0 |\, \lambda \in \mathbb{R}\}$ and define a functional f_0 on L by

$$f_0(\lambda x_0) = \lambda P_U(x_0), \quad \forall\, \lambda \in \mathbb{R}.$$

We have $f_0(\lambda x_0) = P_U(\lambda x_0)$ when $\lambda \geq 0$ and

$$f_0(\lambda x_0) = -P_U((-\lambda)x_0) \leq P_U(\lambda x_0),\ \forall\, \lambda < 0$$

for $0 = P_U((-\lambda)x_0 + \lambda x_0) \leq P_U((-\lambda)x_0) + P_U(\lambda x_0)$. So $f_0(x) \leq P_U(x)$, $\forall\, x \in L$.

We may regard X as the real linear space. Then there exists a $\mathbb{R}$–linear functional f_1 on X such that $f_1(x_0) = f_0(x_0)$ and $f_1(x) \leq P_U(x)$, $\forall\, x \in X$ by Theorem 1.1.3 (1). Consequently, $f_1(x_0) > 1$, $f_1(v) \leq 1$, $\forall\, v \in V$ and $f_1(x) \leq \dfrac{2}{\delta}\|x\|$, $\forall\, x \in X$. Put $f(x) = f_1(x) - i\, f_1(ix)$, $\forall\, x \in X$. It is easy to check that f is a $\mathbb{C}$–linear bounded functional with $\|f\| \leq 2/\delta$ such that $\mathrm{Re}\,(f(x_0)) > 1$ and $\mathrm{Re}\,(f(v)) \leq 1$, $\forall\, v \in V$. $\square$

Let $T \in B(X, Y)$. The range and kernel of T are denoted respectively by $\operatorname{Ran}(T)$ and $\operatorname{Ker} T$. We have the Inverse Operator Theorem as follows:

Theorem 1.1.7. *Let $T \in \mathcal{B}(X, Y)$ with $\operatorname{Ran}(T) = Y$ and $\operatorname{Ker} T = \{0\}$. Then T has a bounded inverse, that is, $T^{-1} \in B(Y, X)$.*

For Banach spaces X, Y, define the linear operations and norm on the product space $X \times Y$ respectively by

$$\lambda\{x_1, y_1\} + \mu\{x_2, y_2\} = \{\lambda x_1 + \mu x_2, \lambda y_1 + \mu y_2\}, \quad \|\{x, y\}\| = \|x\| + \|y\|,$$

for any $\{x_1, y_1\}, \{x_2, y_2\}, \{x, y\} \in X \times Y$ and $\lambda, \mu \in \mathbb{C}$. Then $X \times Y$ is also a Banach space. Let M be a linear subspace of X and $T : M \to Y$ be a linear operator. Denote by $G(T) = \{\{x, Tx\} \in X \times Y \mid x \in M\}$ the graph of T.

Theorem 1.1.8. *Let M be a linear subspace of X and T be a linear mapping from M to Y. If M is closed in X and $G(T)$ is closed in $X \times Y$, then T is bounded.*

Proof. Define the linear mapping $A \colon G(T) \to M$ by $A(\{x, Tx\}) = x$, $\forall x \in M$. Then $\operatorname{Ran}(A) = M$, $\operatorname{Ker} A = \{0\}$ and

$$\|A(\{x, Tx\})\| = \|x\| \leq \|\{x, Tx\}\|, \quad \forall x \in M,$$

i.e., $A \in B(G(T), M)$. Since M and $G(T)$ are all Banach spaces, we have $A^{-1} \in B(M, G(T))$ by Theorem 1.1.7. Thus,

$$\|Tx\| \leq \|\{x, Tx\}\| = \|A^{-1}x\| \leq \|A^{-1}\|\|x\|, \quad x \in M,$$

that is, T is bounded on M. □

Let V be a closed subspace of a normed space W. Two elements x_1 and x_2 in W are called to be equivalent with respect to V, denoted as $x_1 \sim x_2$ if $x_1 - x_2 \in V$. It is easy to check that "$\sim$" is an equivalent relation on W, that is,

(1) $x \sim x$, $\forall x \in W$.
(2) if $x \sim y$, then $y \sim x$, $\forall x, y \in W$.
(3) if $x \sim y$ and $y \sim z$, then $x \sim y$, $\forall x, y, z \in W$.

For $x \in W$, let $[x]$ denote the set $\{y \in W \mid y \sim x\}$ (the equivalence class of x with respect to "$\sim$" in W) and put $X/V = \{[x] \mid x \in W\}$ (the quotient space). Define the linear operation and the norm on W/V respectively by

$$\lambda[x] + \mu[y] = [\lambda x + \mu y], \quad \|[x]\| = \operatorname{dist}(x, V), \ \forall \lambda, \mu \in \mathbb{C}, \ x, y \in W.$$

Obviously, the definition of the linear operation is independent of the choices of x and y and $\|\cdot\|$ is a norm on W/V by Proposition 1.1.1. Thus, W/V is a normed space by above definitions. Moreover, we have

Proposition 1.1.9. *Let V be a closed subspace of a normed space W. Then W/V is a Banach space iff W is a Banach space.*

Definition 1.1.10. A closed subspace M of X is complemented in X if there is a closed subspace N of X such that $M \cap N = \{0\}$ and $X = M + N$.

In this case, we call that X has a decomposition of the oblique direct sum of M and N, written as $X = M \dotplus N$. The closed subspace N is said to be complementary to M and is denoted as M^c.

We have known that not every closed subspace in a Banach space is complemented. Some counter–examples can be seen in [Beauzamy (1985)]. But we still have following special results:

Proposition 1.1.11. *Let M be a closed subspace of X. If $\dim M < \infty$ or $\mathrm{Codim} M \triangleq \dim(X/M) < \infty$, then M is complemented in X.*

Proof. We first assume that $\dim M = n$. Let $\{e_1, \cdots, e_n\}$ be a basis for M. Let M_j be the linear span of $e_1, \cdots, e_{j-1}, e_{j+1}, \cdots, e_n$. Since $e_j \notin M_j$, $j = 1, \cdots, n$, it follows from Corollary 1.1.5 (2) that there exist $f_1, \cdots, f_n \in X^*$ such that $f_j(e_j) = 1$ and $f_j(x) = 0$, $\forall x \in M_j$, that is $f_j(e_i) = \delta_{ij}$, $i, j = 1, \cdots, n$. Set $N = \{x \in X |\, f_j(x) = 0, j = 1, \cdots, n\}$. For any $x \in X$, put $x_1 = \sum\limits_{j=1}^{n} f_j(x) e_j$. It is easy to verify that $x - x_1 \in N$ and $M \cap N = \{0\}$. Thus, $X = M \dotplus N$.

Now assume that $\mathrm{Codim} M = m$. Let $\pi: X \to X/M$ be the canonical mapping. Then there is $\{f_1, \cdots, f_m\} \subset X$ such that $\{\pi(f_1), \cdots, \pi(f_m)\}$ forms a basis for X/M. Put $N = span\{f_1, \cdots, f_m\}$ (the linear span of $f_1, \cdots, f_m$). Then $X = M \dotplus N$. □

Let M be a complemented closed subspace of X. Then $X = M \dotplus M^c$ and hence for any $x \in X$, there is $x_1 \in M$ and $x_2 \in M^c$ uniquely such that $x = x_1 + x_2$. Define the linear operator $P_{M,M^c}: X \to X$ by $P_{M,M^c} x = x_1$. So $x_2 = (I_X - P_{M,M^c})x$. Obviously, $M = \mathrm{Ran}\,(P_{M,M^c})$, $M^c = \mathrm{Ran}\,(I_X - P_{M,M^c})$ and $P^2_{M,M^c} = P_{M,M^c}$, i.e., P_{M,M^c} is an idempotent operator, where I_X is an identity operator on X. We will use I in place of I_X in the following if no confusion occurs.

Proposition 1.1.12. *Keep the symbols as above. The idempotent operator P_{M,M^c} is bounded.*

Proof. Consider the linear operator $j: M \times M^c \to X$ defined by $j(\{x_1, x_2\}) = x_1 + x_2$. From $X = M \dotplus M^c$, we get that $\operatorname{Ran}(j) = X$ and $\operatorname{Ker} j = \{0\}$.

Since X is completed and M, M^c are closed in X, it follows that $M \times M^c$ ($\subset X \times X$) is completed. It follows from Theorem 1.1.7 that $j^{-1}: X \to M \times M^c$ is bounded. Thus,

$$\|P_{M,M^c}x\| + \|(I - P_{M,M^c})x\| = \|j^{-1}x\| \leq \|j^{-1}\|\|x\|, \ \forall\, x \in X$$

and consequently, $P_{M,M^c} \in B(X)$. □

The Hahn–Banach Theorem, the Inverse Operator Theorem and the following Banach–Steinhaus Theorem (or Uniform Boundedness Principle) play important roles in Functional Analysis.

Theorem 1.1.13. *Let* $\{T_\alpha | \alpha \in I\}$ *be a family of operators in* $B(X, Y)$ *with* $\sup\limits_{\alpha \in I} \|T_\alpha x\| < +\infty$ *for every* $x \in X$. *Then* $\sup\limits_{\alpha \in I} \|T_\alpha\| < +\infty$.

Definition 1.1.14. A directed set is a nonempty set I with a relation $\preccurlyeq$ such that

(1) $\alpha \preccurlyeq \alpha$, $\forall\, \alpha \in I$.
(2) If $\alpha, \beta, \gamma \in I$ with $\alpha \preccurlyeq \beta$ and $\beta \preccurlyeq \gamma$, then $\alpha \preccurlyeq \gamma$.
(3) For each pair $\alpha, \beta \in I$, there is a $\gamma \in I$ such that $\alpha \preccurlyeq \gamma$ and $\beta \preccurlyeq \gamma$.

A subset $\{x_\alpha | \alpha \in I\}$ of X is called to be a net if the index set I is a directed set. A net $\{y_\beta | \beta \in J\}$ in X is called to be a subnet of the net $\{x_\alpha | \alpha \in I\}$, if there is a mapping $k: J \to I$ preserving " $\preccurlyeq$ " such that $y_\beta = x_{k(\beta)}$, $\forall\, \beta \in J$ and for any $\alpha \in I$, there is $r \in J$ such that $\alpha \preccurlyeq k(p)$ whenever $r \preccurlyeq p \in J$.

Definition 1.1.15. Let $\{x_\alpha | \alpha \in \Lambda\}$ be a net in X and $x_0 \in X$. We say $\{x_\alpha | \alpha \in \Lambda\}$ converges to x_0 weakly (denoted as $x_\alpha \xrightarrow{w} x_0$, $\alpha \in \Lambda$), if for any $\epsilon > 0$ and any $f \in X^*$, there is a $\gamma \in \Lambda$ such that $|f(x_\alpha) - f(x_0)| < \epsilon$ wherever $\alpha \in \Lambda$ with $\gamma \preccurlyeq \alpha$.

Let $\{f_\lambda | \lambda \in I\}$ be a net in X^* and $f_0 \in X^*$. We say $\{f_\lambda | \lambda \in I\}$ converges to f_0 weakly* (denoted as $f_\lambda \xrightarrow{w^*} f_0$, $\lambda \in I$), if for any $\epsilon > 0$ and any $x \in X$, there is an $\alpha \in I$ such that $|f_\lambda(x) - f_0(x)| < \epsilon$ wherever $\lambda \in I$ with $\alpha \preccurlyeq \lambda$.

Theorem 1.1.16 (Banach–Alaoglu). *Every net* $\{f_\alpha | \alpha \in I\}$ *in* $D(X^*)$ *has a subnet* $\{g_\beta | \beta \in J\}$ *which converges weakly* to an element in* $D(X^*)$.

Applying Theorem 1.1.13 to the weakly (or weakly*) convergent sequences, we have:

Proposition 1.1.17. *Let* $\{x_n\}$ *(resp.* $\{f_n\}$*) be a sequence in* X *(resp.* X^**) such that* $x_n \xrightarrow{w} x_0$ *(resp.* $f_n \xrightarrow{w^*} f_0$*) for some* $x_0 \in X$ *(resp.* $f_0 \in X^*$*). Then* $\sup\limits_{n\geq 1} \|x_n\| < +\infty$ *(resp.* $\sup\limits_{n\geq 1} \|f_n\| < +\infty$*) and* $\|x_0\| \leq \liminf\limits_{n\to\infty} \|x_n\|$ *(resp.* $\|f_0\| \leq \liminf\limits_{n\to\infty} \|f_n\|$*).*

Proposition 1.1.18 (Mazur). *Let* $\{x_n\}$ *be a sequence in* X *such that* $x_n \xrightarrow{w} x_0$ *for some* x_0 *in* X*. Then for any* $\epsilon > 0$*, there are* $x_{n_1}, \cdots, x_{n_k} \subset \{x_1, \cdots, x_n, \cdots\}$ *and non-negative numbers* $\alpha_1, \cdots, \alpha_k$ *with* $\sum\limits_{j=1}^{k} \alpha_j = 1$ *such that* $\|x_0 - \sum\limits_{j=1}^{k} \alpha_j x_{n_j}\| < \epsilon$*.*

Proof. Set $x'_n = x_n - x_1$, $n \geq 1$ and put

$$M = \Big\{ \sum_{j=1}^{n} \lambda_j x'_j | \, \lambda_1, \cdots, \lambda_n \geq 0, \sum_{j=1}^{n} \lambda_j = 1, n \geq 1 \Big\}.$$

Then $x'_n \xrightarrow{w} x_0 - x_1$ and M is convex. We will show that $x_0 - x_1 \in \overline{M}$.

Note that $0 \in \overline{M}$. If $x_0 - x_1 \notin \overline{M}$, then there is $f \in X^*$ such that $\operatorname{Re}(f(x_0 - x_1)) > 1$ and $\operatorname{Re}(f(x)) \leq 1$, $\forall x \in \overline{M}$ by Theorem 1.1.6. Since $x'_n \in \overline{M}$ and $f(x'_n) \to f(x_0 - x_1)$ as $n \to \infty$, we have $\operatorname{Re}(f(x_0 - x_1)) \leq 1$, which contradicts to the assumption $\operatorname{Re}(f(x_0 - x_1)) > 1$. □

Definition 1.1.19. A Banach space X is called to be reflexive, if for every $\phi \in (X^*)^*$, there is $x \in X$ such that $\phi(f) = f(x)$, $\forall f \in X^*$.

We say X is strictly convex if $\|x + y\| < 2$ whenever x, $y \in S(X)$ with $x \neq y$.

We have known that every finite–dimensional Banach space is reflexive; l^p $(1 < p < +\infty)$ is reflexive and strictly convex. Moreover, we have

Theorem 1.1.20 (Eberlein–Shmulyan). *The Banach space* X *is reflexive iff every norm bounded sequence in* X *contains a subsequence which converges weakly to an element of* X*.*

Proposition 1.1.21. *Let* M *be a closed convex subset of* X*. If* X *is reflexive, then for any* $x \in X \backslash M$*, there is* $m \in M$ *such that* $\operatorname{dist}(x, M) = \|x - m\|$*. In addition, if* X *is also strictly convex, then such* m *is uniquely determined by* x*.*

Proof. We can find a sequence $\{y_n\} \subset M$ such that

$$\lim_{n\to\infty} \|x - y_n\| = \text{dist}\,(x, M) = f_M(x).$$

Then $\{\|y_n\|\}$ is bounded. So there is a subsequence $\{y_{n_k}\}$ of $\{y_n\}$ such that $y_{n_k} \xrightarrow{w} m$ for some $m \in X$ by Theorem 1.1.20 and $m \in M$ by Proposition 1.1.18. Thus,

$$\text{dist}\,(x, M) \le \|x - m\| \le \liminf_{n\to\infty} \|x - y_n\| = \text{dist}\,(x, M).$$

Let X be reflexive and strictly convex. If there are m and m' in M such that $m \ne m'$ and $\|x - m\| = \|x - m'\| = \text{dist}\,(x, M)$, then

$$\|(x - m) + (x - m')\| < 2\,\text{dist}\,(x, M).$$

But it is not true for $\frac{1}{2}(m+m') \in M$ and $\|x - \frac{1}{2}(m+m')\| \ge \text{dist}\,(x, M)$. Therefore, m is uniquely determined by x. □

Let $A \in B(X,Y)$. The dual operator $A^*\colon Y^* \to X^*$ is defined by $A^*y^*(x) = y^*(Ax)$, $\forall\, x \in X$ and $y^* \in Y^*$. Clearly, A^* is linear. Moreover, we have

Proposition 1.1.22. *Let Z be a Banach space over $\mathbb{C}$ and let $A, B \in B(X,Y)$, $C \in B(Y,Z)$. Then*

(1) $(A+B)^* = A^* + B^*$, $(\lambda A)^* = \lambda A^*$, $\forall\, \lambda \in \mathbb{C}$.
(2) $(CA)^* = A^*C^*$ *and* $(A^*)^{-1} = (A^{-1})^*$ *when* A^{-1} *exists and is bounded.*
(3) $\|A^*\| = \|A\|$.

Recall that an operator A in $B(X,Y)$ is compact if $\overline{A(D(X))}$ is a compact subset of Y. Let $\mathcal{K}(X,Y)$ denote the set of all compact operators in $B(X,Y)$. Set $\mathcal{K}(X) = \mathcal{K}(X,X)$. We have

Proposition 1.1.23. *Let $A \in B(X,Y)$, $B \in \mathcal{K}(Y,Z)$ and $C \in B(Z,X)$.*

(1) $BA \in \mathcal{K}(X,Z)$ *and* $CB \in \mathcal{K}(Y,X)$.
(2) $\mathcal{K}(X,Y)$ *is a closed subspace of* $B(X,Y)$.
(3) $A \in \mathcal{K}(X,Y)$ *iff* $A^* \in \mathcal{K}(Y^*,X^*)$.
(4) $\text{Ran}\,(I-B)$ *is closed and* $\dim \text{Ker}\,(I-B) < \infty$, $\dim \text{Coker}\,(I-B) < \infty$ *when* $Y = Z$.

Let H be a Hilbert space with the inner product $(\cdot,\cdot)$ and the norm $\|x\| = \sqrt{(x,x)}$, $x \in H$. Let $x, y \in H$. x, y are called to be orthogonal, written as $x \perp y$, if $(x,y) = 0$.

Let M be a subset of H. Put $M^\perp = \{x \in H | x \perp y, \forall y \in M\}$. When M is a closed subspace, we call $M^\perp$ is the orthogonal complement of M. Obviously, $M^\perp$ is a closed subspace of H and $M \cap M^\perp = \{0\}$.

The following results are fundamental important in Hilbert spaces:

Proposition 1.1.24. *Let x, $y \in H$. Then*

(1) $|(x, y)| \leq \|x\|\|y\|$ *(Schwarz Inequality).*
(2) $\|x + y\|^2 = \|x\|^2 + \|y\|^2$, *if $x \perp y$.*
(3) $\|x + y\|^2 + \|x - y\|^2 = 2(\|x\|^2 + \|y\|^2)$ *(Parallelogram Law).*
(4) $(x, y) = \frac{1}{4}\sum_{k=1}^{4} i^k\|x + i^k y\|^2$ *(Polarization Identity).*

Theorem 1.1.25. *Let M be a closed space of H. Then for every $x \in H$, there exists unique $x_0 \in M$ such that $f_M(x) = \|x - x_0\|$ and $x - x_0 \in M^\perp$.*

Theorem 1.1.25 indicates that $H = M \dotplus M^\perp$. In this case, we write it as the form $H = M \oplus M^\perp$, called the orthogonal decomposition of H by M and $M^\perp$. Using Theorem 1.1.25, we can deduce the Riesz representation of bounded linear functionals on H as follows.

Theorem 1.1.26. *Let $f \in H^*$. Then there is a unique $y_f \in H$ such that $\|f\| = \|y_f\|$ and $f(x) = (x, y_f)$, $\forall\, x \in H$.*

In terms of Theorem 1.1.26, we can define the dual operator $A^* : K \to H$ of $A \in B(H, K)$ as $(A^*y, x) = <y, Ax>$, $\forall\, x \in H$ and $y \in K$, where $<\cdot, \cdot>$ is the inner product on K.

Proposition 1.1.27. *Let $A, B \in B(H, K)$, $C \in B(K, L)$. Then*

(1) $(\lambda A + \mu B)^* = \overline{\lambda} A^* + \overline{\mu} B^*$, $\forall\, \lambda$, $\mu \in \mathbb{C}$.
(2) $A^{**} \triangleq (A^*)^* = A$, $(CA)^* = A^*C^*$.
(3) $\|A^*\| = \|A\|$ *and* $\|A^*A\| = \|A\|^2$.

Let $A \in B(H) \triangleq B(H, H)$. A is called to be self–adjoint, normal or unitary respectively if $A^* = A$, $A^*A = AA^*$ or $A^*A = AA^* = I$.

Proposition 1.1.28. *Let $A \in B(H)$. Then $A^* = A$ if and only if $(Ax, x) \in \mathbb{R}$, $\forall\, x \in H$.*

Proof. Clearly, $A^* = A$ implies that

$$(Ax, x) = (x, Ax) = \overline{(Ax, x)}, \quad \forall\, x \in H,$$

that is, $(Ax, x) \in \mathbb{R}$, $\forall\, x \in H$.

Conversely, suppose that $(Ax, x) \in \mathbb{R}$, $\forall\, x \in H$. Then for each pair $x, y \in H$, $(A(x + i^k y), (x + i^k y)) \in \mathbb{R}$, $k = 1, \cdots, 4$. Using the identities

$$(Ax, y) = \frac{1}{4}\sum_{k=1}^{4} i^k (A(x + i^k y), (x + i^k y)),\ \forall\, x, y \in H,$$

we get that

$$\overline{(Ay, x)} = \frac{1}{4}\sum_{k=1}^{4} i^{-k} (A(i^{-k} y + x), (i^{-k} y + x)) = (Ax, y). \tag{1.1.1}$$

(1.1.1) means that $(Ax, y) = (x, Ay) = (A^*x, y)$, $\forall\, x, y \in H$. So $A^* = A$. □

According to Proposition 1.1.28, we can describe the comparasion of two self-adjoint operators in $B(H)$ as follows. Let $A \in \mathcal{B}(H)$ be a self-adjoint operator. A is called to be positive, (in symbol, $A \geq 0$), if $(Ax, x) \geq 0$, $\forall\, x \in H$. Let A_1, A_2 be self-adjoint operators in $\mathcal{B}(H)$. $A_1 \geq A_2$ means that $A_1 - A_2 \geq 0$.

Let M be a closed subspace of H. Theorem 1.1.25 shows that $H = M \oplus M^\perp$. So for any $x \in H$, there are $x_1 \in M$ and $x_2 \in M^\perp$ such that $x = x_1 + x_2$ and $x_1 \perp x_2$. Let P_M be the map of H into itself defined by $P_M x = x_1$. Obviously, P_M is linear, idempotent with $\mathrm{Ran}\,(P_M) = M$ and

$$(P_M x, x) = (x_1, x_1) = \|P_M x\|^2 = \|x\|^2 - \|x_2\|^2 \leq \|x\|^2.$$

Thus, $P_M \in \mathcal{B}(H)$ with $P_M^2 = P_M = P_M^*$. We call that P_M is the orthogonal projection of M or a projection. Clearly, $P_{M^\perp} = I - P_M$.

Proposition 1.1.29. *Let L, M be closed subspaces in H. Then*

(1) *$P_L + P_M$ is the orthogonal projection of $\overline{L + M}$ iff $L \subset M^\perp$.*
(2) *$P_M - P_L$ is an orthogonal projection iff $L \subset M$.*
(3) *$P_M P_L$ is the orthogonal projection of $L \cap M$ iff $P_L P_M = P_M P_L$.*

Proposition 1.1.30. *Let $P \in \mathcal{B}(H)$ be an idempotent operator and put $M = \mathrm{Ran}\,(P)$. Then*

(1) *$P = P_M$ iff $\mathrm{Ran}\,(P^*) \subset \mathrm{Ran}\,(P)$.*
(2) *$(I - P - P^*)^{-1} \in \mathcal{B}(H)$ and $P_M = P(P + P^* - I)^{-1}$.*

Proof. (1) Since $Px = x$, $\forall\, x \in M$, we have $PP_M = P_M$ and $P_M P = P$. Thus, $P_M P P_M = P_M$, $P_{M^\perp} P P_M = 0$, $P_{M^\perp} P P_{M^\perp} = 0$ and

$$P = P_M + P_M P P_{M^\perp}, \quad P^* = P_M + P_{M^\perp} P^* P_M.$$

So for any $x \in H$, there is $y \in H$ such that

$$P^*x = P_M x + P_{M^\perp} P^* P_M x = Py = P_M y + P_M P P_{M^\perp} y$$

and consequently, $P_{M^\perp} P^* P_M x = 0$. Therefore, $P^* = P_M = P$.

(2) Set $N = \mathrm{Ran}\,(I - P)$. Then $P = P_{M,N}$. Since P, $I - P$ are idempotent, we have

$$PP_M = P_M,\ P_M P = P,\ (I - P)P_N = P_N,\ P_N(I - P) = I - P. \quad (1.1.2)$$

By using (1.1.2), we can deduce that

$$(P_N - P_M)(I - P - P^*) = I = (I - P - P^*)(P_N - P_M),$$

that is, $P_N - P_M = (I - P - P^*)^{-1} \in B(H)$. Noting that $PP_N = 0$, we have $P_M = PP_M = -P(I - P - P^*)^{-1}$. □

1.2 Reduced minimum moduli of densely–defined operators

Let M be a subspace of X and $T\colon M \to Y$ be a linear operator. M is called to be the domain of T, denoted by $\mathfrak{D}(T)$. Set $\mathrm{Ker}\,T = \{x \in \mathfrak{D}(T) \,|\, Tx = 0\}$ and $\mathrm{Ran}\,(T) = \{Tx \,|\, x \in \mathfrak{D}(T)\}$. The graph $G(T)$ of T is the subspace $\{\{x, Tx\} \,|\, x \in \mathfrak{D}(T)\}$ of $X \times Y$.

Definition 1.2.1. A linear operator $T\colon \mathfrak{D}(T) \to Y$ is called to be closed if $G(T)$ is closed in $X \times Y$.

Let $\mathcal{D}(X, Y)$ denote the set of all linear operators $T\colon \mathfrak{D}(T) \to Y$ with $\overline{\mathfrak{D}(T)} = X$ and let $\mathcal{C}(X, Y)$ denote the set of all closed operators in $\mathcal{D}(X, Y)$. When $X = Y$, we set $\mathcal{D}(X) = \mathcal{D}(X, X)$ and $\mathcal{C}(X) = \mathcal{C}(X, X)$.

Lemma 1.2.2. *Let $T\colon \mathfrak{D}(T) \to Y$ be a closed operator.*

(1) $\mathrm{Ker}\,T$ *is closed in* X.
(2) *If* $\mathrm{Ker}\,T = \{0\}$, *then the inverse* $T^{-1}\colon \mathrm{Ran}\,(T) \to \mathfrak{D}(T)$ *is a closed operator. In addition, if* $\mathrm{Ran}\,(T)$ *is closed, then* T^{-1} *is bounded.*

Proof. (1) Let $x \in \overline{\mathrm{Ker}\,T}$. Then there is $\{x_n\} \subset \mathrm{Ker}\,T$ such that $x_n \to x$ as $n \to \infty$. Since $\{\{x_n, 0\}\} \subset G(T)$ and $G(T)$ is closed, it follows that $\{x, 0\} \in G(T)$, that is, $x \in \mathfrak{D}(T)$ and $Tx = 0$.

(2) $\mathrm{Ker}\,T = \{0\}$ implies that the inverse $T^{-1}\colon \mathrm{Ran}\,(T) \to \mathfrak{D}(T)$ given by $T^{-1}x = y$ for $Ty = x$, is well–defined and linear. Note that

$$G(T^{-1}) = \{\{y, T^{-1}x\} \,|\, y \in \mathrm{Ran}\,(T)\} \overset{Tx=y}{=\!=\!=} \{\{Tx, y\} \,|\, x \in \mathfrak{D}(T)\}$$

and the mapping $j: Y \times X \to X \times Y$ given by $j(\{y,x\}) = \{x,y\}$ is a homeomorphism. So $j(G(T)) = G(T^{-1})$ is closed when $G(T)$ is closed.

When $\text{Ran}\,(T)$ is closed in Y, $\text{Ran}\,(T)$ is a Banach space. It follows from Theorem 1.1.8 that $T^{-1}: \text{Ran}\,(T) \to \mathfrak{D}(T)$ is bounded. □

Definition 1.2.3. Let $T \in \mathcal{D}(X,Y)$. The conjugate T^* of T is defined as follows.

Put $\mathfrak{D}(T^*) = \{g \in Y^* | g \circ T \text{ continuous on } \mathfrak{D}(T)\}$. For $g \in \mathfrak{D}(T^*)$, let T^* be the operator which sends g to $\widetilde{g \circ T}$, where $\widetilde{g \circ T}$ is the unique continuous linear extension of $g \circ T$ to all of X.

Proposition 1.2.4. *Let $T \in \mathcal{D}(X,Y)$ and $B \in B(Y,X)$, $C \in B(Z,X)$ with* $\text{Ran}\,(C) \subset \mathfrak{D}(T)$. *Then*

(1) *T^* is a closed linear operator.*
(2) *$\mathfrak{D}(T^*)$ is total, that is, for any $y \in Y \backslash \{0\}$, there is $g \in \mathfrak{D}(T^*)$ such that $g(y) \neq 0$, when $G(T)$ is closed.*
(3) *$(BT)^* = T^*B^*$ on $\mathfrak{D}((BT)^*) = \mathfrak{D}(T^*B^*)$.*
(4) *$(TC)^* = C^*T^*$ on $\mathfrak{D}(T^*)$ and $TC \in B(Z,Y)$ if T is a closed operator.*

Proof. (1) It is easy to check that T^* is linear by Definition 1.2.3. Let $\{g,f\} \in \overline{G(T^*)}$. Then there is $\{g_n\} \subset \mathfrak{D}(T^*)$ such that $\|g_n - g\| \to 0$ and $\|T^*g_n - f\| \to 0$ as $n \to \infty$. Thus, for any $x \in \mathfrak{D}(T)$, $g_n(Tx) \to g(Tx)$ and $g_n(Tx) \to f(x)$ and hence $f(x) = g(Tx)$, which means that $g \in \mathfrak{D}(T^*)$ and $T^*(g)(x) = g(Tx) = f(x)$, $\forall\, x \in \mathfrak{D}(T)$. Therefore, $f = T^*g$ for $\overline{\mathfrak{D}(T)} = X$, i.e., T^* is a closed operator.

(2) Let $y_0 \in Y \backslash \{0\}$. Then $\{0, y_0\} \notin G(T)$ and so that there is $h \in (X \times Y)^*$ such that

$$h(\{0,y_0\}) = \text{dist}\,(\{0,y_0\}, G(T)) \neq 0, \;\; h(\{x,Tx\}) = 0, \;\forall\, x \in \mathfrak{D}(T).$$

Set $h_1(x) = h(\{x,0\})$ and $h_2(y) = h(\{0,y\})$, $\forall\,\{x,y\} \in X \times Y$. Then $h_1 \in X^*$, $h_2 \in Y^*$ and $h_1(x) + h_2(y) = h(\{x,y\})$, $\forall\,\{x,y\} \in X \times Y$. So

$$h_1(x) + h_2(Tx) = h(\{x,Tx\}) = 0, \;\forall\, x \in \mathfrak{D}(T), \;\; h_2(y_0) = h(\{0,y_0\}). \tag{1.2.1}$$

From (1.2.1), we get that $h_2 \in \mathfrak{D}(T^*)$ and $h_2(y_0) \neq 0$.

(3) Let $g \in \mathfrak{D}((BT)^*)$. Then for any $x \in \mathfrak{D}(T)$,

$$(BT)^*g(x) = g(BTx) = B^*g(Tx). \tag{1.2.2}$$

(1.2.2) indicates that $B^*g \in \mathfrak{D}(T^*)$ and $(BT)^*g(x) = T^*(B^*g)(x)$. Thus $(BT)^*g = T^*B^*g$ for $\overline{\mathfrak{D}(T)} = X$. On the other hand, let $h \in \mathfrak{D}(T^*B^*)$. Then $B^*h \in \mathfrak{D}(T^*)$ and

$$T^*(B^*h)(x) = B^*h(Tx) = h(BTx), \quad \forall\, x \in \mathfrak{D}(T). \tag{1.2.3}$$

(1.2.3) shows that $h \in \mathfrak{D}((BT)^*)$ and $T^*B^*h(x) = (BT)^*h(x)$, $\forall x \in \mathfrak{D}(T)$. Therefore, $(BT)^* = T^*B^*$ on $\mathfrak{D}((BT)^*) = \mathfrak{D}(T^*B^*)$.

(4) Let $g \in \mathfrak{D}(T^*)$. Since $Cx \in \mathfrak{D}(T)$, $\forall x \in Z$, it follows that

$$(TC)^*g(x) = g(TCx) = T^*g(Cx) = C^*T^*g(x), \quad \forall x \in Z$$

and so that $(TC)^* = C^*T^*$ on $\mathfrak{D}(T^*)$.

Next we show that $G(TC)$ is closed in $Z \times Y$. In fact, let $\{\{z_n, TCz_n\}\} \subset G(TC)$ with $\|z_n - z_0\| \to 0$ and $\|TCz_n - y_0\| \to 0$ as $n \to \infty$ for some $\{z_0, y_0\} \in Z \times Y$. Since C is bounded and T is a closed operator, it follows that $TCz_0 = y_0$. Then, from Theorem 1.1.8, we obtain that $TC \in \mathcal{B}(Z, Y)$ for $\mathfrak{D}(TC) = Z$. □

Corollary 1.2.5. *Let $T \in \mathcal{C}(X, Y)$ and suppose that Y is a reflexive Banach space. Then $T^* \in \mathcal{C}(Y^*, X^*)$.*

Proof. We need only to prove that $\overline{\mathfrak{D}(T^*)} = Y^*$ by Proposition 1.2.4 (1). If $\overline{\mathfrak{D}(T^*)} \neq Y^*$, then there are $g_0 \in Y^* \backslash \overline{\mathfrak{D}(T^*)}$ and $F \in Y^{**}$ such that

$$F(g_0) = \operatorname{dist}(g_0, \overline{\mathfrak{D}(T^*)}) \neq 0, \ F(g) = 0, \ \forall g \in \overline{\mathfrak{D}(T^*)}. \tag{1.2.4}$$

Since Y is reflexive, we can find a nonzero element y_0 in Y such that $F(g) = g(y_0)$, $\forall g \in Y^*$. Thus, $g(y_0) = 0$ by (1.2.4). But it is contrary to Proposition 1.2.4 (2). So $\overline{\mathfrak{D}(T^*)} = Y^*$. □

Remark 1.2.6. Let $T \in \mathcal{D}(H, K)$. The conjugate T^* of T can be defined as follows. Set

$$\mathfrak{D}(T^*) = \{x \in K \,|\, \text{there is } z \in H \text{ such that } < x, Ty >= (z, y), \ \forall y \in \mathfrak{D}(T)\}.$$

For $x \in \mathfrak{D}(T^*)$, the $z \in H$ Such that for every $y \in \mathfrak{D}(T)$, $< x, Ty >= (z, y)$ is unique since $\mathfrak{D}(T)$ is dense in H. Then we put $T^*x = z$ for $x \in \mathfrak{D}(T^*)$. By Proposition 1.2.4 and Corollary 1.2.5, $T^* \in \mathcal{C}(K, H)$.

Define the inner products on $H \times K$ and $K \times H$ by

$$(\{x_1, y_1\}, \{x_2, y_2\}) = (\{y_1, x_1\}, \{y_2, x_2\}) = (x_1, x_2) + < y_1, y_2 >,$$

$\forall x_1, x_2 \in H$, $y_1, y_2 \in K$. Then $H \times K$ and $K \times H$ become Hilbert spaces.

Now we introduce a linear operator $V \colon H \times K \to K \times H$ by $V\{x, y\} = \{-y, x\}$, $\forall x \in H$, $y \in K$. It is easy to check that $V^*\{y, x\} = \{x, -y\}$, $\forall x \in H$, $y \in K$ and $V(M^\perp) = (V(M))^\perp$ for any closed subspace M in $H \times K$ by using $V^*V = I_{H \times K}$ and $VV^* = I_{K \times H}$.

Lemma 1.2.7. *Let $T \in \mathcal{D}(H, K)$. Then $G(T^*) = (V(G(T)))^\perp \subset K \times H$.*

Proof. For any $x \in \mathfrak{D}(T)$ and $y \in \mathfrak{D}(T^*)$, $< Tx, y >= (x, T^*y)$. Thus, $(\{y, T^*y\}, \{-Tx, x\}) =< -Tx, y > +(x, T^*y) = 0$, $\forall x \in \mathfrak{D}(T)$, $y \in \mathfrak{D}(T^*)$ and hence $G(T^*) \subset (V(G(T)))^\perp$.

On the other hand, let $\{z, y\} \in [V(G(T))]^\perp$. Then

$$< z, -Tx > +(y, x) = 0, \quad \forall x \in \mathfrak{D}(T).$$

Consequently, $z \in \mathfrak{D}(T^*)$ and $y = T^*z$, that is, $(V(G(T)))^\perp \subset G(T^*)$. □

Using Lemma 1.2.7, we can deduce following two fundamental results about operators in $\mathcal{C}(H, K)$.

Theorem 1.2.8. *Let $T \in \mathcal{C}(H, K)$. Then*

(1) $T^{**} \triangleq (T^*)^* = T$.
(2) $\mathrm{Ker}\,(I_H + T^*T) = \{0\}$ *and* $(I_H + T^*T)^{-1} \in \mathcal{B}(H)$.

Proof. (1) By Lemma 1.2.7,

$$\begin{aligned} G(T^{**}) &= ((-V^*)(G(T^*)))^\perp = (V^*(G(T^*)))^\perp \\ &= (V^*(V(G(T)))^\perp)^\perp = (V^*V(G(T))^\perp)^\perp = (G(T)^\perp)^\perp \\ &= G(T), \end{aligned}$$

that is, $T^{**} = T$.

(2) Let $x \in \mathfrak{D}(T^*T)$ and $x \in \mathrm{Ker}\,(I_H + T^*T)$. Then $(T^*Tx, x) + (x, x) = 0$. Since $x \in \mathfrak{D}(T)$ and $Tx \in \mathfrak{D}(T^*)$, $(T^*Tx, x) = (Tx, Tx) \geq 0$. So from $(T^*Tx, x) + (x, x) = 0$, we get that $x = 0$, i.e., $\mathrm{Ker}\,(I_H + T^*T) = \{0\}$.

By Lemma 1.2.7,

$$V^*(G(T^*)) = V^*(V(G(T)))^\perp = V^*V(G(T))^\perp = G(T)^\perp.$$

Thus

$$H \times K = G(T) \oplus G(T)^\perp = G(T) \oplus V^*(G(T^*))$$

and hence for each $h \in H$, there is $x \in \mathfrak{D}(T)$ and $y \in \mathfrak{D}(T^*)$ such that $\{h, 0\} = \{x, Tx\} + \{T^*y, -y\}$, that is, $h = x + T^*y$, $Tx = y$. This shows that $x \in \mathfrak{D}(T^*T)$ and $h = (I_H + T^*T)x$. Consequently, $(I_H + T^*T)^{-1}$ is a linear operator from H to $\mathfrak{D}(T^*T)$.

For any x, $y \in H$, put $s = (I_H + T^*T)^{-1}x$, $t = (I_H + T^*T)^{-1}y$. Then s, $t \in \mathfrak{D}(T^*T)$ and

$$\begin{aligned} (x, (I_H + T^*T)^{-1}y) &= ((I_H + T^*T)s, t) = (s, t) + (T^*Ts, t) \\ &= (s, t) + (Ts, Tt) = (s, t) + (s, T^*Tt) \\ &= (s, (I_H + T^*T)t) = ((I_H + T^*T)^{-1}x, y). \end{aligned}$$

Thus $\sup\limits_{y\in D(H)} |(x,(I_H+T^*T)^{-1}y)| \le \|(I_H+T^*T)^{-1}x\|$ for every $x \in H$. Applying Theorem 1.1.13 to $\{\|(I_H+T^*T)^{-1}y\| \,|\, y\in D(H)\}$, we get that $\sup\limits_{y\in D(H)} \|(I_H+T^*T)^{-1}y\| < +\infty$, that is, $(I_H+T^*T)^{-1}$ is bounded. Furthermore, $(I_H+T^*T)^{-1}$ is self-adjoint by the identity

$$(x,(I_H+T^*T)^{-1}y) = ((I_H+T^*T)^{-1}x, y), \quad \forall\, x,\, y\in H. \qquad \square$$

Definition 1.2.9. Let $T\colon \mathcal{D}(T)\to Y$ be a linear operator with $\operatorname{Ker} T$ closed. The reduced minimum modulus $\gamma(T)$ of T is defined by

$$\gamma(T) = \inf\{\|Tx\| \,|\, \operatorname{dist}(x, \operatorname{Ker} T) = 1,\ x\in \mathfrak{D}(T)\}. \tag{1.2.5}$$

We set $\gamma(T)=\infty$ if $T=0$.

Remark 1.2.10. Let $T\in \mathcal{C}(H,K)$ and $x\in\mathfrak{D}(T)$. Then x can be decomposed as $x=x_1+x_2$, where $x_1\in \operatorname{Ker} T$ and $x_2\in(\operatorname{Ker} T)^\perp\cap\mathfrak{D}(T)$. Thus $\operatorname{dist}(x,\operatorname{Ker} T)=\|x_2\|$ and $Tx=Tx_2$ and so that

$$\gamma(T)=\inf\{\|Tx\| \,|\, x\in(\operatorname{Ker} T)^\perp\cap\mathfrak{D}(A),\ \|x\|=1\}.$$

Let $T\in\mathcal{C}(X,Y)\backslash\{0\}$. Then by (1.2.5), we have

$$\|Tx\|\ge\gamma(T)\operatorname{dist}(x,\operatorname{Ker} T), \quad \forall\, x\in\mathfrak{D}(T). \tag{1.2.6}$$

Since $\operatorname{Ker} T$ is closed in X for $T\in\mathcal{C}(X,Y)$ by Lemma 1.2.2, it follows from Proposition 1.1.9 that $X/\operatorname{Ker} T$ is a Banach space. Set

$$[\mathfrak{D}(T)] = \{[x]\in X/\operatorname{Ker} T \,|\, x\in\mathfrak{D}(T)\}.$$

Then $[\mathfrak{D}(T)]$ is dense in $X/\operatorname{Ker} T$ since $\overline{\mathfrak{D}(T)}=X$. Define a linear operator $\hat{T}\colon[\mathfrak{D}(T)]\to Y$ by $\hat{T}([x])=Tx$, $\forall\, x\in\mathfrak{D}(T)$. Clearly, $\hat{T}$ is well-defined and $\mathfrak{D}(\hat{T})=[\mathfrak{D}(T)]$, $\operatorname{Ker}\hat{T}=\{0\}$.

Lemma 1.2.11. *Let $T\in\mathcal{C}(X,Y)\backslash\{0\}$ and let $\hat{T}$ be as above.*

(1) $\hat{T}\in\mathcal{C}(X/\operatorname{Ker} T, Y)$.
(2) $\operatorname{Ran}(T)=\operatorname{Ran}(\hat{T})$ *and* $\gamma(T)=\gamma(\hat{T})$, $\gamma(T^*)=\gamma(\hat{T}^*)$.

Proof. (1) We need only to verify that $G(\hat{T})$ is closed in $(X/\operatorname{Ker} T)\times Y$. Let $\{[x_n]\}\subset\mathfrak{D}(\hat{T})$ and $[x_0]\in X/\operatorname{Ker} T$, $y_0\in Y$ such that

$$\|\hat{T}([x_n])-y_0\|\to 0 \quad\text{and}\quad \|[x_n]-[x_0]\|\to 0 \quad (n\to\infty).$$

Choose $z_n\in\operatorname{Ker} T$ such that

$$\|x_n-x_0-z_n\| < \frac{1}{n} + \operatorname{dist}(x_n-x_0,\operatorname{Ker} T) = \frac{1}{n} + \|[x_n]-[x_0]\|$$

and put $s_n = x_n - z_n$, $n \geq 1$. Then $\|s_n - x_0\| \to 0$ and $\|Ts_n - y_0\| \to 0$ as $n \to \infty$. Since $G(T)$ is closed, $\{x_0, y_0\} \in G(T)$ and hence $\{[x_0], y_0\} \in G(\hat{T})$.

(2) Clearly, $\mathrm{Ran}\,(T) = \mathrm{Ran}\,(\hat{T})$. According to Definition 1.2.9 and the definition of $\hat{T}$, we have

$$\begin{aligned}\gamma(\hat{T}) &= \inf\{\|\hat{T}([x])\| \mid \|[x]\| = 1,\ [x] \in \mathfrak{D}(\hat{T})\}\\ &= \inf\{\|Tx\| \mid \mathrm{dist}\,(x, \mathrm{Ker}\,T) = 1,\ x \in \mathfrak{D}(T)\}\\ &= \gamma(T).\end{aligned}$$

Now for any $g \in \mathfrak{D}(\hat{T}^*)$ and $x \in \mathfrak{D}(T)$, $\hat{T}^*g([x]) = g(\hat{T}([x])) = g(Tx)$. Thus, $g \in \mathfrak{D}(T^*)$ and $T^*g(x) = \hat{T}^*g([x])$. Conversely, for any $g_0 \in \mathfrak{D}(T^*)$ and $x \in \mathfrak{D}(T)$, $T^*g_0(x) = g_0(Tx) = g_0(\hat{T}([x]))$. Noting that $T^*g_0(x - z) = T^*g_0(x)$, $\forall\, z \in \mathrm{Ker}\,T$, we have $|g_0(\hat{T}([x]))| \leq \|T^*g_0\|\|x - z\|$, $\forall\, z \in \mathrm{Ker}\,T$ and consequently, $|g_0(\hat{T}([x]))| \leq \|T^*g_0\|\|[x]\|$. So $g_0 \in \mathfrak{D}(\hat{T}^*)$ and

$$\hat{T}^*g_0([x]) = g_0(\hat{T}([x])) = g_0(Tx) = T^*g_0(x), \quad \forall\, x \in \mathfrak{D}(T),\ g_0 \in \mathfrak{D}(T^*).$$

The above shows that $\mathfrak{D}(T^*) = \mathfrak{D}(\hat{T}^*)$ and $\mathrm{Ker}\,T^* = \mathrm{Ker}\,\hat{T}^*$. Moreover, from $\hat{T}^*g([x]) = T^*g(x) = T^*g(x - z)$, $\forall\, x \in \mathfrak{D}(T)$, $z \in \mathrm{Ker}\,T$ and $g \in \mathfrak{D}(T^*)$, we get that $\|\hat{T}^*g\| = \|T^*g\|$. Therefore, $\gamma(\hat{T}^*) = \gamma(T^*)$. □

Proposition 1.2.12. *Let $T \in \mathcal{C}(X, Y)$. Then* $\mathrm{Ran}\,(T)$ *is closed if and only if $\gamma(T) > 0$.*

Proof. By Lemma 1.2.11 (2), it is equivalent to show that $\mathrm{Ran}\,(\hat{T})$ is closed iff $\gamma(\hat{T}) > 0$.

Firstly, assume $\gamma(\hat{T}) > 0$. Then $\|\hat{T}([x])\| \geq \gamma(\hat{T})\|[x]\|$, $\forall\, [x] \in \mathfrak{D}(\hat{T})$ by (1.2.6). Let $y \in \overline{\mathrm{Ran}\,(\hat{T})}$ and pick $\{[x_n]\} \subset \mathfrak{D}(\hat{T})$ such that $\|\hat{T}([x_n]) - y\| \to 0$ as $n \to \infty$. Then from

$$\|[x_n] - [x_m]\| \leq (\gamma(\hat{T}))^{-1}\|\hat{T}([x_n]) - \hat{T}([x_m])\|, \ \forall n, m \geq 1$$

we obtain that $\{[x_n]\}$ is a Cauchy sequence in $X/\mathrm{Ker}\,T$ and so that there is $x \in X$ such that $\|[x_n] - [x]\| \to 0$ as $n \to \infty$. Thus $\{[x], y\} \in \overline{G(\hat{T})} = G(\hat{T})$ by Lemma 1.2.11 (1), that is, $y = \hat{T}([x])$. So $\mathrm{Ran}\,(\hat{T})$ is closed.

Now suppose that $\mathrm{Ran}\,(\hat{T})$ is closed. Since $\mathrm{Ker}\,\hat{T} = \{0\}$, it follows from Lemma 1.2.2 (2) that $\hat{T}^{-1}\colon \mathrm{Ran}\,(T) \to Y$ is bounded. Thus, for any $[x] \in \mathfrak{D}(\hat{T})$,

$$\|[x]\| = \|\hat{T}^{-1}(\hat{T}([x]))\| \leq (\|\hat{T}^{-1}\|_{\mathrm{Ran}\,(T)})\|\hat{T}([x])\|.$$

This indicates that $\gamma(\hat{T}) \geq (\|\hat{T}^{-1}\|_{\mathrm{Ran}\,(T)})^{-1} > 0$. □

Corollary 1.2.13. *Let $T \in \mathcal{C}(X,Y)$. Suppose that there is a closed subspace M in Y such that* $\mathrm{Ran}\,(T) \cap M = \{0\}$ *and* $\mathrm{Ran}\,(T) + M$ *is closed in* Y. *Then* $\mathrm{Ran}\,(T)$ *is closed.*

Proof. Define a linear operator $W\colon \mathfrak{D}(T) \times M \to X$ by $W(\{x,y\}) = Tx + y$, $\forall\, x \in \mathfrak{D}(T)$ and $y \in M$. Then $\mathrm{Ker}\, W = \{\{x,0\} | \, x \in \mathrm{Ker}\, T\}$ and $\mathrm{Ran}\,(W) = \mathrm{Ran}\,(T) + M$ is closed. Thus, for any $x \in \mathfrak{D}(T)$, $y \in M$,

$$\begin{aligned}\|Tx\| + \|y\| \geq \|W(\{x,y\})\| &\geq \gamma(W)\mathrm{dist}\,(\{x,y\}, \mathrm{Ker}\, W)\\ &= \gamma(W)(\mathrm{dist}\,(x, \mathrm{Ker}\, T) + \|y\|).\end{aligned}$$

So $\gamma(T) \geq \gamma(W)$ and hence $\mathrm{Ran}\,(T)$ is closed when $\mathrm{Ran}\,(W)$ is closed by Proposition 1.2.12. □

Let V be a subspace in X and U be a subspace in X^*. Set

$$V^\perp = \{f \in X^* |\, f(x) = 0,\ x \in V\},\ {}^\perp U = \{x \in X |\, f(x) = 0,\ f \in U\}.$$

Remark 1.2.14. We note that

(1) $\overline{V}^\perp = V^\perp$ and ${}^\perp\overline{U} = {}^\perp U$.
(2) if X is a Hilbert space, then $V^\perp$ is the orthogonal complement of $\overline{V}$ and ${}^\perp U = U^\perp$ by Theorem 1.1.26.
(3) if V is a closed subspace of X, then $\big(X/V\big)^*$ is isomorphic to $V^\perp$. In fact, let $\pi\colon X \to X/V$ be the quotient mapping, i.e., $\pi(x) = [x]$, $\forall\, x \in X$. Then $\|\pi\| \leq 1$. Define linear mapping $\phi\colon \big(X/V\big)^* \to V^\perp$ by $\phi(f) = f \circ \pi$. Clearly, ϕ is bounded and injective.
Now let $g \in V^\perp$ and define $\hat{g}\colon X/V \to \mathbb{C}$ by $\hat{g}([x]) = g(x)$, $\forall\, x \in X$. Since g vanishes on V, it follows that $\hat{g}$ is well–defined and linear. Noting that for any $x \in X$ and $v \in V$,

$$|\hat{g}([x])| = |g(x - v)| \leq \|g\|\|x - v\|,$$

we have $|\hat{g}([x])| \leq \|g\|\|[x]\|$ and consequently, $\|\hat{g}\| \leq \|g\|$. Thus, $\hat{g} \in \big(X/V\big)^*$ and $\phi(\hat{g}) = \hat{g} \circ \pi = g$, that is, ϕ is surjective.

Theorem 1.2.15. *Let $T \in \mathcal{C}(X,Y)$ and $T \neq 0$. Then*

(1) $(\mathrm{Ran}\,(T))^\perp = \mathrm{Ker}\, T^*$ *and* ${}^\perp(\mathrm{Ker}\, T^*) = \overline{\mathrm{Ran}\,(T)}$.
(2) ${}^\perp(\mathrm{Ran}\,(T^*)) \cap \mathfrak{D}(T) = \mathrm{Ker}\, T$ *and* $\overline{\mathrm{Ran}\,(T^*)} \subset (\mathrm{Ker}\, T)^\perp$.

Proof. (1) Let $g \in \mathrm{Ker}\, T^*$. Then for any $x \in \mathfrak{D}(T)$, $g(Tx) = T^*g(x) = 0$, that is, $g \in \mathrm{Ran}\,(T)^\perp$. On the other hand, let $g \in \mathrm{Ran}\,(T)^\perp$. Then $g(Tx) =$

$0, \forall x \in \mathfrak{D}(T)$. Thus, $g \in \mathfrak{D}(T^*)$ and $T^*g(x) = g(Tx) = 0, \forall x \in \mathfrak{D}(T)$. So, $g \in \operatorname{Ker} T^*$ and hence $\operatorname{Ran}(T)^{\perp} = \operatorname{Ker} T^*$.

Let $y \in {}^{\perp}(\operatorname{Ker} T^*)$. If $y \notin \overline{\operatorname{Ran}(T)}$, there is $g_0 \in Y^*$ such that $g_0(y) = 1$ and $g_0(z) = 0, \forall z \in \operatorname{Ran}(T)$. Thus, $g_0(Tx) = 0, \forall x \in \mathfrak{D}(T)$ and hence $g_0 \in \mathfrak{D}(T^*)$ and $T^*g_0(x) = g_0(Tx) = 0, \forall x \in \mathfrak{D}(T)$, i.e., $g_0 \in \operatorname{Ker} T^*$. So, $g_0(y) = 0$. But this contradicts the condition $g_0(y) = 1$. Therefore, ${}^{\perp}(\operatorname{Ker} T^*) \subset \overline{\operatorname{Ran}(T)}$. Conversely, for any $g \in \operatorname{Ker} T^*$ and any $x \in \mathfrak{D}(T)$, we have $g(Tx) = T^*g(x) = 0$, which means that $\operatorname{Ran}(T) \subset {}^{\perp}(\operatorname{Ker} T^*)$. So ${}^{\perp}(\operatorname{Ker} T^*) = \overline{\operatorname{Ran}(T)}$.

(2) Let $x \in \operatorname{Ker} T$. Then for any $f \in \mathfrak{D}(T^*)$, $T^*f(x) = f(Tx) = 0$. Thus, $x \in {}^{\perp}(\operatorname{Ran}(T^*) \cap \mathfrak{D}(T)$. Now let $x \in {}^{\perp}(\operatorname{Ran}(T^*)) \cap \mathfrak{D}(T)$. Then $T^*g(x) = g(Tx) = 0, \forall g \in \mathfrak{D}(T^*)$. Since $\mathfrak{D}(T^*)$ is total in Y^* by Proposition 1.2.4, $g(Tx) = 0, \forall g \in \mathfrak{D}(T^*)$ implies that $Tx = 0$. Therefore, ${}^{\perp}(\operatorname{Ran}(T^*)) \cap \mathfrak{D}(T) = \operatorname{Ker} T$.

For every $g \in \mathfrak{D}(T^*)$ and $x \in \operatorname{Ker} T$, we have $T^*g(x) = g(Tx) = 0$. Thus, $\operatorname{Ran}(T^*) \subset (\operatorname{Ker} T)^{\perp}$ and consequently, $\overline{\operatorname{Ran}(T^*)} \subset (\operatorname{Ker} T)^{\perp}$. □

Let Z be any normed space and $a > 0$. Set $D_Z(a) = \{z \in Z \mid \|z\| \leq a\}$.

Lemma 1.2.16. *Let* $T \in \mathcal{C}(X, Y)$ *with* $\operatorname{Ran}(T^*)$ *closed and* $\operatorname{Ker} T^* = \{0\}$. *Then there is* $r > 0$ *such that* $D_Y(r) \subset T(D_X(2) \cap \mathfrak{D}(T))$.

Proof. By Proposition 1.2.4 (1) and Lemma 1.2.2 (2), $(T^*)^{-1}\colon \operatorname{Ran}(T^*) \to \mathfrak{D}(T^*)$ is bounded. Put $r = \|(T^*)^{-1}\|_{\operatorname{Ran}(T^*)} > 0$. We claim that $D_Y(r) \subset \overline{T(D_X(1) \cap \mathfrak{D}(T))}$.

In fact, if there is $y_0 \in D_Y(r)$ and $y_0 \notin \overline{T(D_X(1) \cap \mathfrak{D}(T))}$, we can find $g_0 \in Y^*$ such that

$$\operatorname{Re}(g_0(y_0)) > 1, \quad \operatorname{Re}(g_0(y)) \leq 1, \ \forall y \in \overline{T(D_X(1) \cap \mathfrak{D}(T))}$$

by Theorem 1.1.6. Put $h(x) = (g_0 \circ T)(x), \forall x \in \mathfrak{D}(T)$. Then, $\operatorname{Re}(h(x)) \leq \|x\|, \forall x \in \mathfrak{D}(T)$ and hence $|\operatorname{Re}(h(x))| \leq \|x\|, \forall x \in \mathfrak{D}(T)$. Noting that

$$h(x) = \operatorname{Re}(h(x)) - i \operatorname{Re}(h(ix)) = |h(x)| \mathrm{e}^{i\theta_x}, \ \forall x \in \mathfrak{D}(T)$$

for some $\theta_x \in \mathbb{R}$, we have

$$|h(x)| = h(\mathrm{e}^{-i\theta_x} x) = \operatorname{Re}(h(\mathrm{e}^{-i\theta_x} x)) \leq \|x\|, \ \forall x \in \mathfrak{D}(T).$$

Thus, $g_0 \in \mathfrak{D}(T^*)$ and $T^*g_0 = h$ with $\|T^*g_0\| \leq 1$. Furthermore, from $g_0 = (T^*)^{-1}T^*g_0$, we get

$$1 < \operatorname{Re}(g_0(y_0)) \leq \|g_0\| \|y_0\| \leq \|(T^*)^{-1}\|_{\operatorname{Ran}(T^*)} \|T^*g_0\| \|y_0\| \leq r^{-1}\|y_0\|,$$

i.e., $\|y_0\| > r$, which contradicts the assumption that $y_0 \in D_Y(r)$.

Let $y \in D_Y(r)$. Since $D_Y(2^{-n}r) \subset \overline{T(D_X(2^{-n}) \cap \mathfrak{D}(T))}$, $n = 0, 1, \cdots$, it follows that there is $x_1 \in D_X(1) \cap \mathfrak{D}(T)$ such that $\|y - Tx_1\| < 2^{-1}r$ and then there exists $x_2 \in D_X(2^{-1}) \cap \mathfrak{D}(T)$ such that $\|y - Tx_1 - Tx_2\| < 2^{-2}r$. In this way, we can find $x_n \in D_X(2^{-n}) \cap \mathfrak{D}(T)$ such that

$$\|y - Tx_1 - \cdots - Tx_n\| < 2^{-n}r, \; n = 1, 2, \cdots.$$

Put $s_n = \sum_{k=1}^{n} x_k$, $n = 1, 2, \cdots$. Then $\{s_n\} \subset \mathfrak{D}(T)$ and $\|y - Ts_n\| < 2^{-n}r$, $n \geq 1$. Note that $\|s_n\| \leq 2^{-(n-1)}$, $n \geq 1$. Thus, $\{s_n\}$ converges to an element x_0 in X with $\|x_0\| \leq 2$. Therefore, $\{x_0, y\} \in \overline{G(T)} = G(T)$, that is, $x_0 \in D_X(2) \cap \mathfrak{D}(T)$ and $Tx_0 = y$. □

Theorem 1.2.17. *Let $T \in \mathcal{C}(X, Y)$ and $T \neq 0$. Then following statements are equivalent*

(1) $\mathrm{Ran}\,(T)$ *is closed in* Y.
(2) $\mathrm{Ran}\,(T^*) = (\mathrm{Ker}\, T)^{\perp}$.
(3) $\mathrm{Ran}\,(T^*)$ *is closed in* X^*.

Proof. (1)⇒(2) By Proposition 1.2.12, $\gamma(T) > 0$. Let $f \in (\mathrm{Ker}\, T)^{\perp}$ and put $g(Tx) = f(x)$, $\forall x \in \mathfrak{D}(T)$. Then g is well–defined for $f(x) = 0$, $\forall x \in \mathrm{Ker}\, T$ and g is also linear on $\mathrm{Ran}\,(T)$. Noting that for any $z \in \mathrm{Ker}\, T$,

$$|f(x)| = |f(x - z)| \leq \|f\|\|x - z\|, \quad \forall x \in \mathfrak{D}(T),$$

we have

$$|f(x)| \leq \|f\| \mathrm{dist}\,(x, \mathrm{Ker}\, T) \leq \|f\|(\gamma(T))^{-1}\|Tx\|, \quad \forall x \in \mathfrak{D}(T)$$

and consequently, $|g(y)| \leq \|f\|(\gamma(T))^{-1}\|y\|$, $\forall y \in \mathrm{Ran}\,(T)$. Thus, there is $h \in Y^*$ such that $h(y) = g(y)$, $\forall y \in \mathrm{Ran}\,(T)$ and $\|h\| \leq \|f\|(\gamma(T))^{-1}$ by Proposition 1.1.4. Now $h(Tx) = g(Tx) = f(x)$, $\forall x \in \mathfrak{D}(T)$ indicates that $h \in \mathfrak{D}(T^*)$ and $T^*h(x) = h(Tx) = f(x)$, $\forall x \in \mathfrak{D}(T)$. Thus, $(\mathrm{Ker}\, T)^{\perp} \subset \mathrm{Ran}\,(T^*)$ and hence $\mathrm{Ran}\,(T^*) = (\mathrm{Ker}\, T)^{\perp}$ by Theorem 1.2.15 (2).

The implication (2)⇒(3) is obvious.

(3)⇒(1) Let T_1 denote the operator $T\colon \mathfrak{D}(T) \to \overline{\mathrm{Ran}\,(T)} = Y_1$. Clearly, $G(T_1) = G(T)$ and $\mathrm{Ker}\, T_1^* = (\mathrm{Ran}\,(T_1))^{\perp} = \{0\}$ by Theorem 1.2.15 (1).

For any $g \in \mathfrak{D}(T_1^*) \subset Y_1^*$, there is $h \in Y^*$ which extends g by Proposition 1.1.4. Thus, for any $x \in \mathfrak{D}(T)$, $T_1^*g(x) = g(Tx) = h(Tx)$ and so that $h \in \mathfrak{D}(T^*)$ with $T_1^*g = T^*h$. Therefore, $\mathrm{Ran}\,(T_1^*) \subset \mathrm{Ran}\,(T^*)$. Now let $g \in \mathfrak{D}(T^*) \subset Y^*$ and g_1 be the restriction of g on Y_1. Then

$$T^*g(x) = g(Tx) = g_1(T_1x) = T_1^*g_1(x), \; \forall x \in \mathfrak{D}(T) = \mathfrak{D}(T_1).$$

Thus, $T^*g = T_1^*g_1 \in \mathrm{Ran}\,(T_1^*)$ and consequently, $\mathrm{Ran}\,(T_1^*) = \mathrm{Ran}\,(T^*)$ is closed in X^*. Then $(T_1^*)^{-1}\colon \mathrm{Ran}\,(T_1^*) \to X^*$ is bounded by Lemma 1.2.2 and moreover, $D_{Y_1}(r) \subset T_1(D_X(2) \cap \mathfrak{D}(T_1))$ by Lemma 1.2.16, where $r = \|(T_1^*)^{-1}\|^{-1}$. Therefore, for any $x \in \mathfrak{D}(T)$ with $x \notin \mathrm{Ker}\,T$, there is $z \in D_X(2) \cap \mathfrak{D}(T_1)$ such that $r\dfrac{T_1x}{\|T_1x\|} = T_1z$. Set $y = x - r^{-1}\|T_1x\|z$. Then $y \in \mathrm{Ker}\,T$ and

$$\begin{aligned}\mathrm{dist}\,(x, \mathrm{Ker}\,T) &= \mathrm{dist}\,(r^{-1}\|Tx\|z, \mathrm{Ker}\,T)\\ &= r^{-1}\|Tx\|\mathrm{dist}\,(z, \mathrm{Ker}\,T) \le r^{-1}\|Tx\|\|z\|\\ &\le 2r^{-1}\|Tx\|.\end{aligned}$$

The above means that $\gamma(T) \ge 2^{-1}r > 0$. It follows from Proposition 1.2.12 that $\mathrm{Ran}\,(T)$ is closed in Y. □

Corollary 1.2.18. *Let $T \in \mathcal{C}(X, Y)$ and $T \ne 0$. Then $\gamma(T) = \gamma(T^*)$.*

Proof. By Proposition 1.2.12 and Theorem 1.2.17, we have $\gamma(T) = 0$ iff $\gamma(T^*) = 0$.

Assume that $\gamma(T) > 0$. Then $\mathrm{Ran}\,(T)$ is closed by Proposition 1.2.12. Consider $T_1\colon \mathfrak{D}(T_1) \to \mathrm{Ran}\,(T) = Y_1$ given by $T_1x = Tx$, $\forall\, x \in \mathfrak{D}(T_1) = \mathfrak{D}(T)$ and assume that $\mathrm{Ker}\,T_1 = \{0\}$ by Lemma 1.2.11.

We first prove that $\gamma(T_1^*) = \gamma(T^*)$. Let $g_1 \in \mathfrak{D}(T_1^*)$ and $h \in Y^*$ such that $g_1 = h|_{Y_1}$. Then by the proof (3)⇒(1) of Theorem 1.2.17, $h \in \mathfrak{D}(T^*)$ and $T^*h = T_1^*g_1$. Noting that for any $x \in \mathfrak{D}(T)$ and $\phi \in \mathrm{Ker}\,T^*$, $\phi(Tx) = T^*\phi(x) = 0$, we have

$$|g_1(Tx)| = |h(Tx)| = |(h - \phi)(Tx)| \le \|h - \phi\|\|Tx\|$$

and hence $\mathrm{dist}\,(h, \mathrm{Ker}\,T^*) \ge \|g_1\|_{Y_1}$. Therefore,

$$\|T_1^*g_1\| = \|T^*h\| \ge \gamma(T^*)\mathrm{dist}\,(h, \mathrm{Ker}\,T^*) \ge \gamma(T^*)\|g_1\|_{Y_1}. \qquad (1.2.7)$$

Since $\mathrm{Ker}\,T_1^* = \mathrm{Ran}\,(T_1)^{\perp} = \{0\}$, it follows from 1.2.7 that $\gamma(T_1^*) \ge \gamma(T^*)$.

Now let $g \in \mathfrak{D}(T^*)$ and put $g_1 = g|_{Y_1}$. Then $g_1 \in \mathfrak{D}(T_1^*)$ and $T_1^*g_1 = T^*g$. By Proposition 1.1.4, there exists $h \in Y^*$ such that

$$h(y) = g_1(y),\ \forall\, y \in Y_1 \text{ and } \|h\| = \|g_1\|_{Y_1}.$$

Then $h \in \mathfrak{D}(T^*)$ and $T^*h = T_1^*g_1 = T^*g$, i.e., $h - g \in \mathrm{Ker}\,T^*$ and moreover, for each $g \in \mathfrak{D}(T^*)$,

$$\begin{aligned}\|T^*g\| &= \|T_1^*g_1\| \ge \gamma(T_1^*)\|g_1\|_{Y_1} = \gamma(T_1^*)\|h\|\\ &\ge \gamma(T_1^*)\mathrm{dist}\,(h, \mathrm{Ker}\,T^*) = \gamma(T_1^*)\mathrm{dist}\,(g, \mathrm{Ker}\,T^*),\end{aligned}$$

which means that $\gamma(T^*) \geq \gamma(T_1^*)$.

Next, we show that $\gamma(T) = \gamma(T_1) = \gamma(T_1^*)$. Clearly, $\gamma(T) = \gamma(T_1)$. Since $T_1 \in \mathcal{C}(X, Y_1)$ with $\operatorname{Ker} T_1 = \{0\}$ and $\operatorname{Ran}(T_1) = Y_1$, it follows from Lemma 1.2.2 that $B = T_1^{-1} \colon Y_1 \to \mathfrak{D}(T_1)$ is bounded. From $BT_1 = I_X$ on $\mathfrak{D}(T_1)$ and $T_1 B = I_{Y_1}$, we get that $\operatorname{Ran}(B^*) \subset \mathfrak{D}(T_1^*)$, $T_1^* B^* = I_{X^*}$ and $B^* T_1^* = I_{(Y_1)^*}$ on $\mathfrak{D}(T_1^*)$ by Proposition 1.2.4.

From $BT_1(x) = x$, $\forall\, x \in \mathfrak{D}(T)$ and $T_1 B(y) = y$, $\forall\, y \in Y$, we get that

$$\|x\| = \|BT_1(x)\| \leq \|B\|\|T_1 x\|, \quad \|y\| = \|T_1 B(y)\| \geq \gamma(T_1)\|By\|.$$

Thus $\gamma(T) = \|B\|^{-1}$.

Similarly, from $T_1^* B^* = I_{X^*}$ and $B^* T_1^* = I_{(Y_1)^*}$ on $\mathfrak{D}(T_1^*)$, we have $\gamma(T^*) = \|B^*\|^{-1}$ by above argument.

Therefore, $\gamma(T^*) = \gamma(T_1^*) = \|B^*\|^{-1} = \|B\|^{-1} = \gamma(T_1) = \gamma(T)$. □

1.3 The gap function of subspaces

For the Banach space X, let $\operatorname{Sb}(X)$ denote the set of all subspaces in X.

Definition 1.3.1. The function $\delta(\cdot, \cdot)$ on $\operatorname{Sb}(X)$ is defined by

$$\delta(M, N) = \begin{cases} \sup\{f_N(x) \mid x \in M,\ \|x\| = 1\} & M \neq \{0\} \\ 0 & M = \{0\} \end{cases}$$

$\forall\, M,\, N \in \operatorname{Sb}(X)$.

Let $M,\, N \in \operatorname{Sb}(X)$. Put $\hat{\delta}(M, N) = \max\{\delta(M, N), \delta(N, M)\}$. $\hat{\delta}(M, N)$ is called the gap of M and N.

Proposition 1.3.2. *Let L, M, $N \in \operatorname{Sb}(X)$. Then*

(1) $0 \leq \delta(M, N) \leq 1$, $0 \leq \hat{\delta}(M, N) \leq 1$ *and* $\hat{\delta}(M, N) = \hat{\delta}(N, M)$.
(2) $\delta(\overline{M}, \overline{N}) = \delta(M, N)$, $\hat{\delta}(\overline{M}, \overline{N}) = \hat{\delta}(M, N)$.
(3) $\delta(M, N) = 0$ *iff* $\overline{M} \subset \overline{N}$ *and* $\hat{\delta}(M, N) = 0$ *iff* $\overline{M} = \overline{N}$.
(4) $\delta(L, N) \leq \delta(L, M) + (1 + \delta(L, M))\delta(M, N)$.
(5) $\hat{\delta}(L, N) \leq 2(\hat{\delta}(L, M) + \hat{\delta}(M, N))$.

Proof. (1) is obvious.

(2) We have $\operatorname{dist}(x, \overline{N}) = \operatorname{dist}(x, N)$, $\forall\, x \in X$ by Proposition 1.1.1 (2). Thus, $\delta(M, N) \leq \delta(\overline{M}, \overline{N})$.

Now let $\bar{x} \in \overline{M}$ with $\|\bar{x}\| = 1$. Then for any $\epsilon \in (0, 1)$, there is $x_0 \in M$ such that $\|\bar{x} - x_0\| < \epsilon$. Put $x = \dfrac{x_0}{\|x_0\|}$. Then $x \in M$ with $\|x\| = 1$ and

$\|\bar{x} - x\| < 2\epsilon$. Thus, by Proposition 1.1.1 (5),

$$\text{dist}\,(\bar{x}, \overline{N}) \leq \text{dist}\,(x, \overline{N}) + \|\bar{x} - x\| < \delta(M, N) + 2\epsilon$$

and hence $\delta(\overline{M}, \overline{N}) \leq \delta(M, N)$ by letting $\epsilon \to 0$. Therefore, $\delta(\overline{M}, \overline{N}) = \delta(M, N)$ and $\hat{\delta}(\overline{M}, \overline{N}) = \hat{\delta}(M, N)$.

(3) $\delta(M, N) = 0$ implies that $\text{dist}\,(x, N) = 0$, $\forall\, x \in \overline{M}$ with $\|x\| = 1$ by (2). Thus, $x \in \overline{N}$ and consequently, $\overline{M} \subset \overline{N}$. Conversely, if $\overline{M} \subset \overline{N}$, then $\text{dist}\,(x, \overline{N}) = 0$, $\forall\, x \in \overline{M}$. So that $\delta(\overline{M}, \overline{N}) = 0$.

Since $\hat{\delta}(M, N) = 0$ iff $\delta(M, N) = \delta(N, M) = 0$, we have $\hat{\delta}(M, N) = 0$ iff $\overline{M} \subset \overline{N}$ and $\overline{N} \subset \overline{M}$, i.e., $\overline{M} = \overline{N}$ by above argument.

(4) For any $x \in L$, $y \in M$ and $z \in N$, we have

$$\text{dist}\,(x, N) \leq \|x - z\| \leq \|x - y\| + \|y - z\|. \tag{1.3.1}$$

(1.3.1) indicates that

$$\begin{aligned} \text{dist}\,(x, N) &\leq \|x - y\| + \text{dist}\,(y, N) \leq \|x - y\| + \|y\|\delta(M, N) \\ &\leq \|x - y\| + (\|x - y\| + \|x\|)\delta(M, N). \end{aligned}$$

Thus, $\text{dist}\,(x, N) \leq \text{dist}\,(x, M) + (\|x\| + \text{dist}\,(x, M))\delta(M, N)$ and consequently, the assertion follows.

(5) From (4), we get that

$$\begin{aligned} \delta(L, N) &\leq \delta(L, M) + 2\delta(M, N) \leq 2(\delta(L, M) + \delta(M, N)), \\ \delta(N, L) &\leq \delta(N, M) + 2\delta(M, L) \leq 2(\delta(M, L) + \delta(N, M)). \end{aligned}$$

Therefore, $\hat{\delta}(L, N) \leq 2(\hat{\delta}(L, M) + \hat{\delta}(M, N))$. □

Proposition 1.3.3. *Let M, $N \in \text{Sb}(X)$.*

(1) *If $\overline{N} \subset \overline{M}$ and $\overline{N} \neq \overline{M}$, then $\delta(M, N) = 1$.*
(2) *If M and N are of finite-dimensional and $\dim M > \dim N$, then $\delta(M, N) = 1$.*

Proof. (1) Let $x_0 \in \overline{M} \backslash \overline{N}$. Then $d = \text{dist}\,(x_0, \overline{N}) > 0$. Thus, for any $\epsilon > 0$, there is $y_\epsilon \in \overline{N}$ such that $d = \text{dist}\,(x_0, N) > \|x_0 - y_\epsilon\| - \epsilon$. Put $x_\epsilon = \dfrac{x_0 - y_\epsilon}{\|x_0 - y_\epsilon\|}$. Then $x_\epsilon \in \overline{M}$ with $\|x_\epsilon\| = 1$ and for any $z \in \overline{N}$,

$$\|x_\epsilon - z\| = \frac{\|x_0 - y_\epsilon - \|x_0 - y_\epsilon\|z\|}{\|x_0 - y_\epsilon\|} > \frac{d}{d + \epsilon} = 1 - (d + \epsilon)^{-1}\epsilon.$$

This means that $\delta(M, N) = \delta(\overline{M}, \overline{N}) = 1$ by Proposition 1.3.2.

(2) Put $X_0 = M + N$. Then X_0 is a finite–dimensional Banach space. First assume that X_0 is strictly convex. Then for any $x \in M$, there is a

unique $\rho(x) \in N$ such that $\operatorname{dist}(x, N) = \|x - \rho(x)\|$ by Proposition 1.1.21. Then $\rho\colon M \to N$ is a mapping with $\|\rho(x)\| \leq 2\|x\|$, $\forall\, x \in M$. Since

$$\begin{aligned}\|x - (-\rho(-x))\| = \| - x - \rho(-x)\| &= \operatorname{dist}(-x, N) \\ &= \operatorname{dist}(x, N) = \|x - \rho(x)\|, \quad \forall\, x \in M,\end{aligned}$$

we have $\rho(x) = -\rho(-x)$, $\forall\, x \in M$ by the uniqueness of $\rho(x)$.

Let $x_0 \in M$ and $\{x_n\} \subset M$ such that $\|x_n - x_0\| \to 0$. Passing to a subsequence if necessary, we may assume that $\rho(x_n) \to y_0$ as $n \to \infty$ for some $y_0 \in N$ since $\{\rho(x_n)\}$ is bounded in N and N is of finite–dimensional.

Noting that $|\operatorname{dist}(x_n, N) - \operatorname{dist}(x_0, N)| \leq \|x_n - x_0\|$, $\forall\, n \in \mathbb{N}$ by Proposition 1.1.1 (5), we get that $\|x_0 - y_0\| = \|x_0 - \rho(x_0)\|$ and hence $y_0 = \rho(x_0)$. This indicates that ρ is continuous on M.

Let $S_M(0) = \{x \in M |\, \|x\| = 1\}$. Since $\rho\colon S_M(0) \to N$ is continuous and $\rho(-x) = -\rho(x)$, $\forall\, x \in S_M(0)$, it follows from Borsuk's Antipodal Theorem (Theorem A.1.3) that there exists $m \in M$ with $\|m\| = 1$ such that $\rho(m) = 0$. Thus, $\operatorname{dist}(m, N) = \|m - \rho(m)\| = 1$.

In general, we can choose a Hilbert norm $\|\cdot\|_H$ on X_0 and define strictly convex norms $\|\cdot\|_n = \|\cdot\| + \dfrac{1}{n}\|\cdot\|_H$ on X_0. Then there is a sequence $\{m_n\} \subset M$ such that

$$\|m_n\|_n = \inf_{z \in N} \|m_n - z\|_n = 1, \quad \forall\, n \in N$$

by above argument. Clearly, $\|m_n\| \leq 1$, $\forall\, n \in \mathbb{N}$. Passing to a subsequence if necessary, we can assume that $\{m_n\}$ is a convergent sequence. Let $m \in M$ such that $\|m_n - m\| \to 0$. Then $\|m_n\|_n \to \|m\|$ and $\|m_n - m\|_n \to 0$ as $n \to \infty$ since the norm $\|\cdot\|_H$ is equivalent to $\|\cdot\|$. So $\|m\| = 1$ and

$$1 = \lim_{n \to \infty} \inf_{z \in N} \|m_n - z\|_n = \inf_{z \in N} \|m - z\| = \operatorname{dist}(m, N),$$

that is, $\delta(M, N) = 1$. □

Proposition 1.3.4. *Let M and N be closed subspaces in X. Then*

$$\delta(M, N) = \delta(N^{\perp}, M^{\perp}), \quad \hat{\delta}(M, N) = \hat{\delta}(M^{\perp}, N^{\perp}).$$

Proof. Let $x \in M$ with $\|x\| = 1$. If $x \notin N$, it follows from Corollary 1.1.5 (2) that there is $f \in X^*$ such that

$$f(y) = 0, \ \forall\, y \in N, \ f(x) = 1, \ \|f\| = 1/\operatorname{dist}(x, N).$$

Set $g = \operatorname{dist}(x, N) f$. Then $g \in N^{\perp}$ with $g(x) = \operatorname{dist}(x, N)$ and $\|g\| = 1$. Thus, for any $h \in M^{\perp}$, $\operatorname{dist}(x, N) = \|(g - h)(x)\| \leq \|g - h\|$ and hence $\operatorname{dist}(x, N) \leq \operatorname{dist}(g, M^\perp) \leq \delta(N^\perp, M^\perp)$, which gives $\delta(M, N) \leq \delta(N^\perp, M^\perp)$.

Let $f \in N^\perp$ with $\|f\| = 1$ and let $g = f|_M$ (the restriction of f on M). Then $g \in M^*$ with $\|g\|_M = \sup\{|f(x)|| x \in D(M)\}$. By Proposition 1.1.4, there is $\tilde{g} \in X^*$ such that $\tilde{g}|_M = g$ and $\|\tilde{g}\| = \|g\|_M$. Therefore, $\tilde{g} - f \in M^\perp$ and consequently,

$$\operatorname{dist}(f, M^\perp) = \operatorname{dist}(\tilde{g}, M^\perp) \leq \|\tilde{g}\| = \|g\|_M.$$

Since for any $y \in N$ with $\|y\| = 1$ and $x \in D(M)$,

$$|f(x)| = |f(x - y)| \leq \|f\|\|x - y\| = \|x - y\|,$$

we have $|f(x)| \leq \operatorname{dist}(x, N) \leq \delta(M, N)$. So $\operatorname{dist}(f, M^\perp) \leq \|g\|_M \leq \delta(M, N)$ and hence $\delta(N^\perp, M^\perp) \leq \delta(M, N)$.

The above shows that $\hat{\delta}(M, N) = \hat{\delta}(M^\perp, N^\perp)$. □

Let $T \in \mathcal{C}(X, Y)$ and $A \in \mathcal{B}(X, Y)$. Put $\bar{T} = T + A$. Then $\mathfrak{D}(\bar{T}) = \mathfrak{D}(T)$ and $G(\bar{T})$ is closed in $X \times Y$, i.e., $\bar{T} \in \mathcal{C}(X, Y)$.

Lemma 1.3.5. *Let $T \in \mathcal{C}(X, Y)$ and $A \in B(X, Y)$. Put $\bar{T} = T + A$. Then*

(1) $\gamma(T)\delta(\operatorname{Ker}\bar{T}, \operatorname{Ker}T) \leq \|A\|$.
(2) $\gamma(T)\delta(\operatorname{Ran}(T), \operatorname{Ran}(\bar{T})) \leq \|A\|$.

Proof. (1) From $\|Tx\| \geq \gamma(T)\operatorname{dist}(x, \operatorname{Ker}T)$, $x \in \mathfrak{D}(T)$, we get that for any $x \in \operatorname{Ker}\bar{T}$ with $\|x\| = 1$,

$$\gamma(T)\operatorname{dist}(x, \operatorname{Ker}T) \leq \|(\bar{T} - A)x\| = \|Ax\| \leq \|A\|.$$

Consequently, $\gamma(T)\delta(\operatorname{Ker}\bar{T}, \operatorname{Ker}T) \leq \|A\|$.

(2) For any $u \in \operatorname{Ran}(T)$ with $\|u\| = 1$, take $x \in \mathfrak{D}(T)$ such that $u = Tx$. Then $x \neq 0$ and for any $z \in \operatorname{Ker}T$,

$$\operatorname{dist}(u, \operatorname{Ran}(\bar{T})) \leq \|u - \bar{T}(x - z)\| = \|T(x - z) - \bar{T}(x - z)\| \leq \|A\|\|x - z\|.$$

Thus, $\operatorname{dist}(u, \operatorname{Ran}(T)) \leq \|A\|\operatorname{dist}(x, \operatorname{Ker}T)$ and hence

$$\gamma(T)\operatorname{dist}(u, \operatorname{Ran}(T)) \leq \|A\|\|Tx\| = \|A\|\|u\| = \|A\|.$$

So $\gamma(T)\delta(\operatorname{Ran}(T), \operatorname{Ran}(\bar{T})) \leq \|A\|$. □

Lemma 1.3.6. *Let T, $\bar{T} \in \mathcal{C}(X, Y)$ with $\mathfrak{D}(T) = \mathfrak{D}(\bar{T})$. Assume that there are constants $\lambda > 0$ and $\mu \in \mathbb{R}$ such that*

$$\|\bar{T}x\| \geq \lambda\|Tx\| + \mu\|x\|, \quad \forall x \in \mathfrak{D}(T). \tag{1.3.2}$$

Then

$$\gamma(\bar{T}) \geq \lambda\gamma(T)(1 - 2\delta(\operatorname{Ker} T, \operatorname{Ker} \bar{T})) + \mu. \tag{1.3.3}$$

Especially, if $X = H$ and $Y = K$, then (1.3.3) can be improved as

$$\gamma(\bar{T}) \geq \lambda\gamma(T)[1 - \delta^2(\operatorname{Ker} T, \operatorname{Ker} \bar{T})]^{1/2} + \mu \tag{1.3.4}$$

$$\geq \lambda\gamma(T)[1 - \delta(\operatorname{Ker} T, \operatorname{Ker} \bar{T})] + \mu. \tag{1.3.5}$$

Proof. By the definition of $\gamma(\bar{T})$, we can pick $\{x_n\} \subset \mathfrak{D}(T)$ such that $\|\bar{T}x_n\| \to \gamma(\bar{T})$ as $n \to \infty$ and $\operatorname{dist}(x_n, \operatorname{Ker}\bar{T}) = 1$, $n \geq 1$. Choose $\{y_n\} \subset \operatorname{Ker}\bar{T}$ such that $1 \leq \|x_n - y_n\| < 1 + n^{-1}$, $n \in \mathbb{N}$. Then for any $n \geq 1$,

$$\begin{aligned}\|\bar{T}x_n\| &= \|\bar{T}(x_n - y_n)\| \geq \lambda\|T(x_n - y_n)\| + \mu\|x_n - y_n\| \\ &\geq \lambda\gamma(T)\operatorname{dist}(x_n - y_n, \operatorname{Ker} T) + \mu\|x_n - y_n\|.\end{aligned}$$

Pick $\{z_n\} \subset \operatorname{Ker} T$ such that $\|x_n - y_n - z_n\| < \operatorname{dist}(x_n - y_n, \operatorname{Ker} T) + n^{-1}$, $n \in \mathbb{N}$. Then $\|x_n - y_n - z_n\| < 1 + 2n^{-1}$ and so that $\|z_n\| < 2 + 3n^{-1}$, $n \geq 1$. Furthermore, for any $z \in \operatorname{Ker} T$ and $n \geq 1$,

$$\begin{aligned}\operatorname{dist}(x_n - y_n, \operatorname{Ker} T) &> \|(x_n - z) - (y_n + z_n - z)\| - \frac{1}{n} \\ &\geq \|x_n - z\| - \|y_n + z_n - z\| - \frac{1}{n} \\ &\geq 1 - \frac{1}{n} - \|z_n - (z - y_n)\|.\end{aligned}$$

Thus, for any $n \in \mathbb{N}$,

$$\operatorname{dist}(x_n - y_n, \operatorname{Ker} T) \geq 1 - \frac{1}{n} - \operatorname{dist}(z_n, \operatorname{Ker}\bar{T}).$$

Note that, for any $n \geq 1$,

$$\operatorname{dist}(z_n, \operatorname{Ker}\bar{T}) \leq \|z_n\|\delta(\operatorname{Ker} T, \operatorname{Ker}\bar{T}) < \big(2 + \frac{3}{n}\big)\delta(\operatorname{Ker} T, \operatorname{Ker}\bar{T}).$$

So let $n \to \infty$, we get that

$$\gamma(\bar{T}) \geq \lambda\gamma(T)(1 - 2\delta(\operatorname{Ker} T, \operatorname{Ker}\bar{T})) + \mu.$$

Now let $T, \bar{T} \in C(X, Y)$ with $\mathfrak{D}(T) = \mathfrak{D}(\bar{T})$ satisfying (1.3.2). Let $x \in (\operatorname{Ker}\bar{T})^{\perp}$ with $\|x\| = 1$. Write x as $x = x_1 + x_2$, where $x_1 \in \operatorname{Ker} T$ and $x_2 \in (\operatorname{Ker} T)^{\perp}$ (Theorem 1.1.20). Then

$$\operatorname{dist}(x, \operatorname{Ker} T) = \|x_2\| = (1 - \|x_1\|^2)^{1/2}, \quad \|x_1\| = \operatorname{dist}(x, (\operatorname{Ker} T)^{\perp})$$

and so that

$$\begin{aligned}\|\bar{T}x\| &\geq \lambda\|Tx\| + \mu \geq \lambda\gamma(T)\operatorname{dist}(x, \operatorname{Ker} T) + \mu \\ &\geq \lambda\gamma(T)[1 - (\operatorname{dist}(x, (\operatorname{Ker} T)^{\perp}))^2]^{1/2} + \mu.\end{aligned}$$

Since $\gamma(\bar{T}) = \inf\{\|\bar{T}x\| \,|\, x \in (\operatorname{Ker}\bar{T})^{\perp}, \|x\| = 1\}$ and

$$\delta((\operatorname{Ker}\bar{T})^{\perp}, (\operatorname{Ker}T)^{\perp}) = \delta(\operatorname{Ker}T, \operatorname{Ker}\bar{T})$$

by Proposition 1.3.4, it follows that

$$\gamma(\bar{T}) \geq \lambda\gamma(T)[1 - \delta^2(\operatorname{Ker}T, \operatorname{Ker}\bar{T})]^{1/2} + \mu.$$

Since $1 - \delta^2(\operatorname{Ker}T, \operatorname{Ker}\bar{T}) \geq [1 - \delta(\operatorname{Ker}T, \operatorname{Ker}\bar{T})]^2$, we can get that

$$\gamma(\bar{T}) \geq \lambda\gamma(T)[1 - \delta(\operatorname{Ker}T, \operatorname{Ker}\bar{T})] + \mu.$$

□

Let M, N be closed subspaces in the Hilbert space H and let P_M, P_N be orthogonal projections of H onto M and N respectively.

Lemma 1.3.7. *Let M, N and P_M, P_N be as above.*

(1) $\delta(M, N) < 1$ *implies that* $M \cap N^{\perp} = \{0\}$;
(2) $\delta(M, N) = \|(I - P_N)P_M\| \leq \|P_M - P_N\|$;
(3) $\hat{\delta}(M, N) = \|P_M - P_N\|$.

Proof. (1) If $M \cap N^{\perp} \neq \{0\}$, we can pick $x \in M \cap N^{\perp}$ such that $\|x\| = 1$. Thus, $\operatorname{dist}(x, N) = \|x\| = 1$ and hence $\delta(M, N) = 1$, which contradicts to the assumption that $\delta(M, N) < 1$.

(2) Since for any $z \in H$, $\operatorname{dist}(z, N) = \|z - P_N z\|$, we have

$$\|(I - P_N)P_M x\| = \operatorname{dist}(P_M x, N) \leq \|P_M x\|\delta(M, N), \ \forall x \in H.$$

Thus, $\|(I - P_N)P_M\| \leq \delta(M, N)$.

On the other hand, for any $z \in M$ with $\|z\| = 1$, we have

$$\begin{aligned}\operatorname{dist}(z, N) &= \|(I - P_N)P_M z\| \leq \|(I - P_N)P_M\| \\ &= \|(P_M - P_N)P_M\| \leq \|P_M - P_N\|\end{aligned}$$

and consequently, $\delta(M, N) \leq \|(I - P_N)P_M\| \leq \|P_M - P_N\|$.

(3) Clearly, $\hat{\delta}(M, N) \leq \|P_M - P_N\|$ by (2). Now let $\delta = \hat{\delta}(M, N)$. Then $\|P_M(I - P_N)\| = \|(I - P_N)P_M\| \leq \delta$ and $\|(I - P_M)P_N\| \leq \delta$. So for any $x \in H$, we get that

$$\begin{aligned}\|(P_M - P_N)x\|^2 &= \|P_M(I - P_N)x - (I - P_M)P_N x\|^2 \\ &= \|P_M(I - P_N)x\|^2 + \|(I - P_M)P_N x\|^2 \\ &\leq \|P_M(I - P_N)\|^2\|(I - P_N)x\|^2 + \|(I - P_M)P_N\|^2\|P_N x\|^2 \\ &\leq \delta^2\|(I - P_N)x\|^2 + \delta^2\|P_N x\|^2 \\ &= \delta^2\|x\|^2.\end{aligned}$$

So $\|P_M - P_N\| \leq \delta$ and the assertion follows. □

Theorem 1.3.8. *Let M and N be closed subspaces of H. Suppose that $M \cap N^\perp = \{0\}$. Then $\delta(M,N) = \|P_M - P_N\|$.*

Proof. If $\delta(M,N) = 1$, then $\|P_M - P_N\| = \hat{\delta}(M,N) = 1$ by Lemma 1.3.7. Thus, $\delta(M,N) = \|P_M - P_N\| = 1$.

Now assume that $\delta = \delta(M,N) < 1$. Define an operator $A\colon M \to N$ by $Ax = P_N x$, $\forall\, x \in M$. By Lemma 1.3.7 (2),

$$\|(I - P_N)x\| = \|(I - P_N)P_M x\| \le \delta\|x\|, \ \forall\, x \in M.$$

Thus, $\|Ax\| \ge \|x\| - \|(I - P_N)x\| \ge (1-\delta)\|x\|$, $\forall\, x \in M$ and consequently, $\operatorname{Ker} A = \{0\}$ and $\operatorname{Ran}(A)$ is closed in N. Let $z \in N \cap (\operatorname{Ran}(A))^\perp$. Then

$$0 = (z, P_N x) = (P_N z, x) = (z, x), \ \forall\, x \in M.$$

Thus, $z \in N \cap M^\perp$. Since $N \cap M^\perp = \{0\}$, it follows that $N \cap (\operatorname{Ran}(A))^\perp = \{0\}$. Consequently, $\operatorname{Ran}(A) = N$ by Theorem 1.1.25.

For any $z \in H$, choose $z_0 \in M$ such that $P_N z = Az_0 = P_N z_0$. Suppose that $P_N z \ne 0$. Then $z_0 \ne 0$. Thus,

$$\begin{aligned}
\|(I - P_M)P_N z\|^2 &= \|(I - P_M)P_N z_0\|^2 \\
&= \Big\|(I - P_M)\Big(P_N z_0 - \frac{\|P_N z_0\|^2}{\|z_0\|^2} z_0\Big)\Big\|^2 \\
&\le \Big\|P_N z_0 - \frac{\|P_N z_0\|^2}{\|z_0\|^2} z_0\Big\|^2 \\
&= \frac{\|P_N z_0\|^2}{\|z_0\|^2}(\|z_0\|^2 - \|P_N z_0\|^2).
\end{aligned}$$

Note that

$$\begin{aligned}
\|z_0\|^2 = \|(I - P_N)z_0 + P_N z_0\|^2 &= \|(I - P_N)z_0\|^2 + \|P_N z_0\|^2 \\
\le \|(I - P_N)P_M z_0\|^2 + \|P_N z_0\|^2 &\le \delta^2\|z_0\|^2 + \|P_N z_0\|^2.
\end{aligned}$$

We have

$$\|(I - P_M)P_N z\|^2 \le \frac{\|P_N z_0\|^2}{\|z_0\|^2}\delta^2\|z_0\|^2 = \delta^2\|P_N z\|^2. \tag{1.3.6}$$

Since (1.3.6) is also true when $P_N z = 0$, we have $\|(I - P_M)P_N\| \le \delta$, that is, $\hat{\delta}(M,N) \le \delta$ by Lemma 1.3.7 (2). Finally, we get that $\|P_M - P_N\| = \delta(M,N)$ by Lemma 1.3.7 (2) and (3). □

Combining Theorem 1.3.8 with Lemma 1.3.5 and Proposition 1.2.12, we have following useful results.

Corollary 1.3.9. *Let $T \in C(H,K)$ and $A \in B(H,K)$. Put $\bar{T} = T + A$. Assume that* $\operatorname{Ran}(T)$ *and* $\operatorname{Ran}(\bar{T})$ *are all closed.*

(1) *If* $\operatorname{Ker} T \cap (\operatorname{Ker} \bar{T})^{\perp} = \{0\}$ *and* $\operatorname{Ker} \bar{T} \cap (\operatorname{Ker} T)^{\perp} = \{0\}$, *then*

$$\|P_{\operatorname{Ker} \bar{T}} - P_{\operatorname{Ker} T}\| \leq \|A\| \min\{\gamma(\bar{T})^{-1}, \gamma(T)^{-1}\};$$

(2) *If* $\operatorname{Ran}(\bar{T}) \cap (\operatorname{Ran}(T))^{\perp} = \{0\}$ *and* $\operatorname{Ran}(T) \cap (\operatorname{Ran}(\bar{T}))^{\perp} = \{0\}$, *then*

$$\|P_{\operatorname{Ran}(\bar{T})} - P_{\operatorname{Ran}(T)}\| \leq \|A\| \min\{\gamma(\bar{T})^{-1}, \gamma(T)^{-1}\}.$$

1.4 Banach algebras

An algebra is a complex liner space $\mathcal{A}$, together with a mapping $\mathcal{A} \times \mathcal{A} \to \mathcal{A}$ (called a multiplication), which satisfies following conditions:

(1) $x(y+z) = xy + xz, (y+z)x = yx + zx, \forall\, x,\, y,\, z \in \mathcal{A}$.
(2) $\lambda(xy) = (\lambda x)y = x(\lambda y), \forall\, x,\, y \in \mathcal{A},\ \lambda \in \mathbb{C}$.
(3) $(xy)z = x(yz), \forall\, x,\, y,\, z \in \mathcal{A}$.

Definition 1.4.1. Let $\mathcal{A}$ be an algebra. If $\mathcal{A}$ is a Banach space and whose norm $\|\cdot\|$ satisfies the condition:

$$\|xy\| \leq \|x\|\|y\|, \quad \forall\ x,\ y, \in \mathcal{A},$$

we call $\mathcal{A}$ is a Banach algebra.

Let $\mathcal{A}$ be a Banach algebra. An element e in $\mathcal{A}$ is called to be a unit if $ex = xe = x$, $\forall\, x \in \mathcal{A}$. Clearly, the unit e is unique. $\mathcal{A}$ is called to be commutative, if $xy = yx, \forall x, y \in \mathcal{A}$.

Example 1.4.2.

(1) Let X be a Banach space. Then $B(X)$ is a Banach algebra with the unit $I = I_X$ under the norm

$$\|T\| = \sup\{\|x\| \,|\, x \in D(X)\}, \quad \forall T \in B(X).$$

(2) Let M be a compact Hausdorff space and let $C(M)$ be the set of all continuous functions from M to $\mathbb{C}$. Define the linear operation, multiplication and norm on $C(M)$ respectively, by

$$(\lambda f + \mu g)(x) = \lambda f(x) + \mu g(x),\ \forall\, \lambda, \mu \in \mathbb{C},\ f,\, g \in C(M),\ x \in M,$$
$$(fg)(x) = f(x)g(x), \quad \|f\| = \sup_{x \in M} |f(x)|,\ \forall\, f,\, g \in C(M),\ x \in M$$

($|f(M)|$ is bounded on M). Then $C(M)$ becomes a commutative Banach algebra with unit 1.

(3) Let $\mathcal{A}$ be a Banach algebra. Then the matrix algebra $M_n(\mathcal{A})$ of matrices of order n over $\mathcal{A}$ is also a Banach algebra under the norm

$$\|(a_{ij})_{n\times n}\| = \sum_{i,j=1}^{n} \|a_{ij}\|, \ \forall\,(a_{ij})_{n\times n} \in M_n(\mathcal{A}).$$

Remark 1.4.3. Let $\mathcal{A}$ be a Banach algebra.

(1) The mapping $\{x, y\} \mapsto xy$ of $\mathcal{A} \times \mathcal{A}$ to $\mathcal{A}$ is continuous. In fact, for $x_0, y_0, x, y \in \mathcal{A}$, we have

$$\|xy - x_0y_0\| = \|(x - x_0)y + x_0(y - y_0)\| \le \|x - x_0\|\|y\| + \|x_0\|\|y - y_0\|.$$

So if $\|x - x_0\|$ and $\|y - y_0\|$ are small enough, then $\|xy - x_0y_0\|$ will tend to zero.

(2) If $\mathcal{A}$ has the unit e and $\|e\| \neq 1$, then we can define a new norm $\|x\|_0 = \sup\{\|xy\| \,|\, y \in D(A)\}$ on $\mathcal{A}$ such that $(\mathcal{A}, \|\cdot\|_0)$ is a Banach algebra with $\|e\|_0 = 1$. In fact, from $\|xy\| \le \|x\|\|y\|$, $\forall\, x,\, y \in \mathcal{A}$, we get that $\|x\|_0 \le \|x\|$, $\forall\, x \in \mathcal{A}$ and from $\|xy\| \le \|x\|_0$, $\forall\, x \in \mathcal{A}$ and $y \in D(A)$, we have $\|xy\| \le \|x\|_0\|y\|$, $\forall\, x,\, y \in \mathcal{A}$. Then for any $x,\, y,\, z \in \mathcal{A}$,

$$\begin{aligned} \|(x + y)z\| &\le \|xz\| + \|yz\| \le (\|x\|_0 + \|y\|_0)\|z\| \\ \|(xy)z\| &\le \|x\|_0\|yz\| \le \|x\|_0\|y\|_0\|z\| \end{aligned}$$

and hence

$$\|x + y\|_0 \le \|x\|_0 + \|y\|_0, \quad \|xy\|_0 \le \|x\|_0\|y\|_0.$$

Thus, from $\|e\|^{-1}\|x\| \le \|x\|_0 \le \|x\|$, $\forall\, x \in \mathcal{A}$, we get that $(\mathcal{A}, \|\cdot\|_0)$ is a Banach algebra with $\|e\|_0 = 1$. Therefore, for a Banach algebra with a unit e, we always assume that $\|e\| = 1$. For convenience, we let 1 denoted the unit e of $\mathcal{A}$.

(3) If $\mathcal{A}$ has no unit, we set $\tilde{\mathcal{A}} = \{x + \lambda \,|\, x \in \mathcal{A}, \lambda \in \mathbb{C}\}$, in which the linear operation and multiplication are given respectively, by

$$\begin{aligned} &\alpha(x + \lambda) + \beta(y + \mu) = (\alpha x + \beta y) + (\alpha\lambda + \beta\mu), \\ &\qquad (x + \lambda)(y + \mu) = (xy + \lambda y + \mu x) + \lambda\mu, \ \forall\, x + \lambda, y + \mu \in \tilde{\mathcal{A}}. \end{aligned}$$

It is easy to check that under the norm $\|x + \lambda\| = \|x\| + |\lambda|$, $\forall\, x + \lambda \in \tilde{\mathcal{A}}$, $\tilde{\mathcal{A}}$ becomes a Banach algebra with the unit 1.

Let $\mathcal{A}$ be a unital Banach algebra and let $x \in \mathcal{A}$. If there is $y \in \mathcal{A}$ such that $xy = yx = 1$, we call that x is invertible in $\mathcal{A}$ and y is the inverse of x. The inverse of x is unique. So we let x^{-1} denote y.

Let $GL(\mathcal{A})$ denote the set of all invertible elements in $\mathcal{A}$. Clearly, if $x, y \in GL(\mathcal{A})$, then $x^{-1}, y^{-1}, xy \in GL(\mathcal{A})$ and $(xy)^{-1} = y^{-1}x^{-1}$. Thus $GL(\mathcal{A})$ is a group with the unit 1.

Definition 1.4.4. Let $\mathcal{A}$ be a Banach algebra and $x \in \mathcal{A}$.

(1) If $\mathcal{A}$ is unital, the spectrum $\sigma(x)$ of x is the set $\{\lambda \in \mathbb{C} | \lambda - x \notin GL(\mathcal{A})\}$.
(2) If $\mathcal{A}$ is non–unital, the spectrum $\sigma(x)$ of x is defined by

$$\sigma(x) = \{\lambda \in \mathbb{C} | \lambda - x \notin GL(\tilde{\mathcal{A}})\}.$$

Obviously, in this case, $0 \in \sigma(x)$.

The following is the main result about the spectrum of an element in a Banach algebra:

Theorem 1.4.5. *Let $\mathcal{A}$ be a Banach algebra and $x \in \mathcal{A}$. Then*

(1) *$\sigma(x)$ is nonempty and compact in $\mathbb{C}$.*
(2) $\max\{|\lambda| | \lambda \in \sigma(x)\} = \lim\limits_{n\to\infty} \|x^n\|^{\frac{1}{n}}$, *where* $\max\{|\lambda| | \lambda \in \sigma(x)\}$ *is called the spectral radius of x and is denoted by $r(x)$.*

To prove this theorem, we need two lemmas as follows.

Lemma 1.4.6. *Let $\mathcal{A}$ be a Banach algebra and $a \in \mathcal{A}$. Then* $\lim\limits_{n\to\infty} \|a^n\|^{\frac{1}{n}} = \inf\limits_{n\geq 1} \|a^n\|^{\frac{1}{n}} \triangleq \mu(a)$.

Proof. For any $\epsilon > 0$, there is $m \in \mathbb{N}$ such that $\|a^m\|^{\frac{1}{m}} < \mu(a) + \epsilon$. Thus, for any $n \in \mathbb{N}$, there are positive integer numbers p, q with $0 \leq q \leq m-1$ such that $n = pm + q$. Therefore,

$$\begin{aligned}\|a^n\|^{\frac{1}{n}} &= \|a^{pm}a^q\|^{\frac{1}{n}} \leq \|a^m\|^{\frac{p}{n}} \|a\|^{\frac{q}{n}} \\ &< (\mu(a)+\epsilon)^{\frac{mp}{n}} \|a\|^{\frac{q}{n}} \to \mu(a)+\epsilon, (n \to \infty).\end{aligned}$$

for $\dfrac{mp}{n} \to 1$ and $\dfrac{q}{n} \to 0$ as $n \to \infty$, that is, $\overline{\lim\limits_{n\to\infty}} \|a^n\|^{\frac{1}{n}} \leq \mu(a) + \epsilon$.

On the other hand, since $\mu(a) \leq \|a^n\|^{\frac{1}{n}}$, $n \leq 1$, we have

$$\mu(a) \leq \underline{\lim_{n\to\infty}} \|a^n\|^{\frac{1}{n}} \leq \overline{\lim_{n\to\infty}} \|a^n\|^{\frac{1}{n}} \leq \mu(a) + \epsilon.$$

and hence $\lim\limits_{n\to\infty} \|a^n\|^{\frac{1}{n}} = \mu(a)$ by letting $\epsilon \to 0^+$. □

Lemma 1.4.7. *Let $\mathcal{A}$ be a unital Banach algebra and let $a \in \mathcal{A}$.*

(1) *If $a \in GL(A)$, then $\mu(a) \geq \|a^{-1}\|^{-1} > 0$.*

(2) *If* $\mu(a) < 1$, *then* $1 - a \in GL(A)$ *with* $(1-a)^{-1} = \sum\limits_{n=0}^{\infty} a^n$.

(3) *If* $\|a\| < 1$, *then* $1 - a \in GL(A)$ *and* $\|(1-a)^{-1}\| \leq (1 - \|a\|)^{-1}$.

Proof. (1) Since $a^n(a^{-1})^n = 1$, it follows that

$$1 = \|a^n(a^{-1})^n\| \leq \|a^n\|\|(a^{-1})^n\| \leq \|a^n\|\|a^{-1}\|^n$$

and so that $\mu(a) \geq \|a^{-1}\|^{-1}$.

(2) Take $\epsilon = (1 - \mu(a))/2$. Then $\mu(a) + \epsilon < 1$ and $0 \leq \|a^n\|^{\frac{1}{n}} - \mu(a) < \epsilon$ for n large enough. Put $a_n = \sum\limits_{k=0}^{n} a^k$, where $a^0 = 1$. Then from

$$\|a_n - a_m\| \leq \sum_{k=m+1}^{n} \|a^k\| \text{ and } \|a^n\| < (\mu(a) + \epsilon)^n, \ \forall n > m$$

for m large enough, we get that $\{a_n\}$ is a cauchy sequence in $\mathcal{A}$. So there is $b \in \mathcal{A}$ such that $\|a_n - b\| \to 0$ $(n \to \infty)$. Noting that

$$(1-a)a_n = \sum_{k=0}^{n} a^k - \sum_{k=0}^{n} a^{k+1} = 1 - a^{n+1} = a_n(1-a), \ \lim_{n\to\infty} \|a^n\| = 0,$$

we have $(1-a)b = b(1-a) = 1$, that is, $1 - a \in GL(\mathcal{A})$ with $(1-a)^{-1} = b = \sum\limits_{n=0}^{\infty} a_n$.

(3) Since $\mu(a) \leq \|a\| < 1$, it follows from (2) that $1 - a \in GL(\mathcal{A})$ with $(1-a)^{-1} = \sum\limits_{n=0}^{\infty} a^n$ and

$$\|(1-a)^{-1}\| \leq \sum_{n=0}^{\infty} \|a^n\| \leq \sum_{n=0}^{\infty} \|a\|^n = \frac{1}{1 - \|a\|}.$$

□

Corollary 1.4.8. *Let $\mathcal{A}$ be a unital Banach algebra. Let $a \in GL(\mathcal{A})$ and $b \in \mathcal{A}$ with $\|b - a\| < \|a^{-1}\|^{-1}$. Then $b \in GL(\mathcal{A})$ and*

$$\|b^{-1} - a^{-1}\| \leq \frac{\|a^{-1}\|^2\|b-a\|}{1 - \|a^{-1}\|\|b-a\|}.$$

Proof. From $\|b - a\| < \|a^{-1}\|^{-1}$, we get that

$$\|1 - a^{-1}b\| = \|a^{-1}(a+b)\| \leq \|a^{-1}\|\|b-a\| < 1.$$

Thus $a^{-1}b \in GL(\mathcal{A})$ and hence $b \in GL(\mathcal{A})$ by Lemma 1.4.7 and moreover,

$$b^{-1}a = (a^{-1}b)^{-1} = 1 + \sum_{n=1}^{\infty}(1 - a^{-1}b)^n,$$

$$b^{-1} - a^{-1} = \Big(\sum_{n=1}^{\infty}(1 - a^{-1}b)^n\Big)a^{-1}.$$

thus,

$$\|b^{-1} - a^{-1}\| \le \frac{\|a^{-1}\|\|1 - a^{-1}b\|}{1 - \|1 - a^{-1}b\|} \le \frac{\|a^{-1}\|^2\|a - b\|}{1 - \|a^{-1}\|\|a - b\|}. \qquad \square$$

From Corollary 1.4.8, we have

Corollary 1.4.9. *Let $\mathcal{A}$ be a unital Banach algebra. Then*

(1) *$GL(\mathcal{A})$ is an open subset of $\mathcal{A}$.*
(2) *The mapping $x \mapsto x^{-1}$ is a homeomorphism from $GL(\mathcal{A})$ onto itself.*

Corollary 1.4.10. *Let $\mathcal{A}$ be a unital Banach algebra and let $a \in \mathcal{A}$.*

(1) *If $\lambda_0 \in \mathbb{C}$ and $|\lambda_0| > \mu(a)$, then $\lambda_0 - a \in GL(\mathcal{A})$ and $(\lambda_0 - a)^{-1} = \sum\limits_{n=0}^{\infty} \lambda_0^{-(n+1)} a^n$.*
(2) *If $\lambda_0 \in \mathbb{C}\backslash\sigma(a)$, then for any $\lambda \in \mathbb{C}$ with $|\lambda - \lambda_0| < \dfrac{1}{\mu((\lambda_0 - a)^{-1})}$, $\lambda - a \in GL(\mathcal{A})$ and $(\lambda - a)^{-1} = \sum\limits_{n=0}^{\infty} (-1)^n (\lambda_0 - a)^{-(n+1)} (\lambda - \lambda_0)^n$.*

Proof. (1) Since $\mu(\lambda_0^{-1}a) = |\lambda_0|^{-1}\mu(a) < 1$, it follows from Lemma 1.4.7 that $\lambda_0 - a = \lambda_0(1 - \lambda_0^{-1}a) \in GL(\mathcal{A})$ and

$$(\lambda_0 - a)^{-1} = \lambda_0^{-1}(1 - \lambda_0^{-1}a)^{-1} = \sum_{n=0}^{\infty} \lambda_0^{-(n+1)} a^n.$$

(2) Noting that $\mu((\lambda - \lambda_0)(\lambda_0 - a)^{-1}) = |\lambda_0 - \lambda|\mu((\lambda_0 - a)^{-1}) < 1$, we have $\lambda - a = (\lambda_0 - a)((\lambda - \lambda_0)(\lambda_0 - a)^{-1} + 1) \in GL(\mathcal{A})$ and

$$\begin{aligned}(\lambda - a)^{-1} &= ((\lambda - \lambda_0)(\lambda_0 - a)^{-1} + 1)^{-1}(\lambda_0 - a)^{-1} \\ &= \sum_{n=0}^{\infty} (-1)^n (\lambda_0 - a)^{-(n+1)} (\lambda - \lambda_0)^n\end{aligned}$$

by Lemma 1.4.7 (2). $\square$

Now we give the prove of Theorem 1.4.5 as follows.

(1) By Corollary 1.4.10 (2), $\mathbb{C}\backslash\sigma(a)$ is open in $\mathbb{C}$. Noting that $|\lambda| \le \mu(a)$, whenever $\lambda \in \sigma(a)$ by Corollary 1.4.10 (1), we get that $\sigma(a)$ is bounded and closed in $\mathbb{C}$.

Let $\phi \in \mathcal{A}^*$. Then $g_\phi(\lambda) = \phi((\lambda - a)^{-1})$ is analytic on $\mathbb{C}\backslash\sigma(a)$ by Corollary 1.4.10 (2). If $\sigma(a) = \emptyset$, then $g_\phi(\lambda)$ is analytic on $\mathbb{C}$. Since for any $\lambda \in \mathbb{C}$ with $|\lambda| > \|a\| \ge \mu(a)$,

$$(\lambda - a)^{-1} = \sum_{n=0}^{\infty} \lambda^{-(n+1)} a^n, \quad \|(\lambda - a)^{-1}\| \le \sum_{n=0}^{\infty} \frac{\|a\|^n}{|\lambda|^{n+1}} = \frac{1}{|\lambda| - \|a\|}.$$

By Corollary 1.4.10 (1) and Lemma 1.4.7 (3), it follows that $\lim\limits_{\lambda\to\infty} g_\phi(\lambda) = 0$. Thus, $g_\phi(\lambda) = 0$, $\forall\,\lambda \in \mathbb{C}$ by Liouville's Theorem and hence $(\lambda - a)^{-1} = 0$, $\forall\,\lambda \in \mathbb{C}$. But this is impossible. So $\sigma(a) \neq \emptyset$.

(2) We have got that $r(a) \leq \mu(a)$. If $r(a) < \mu(a)$, we can choose $\epsilon > 0$ such that $r(a) + \epsilon < \mu(a)$. Then $|\lambda| < \mu(a) - \epsilon$, $\forall\,\lambda \in \sigma(a)$. Put

$$\Gamma = \{\lambda \in \mathbb{C} \,|\, |\lambda| = \mu(a) - \epsilon\},\ \Gamma' = \{\lambda \in \mathbb{C} \,|\, |\lambda| = \mu(a) + \epsilon\}.$$

Suppose that Γ, Γ' are oriented counter–clockwise. Let $f \in \mathcal{A}^*$. By Corollary 1.4.10 (2), $(\lambda - a)^{-1} \in GL(\mathcal{A})$ when $|\lambda| > r(a)$ and $g_f(\lambda) = f((\lambda - a)^{-1})$ is analytic on the domain $\{\lambda \in \mathbb{C} \,|\, |\lambda| > r(a)\}$ which contains Γ and Γ'. Thus by Canchy's Theorem,

$$\oint_\Gamma f((\lambda - a)^n)\lambda^n \mathrm{d}\,\lambda = \oint_{\Gamma'} f((\lambda - a)^n)\lambda^n \mathrm{d}\,\lambda, \quad \forall n \geq 0. \tag{1.4.1}$$

Since $f((\lambda - a)^{-1}) = \sum\limits_{n=0}^{\infty} f(a^n)\lambda^{-(n+1)}$, $\lambda \in \Gamma'$ by Corollary 1.4.10 (1), it follows from (1.4.1) that for $n \in \mathbb{N}$,

$$|f(a^n)| \leq \frac{1}{2\pi}\oint_\Gamma |f((\lambda - a)^{-1})||\lambda|^n |\mathrm{d}\,\lambda| \leq (\mu(a) - \epsilon)^{n+1}\|f\| \max_{\lambda\in\Gamma} \|(\lambda - a)^{-1}\|.$$

So $\|a^n\| \leq (\mu(a) - \epsilon)^{n+1} \max\limits_{\lambda\in\Gamma} \|(\lambda - a)^{-1}\|$, $n \geq 1$ by Corollary 1.1.5 (1) and hence $\mu(a) = \lim\limits_{n\to\infty} \|a^n\|^{\frac{1}{n}} \leq \mu(a) - \epsilon$. But this is impossible. Therefore $r(a) = \mu(a)$.

Proposition 1.4.11. *Let $\mathcal{A}$ be a Banach algebra and let $a, b \in \mathcal{A}$. If $ab = ba$, then*

$$r(ab) \leq r(a)r(b) \text{ and } r(a + b) \leq r(a) + r(b).$$

Proof. Since $\|(ab)^n\| = \|a^n b^n\| \leq \|a^n\|\|b^n\|$, we have $r(ab) \leq r(a)r(b)$.

Now choose real numbers α, β such that $r(a) < \alpha$ and $r(b) < \beta$. Set $x = \alpha^{-1}a$ and $y = \beta^{-1}b$. Then

$$\|(a + b)^n\| = \Big\|\sum_{k=0}^{n} \binom{n}{k} a^k b^{n-k}\Big\| \leq \sum_{k=0}^{n} \binom{n}{k} \alpha^k \beta^{n-k}\|x^k\|\|y^{n-k}\|.$$

For each n, choose $n' \in \mathbb{N}$ with $0 \leq n' \leq n$ such that $\|x^{n'}\|\|y^{n-n'}\| = \max\limits_{0\leq k\leq n} \|x^k\|\|y^{n-k}\|$. Then

$$r(a + b) \leq (\alpha + \beta)\|x^{n'}\|^{\frac{1}{n}}\|y^{n-n'}\|^{\frac{1}{n}}.$$

Pick a sequence $\{n'_m\}$ such that $\delta = \lim_{m\to\infty} \frac{n'_m}{n_m}$ exists. Note that $0 \le \delta \le 1$ and that $1-\delta = \lim_{m\to\infty} \frac{n_m - n'_m}{n_m}$. If $\delta \ne 0$, then $\lim_{m\to\infty} n'_m = \infty$ and

$$\lim_{m\to\infty} \|x^{n'_m}\|^{\frac{1}{n_m}} = \lim_{m\to\infty} \left(\|x^{n'_m}\|^{\frac{1}{n'_m}}\right)^{\frac{n'_m}{n_m}} = (r(x))^{\delta} \le 1$$

for $r(x) = \alpha^{-1} r(a) < 1$. If $\delta = 0$, then

$$\limsup_{m\to\infty} \|x^{n'_m}\|^{\frac{1}{n_m}} \le \lim_{m\to\infty} \|a\|^{\frac{n'_m}{n_m}} \le 1.$$

Therefore, in either case, $\limsup_{m\to\infty} \|x^{n'_m}\|^{\frac{1}{n_m}} \le 1$. Similarly we obtain $\limsup_{m\to\infty} \|y^{n-n'_m}\|^{\frac{1}{n_m}} \le 1$. It follows that $r(a+b) \le \alpha + \beta$. Since this holds for all $\alpha > r(a)$ and $\beta > r(b)$, we have $r(a+b) \le r(a) + r(b)$. □

Let $a \in \mathcal{A}$. a is quasi–nilpotent, if $\sigma(a) = \{0\}$ or equivalently, $r(a) = 0$. Let $QN(\mathcal{A})$ denote the set of all quasi–nilpotent elements in $\mathcal{A}$. We have

Corollary 1.4.12. *Let a, $b \in QN(\mathcal{A})$ and $c \in \mathcal{A}$.*

(1) *If $ac = ca$, then $ac \in QN(\mathcal{A})$.*
(2) *If $ab = ba$, then $a + b \in QN(\mathcal{A})$.*
(3) *If $ab = 0$, then $a + b \in QN(\mathcal{A})$.*

Proof. By Proposition 1.4.11, we have

$$r(ac) \le r(a)r(c) = 0 \text{ and } r(a+b) \le r(a) + r(b) = 0.$$

Thus, we obtain (1) and (2).

To prove (3), let $\lambda \in \mathbb{C}\backslash\{0\}$. Since both $\lambda - a$ and $\lambda - b$ are all invertible in $\tilde{\mathcal{A}}$ and $ab = 0$, we get that

$$\lambda - (a+b) = \lambda^{-1}(\lambda - a)(\lambda - b)$$

is invertible in $\tilde{\mathcal{A}}$. So $a + b \in QN(\mathcal{A})$. □

Let X, Y be Banach spaces and let $S \in B(X,Y)$ and $T \in B(Y,X)$. Then $TS \in B(X)$, $ST \in B(Y)$. So $\sigma(TS) \ne \emptyset$ and $\sigma(ST) \ne \emptyset$ by Theorem 1.4.5. Moreover, we have

Proposition 1.4.13. *Let $S \in B(X,Y)$ and $T \in B(Y,X)$. Then*

$$\sigma(TS)\backslash\{0\} = \sigma(ST)\backslash\{0\}.$$

Proof. Let $\lambda \in \sigma(TS)$ and $\lambda \neq 0$. If $\lambda \notin \sigma(ST)$, then

$$(\lambda I_Y - ST)R = R(\lambda I_Y - ST) = I_Y$$

for some $R \in B(Y)$ and so that $STR = RST = \lambda R - I_Y$. Thus,

$$(\lambda I_X - TS)(I_X + TRS) = \lambda I_Y + \lambda TRS - TS - T(\lambda R - I_Y)S = \lambda I_X,$$
$$(I_X + TRS)(\lambda I_X - TS) = \lambda I_X - TS + \lambda TRS - T(\lambda R - I_Y)S = \lambda I_X.$$

and consequently, $\lambda I_X - TS$ is invertible in $B(X)$ with $(\lambda I_X - TS)^{-1} = \lambda^{-1}(I_X + TRS)$. But it contradicts to the assumption that $\lambda \in \sigma(TS)$. So $\sigma(TS)\backslash\{0\} \subset \sigma(ST)\backslash\{0\}$. Similarly, we have $\sigma(ST)\backslash\{0\} \subset \sigma(TS)\backslash\{0\}$. □

Analog to Proposition 1.4.13, in a Banach algebra $\mathcal{A}$, we also have

Proposition 1.4.14. *Let $\mathcal{A}$ be a Banach algebra and $a, b \in \mathcal{A}$. Then $\sigma(ab)\backslash\{0\} = \sigma(ba)\backslash\{0\}$.*

Let X be a Banach space and let L be a smooth simple curve in $\mathbb{C}$ with finite length $|L|$ and the positive orientation as in Complex Analysis. Let $f\colon L \to X$ be a continuous mapping. For each $\phi \in X^*$, we consider the integral of the continuous function $\lambda \mapsto \phi(f(\lambda))$, $\forall\, \lambda \in L$

$$\int_L \phi(f(\lambda))\mathrm{d}\,\lambda = F(\phi).$$

Clearly, $F(\phi)$ is linear functional on X^* and

$$|F(\phi)| \leq |L| \max_{\lambda \in L} |\phi(f(\lambda))| \leq (|L| \max_{\lambda \in L} \|f(\lambda)\|)\|\phi\|.$$

Thus $F \in (X^*)^*$. On the other hand, since $f(\lambda)$ is continuous on L, it follows that for some $\{\lambda_o, \cdots, \lambda_n\} \subset L$ with $\Delta\lambda = \max\limits_{0 \leq j \leq n} |\lambda_{j+1} - \lambda_j| \to 0$, $\sum\limits_{j=0}^{n} f(\lambda_j)(\lambda_{j+1} - \lambda_j)$ converges to an element b in X. So for every $\phi \in X^*$,

$$\phi(b) = \lim_{\Delta\lambda \to 0} \sum_{=0}^{n} \phi(f(\lambda_j))(\lambda_{j+1} - \lambda_j) = \int_L \phi(f(\lambda))\mathrm{d}\,\lambda = F(\phi).$$

In this case, we denote F by b symbolically. b is given by $b = \displaystyle\int_L f(\lambda)\mathrm{d}\,\lambda$.

Now let $\mathcal{A}$ be a united Banach algebra and let $a \in \mathcal{A}$. Let f be a holomorphic function in an open neighborhood Ω of $\sigma(a)$ and C be a smooth simple closed curve in Ω enclosing $\sigma(a)$ and oriented counter–clockwise. For each $\phi \in \mathcal{A}^*$, we can find $G \in \mathcal{A}^{**}$ and $y \in \mathcal{A}$ such that

$$G(\phi) = \frac{1}{2\pi i} \oint_C \phi((\lambda - a)^{-1} f(\lambda))\mathrm{d}\,\lambda = \phi(y)$$

by above argument. According to Cauchy's Theorem for contour integrals, G does not depend on the choice of the curve C, but only the function f. Therefore, we denote G by $f(a)$, that is, $f(a) = \frac{1}{2\pi i} \oint_C f(\lambda)(\lambda - a)^{-1} \mathrm{d}\,\lambda$.

Proposition 1.4.15. *Let $\mathcal{A}$ be a unital Banach algebra and let $a \in \mathcal{A}$. Let Ω be an open neighborhood of $\sigma(a)$.*

(1) *Let f, g be holomorphic on Ω. Then*
$$(f+g)(a) = f(a) + g(a), \quad (fg)(a) = f(a)g(a).$$
(2) *If $f \equiv 1$, then $f(a) = 1$; if $f(\lambda) \equiv \lambda$, $\forall\, \lambda \in \Omega$, then $f(a) = a$. Let $\Omega = D(0,r) = \{\lambda \in \mathbb{C} \,|\, |\lambda| < r\}$ for some $r > 0$ and if $f(\lambda) = \sum_{n=0}^{\infty} a_n \lambda^n$, $\forall\, \lambda \in \Omega$, then $f(a) = \sum_{n=0}^{\infty} a_n a^n$.*
(3) *Let f be holomorphic on Ω. Then $\sigma(f(a)) = f(\sigma(a))$.*
(4) *Let f be holomorphic on Ω and g be holomorphic on a neighborhood of $f(\sigma(a))$. Then $(g \circ f)(a) = g(f(a))$.*

Proof. (1) Clearly, $(f+g)(a) = f(a) + g(a)$.

Let C_1, C_2 be smooth simple closed curve in Ω enclosing $\sigma(a)$ and oriented counter–clockwise such that C_2 lies completely inside the curve C_1. We have then

$$\begin{aligned}
f(a)g(a) &= \Big(\frac{1}{2\pi i} \oint_{C_1} f(\lambda)(\lambda - a)^{-1} \mathrm{d}\,\lambda\Big)\Big(\frac{1}{2\pi i} \oint_{C_2} g(\mu)(\mu - a)^{-1} \mathrm{d}\,\mu\Big) \\
&= -\frac{1}{4\pi^2} \oint_{C_1} \oint_{C_2} f(\lambda) g(\mu)(\lambda - a)^{-1}(\mu - a)^{-1} \mathrm{d}\,\lambda \mathrm{d}\,\mu \\
&= -\frac{1}{4\pi^2} \oint_{C_1} \oint_{C_2} \frac{f(\lambda) g(\mu)}{\lambda - \mu}((\mu - a)^{-1} - (\lambda - a)^{-1}) \mathrm{d}\,\lambda\, \mathrm{d}\,\mu \\
&= \frac{1}{2\pi i} \oint_{C_2} \frac{1}{2\pi i}\Big(\oint_{C_1} \frac{f(\lambda)}{\lambda - \mu} \mathrm{d}\,\lambda\Big) g(\mu)(\mu - a)^{-1} \mathrm{d}\,\mu \\
&\quad + \frac{1}{4\pi^2} \oint_{C_1} \Big(\oint_{C_2} \frac{g(\mu)}{\lambda - \mu} \mathrm{d}\,\mu\Big) f(\lambda)(\lambda - a)^{-1} \mathrm{d}\,\lambda.
\end{aligned}$$

Because $\frac{g(\mu)}{\lambda - \mu}$ is holomorphic inside the C_2 when λ lies on C_1, $\oint_{C_2} \frac{g(\mu)}{\lambda - \mu} \mathrm{d}\,\mu = 0$. Note that $\frac{1}{2\pi i} \oint_{C_1} \frac{f(\lambda)}{\lambda - \mu} \mathrm{d}\,\lambda = f(\mu)$ when μ lies on C_2. Thus,

$$f(a)g(a) = \frac{1}{2\pi i} \oint_{C_2} f(\mu) g(\mu)(\mu - a)^{-1} \mathrm{d}\,\mu = (fg)(a).$$

(2) Choose $\epsilon > 0$ such that $C = \{\lambda \in \mathbb{C} \mid |\lambda| = \|a\| + \epsilon\}$ (with positive orientation) is contained in Ω. Since

$$(\lambda - a)^{-1} = \lambda^{-1}(1 - \lambda^{-1}a)^{-1} = \sum_{n\to 0}^{\infty} a^n \lambda^{-(n+1)}$$

is uniformly on C and $f(\lambda)$ is analytic on $\{\lambda \in \mathbb{C} \mid |\lambda| \leq \|a\| + \epsilon\}$, we have

$$\begin{aligned} f(a) &= \frac{1}{2\pi i}\oint_C f(\lambda)\Big(\sum_{n=0}^{\infty} a^n \lambda^{-(n+1)}\Big) \mathrm{d}\,\lambda \\ &= \sum_{n=0}^{\infty} a^n \big(\frac{1}{2\pi i}\oint_c f(\lambda)\lambda^{-(n+1)} \mathrm{d}\,\lambda\big) = \sum_{n=0}^{\infty} \frac{a^n}{n!} f^{(n)}(0) \\ &= \sum_{n=0}^{\infty} a_n a^n \end{aligned}$$

and so that $f(a) = 1$ when $f(\lambda) \equiv 1$ and $f(a) = a$ when $f(\lambda) \equiv \lambda$, $\forall\, \lambda \in \Omega$.

(3) Let $b = f(a)$. If $\mu \notin f(\sigma((a))$, then $h(\lambda) = (\mu - f(\lambda))^{-1}$ is holomorphic on a neighborhood of $\sigma(a)$. Then

$$(\mu - b)h(a) = h(a)(\mu - b) = (\mu - f)^{-1}(\mu - f)(a) = 1$$

by (1) and (2). Thus $\mu \notin \sigma(f(a))$, i.e., $\sigma(f(a)) \subset f(\sigma(a))$. If $\mu \in f(\sigma(a))$, then there is $\lambda_0 \in \sigma(a)$ such that $\mu = f(\lambda_0)$. Set

$$h(\lambda) = \begin{cases} \dfrac{f(\lambda) - \mu}{\lambda - \lambda_0} & \lambda \neq \lambda_0 \\ f'(\lambda_0) & \lambda = \lambda_0 \end{cases}, \qquad \forall\, \lambda \in \Omega.$$

Then $h(\lambda)$ is holomorphic on Ω with $f(\lambda) - \mu = (\lambda - \lambda_0)h(\lambda)$, $\forall\, \lambda \in \Omega$. Thus $\mu - b = (\lambda_0 - a)h(a) = h(a)(\lambda_0 - a)$ and consequently, $\mu \in \sigma(b)$ (for $\lambda_0 \in \sigma(a)$).

(4) Choose smooth simple closed curves C_1 and C_2 with positive orientation such that C_1 encloses $f(\sigma(a))$ and is contained in the domain of g and C_2 encloses $f^{-1}(C_1)$ and is contained in Ω. Then

$$\begin{aligned} (g \circ f)(a) &= \frac{1}{2\pi i}\oint_{C_1} (g \circ f)(\lambda)(\lambda - a)^{-1} \mathrm{d}\,\lambda \\ &= \big(\frac{1}{2\pi i}\big)^2 \oint_{C_1} \Big(\oint_{C_2} \frac{g(\mu)}{\mu - f(\lambda)} \mathrm{d}\,\mu\Big)(\lambda - a)^{-1} \mathrm{d}\,\lambda \\ &= \frac{1}{2\pi i}\oint_{C_2} g(\mu)\Big(\frac{1}{2\pi i}\oint_{C_1} (\mu - f(\lambda))^{-1}(\lambda - a)^{-1} \mathrm{d}\,\lambda\Big) \mathrm{d}\,\mu \\ &= \frac{1}{2\pi i}\oint_{C_2} g(\mu)(\mu - f(a))^{-1} \mathrm{d}\,\mu \\ &= g(f(a)). \end{aligned}$$

□

Proposition 1.4.16. *Let $\mathcal{A}$ be a unital Banach algebra and let $GL_0(\mathcal{A})$ denote the connected component of* 1 *in $GL(\mathcal{A})$. Then*

(1) *$GL_0(\mathcal{A})$ is a normal subgroup of $GL(\mathcal{A})$.*
(2) *$GL_0(\mathcal{A}) = \{e^{a_1} \cdots e^{a_n} \,|\, a_1, \cdots, a_n \in \mathcal{A},\, n \in \mathbb{N}\}$.*

Proof. (1) Given $a \in GL_0(\mathcal{A})$. Set $U_a = U(a, \|a^{-1}\|^{-1})$. For each $b \in U_a$, put $f(t) = ta + (1-t)b$, $\forall t \in [0,1]$. Since

$$\|a - f(t)\| = (1-t)\|a - b\| \le \|a - b\| < \frac{1}{\|a^{-1}\|}, \quad \forall t \in [0,1],$$

it follows from Corollary 1.4.8 and Corollary 1.4.9 that $f\colon [0,1] \to GL(\mathcal{A})$ is a continuous mapping. Note that $f(0) = b$ and $f(1) = a$. So $b \in GL_0(\mathcal{A})$ and hence $GL_0(\mathcal{A})$ is open in $GL(\mathcal{A})$.

Let $\{x_n\} \subset GL_0(\mathcal{A})$ and $x_0 \in GL(\mathcal{A})$ with $\|x_n - x_0\| \to 0$ as $n \to \infty$. Choose $n_0 \in \mathbb{N}$ such that $\|x_{n_0} - x_0\| < \|x_0^{-1}\|^{-1}$. Then $x_0 \in GL_0(\mathcal{A})$ by above argument. Thus, $GL_0(\mathcal{A})$ is also closed in $GL(\mathcal{A})$.

Let $a \in GL(\mathcal{A})$ and let C_a be a connected component of a in $GL(\mathcal{A})$. Since $x \mapsto ax$ is a homeomorphism from $GL(\mathcal{A})$ onto itself, we get that $a(GL_0(\mathcal{A})$ is clopen and $a^{-1}C_a$ is connected. So $a \in a(GL_0(\mathcal{A})) \subset C_a$ and $1 \in a^{-1}C_a \subset GL_0(\mathcal{A})$, that is, $C_a = a(GL_0(\mathcal{A})$.

Similarly, we can obtain $C_a = (GL_0(\mathcal{A}))a$. Therefore, $GL_0(\mathcal{A})$ is a normal subgroup of $GL(\mathcal{A})$.

(2) Put $G(\mathcal{A}) = \{\mathrm{e}^{a_1} \cdots \mathrm{e}^{a_n} \,|\, a_1, \cdots, a_n \in \mathcal{A},\, n \in \mathbb{N}\}$. Let $a_1, \cdots, a_n \in \mathcal{A}$ and put $g(t) = \mathrm{e}^{ta_1} \cdots \mathrm{e}^{ta_n}$, $\forall t \in [0,1]$. Then $f\colon [0,1] \to GL(\mathcal{A})$ is continuous and $f(0) = 1$, $f(1) = \mathrm{e}^{a_1} \cdots \mathrm{e}^{a_n}$. Thus, $G(\mathcal{A}) \subset GL_0(\mathcal{A})$.

Let $x \in G(\mathcal{A})$. Then for any $y \in O(x, \|x^{-1}\|^{-1})$, $\|x^{-1}y - 1\| < 1$. So $\sigma(x^{-1}y) \subset \{z \in \mathbb{C} \,|\, |z-1| < 1\} = U$. Since $\mathrm{In}\lambda$ is holomorphic on U, $z = \mathrm{In}(x^{-1}y) \in \mathcal{A}$. Noting that $\lambda = \mathrm{e}^{\mathrm{In}\lambda}$, we have $x^{-1}y = \mathrm{e}^z$ by Proposition 1.4.15 (4). Thus $y = x\mathrm{e}^z \in G(\mathcal{A})$, that is, $G(\mathcal{A})$ is open in $GL_0(\mathcal{A})$. To prove $G(\mathcal{A})$ is closed in $GL_0(\mathcal{A})$, let $\{x_n\} \subset G(\mathcal{A})$ and $x_0 \in GL_0(\mathcal{A})$ with $\|x_n - x_0\| \to 0$ as $n \to \infty$. Choose $n_1 \in \mathbb{N}$ such that $\|x_{n_1} - x_0\| < \|x_0^{-1}\|^{-1}$. Put $z_0 = \mathrm{In}(x_{n_1}x_0^{-1})$. Then $x_{n_1}x_0^{-1} = \mathrm{e}^{z_0}$ and hence $x_0 = \mathrm{e}^{-z_0}x_{n_1} \in G(\mathcal{A})$. Since $GL_0(\mathcal{A})$ is connected, we have $G(\mathcal{A}) = GL_0(\mathcal{A})$. □

Proposition 1.4.17. *Let $\mathcal{A}$ be a unital Banach algebra and $a \in \mathcal{A}$. Suppose that* $\mathrm{Re}\,\lambda < 0$ *for every $\lambda \in \sigma(a)$. Then*

$$a^{-1} = \int_0^{+\infty} e^{ta}\mathrm{d}\,t \triangleq \lim_{A \to +\infty} \int_0^A e^{ta}\mathrm{d}\,t.$$

Proof. By assumption, $0 \notin \sigma(a)$, i.e., $a \in GL(\mathcal{A})$. Let C be a smooth simple curve in an open neighborhood of $\sigma(a)$ in the left half plane, which encloses $\sigma(a)$ and is oriented counter–clockwise. Then

$$e^{ta} = \frac{1}{2\pi i}\oint_C e^{t\lambda}(\lambda - a)^{-1}\mathrm{d}\,\lambda \text{ and}$$

$$\int_0^A e^{ta}\mathrm{d}\,t = \frac{1}{2\pi i}\oint_C (\lambda - a)^{-1}\mathrm{d}\,\lambda \int_0^A e^{t\lambda}\mathrm{d}\,\lambda = \frac{1}{2\pi i}\oint_C \frac{(1-e^{t\lambda})}{\lambda}(\lambda - a)^{-1}\mathrm{d}\,\lambda.$$

Noting that C is compact and $\lambda \mapsto \mathrm{Re}\,\lambda$ is continuous, we can find $\lambda_0 \in C$ such that $\max\limits_{\lambda \in C}(\mathrm{Re}\,\lambda) = \mathrm{Re}\,\lambda_0 < 0$. Thus

$$\begin{aligned}\Big\|\int_0^A e^{ta}\mathrm{d}\,t - a^{-1}\Big\| &\le \frac{1}{2\pi}\oint_C \frac{|e^{\lambda t}|}{|\lambda|}\|(\lambda - a)^{-1}\||\mathrm{d}\,\lambda| \\ &\le \frac{1}{2\pi}\frac{|C|}{\min\limits_{\lambda\in C}|\lambda|}\max_{\lambda\in C}\|(\lambda - a)^{-1}\|e^{t(\mathrm{Re}\,\lambda_0)}\end{aligned}$$

and consequently, $a^{-1} = \displaystyle\int_0^{+\infty} e^{ta}\mathrm{d}\,t$ for $\lim\limits_{t\to+\infty} e^{t(\mathrm{Re}\,\lambda_0)} = 0$. □

Let $\mathcal{A}$ be a unital commutative Banach algebra. A multiplicative linear functional φ on $\mathcal{A}$ is a mapping from $\mathcal{A}$ to $\mathbb{C}$ such that for any $x,\, y \in \mathcal{A}$ and $\lambda,\, \mu, \in \mathbb{C}$,

$$\varphi(\lambda x + \mu y) = \lambda\varphi(x) + \mu\varphi(y),\ \ \varphi(xy) = \varphi(x)\varphi(y).$$

Let $M_{\mathcal{A}}$ denote the set of all nonzero multiplicative linear functionals on $\mathcal{A}$.

Proposition 1.4.18. *Let $\mathcal{A}$ be a unital commutative Banach algebra and let $a \in \mathcal{A}$. Then for every $\lambda \in \sigma(a)$, there is a $\phi \in M_{\mathcal{A}}$ such that $\phi(a) = \lambda$.*

Proof. Set $I(a) = \overline{\{b(\lambda - a) |\, b \in \mathcal{A}\}}$. Then $I(a)$ is a closed ideal of $\mathcal{A}$ and $1 \notin I(a)$ for $\lambda \in \sigma(a)$. Then by Zorn's Lemma, there is a maximal ideal $\mathcal{J}$ containing $I(a)$ in $\mathcal{A}$ and $\mathcal{J} \ne \mathcal{A}$. It is easy to check that $\mathcal{J}$ is closed. Then $\{xy + k \,|\, x,\, y \in \mathcal{A},\ k \in \mathcal{J}\} = \mathcal{A}$ and $\mathcal{A}/\mathcal{J}$ is a field. On the other hand, the Banach space $\mathcal{A}/\mathcal{J}$ is also a commutative Banach algebra under the multiplication $[x][y] = [xy]$ and norm $\|[x]\| = \mathrm{dist}\,(x, \mathcal{J})$, $\forall\, x,\, y \in \mathcal{A}$.

Let $x \in \mathcal{A}\backslash\mathcal{J}$. Then $\sigma([x]) \ne \emptyset$ in $\mathcal{A}/\mathcal{J}$ by Theorem 1.4.5. Since $\mathcal{A}/\mathcal{J}$ is a field, we must have $\sigma([x]) = \{\lambda_x\}$ and $[x] = \lambda_x[1]$. Thus $\hat{\phi}([x]) = \lambda_x$ defines a multiplicative linear functional on $\mathcal{A}/\mathcal{J}$ and hence $\phi(x) = \hat{\phi}([x])$, $\forall\, x \in \mathcal{A}$ gives a multiplicative linear functional on $\mathcal{A}$, which satisfies the conditions that $\phi(1) = 1$ and $\phi(a) = \lambda$. □

Lemma 1.4.19. *Let $\mathcal{A}$ be a unital commutative Banach algebra. Then*

(1) *every element φ in $M_{\mathcal{A}}$ is continuous on $\mathcal{A}$ with $\varphi(1) = \|\varphi\| = 1$.*
(2) *$M_{\mathcal{A}}$ is w^*–compact in $D(\mathcal{A}^*)$.*

Proof. (1) Since $\varphi \not\equiv 0$, we have $\varphi(x_0) \neq 0$ for some $x_0 \in \mathcal{A}$. Then from $1\dot{x}_0 = x_0$ and $\varphi(x_0) = \varphi(1 \cdot x_0) = \varphi(1)\varphi(x_0)$, we get that $\varphi(1) = 1$.

Now for every $a \in \mathcal{A}$, we have $\varphi(a) \in \sigma(a)$. Otherwise, there is $b \in \mathcal{A}$ such that $b(\varphi(a) \cdot 1 - a) = 1$. In this case,

$$1 = \varphi(1) = \varphi(b(\varphi(a) \cdot 1 - a)) = \varphi(b)(\varphi(a) - \varphi(a)) = 0.$$

but it is impossible. Therefore, $|\varphi(a)| \leq \|a\|$, that is, φ is continuous with $\|\varphi\| \leq 1$. From $1 = |\varphi(1)| \leq \|\varphi\|\|1\|$, we finally get that $\|\varphi\| = 1$.

(2) Let $\{\varphi_\alpha\} \subset M_{\mathcal{A}}$ be a net and $\varphi \in \mathcal{A}^*$ such that $\varphi_\alpha \xrightarrow{w^*} \varphi$. Then $1 = \varphi_\alpha(1) \to \varphi(1)$, i.e., $\varphi(1) = 1$ and

$$\varphi_\alpha(x) \to \varphi(x), \;\; \varphi_\alpha(y) \to \varphi(y), \;\; \varphi_\alpha(xy) \to \varphi(xy),$$

$\forall\, x,\, y \in \mathcal{A}$. Since $\varphi_\alpha(xy) = \varphi_\alpha(x)\varphi_\alpha(y)$, we get that $\varphi(xy) = \varphi(x)\varphi(y)$, i.e., $\varphi \in M_{\mathcal{A}}$. This means that $M_{\mathcal{A}}$ is w^*–closed in $D(A^*)$. Noting that $D(\mathcal{A}^*)$ is w^*–compact by Theorem 1.1.16, $M_{\mathcal{A}}$ is also w^*–compact. □

Definition 1.4.20. Let $\mathcal{A}$ be a commutative unital Banach algebra. $M_{\mathcal{A}}$ is called the spectral space of $\mathcal{A}$. The mapping $\Gamma_{\mathcal{A}} \colon \mathcal{A} \to C(M_{\mathcal{A}})$ given by

$$\Gamma_{\mathcal{A}}(a)(\varphi) = \hat{a}(\varphi) \triangleq \varphi(a), \quad \forall\, \varphi \in M_{\mathcal{A}},$$

is called the Gelfand transformation.

According to Lemma 1.4.19 and Definition 1.4.20, we have

Proposition 1.4.21. *Let $\mathcal{A}$ be a unital commutative Banach algebra. Then $\Gamma_{\mathcal{A}}$ is a homomorphism with $\Gamma_{\mathcal{A}}(1) = 1$ and*

$$\|\Gamma_{\mathcal{A}}(a)\| \triangleq \max_{\varphi \in M_{\mathcal{A}}} |\Gamma_{\mathcal{A}}(a)(\varphi)| \leq \|a\|, \; \forall\, a \in \mathcal{A}.$$

Moreover, $\Gamma_{\mathcal{A}}$ is injective when $\mathcal{A}$ is semi–simple (i.e., $\bigcap_{\varphi \in M_A} \operatorname{Ker} \varphi = \{0\}$).

Proposition 1.4.22. *Let $\mathcal{A}$ be a unital semi–simple commutative Banach algebra. For $A = (a_{ij})_{n \times n} \in M_n(\mathcal{A})$, put*

$$\hat{A}(\phi) = (\phi(a_{ij}))_{n \times n} = (\hat{a}_{ij}(\phi))_{n \times n}, \;\; \forall\, \phi \in M_{\mathcal{A}}.$$

Then A is invertible in $M_n(\mathcal{A})$ iff $\hat{A}(\phi) \in GL(M_n(\mathbb{C}))$, $\forall\, \phi \in M_{\mathcal{A}}$.

Proof. If A is invertible in $M_n(\mathcal{A})$, then there is $B \in M_n(\mathcal{A})$ such that $AB = BA = 1_n$. Thus, $\hat{A}(\phi)\hat{B}(\phi) = \hat{B}(\phi)\hat{A}(\phi) = I_n$, $\forall\, \phi \in M_{\mathcal{A}}$ and hence $\hat{A}(\phi) \in GL(M_n(\mathbb{C}))$, $\forall\, \phi \in M_{\mathcal{A}}$.

Now assume that for each $\phi \in M_{\mathcal{A}}$, $\hat{A}(\phi) = (\hat{a}_{ij}(\phi))_{n\times n}$ is invertible in $M_n(\mathbb{C})$. Then $\det(\hat{A})(\phi) = \det(\hat{A}(\phi)) \neq 0$, $\forall\, \phi \in M_{\mathcal{A}}$.

For i, $j = 1, \cdots, n$, let $M_{ij}(\hat{A})$ denote the $(n-1) \times (n-1)$ minor of $\hat{A}$ formed by deleting the ith row and jth column of $\hat{A}$. Put $\hat{A}_{ij} = (-1)^{i+j} M_{ij}(\hat{A})$ and $\mathrm{adj}(\hat{A}) = (\hat{A}_{ij})^T_{n\times n}$ (the transport of $(\hat{A}_{ij})_{n\times n}$), that is, $\mathrm{adj}(\hat{A})$ is the adjoint matrix of $\hat{A}$. Then

$$\hat{A}(\mathrm{adj}(\hat{A})) = (\mathrm{adj}(\hat{A}))\hat{A} = \det(\hat{A})\, I_n \tag{1.4.2}$$

by Laplace's Theorem. Let $c \in \mathcal{A}$ and $B \in M_n(\mathcal{A})$ such that $\det(\hat{A}) = \hat{c}$ and $\mathrm{adj}(\hat{A}) = \hat{B}$. Since $\mathcal{A}$ is semi–simple, it follows from (1.4.2) that $AB = BA = \mathrm{diag}(c, \cdots, c)$. Note that $\hat{c}(\phi) \neq 0$, $\forall\, \phi \in M_{\mathcal{A}}$ implies that $c \in GL(\mathcal{A})$ by Proposition 1.4.18. So A is invertible in $M_n(\mathcal{A})$. □

1.5 C^*–algebras

Let $\mathcal{A}$ be a Banach algebra. An involution is a continuous mapping $a \mapsto a^*$ of $\mathcal{A}$ into itself such that for all a, $b \in \mathcal{A}$ and $\alpha, \beta \in \mathbb{C}$, the following conditions hold:

$$(a^*)^* = a, \;\; (ab)^* = b^*a^*, \;\; (\alpha a + \beta b)^* = \bar{\alpha} a^* + \bar{\beta} b^*.$$

Definition 1.5.1. A C^*–algebra $\mathcal{A}$ is a Banach algebra with involution such that $\|a^*a\| = \|a\|^2$ for every $a \in \mathcal{A}$.

Clearly, by Definition 1.5.1, we have $\|a^*\| = \|a\|, \forall\, a \in \mathcal{A}$. Moreover, if $\mathcal{A}$ has the unit 1, then $1^* = 1^*1 = (1^*1)^* = (1^*)^* = 1$ and $\|1\|^2 = \|1^* \cdot 1\| = \|1\|^2$, i.e., $\|1\| = 1$.

Example 1.5.2.

(1) Let H be Hilbert space. Then $B(H)$ is a C^*–algebra with the unit I_H, where A^* is the adjoint of A, $\forall\, A \in B(H)$, for $\|A^*A\| = \|A\|^2$ by Proposition 1.1.27.
(2) Let $K(H)$ be the set of all compact operations in $B(H)$. Then by Proposition 1.1.23, $\mathcal{K}(H)$ is the closed subalgebra of $B(H)$ such that $A^* \in \mathcal{K}(H)$ whenever $A \in \mathcal{K}(H)$. Thus $\mathcal{K}(H)$ is a C^*-algebra without a unit when $\dim H = \infty$.

(3) Let M be a compact Hausdorff space. $C(M)$ (cf. Example 1.4.2 (2)) is a commutative C^*-algebra with unit 1, where the involution is given by $f^*(x) = \overline{f(x)}, \forall f \in C(M), \forall x \in M$.

(4) Let M be locally compact but not compact, then $C_0(M)$ (the algebra of continuous functions on M vanishing at infinity) is a commutative C^*-algebra without a unit, where the norm and the involution on $C_0(M)$ are given, respectively, by

$$\|f\| = \sup_{x \in M} |f(x)|, \; f^*(x) = \overline{f(x)}, \; \forall f \in C_0(M), \; \forall x \in M.$$

Proposition 1.5.3. *Let $\mathcal{A}$ be a non-unital $C^* -$ algebra. Then there is a unital C^*-algebra $\tilde{\mathcal{A}}$ such that $\mathcal{A}$ is a closed ideal of $\tilde{\mathcal{A}}$ and $\tilde{\mathcal{A}}/\mathcal{A} = \mathbb{C}$.*

Proof. Put $\tilde{\mathcal{A}} = \{a + \lambda | \, a \in \mathcal{A}, \lambda \in \mathbb{C}\}$. Similar to Remark 1.4.3 (3), $\tilde{\mathcal{A}}$ is an algebra with unit 1 under the linear operation and multiplication given respectively by

$$\alpha(a+\lambda)+\beta(b+\mu) = (\alpha a+\beta b)+(\alpha\lambda+\beta\mu), \; (a+\lambda)(b+\mu) = (ab+\lambda b+\mu a)+\lambda\mu,$$

$\forall \, a + \lambda, \; b + \mu \in \tilde{\mathcal{A}}$. Define the involution on $\tilde{\mathcal{A}}$ by $(a + \lambda)^* = a^* + \bar{\lambda}$ and the norm on $\tilde{\mathcal{A}}$ by

$$\|a + \lambda\| = \sup\{\|ay + \lambda y\| \, | \, y \in D(A)\}, \quad \forall \, a \in \mathcal{A}, \; \lambda \in \mathbb{C}.$$

So $\|ay + \lambda y\| \le \|a + \lambda\| \|y\|$ and hence

$$\begin{aligned} \|ay + \lambda y\|^2 &= \|(ay + \lambda y)^*(ay + \lambda y)\| = \|y^*(a + \lambda)^*(ay + \lambda y)\| \\ &\le \|y^*\| \|(a + \lambda)^*(a + \lambda)y\| \le \|y\|^2 \|(a + \lambda)^*(a + \lambda)\|, \end{aligned}$$

$\forall \, y \in \mathcal{A}, \; a + \lambda \in \tilde{\mathcal{A}}$. Consequently, $\|a + \lambda\|^2 \le \|(a + \lambda)^*(a + \lambda)\|$.

Let $a + \lambda, \; b + \mu \in \tilde{\mathcal{A}}$. Then for any $z \in \mathcal{A}$,

$$\|(a + \lambda)(b + \mu)z\| \le \|a + \lambda\| \|(b + \mu)z\| \le \|a + \lambda\| \|b + \mu\| \|z\|. \qquad (1.5.1)$$

From (1.5.1), we have $\|(a + \lambda)(b + \mu)\| \le \|a + \lambda\| \|b + \mu\|$. Thus

$$\|a + \lambda\|^2 \le \|(a + \lambda)^*(a + \lambda)\| \le \|(a + \lambda)^*\| \|a + \lambda\|$$

and so that

$$\|a + \lambda\| \le \|(a + \lambda)^*\| \le \|(a^* + \bar{\lambda})^*\| = \|a + \lambda\|,$$

i.e., $\|(a + \lambda)^*\| = \|a + \lambda\|$. Therefore $\|a + \lambda\|^2 = \|(a + \lambda)^*(a + \lambda)\|$.

Noting that $\sup\{\|ab\| \, | \, b \in D(\mathcal{A})\} = \|a\|$. So $\mathcal{A}$ is completed in $\tilde{\mathcal{A}}$ under the norm of $\tilde{\mathcal{A}}$ and hence $\mathcal{A}$ is closed in $\tilde{\mathcal{A}}$. Since $\tilde{\mathcal{A}}/\mathcal{A} = \mathbb{C}$ is completed, we obtain that $\tilde{\mathcal{A}}$ is completed by Proposition 1.1.9. □

Definition 1.5.4. Let a be an element in a C^*–algebra $\mathcal{A}$. If $a^* = a$, a is called to be self–adjoint; if $a^*a = aa^*$, a is called to be normal. If $aa^*a = a$, a is called the partial isometry.

When $\mathcal{A}$ is unital, a is unitary if $a^*a = aa^* = 1$; a is called the isometry (or co–isometry) if $a^*a = 1$ (or $aa^* = 1$).

Let $\mathcal{A}_{sa}$ (resp. $U(\mathcal{A})$) denote the set of all self–adjoint elements (resp. unitary elements) in $\mathcal{A}$.

Proposition 1.5.5. *Let $\mathcal{A}$ be a C^*–algebra and $a \in \mathcal{A}$.*

(1) *If $\mathcal{A}$ is unital and a is invertible, then a^* is invertible and $(a^*)^{-1} = (a^{-1})^*$.*
(2) *$\sigma(a^*) = \{\bar{\lambda} \in \mathbb{C} \,|\, \lambda \in \sigma(a)\}$. If $\mathcal{A}$ is unital and a is invertible in $\mathcal{A}$, then $\sigma(a^{-1}) = \{\lambda^{-1} \,|\, \lambda \in \sigma(a)\}$.*
(3) *$a = x + i\,y$, where $x, y \in \mathcal{A}_{sa}$.*
(4) *$\|a\| = r(a)$ when a is normal.*
(5) *$\|a\| = 1$ and $\sigma(a) \subset \mathbf{S}^1$ if $a \in U(\mathcal{A})$ when $\mathcal{A}$ is unital.*
(6) *a is a partial isometry iff $(a^*a)^2 = a^*a$ or $(aa^*)^2 = aa^*$.*

Proof. (1) From $aa^{-1} = a^{-1}a = 1$, we get that $1 = (a^{-1})^*a^* = a^*(a^{-1})^*$. Consequently, $a^* \in GL(\mathcal{A})$ and $(a^*)^{-1} = (a^{-1})^*$.

(2) If $\lambda \notin \sigma(a)$, then $(\lambda - a)^* \in GL(\tilde{\mathcal{A}})$ and consequently, $\bar{\lambda} \notin \sigma(a^*)$. On the other hand, if $\bar{\lambda} \notin \sigma(a^*)$, then $\lambda - a = (\bar{\lambda} - a^*)^* \in GL(\tilde{\mathcal{A}})$. This proves $\sigma(a^*) = \{\bar{\lambda} \in \mathbb{C} \,|\, \lambda \in \sigma(a)\}$.

When $0 \notin \sigma(a)$, we have $\lambda - a = \lambda a(a^{-1} - \lambda^{-1})$, $\forall \lambda \in \sigma(a)$. The assertion follows.

(3) Put $x = \dfrac{a^* + a}{2}$ and $y = \dfrac{a - a^*}{2i}$. Then $x, y \in \mathcal{A}_{sa}$ and $a = x + iy$.

(4) If $a^*a = aa^*$, then

$$\|a\|^4 = \|a^*a\|^2 = \|a^*aa^*a\| = \|(a^2)^*a^2\| = \|a^2\|^2.$$

So $\|a^2\| = \|a\|^2$. By induction, we have $\|a^{2^n}\| = \|a\|^{2^n}$, $\forall n \in \mathbb{N}$. So $r(a) = \lim\limits_{n\to\infty} \|a^{2^n}\|^{\frac{1}{2^n}} = \|a\|$.

(5) When $a \in U(\mathcal{A})$, $\|a\|^2 = \|a^*a\| = 1$. So $\|a\| = 1$.

Let $\lambda \in \sigma(a)$. If $|\lambda| > 1$, then $\|\lambda^{-1}a\| < 1$. So $\lambda - a = \lambda(1 - \lambda^{-1}a)$ is invertible by Lemma 1.4.7 (3); If $|\lambda| < 1$, then $\|\lambda a^*\| < 1$. Thus, $\lambda - a = a(\lambda a^* - 1)$ is invertible. But these two cases contradict with the assumption that $\lambda \in \sigma(a)$. Therefore, $\sigma(a) \subset \mathbf{S}^1$.

(6) If $aa^*a = a$, then $aa^*aa^* = aa^*$ and $a^*aa^*a = a^*a$. Conversely, if $(a^*a)^2 = a^*a$, then

$$\|a - aa^*a\|^2 = \|(a^* - a^*aa^*)(a - aa^*a)\| = 0$$

and consequently, $a = aa^*a$. If $(aa^*)^2 = aa^*$, then from

$$\|a - aa^*a\|^2 = \|(a - aa^*a)(a^* - a^*aa^*)\| = 0,$$

we obtain $a = aa^*a$. □

Let $\mathcal{A}$ be a unital C^*–algebra. Let $U_0(\mathcal{A})$ denote the connected component of 1 in $U(\mathcal{A})$. Similar to Proposition 1.4.16, we have

Proposition 1.5.6. *Let $\mathcal{A}$ be a unital C^*–algebra. Then $U_0(\mathcal{A})$ is a normal subgroup of $U(\mathcal{A})$ and $U_0(\mathcal{A}) = \{e^{ia_1} \cdots e^{ia_n} \,|\, a_1, \cdots, a_n \in \mathcal{A}_{sa},\, n \in \mathbb{N}\}$.*

Proposition 1.5.7. *Let $\mathcal{A}$ be a C^*–algebra and $a \in \mathcal{A}_{sa}$. Then $\sigma(a) \subset \mathbb{R}$.*

Proof. Suppose that $\mathcal{A}$ is unital. If $\mathcal{A}$ is non–unital, we replace $\mathcal{A}$ by $\tilde{\mathcal{A}}$.

Put $f(z) = \sum\limits_{n=0}^{\infty} \dfrac{(iz)^n}{n!} = \mathrm{e}^{iz}$ and $g(z) = \sum\limits_{n=0}^{\infty} \dfrac{(-iz)^n}{n!} = \mathrm{e}^{-iz}$, $z \in \mathbb{C}$. Then

$$f(a)g(a) = g(a)f(a) = (fg)(a) = 1$$

by Proposition 1.4.15. Since $(f(a))^* = g(a)$, $u = f(a)$ is unitary in $\mathcal{A}$. Thus, $f(\sigma(a)) = \sigma(f(a)) \subset \mathbf{S}^1$ by Proposition 1.4.15 (3) and Proposition 1.5.5 (4). Consequently, for any $\lambda + i\mu \in \sigma(a)$, we have $|\mathrm{e}^{i(\lambda+i\mu)}| = 1$. Hence $\mu = 0$. □

Let $\mathcal{B}$ be a C^*–subalgebra of $\mathcal{A}$. Suppose that 1 is the unit of $\mathcal{A}$ and e is the unit of $\mathcal{B}$. Let $a \in \mathcal{B}$ and set

$$\sigma_{\mathcal{B}}(a) = \{\lambda \in \mathbb{C} \,|\, \lambda e - a \notin GL(\mathcal{B})\},\ \sigma_{\mathcal{A}}(a) = \{\lambda \in \mathbb{C} \,|\, \lambda - a \notin GL(\mathcal{A})\} \triangleq \sigma(a).$$

If $\mathcal{A}$ (or $\mathcal{B}$) is non–unital, we set $\sigma_{\mathcal{A}}(a) = \sigma_{\tilde{\mathcal{A}}}(a)$ (or $\sigma_{\mathcal{B}}(a) = \sigma_{\tilde{\mathcal{B}}}(a)$. Clearly, $\sigma_{\mathcal{A}}(a) \backslash \{0\} \subset \sigma_{\mathcal{B}}(a) \backslash \{0\}$. Furthermore, we have

Corollary 1.5.8. *Let $\mathcal{B}$ be a C^*–subalgebra of the unital C^*–algebra $\mathcal{A}$. Let $a \in \mathcal{B}$. If $\mathcal{B}$ and $\mathcal{A}$ have the same unit 1, then $\sigma_{\mathcal{B}}(a) = \sigma_{\mathcal{A}}(a)$. If $\mathcal{B}$ is no unital, then $\sigma_{\mathcal{B}}(a) \backslash \{0\} = \sigma_{\mathcal{A}}(a) \backslash \{0\}$.*

Proof. Suppose that $\mathcal{A}$ and $\mathcal{B}$ have the unit 1.

Since $\sigma_{\mathcal{A}}(a) \subset \sigma_{\mathcal{B}}(a)$, we have to prove $\sigma_{\mathcal{B}}(a) \subset \sigma(a)$. Let $\lambda \in \sigma_{\mathcal{B}}(a)$. If $\lambda \notin \sigma_{\mathcal{A}}(a)$. Then $\lambda - a$ and $(\lambda - a)^*$ are all invertible in $\mathcal{A}$. Put $b = (\lambda - a)^*(\lambda - a)$ and $c = (\lambda - a)(\lambda - a)^*$. Then b, $c \in GL(\mathcal{A})$. Since $\lambda \in \sigma_{\mathcal{B}}(a)$, it follows that $0 \in \sigma_{\mathcal{B}}(b) \cup \sigma_{\mathcal{B}}(c)$.

Assume that $0 \in \sigma_{\mathcal{B}}(b)$. Choose $\mu \in \mathbb{R}\backslash\{0\}$ such that $|\mu| < \dfrac{1}{2\|b^{-1}\|}$. Thus, $i\,\mu - b = -b(1 - i\mu b^{-1})$ is invertible and

$$\|(i\mu - b)^{-1}\| = \|b^{-1}(1 - i\mu b)^{-1}\| \le \|b^{-1}\|\Big(\sum_{n=0}^{\infty} |\mu|\|b^{-1}\|\Big)$$
$$\le \frac{\|b^{-1}\|}{1 - |\mu|\|b^{-1}\|} < 2\|b^{-1}\|.$$

Since $b^* = b$ and $i\mu \notin \mathbb{R}$, $i\mu - b$ is invertible in $\mathcal{B}$ by Proposition 1.5.7. Note that

$$\|b(i\mu - b)^{-1} + 1\| = |\mu|\|(i\mu - b)^{-1}\| < 2|\mu|\|b^{-1}\| < 1.$$

So $b(i\mu - b)^{-1}$ is invertible in $\mathcal{B}$ and hence $0 \notin \sigma_{\mathcal{B}}(a)$, a contradiction.

Similarly, if $0 \in \sigma_{\mathcal{B}}(c)$, we also get a contradiction. All these prove that $\lambda \in \sigma_{\mathcal{A}}(a)$.

If $\mathcal{B}$ is no unital, we consider $\tilde{\mathcal{B}}$. Above argument shows that $\sigma_{\mathcal{B}}(a) = \sigma(a)$ and $\sigma_{\mathcal{B}}(a)\backslash\{0\} = \sigma(a)\backslash\{0\}$. □

Remark 1.5.9. By means of the proof of Corollary 1.5.8, we have:

(1) if both $\mathcal{A}$ and $\mathcal{B}$ are all non–unital, $\sigma_{\mathcal{A}}(a) = \sigma_{\mathcal{B}}(a)$;
(2) if one of $\mathcal{A}$ and $\mathcal{B}$ is non–unital, then $\sigma_{\mathcal{A}}(a)\backslash\{0\} = \sigma_{\mathcal{B}}(a)\backslash\{0\}$.

Let $\mathfrak{P}(\mathbb{R})$ denote the set of polynomials of real variable with complex coefficients and put $\mathfrak{P}_0(\mathbb{R}) = \{p \in \mathfrak{P}(\mathbb{R})|\, p(0) = 0\}$. Let $\mathcal{A}$ be a C^*–algebra and $a \in \mathcal{A}_{sa}$. Put $C^*(a) = \overline{span}\{p(a)|\, p \in \mathfrak{P}_0(\mathbb{R})\}$ (the C^*–subalgebra of $\mathcal{A}$ generated by a). Then $C^*(a)$ is a commutative C^*–algebra without a unit.

Theorem 1.5.10. *Let $\mathcal{A}$ be a C^*–algebra and $a \in \mathcal{A}_{sa}$. Then there is an isomorphism $\Gamma\colon C^*(a) \to C_0(\sigma(a)\backslash\{0\})$ such that for any $\lambda \in \sigma(a)\backslash\{0\}$ and any $x \in C^*(a)$,*

$$\Gamma(a)(\lambda) = \lambda, \quad \Gamma(x^*) = (\Gamma(x))^*, \quad \|\Gamma(x)\| = \|x\|.$$

Proof. Put $\mathcal{B} = C^*(a)$. Then $\sigma_{\mathcal{B}}(a)\backslash\{0\} = \sigma(a)\backslash\{0\}$ by Remark 1.5.9. Define a mapping $\Gamma_0\colon \{p(a)|\, p \in \mathfrak{P}_0(\mathbb{R})\} \to \mathfrak{P}_0(\mathbb{R})$ by

$$\Gamma_0(p(a))(\lambda) = p(\lambda), \quad \lambda \in \sigma(a)\backslash\{0\}.$$

Since $p(a)$ is normal for every $p \in \mathfrak{P}_0(\mathbb{R})$, it follow from Proposition 1.5.5 that $\|p(a)\| = \max\{|\mu|\,|\, \mu \in \sigma(p(a))\}$. Noting that $\sigma(p(a)) = p(\sigma(a))$ by Proposition 1.4.15, we have $\|p(a)\| = \max\limits_{\lambda \in \sigma(a)\backslash\{0\}} |p(\lambda)| = \|p\|$. This indicates that Γ_0 is a well-defined homomorphism such that $\|\Gamma_0(p(a))\| = \|p\|$, $\forall\, p \in \mathfrak{P}_0(\mathbb{R})$. Furthermore, $\Gamma_0((p(a))^*) = p^*$, $\forall\, p \in \mathfrak{P}_0(\mathbb{R})$.

Let $x \in C^*(a)$. Then there is $\{p_n\} \in \mathfrak{P}_0(\mathbb{R})$ such that $\|x - p_n(a)\| \to 0$ as $n \to \infty$. Since

$$\|p_n - p_m\| = \|\Gamma_0(p_n(a)) - \Gamma_0(p_m(a))\| = \|p_n(a) - p_m(a)\|,\ \forall\, m,\, n \in \mathbb{N},$$

it follows that $\{p_n\}$ is a Cauchy sequence in $C_0(\sigma(a)\backslash\{0\})$. Thus we can extend Γ_0 to $C^*(a)$ by setting $\Gamma(x) = \lim_{n\to\infty} \Gamma_0(p_n(a))$. It is easy to check that Γ is a $*$-homomorphism with

$$\|\Gamma(x)\| = \|x\|,\ \forall\, x \in C^*(a) \text{ and } \Gamma(a)(\lambda) = \lambda,\ \forall\, \lambda \in \sigma(a)\backslash\{0\}.$$

To prove Γ is surjective, let $g \in C_0(\sigma(a)\backslash\{0\})$. Then we can find a sequence $\{q_n\} \in \mathfrak{P}_0(\mathbb{R})$ such that $\|g - q_n\| \to 0$ when $n \to \infty$ by Weierstrass Theorem. So $\{q_n(a)\}$ is a Cauchy sequence in $C^*(a)$ and hence there exists $y \in C^*(a)$ such that $\|y - q_n(a)\| \to 0$ as $n \to \infty$. Therefore,

$$\Gamma(y)(\lambda) = \lim_{n\to\infty} \Gamma(q_n(a))(\lambda) = \lim_{n\to\infty} q_n(\lambda) = g(\lambda),\ \forall\, \lambda \in \sigma(a)\backslash\{0\},$$

that is, $g = \Gamma(y)$. □

According to Theorem 1.5.10, we have the following definition.

Definition 1.5.11. Let $\mathcal{A}$ be a C^*-algebra and let $a \in \mathcal{A}_{sa}$. Let $f \in C_0(\sigma(a)\backslash\{0\})$. Then $f(a)$ is defined as the preimage of f under Γ, that is, $f(a) = \Gamma^{-1}(\{f\})$.

Proposition 1.5.12. *Let $\mathcal{A}$ be a C^*-algebra and let $a,\, b \in \mathcal{A}_{sa}$. Let $f \in C_0(\sigma(a)\backslash\{0\})$. Then $\sigma(f(a))\backslash\{0\} = f(\sigma(a)\backslash\{0\})$.*

Proof. Let $\mathcal{B} = C^*(a)$ and $\mathcal{C} = C_0(\sigma(a)\backslash\{0\})$. We extend $\Gamma\colon \mathcal{B} \to \mathcal{C}$ in Theorem 1.5.10 to $\Gamma^+\colon \tilde{\mathcal{B}} \to \tilde{\mathcal{C}}$ by $\Gamma^+(\lambda + x) = \lambda + \Gamma(x)$, $\forall\, \lambda \in \mathbb{C},\ x \in \mathcal{B}$. It is easy to verify that Γ^+ is a $*$-homomorphism and is also surjective. Let $\lambda + x \in \tilde{\mathcal{B}}$. For any $b \in \mathcal{B}$, put $g = \Gamma(b) \in \mathcal{C}$. Then $\|g\| = \|b\|$ and

$$\|(\lambda + \Gamma(a))g\| = \|\Gamma((\lambda + a)b)\| = \|(\lambda + a)b\| \le \|\lambda + a\|\|g\|$$
$$\|(\lambda + a)b\| = \|(\lambda + \Gamma(a))g\| \le \|\lambda + \Gamma(a)\|\|g\| = \|\lambda + \Gamma(a)\|\|b\|.$$

and so that $\|\Gamma^+(\lambda + a)\| = \|\lambda + a\|$. Thus, Γ^+ is a $*$-isomorphism from $\tilde{\mathcal{B}}$ onto $\tilde{\mathcal{C}}$ and consequently, $\sigma_{\mathcal{B}}(f(a)) = \sigma_{\mathcal{C}}(f)$ for $f = \Gamma(f(a))$.

Obviously, $f(\sigma(a)) \subset \sigma_{\mathcal{C}}(f)$. Now suppose that $\lambda \in \sigma_{\mathcal{C}}(f)$ and $\lambda \notin f(\sigma(a))$. Since $\sigma(a)$ is compact and f is continuous on $\sigma(a)$, it follows that $f(\sigma(a))$ is compact in $\mathbb{C}$. So $\operatorname{dist}(\lambda, f(\sigma(a))) = d > 0$ and hence $f - \lambda \in GL(\tilde{\mathcal{C}})$ with $(f - \lambda)^{-1}(\mu) = \dfrac{1}{f(\mu) - \lambda}$, $\forall\, \mu \in \sigma(a)$, a contradiction. Therefore, $\sigma(f(a))\backslash\{0\} = f(\sigma(a))\backslash\{0\}$ by Corollary 1.5.8. □

Definition 1.5.13. Let $\mathcal{A}$ be a C^*–algebra. An element $a \in \mathcal{A}_{sa}$ is positive if $\sigma(a) \subset \mathbb{R}_+$. This is denoted by $a \geq 0$. Let $\mathcal{A}_+$ denote the set of all positive elements in $\mathcal{A}$.

Proposition 1.5.14. *Let $\mathcal{A}$ be a C^*–algebra and $a \in \mathcal{A}_{sa}$. Then the following statements are equivalent:*

(1) $a \geq 0$.
(2) $a = b^2$ *for some* $b \in \mathcal{A}_{sa}$.
(3) $\|t - a\| \leq t$ *for all* $t \geq \|a\|$.
(4) $\|t - a\| \leq t$ *for some* $t \geq \|a\|$.

Proof. (1)⇒(2) Let $f(t) = t^{1/2}$, $t \in \sigma(a)\backslash\{0\}$. Set $b = f(a)$. Then $b \in \mathcal{A}_{sa}$ and $b^2 = a$ by Theorem 1.5.10 and Definition 1.5.11.

(2)⇒(1) By Proposition 1.5.12 (1), we have

$$\sigma(a)\backslash\{0\} = \{\lambda^2 \,|\, \lambda \in \sigma(b)\backslash\{0\}\} \subset \mathbb{R}_+.$$

Thus, $a \geq 0$.

(1)⇒(3) Since $r(a) = \|a\|$, it follows that for all $t \geq \|a\| \geq \lambda$, $\forall\, \lambda \in \sigma(a)$. Thus, $\sigma(t - a) = \{t - \lambda \,|\, \lambda \in \sigma(a)\} \subset [0, t]$ and hence $\|t - a\| \leq t$ by Proposition 1.5.5 (3).

(3)⇒(4) is clear.

(4)⇒(1) We have $t \geq \|a\| \geq |\lambda| \geq \lambda$, $\forall\, \lambda \in \sigma(a)$ and

$$t \geq \|t - a\| = \max\{|t - \lambda| \,|\, \lambda \in \sigma(a)\} \geq |t - \lambda|, \ \forall\, \lambda \in \sigma(a).$$

Thus, $t \geq t - \lambda$ and hence $\lambda \geq 0$, $\forall\, \lambda \in \sigma(a)$. So, $a \geq 0$. □

Corollary 1.5.15. *Let $\mathcal{A}$ be a C^*–algebra. Then $\mathcal{A}_+$ is closed in $\mathcal{A}$ and*

$$\mathcal{A}_+ = \mathcal{A}_+ + \mathcal{A}_+ \triangleq \{a + b \,|\, a,\, b \in \mathcal{A}_+\}.$$

Proof. Let $\{a_n\} \subset \mathcal{A}_+$ and $a \in \mathcal{A}$ such that $\|a_n - a\| \to 0$ as $n \to \infty$. Then $a \in \mathcal{A}_{sa}$ and there is $M > 0$ such that $\|a_n\| \leq M$, $\forall\, n \in \mathbb{N}$. Thus, $\|M - a_n\| \leq M$, $\forall\, n \in \mathbb{N}$ by Proposition 1.5.14 and consequently, $\|a\| \leq M$ and $\|M - a\| \leq M$. So $a \in \mathcal{A}_+$ by using Proposition 1.5.14 again.

Clearly, $\mathcal{A}_+ \subset \mathcal{A}_+ + \mathcal{A}_+$. Now let $a,\, b \in \mathcal{A}_+$. and put $t = \|a\| + \|b\|$. Note that

$$\big\|\|a\| - a\big\| \leq \|a\|, \quad \big\|\|b\| - b\big\| \leq \|b\|$$

by Proposition 1.5.14, we have $\|a + b\| \leq t$ and

$$\|t - (a + b)\| = \big\|\|a\| - a + \|b\| - b\big\| \leq \big\|\|a\| - a\big\| + \big\|\|b\| - b\big\| \leq \|a\| + \|b\| = t,$$

which means that $a + b \in \mathcal{A}_+$ by Proposition 1.5.14. □

Let $\mathcal{A}$ be a C^*–algebra and let $a \in \mathcal{A}_{sa}$, $b \in \mathcal{A}_+$. Put

$$f_+(t) = \max\{t, 0\} \text{ and } f_-(t) = -\min\{t, 0\}, \quad \forall\, t \in [-\|a\|, \|a\|].$$

Then f_+, $f_- \in (C_0([-\|a\|, \|a\|]\backslash\{0\}))_+$ and $f_+f_- = 0$, $f_+(t) - f_-(t) = t$, $\forall\, t \in [-\|a\|, \|a\|]$. Put $a_+ = f_+(a)$, $a_- = f_-(a) \in \mathcal{A}_+$. Then $a_+a_- = 0$ and $a = a_+ - a_-$.

For $n \in \mathbb{N}$ with $n \geq 2$, put $g_n(t) = t^{1/n}$, $\forall\, t \in [0, \|b\|]$. Then $g_n \in (C_0((0, \|b\|]))_+$. Set $b^{1/n} = g_n(b)$. Then $b^{1/n} \in \mathcal{A}_+$ and $(b^{1/n})^n = b$ by Theorem 1.5.10.

Proposition 1.5.16. *Let $(H, (\cdot, \cdot))$ and $(K, <\cdot, \cdot>)$ be Hilbert spaces. Let $T \in B(H, K)$. Then*

(1) $T^*T \in B(H)_+$.

(2) *There is a unique operator U in $B(H, K)$ such that*

$$U^*U = P_{\overline{\mathrm{Ran}\,(T^*)}}, \; UU^* = P_{\overline{\mathrm{Ran}\,(T)}} \text{ and } T = U|T|. \tag{1.5.2}$$

In this case, we call $T = U|T|$ is the polar decomposition of T.

Proof. (1) Clearly, $T^*T \in B(H)_{sa}$. Note that for any $\lambda > 0$ and any $\xi \in H$,

$$\|(\lambda I + T^*T)\xi\|^2 = \|T^*T\xi\|^2 + \lambda^2\|\xi\|^2 + 2\lambda\|T\xi\|^2 \geq \lambda^2\|\xi\|^2.$$

So $\mathrm{Ker}\,(\lambda I + T^*T) = \{0\}$ and $\mathrm{Ran}\,(\lambda I + T^*T)$ is closed in H and consequently, $\mathrm{Ran}\,(\lambda I + T^*T) = (\mathrm{Ker}\,(\lambda I + T^*T))^\perp = H$. Then $\lambda I + T^*T \in GL(B(H))$ by Theorem 1.1.7. So $\sigma(T^*T) \subset \mathbb{R}_+$, i.e., $T^*T \in B(H)_+$.

(2) Set $|T| = (T^*T)^{1/2} \in B(H)_+$. Since for any $x \in H$,

$$\|Tx\|^2 = <Tx, Tx> = (T^*Tx, x) = (|T|^2x, x) = \||T|x\|^2, \tag{1.5.3}$$

we can define a linear operator $U_0 \colon \mathrm{Ran}\,(|T|) \to \mathrm{Ran}\,(T)$ by $U_0(|T|x) = Tx$, $\forall\, x \in H$. Thus $\|U_0y\| = \|y\|$, $\forall\, y \in \mathrm{Ran}\,(|T|)$ by (1.5.3) and hence U_0 can be extended to be a linear operator U_1 from $\overline{\mathrm{Ran}\,(|T|)}$ to $\overline{\mathrm{Ran}\,(T)}$ such that $\|U_1y\| = \|y\|$, $\forall\, y \in \overline{\mathrm{Ran}\,(|T|)}$.

Put $U = P_{\overline{\mathrm{Ran}\,(T)}}U_1P_{\overline{\mathrm{Ran}\,(T^*)}}$. Using the facts that

$$\overline{\mathrm{Ran}\,(T^*)} = (\mathrm{Ker}\,T)^\perp = (\mathrm{Ker}\,|T|)^\perp = \overline{\mathrm{Ran}\,(|T|)},$$

it is easy to check that U satisfies (1.5.2).

If there is another $V \in B(H, K)$ such that $T = V|T|$ and $V^*V = U^*U$, $VV^* = UU^*$. Then $V\xi = U\xi$, $\forall\, \xi \in \overline{\mathrm{Ran}\,(|T|)}$. From

$$V^*V = U^*U = P_{\overline{\mathrm{Ran}\,(T^*)}} = I_H - P_{\mathrm{Ker}\,T},$$

we get that $\mathrm{Ker}\,U = \mathrm{Ker}\,V = \mathrm{Ker}\,T$. Thus, if $\xi \in \mathrm{Ker}\,(|T|)$, then $U\xi = V\xi = 0$. Since $H = \overline{\mathrm{Ran}\,(|T|)} \oplus \mathrm{Ker}\,(|T|)$, it follows from above argument that $V = U$. $\square$

We can generalize Proposition 1.5.16 (1) to the set of C^*–algebras as follows.

Theorem 1.5.17. *Let $\mathcal{A}$ be a C^*–algebra and let $a \in \mathcal{A}_{sa}$. Then $a \in \mathcal{A}_+$ iff $a = b^*b$ for some $b \in \mathcal{A}$.*

Proof. If $a \in \mathcal{A}_+$, then $a = b^2 = bb$ for some $b \in \mathcal{A}_{sa}$ by Proposition 1.5.14 (2).

Conversely, suppose that $a = b^*b$ for some $b \in \mathcal{A}$. Then $a^* = a$. Set $a = a_+ - a_-$. We have

$$-(b(a_-)^{1/2})^*(b(a_-)^{1/2}) = -(a_-)^{1/2}(a_+ - a_-)(a_-)^{1/2} = (a_-)^2 \in \mathcal{A}_+.$$

Put $b(a_-)^{1/2} = x_1 + ix_2$, $x_1, x_2 \in \mathcal{A}_{sa}$. Then

$$(b(a_-)^{1/2})(b(a_-)^{1/2})^* = 2(x_1^2 + x_2^2) - (b(a_-)^{1/2})^*(b(a_-)^{1/2}) \in \mathcal{A}_+$$

by Corollary 1.5.15. Since

$$\sigma((b(a_-)^{1/2})(b(a_-)^{1/2})^*)\backslash\{0\} = \sigma((b(a_-)^{1/2})^*(b(a_-)^{1/2}))\backslash\{0\}$$

by Proposition 1.4.14, we get that $\sigma(-(a_-)^2) \subset \mathbb{R}_+$. Thus $\sigma((a_-)^2) = \{0\}$ and hence $a_- = 0$. $\square$

Definition 1.5.18. Let $\mathcal{A}$ be a C^*–algebra and let $a, b \in \mathcal{A}_{sa}$. We write $a \leq b$ (or $b \geq a$) if $a - b \geq 0$.

Proposition 1.5.19. *Let $\mathcal{A}$ be a C^*–algebra.*

(1) $\mathcal{A}_+ = \{a^*a \,|\, a \in \mathcal{A}\}$.
(2) *Let $a, b \in \mathcal{A}_{sa}$ and $c \in \mathcal{A}$. Then $a \leq b$ implies that $c^*ac \leq c^*bc$.*
(3) *If $a, b \in \mathcal{A}_+$ and $a \leq b$, then $\|a\| \leq \|b\|$.*
(4) *If $\mathcal{A}$ is unital and $a, b \in \mathcal{A}_+$ are invertible, then $a \leq b$ implies that $0 < b^{-1} \leq a^{-1}$.*

Proof. (1) follows from Theorem 1.5.17 and Proposition 1.5.14.

(2) Pick $d \in \mathcal{A}$ such that $b - a = d^*d$. Then $c^*bc - c^*ac = (dc)^*(dc)$ and hence $c^*ac \leq c^*bc$ by Theorem 1.5.17.

(3) Since $b \leq \|b\|$ in $\tilde{\mathcal{A}}$, we have

$$\|b\| - a = (\|b\| - b) + (b - a) \in \mathcal{A}_+ + \mathcal{A}_+ = \mathcal{A}_+$$

by Corollary 1.5.15. So for any $\lambda \in \sigma(a)$, $\|b\| - \lambda \geq 0$ and consequently, $\|a\| \leq \|b\|$.

(4) Since $(b^{1/2})^2 = b$ is invertible in $\mathcal{A}$, it follows that $b^{1/2}$ is also invertible in $\mathcal{A}$. So by Proposition 1.5.5 (2), $(b^{1/2})^{-1} \in \mathcal{A}_+$. Thus

$$(b^{1/2})^{-1}a(b^{1/2})^{-1} \leq (b^{1/2})^{-1}b^{1/2}b^{1/2}(b^{1/2})^{-1} = 1$$

and hence

$$\begin{aligned}a^{1/2}b^{-1}a^{1/2} &= a^{1/2}(b^{1/2})^{-1}(b^{1/2})^{-1}a^{1/2}\\ &\leq \|a^{1/2}(b^{1/2})^{-1}(b^{1/2})^{-1}a^{1/2}\|\\ &= \|(b^{1/2})^{-1}a(b^{1/2})^{-1}\| \leq 1.\end{aligned}$$

Finally, $b^{-1} = (a^{1/2})^{-1}a^{1/2}b^{-1}a^{1/2}(a^{1/2})^{-1} \leq (a^{1/2})^{-1}(a^{1/2})^{-1} = a^{-1}$. □

Corollary 1.5.20. *Let $\mathcal{A}$ be a unital C^*-algebra and let a, $b \in \mathcal{A}_+$. If b is invertible in $\mathcal{A}$, then $a + b \in GL(\mathcal{A})$.*

Proof. Put $c = a + b \in \mathcal{A}_+$. Then

$$(b^{1/2})^{-1}c(b^{1/2})^{-1} = (b^{1/2})^{-1}a(b^{1/2})^{-1} + 1.$$

Since $(b^{1/2})^{-1}a(b^{1/2})^{-1} \in \mathcal{A}_+$ by Proposition 1.5.19 (2) and

$$\sigma((b^{1/2})^{-1}c(b^{1/2})^{-1}) = \{\lambda + 1 \mid \lambda \in \sigma((b^{1/2})^{-1}a(b^{1/2})^{-1})\},$$

$(b^{1/2})^{-1}c(b^{1/2})^{-1}$ is invertible in $\mathcal{A}$ and so is the positive element c. □

Corollary 1.5.21. *Let H be a Hilbert space and P be an idempotent operator in $B(H)$. Then P is a projection iff $\|P\| = 1$.*

Proof. When P is a projection, $\|P\| = 1$. Conversely, put $M = \operatorname{Ran}(P)$. Since $Px = x$, $\forall x \in M$, we have $PP_M = P_M$ and $P_MP = P$. Thus, $P_MPP_M = P_M$, $P_{M^\perp}PP_M = 0$, $P_{M^\perp}PP_{M^\perp} = 0$ and

$$P = P_M + P_MPP_{M^\perp}, \quad PP^* = P_M + P_MPP_{M^\perp}P^*P_M. \tag{1.5.4}$$

Note that $PP^* \in B(H)_+$ by Proposition 1.5.19 (1) and $\|PP^*\| = \|P\|^2 = 1$. So $PP^* \leq I$ and hence $P_MPP_{M^\perp}P^*P_M \leq I - P_M$ by (1.5.4). But this means that $P_MPP_{M^\perp}P^*P_M = 0$. Therefore, $P = P_M$ by (1.5.4). □

Definition 1.5.22. An approximate unit for a C*-algebra $\mathcal{A}$ is an increasing net $\{u_\lambda\}$ of positive elements in the closed unit ball $D(\mathcal{A})$ of $\mathcal{A}$ such that $a = \lim\limits_\lambda a\,u_\lambda$ or equivalently, $a = \lim\limits_\lambda u_\lambda a$ for all $a \in \mathcal{A}$.

Theorem 1.5.23. *Every C^*-algebra admits an approximate unit.*

Proof. Let $\mathcal{A}$ be a C^*-algebra. Put $\Lambda = \{a \in \mathcal{A}_+ \mid \|a\| < 1\}$ and $u_\lambda = \lambda$, $\forall \lambda \in \Lambda$. We will show that $\{u_\lambda\}_\Lambda$ is an approximate unit for $\mathcal{A}$.

Let a, $b \in \Lambda$ and put $f_\alpha(t) = \dfrac{1}{\alpha}\Big(1 - \dfrac{1}{1+\alpha t}\Big)$, $t \in (-\alpha^{-1}, +\infty)$, $\alpha > 0$. Then for any x, $y \in \mathcal{A}_+$, $x \leq y$ implies that $(1 + \alpha y)^{-1} \leq (1 + \alpha x)^{-1}$ by Corollary 1.5.20 and Proposition 1.5.19 (4). Thus, $f_\alpha(x) \leq f_\alpha(y)$, $\alpha > 0$.

Since $a, b \in \mathcal{A}_+$ and $\|a\| < 1$, $\|b\| < 1$, we have

$$1 - a = (1 - \|a\|) + (\|a\| - a)$$

is invertible in $\tilde{\mathcal{A}}$ by Corollary 1.5.20. Similarly, $1 - b$ is also invertible in $\tilde{\mathcal{A}}$. Put $s = (1-a)^{-1}a$, $t = (1-b)^{-1}b$. Then $s, t \in \mathcal{A}_+$ and $1+s$, $1+t$, $1+s+t$ are all invertible in $\tilde{\mathcal{A}}$. Set $w = f_1(1+s+t)$. Then $w \in \mathcal{A}_+$ with $\|w\| < 1$ and moreover, $w \geq f_1(s) = s(1-s)^{-1} = a$ and $w \geq f_1(t) = b$. Therefore, $\{u_\lambda\}_\Lambda$ is an increasing net.

Let $x \in \mathcal{A}_+$ and put $g(u) = \|x(1-u)x\|$, $u \in \Lambda$. When $v \in \Lambda$ and $v \leq u$, we have $x(1-u)x \leq x(1-v)x$ and $g(u) \leq g(v)$ by Proposition 1.5.19 (3). Thus, $\{g(\lambda)\}_\Lambda$ is a decreasing net.

For $\alpha > 0$, $x \in \mathcal{A}_+ \backslash \{0\}$, we have $1 + \alpha x \geq 1$ and $1 - (1+\alpha x)^{-1} \geq 0$ by Proposition 1.5.19 (4). Furthermore,

$$\|1 - (1+\alpha x)^{-1}\| = \max\{1 - (1+\alpha\lambda)^{-1} \,|\, \lambda \in \sigma(x)\} < 1.$$

So $\alpha f_\alpha(x) \in \Lambda$, $\forall\, \alpha > 0$ and consequently,

$$g(\alpha f_\alpha(x)) = \|x(1 - \alpha f_\alpha(x))x\| = \|(1+\alpha x)^{-1}x^2\| \leq \frac{\|x\|}{\alpha}. \tag{1.5.5}$$

(1.5.5) indicates that for any $\epsilon > 0$, there is $\alpha_0 > 0$ such that $\|x(1-u)x\| < \epsilon^2$, $\forall\, u \in \Lambda$ with $u \geq \alpha_0 f_{\alpha_0}(x)$. Noting that

$$(1-u)^2 = (1-u)^{1/2}(1-u)(1-u)^{1/2} \leq 1 - u$$

when $u \in \Lambda$, we have $x(1-u)^2x \leq x(1-u)x$ and $\|x(1-u)^2x\| \leq \|x(1-u)x\|$. Thus,

$$\|(1-u)x\|^2 = \|x(1-u)^2x\| \leq \|x(1-u)x\| < \epsilon^2, \ \forall\, u \geq \alpha_0 f_{\alpha_0}(x),$$

that is, $\lim\limits_{\lambda \in \Lambda} \|(1-u_\lambda)x\| = 0$.

Now for any $x \in \mathcal{A}$, we have $x = x_1 + ix_2$, $x_1, x_2 \in \mathcal{A}_{sa}$ and

$$x_1 = (x_1)_+ - (x_1)_-, \quad x_2 = (x_2)_+ - (x_2)_-,$$

where $(x_1)_+, (x_1)_-, (x_2)_+$ and $(x_2)_-$ are all positive. This implies that for each $x \in \mathcal{A}$, $\lim\limits_{\lambda \in \Lambda} \|(1-u_\lambda)x\| = \lim\limits_{\lambda \in \Lambda} \|x(1-u_\lambda)\| = 0$. □

Corollary 1.5.24. *A separable C^*–algebra has a sequential approximate unit.*

Proof. Let $\{x_n\}$ be a countable dense subset of a C^*–algebra $\mathcal{A}$ and let $\{u_\lambda\}$ be an approximate unit for $\mathcal{A}$. A simple induction argument shows that there exist $i_1 \leq i_2 \leq \cdots$ such that $\|x_k u_i - x_k\| < 1/2^n$ for $i \leq k \leq n$ and $i \geq i_n$. Let $v_n = u_{i_n}$. Then for all $k \in \mathbb{N}$, $\|x_k v_n - x_k\| \to 0$ as $n \to \infty$. Since $\{v_n\}$ is bounded, $\|xv_n - x\| \to 0$ as $n \to \infty$ for all $x \in \mathcal{A}$. □

Theorem 1.5.25. *Let $\mathcal{J}$ be a non-zero closed ideal of a C^*-algebra $\mathcal{A}$. Then*

(1) *$\mathcal{J}$ is a C^*-algebra.*
(2) $\operatorname{dist}(x, \mathcal{J}) = \lim_\lambda \|x(1-u_\lambda)\| = \lim_\lambda \|(1-u_\lambda)x\|$ *for every $x \in \mathcal{A}$, where $\{u_\lambda\}$ is an approximate unit of $\mathcal{J}$.*
(3) *$\mathcal{A}/\mathcal{J}$ is a C^*-algebra equipped with operations:*

$$\alpha[x]+\beta[y] = [\alpha x+\beta y],\ [x][y] = [xy],\ [x]^* = [x^*],\ \forall\, x, y \in \mathcal{A},\ \alpha, \beta \in \mathbb{C}$$

and the norm $\|[x]\| = \operatorname{dist}(x, \mathcal{J})$, $\forall\, x \in \mathcal{A}$.

Proof. (1) We need to prove that if $x \in \mathcal{J}$, then $x^* \in \mathcal{J}$. Put $\mathcal{J}^* = \{a^* | \, a \in \mathcal{J}\}$ and $\mathcal{I} = \mathcal{J} \cap \mathcal{J}^*$. Then $\mathcal{I}$ is a C^*-subalgebra of $\mathcal{A}$ and moreover, $\mathcal{I} \neq \{0\}$ provide $\mathcal{J} \neq \{0\}$ for $a \in \mathcal{J}\backslash\{0\}$, $a^*a \in \mathcal{I}$ and $a^*a \neq 0$.

Let $\{u_\lambda\}$ be an approximate unit for $\mathcal{I}$ and let $x \in \mathcal{J}$. Then $x^*x \in \mathcal{I}$ and hence $\|x^*x(1-u_\lambda)\| \to 0$. Thus, from

$$\|x - xu_\lambda\|^2 = \|(1-u_\lambda)x^*x(1-u_\lambda)\| \leq \|1-u_\lambda\|\|x^*x(1-u_\lambda)\|,$$

we get that $\lim_\lambda \|x - xu_\lambda\| = \lim_\lambda \|x^* - u_\lambda x^*\| = 0$. Note that $u_\lambda \in \mathcal{I} \subset \mathcal{J}$. Therefore, $u_\lambda x^* \in \mathcal{J}$ and hence $x^* \in \mathcal{J}$ and $\{u_\lambda\}$ is also an approximate unit for $\mathcal{J}$.

(2) Let $x \in \mathcal{A}$ and set $\alpha = \operatorname{dist}(x, \mathcal{J}) = \operatorname{dist}(x^*, \mathcal{J})$. Let $\{u_\lambda\}$ be an approximate unit for $\mathcal{J}$. It follows from Proposition 1.5.19 that

$$\|x(1-u_\lambda)x^*\| \leq \|x(1-u_\mu)x^*\|, \quad \lambda \geq \mu.$$

Thus $\beta = \lim_\lambda \|x(1-u_\lambda)x^*\|$ exists. Since $0 \leq 1-u_\lambda \leq 1$, we have

$$\alpha^2 \leq \|x(1-u_\lambda)\|^2 = \|x(1-u_\lambda)^2x^*\| \leq \|x(1-u_\lambda)x^*\| \tag{1.5.6}$$

and so that $\alpha^2 \leq \beta$ by (1.5.6). On the other hand, for any $\epsilon > 0$, we can find $b_0 \in \mathcal{J}$ such that $\alpha + \epsilon < \|x + b_0\|$. Then

$$\begin{aligned}(\alpha+\epsilon)^2 &> \|x+b_0\|\|1-u_\lambda\|\|(x+b_0)^*\| \geq \|(x+b_0)(1-u_\lambda)(x+b_0)^*\| \\ &\geq \|x(1-u_\lambda)x^*\| - \|b_0(1-u_\lambda)(x+b_0)^*\| - \|(x+b_0)(1-u_\lambda)b_0^*\|.\end{aligned}$$

Since $\|b_0(1-u_\lambda)\| \to 0$ and $\|(1-u_\lambda)b_0^*\| \to 0$, we get that $(\alpha+\epsilon)^2 \geq \beta$ from above inequality and consequently, $\alpha^2 \geq \beta$ by letting $\epsilon \to 0^+$. Finally, we obtain the assertion from (1.5.6).

(3) We need to verify that $\|[a][b]\| \leq \|[a]\|\|[b]\|$ and $\|[a]\|^2 = \|[a]^*[a]\|$, $\forall\, a, b \in \mathcal{A}$. Let $a, b \in \mathcal{A}$. Note that for any $k_1, k_2 \in \mathcal{J}$, ak_2, k_1b, $k_1k_2 \in \mathcal{J}$. So we have

$$\begin{aligned}\|[a][b]\| &\leq \|ab - ak_2 - k_2b - k_1k_2\| = \|(a-k_1)(b-k_2)\| \\ &\leq \|a-k_1\|\|b-k_2\|\end{aligned}$$

and hence $\|[a][b]\| \le \|[a]\|[b]\|$. Thus, $\|[a]^*[a]\| \le \|[a]^*\|[a]\| \le \|[a]\|^2$.

Let $\{u_\lambda\}$ be an approximate unit for $\mathcal{J}$. Then

$$\|[a]^*[a]\| = \|[a^*a]\| = \lim_\lambda \|a^*a(1-u_\lambda)\|$$

by (2). Noting that

$$\|a^*a(1-u_\lambda)\| \ge \|(1-u_\lambda)a^*a(1-u_\lambda)\| = \|a(1-u_\lambda)\|^2,$$

we have $\|[a]\|^2 \ge \|[a]^*[a]\| \ge \|[a]\|^2$. □

Definition 1.5.26. Let $\mathcal{A}$ and $\mathcal{B}$ be C^*–algebras. A mapping $\phi\colon \mathcal{A} \to \mathcal{B}$ is called a $*$–homomorphism if ϕ is an algebraic homomorphism and $\phi(a^*) = (\phi(a))^*$, $\forall\, a \in \mathcal{A}$.

The $*$–homomorphism $\phi\colon \mathcal{A} \to \mathcal{B}$ is called to be isomorphic if $\operatorname{Ker}\phi = \{0\}$ and $\operatorname{Ran}(\phi) = \mathcal{B}$.

Proposition 1.5.27. *Let ϕ be a $*$–homomorphism between C^*–algebras $\mathcal{A}$ and $\mathcal{B}$. Then*

(1) $\|\phi(a)\| \le \|a\|$, $\forall\, a \in \mathcal{A}$.
(2) $\phi(f(a)) = f(\phi(a))$ *when* $a \in \mathcal{A}_{sa}$ *and* $f \in C_0(\sigma(a)\backslash\{0\})$.
(3) $\|\phi(a)\| = \|a\|$, $\forall\, a \in \mathcal{A}$ *when ϕ is injective.*
(4) $\operatorname{Ran}(\phi) = \phi(\mathcal{A})$ *is a C^*–subalgebra of $\mathcal{B}$.*

Proof. We extend ϕ to a $*$–homomorphism $\phi^+\colon \tilde{\mathcal{A}} \to \tilde{\mathcal{B}}$ by $\phi^+(a+\lambda) = \phi(a) + \lambda$, $\forall\, a \in \mathcal{A}$, $\lambda \in \mathbb{C}$.

(1) Let $a \in \mathcal{A}$. Since $a^*a \le \|a^*a\|$, it follows from Theorem 1.5.23 that $\|a\|^2 - a^*a = b^*b$ for some $b \in \mathcal{A}$. Thus

$$\|a\|^2 - \phi(a^*a) = \phi^+(b^*b) = (\phi^+(b))^*(\phi^+(b)) \in \mathcal{B}_+$$

and consequently, $(\phi(a))^*\phi(a) \le \|a\|^2$. Therefore $\|\phi(a)\| \le \|a\|$ by Proposition 1.5.19 (3).

(2) Let $a \in \mathcal{A}_{sa}$ and put $b = \phi(a)$. Obviously, $\sigma(b) \subset \sigma(a)$ and $f \in C_0(\sigma(b)\backslash\{0\})$. Choose $\{p_n\} \subset \mathfrak{P}_0(\mathbb{R})$ such that $\|f - p_n\| \to 0$ as $n \to \infty$. Since $p_n(b) = \phi(p_n(a))$, it follows from Theorem 1.5.10 that $f(b) = \phi(f(a))$.

(3) If there is $a_0 \in \mathcal{A}$ such that $\|\phi(a_0)\| < \|a_0\|$, then $\|\phi(a_0^*a_0)\| < \|a_0^*a_0\|$. Set $d = a_0^*a_0$. Pick a continuous function f_0 on $[0, \|d\|]$ such that $f_0(\|d\|) = 1$ and $f_0(t) = 0$, $\forall\, t \in [0, \|\phi(d)\|]$. Then $f_0(\phi(d)) = 0$ and $\phi(f_0(d)) = f_0(\phi(d)) = 0$ by (2). Since ϕ is injective, we have $f_0(d) = 0$. But it is impossible for $\|d\| \in \sigma(d)$ and $f_0(\|d\|) = 1$.

(4) Put $\mathcal{J} = \operatorname{Ker}\phi$. Since ϕ is continuous by (1), it follows that $\mathcal{J}$ is a closed ideal of $\mathcal{A}$ and $\mathcal{A}/\mathcal{J}$ is a C^*–algebras by Theorem 1.5.25 (3).

Define a $*$–homomorphism $\hat{\phi}\colon \mathcal{A}/\mathcal{J} \to \mathcal{B}$ by $\hat{\phi}([a]) = \phi(a)$, $\forall\, a \in \mathcal{A}$. Since $\operatorname{Ker}\hat{\phi} = \{0\}$, we have

$$\|\phi(a)\| = \|\hat{\phi}([a])\| = \|[a]\| = \operatorname{dist}(a, \operatorname{Ker}\phi), \ \forall\, a \in \mathcal{A}$$

by (3) and so that $\gamma(\phi) = 1$, that is, $\operatorname{Ran}(\phi)$ is closed in $\mathcal{B}$. Therefore, $\phi(\mathcal{B})$ is a C^*–subalgebra of $\mathcal{B}$. □

Definition 1.5.28. Let $\mathcal{B}$ be C^*–subalgebra of a C^*–algebra $\mathcal{A}$. We say that $\mathcal{B}$ is hereditary if for any $a \in \mathcal{A}_+$ and $b \in \mathcal{B}_+$, $a \le b$ implies that $a \in \mathcal{B}$.

Obviously, 0 and $\mathcal{A}$ are hereditary C^*–subalgebras of $\mathcal{A}$. Any intersection of hereditary C^*–subalgebras is hereditary. Moreover, we have the following:

Proposition 1.5.29. *Let $\mathcal{A}$ be a C^*–algebra.*

(1) *Every closed ideal of $\mathcal{A}$ is hereditary.*
(2) *$\overline{a\mathcal{A}a^*} = \overline{span}\{axa^* \mid x \in \mathcal{A}\}$ is hereditary.*
(3) *If $\mathcal{B}$ is a hereditary C^*–subalgebra of $\mathcal{A}$, then*

$$\mathcal{B}\,\mathcal{A}\,\mathcal{B} = span\{axb \mid a, b \in \mathcal{B},\ x \in \mathcal{A}\} \subset \mathcal{B}.$$

Proof. (1) Let $\mathcal{J}$ be a closed ideal of $\mathcal{A}$. Then $\mathcal{J}$ is a C^*–subalgebra of $\mathcal{A}$ by Theorem 1.5.25. Let $\{u_\lambda\}$ be an approximate unit for $\mathcal{J}$. Let $a \in \mathcal{A}_+$ and $b \in \mathcal{B}_+$ with $a \le b$. Then

$$\|(1-u_\lambda)a(1-u_\lambda)\| \le \|(1-u_\lambda)b(1-u_\lambda)\| \le \|1-u_\lambda\|\|b(1-u_\lambda)\|.$$

Note that

$$\|a(1-u_\lambda)\|^2 \le \|a^{1/2}\|^2\|a^{1/2}(1-u_\lambda)\|^2 = \|a\|\|(1-u_\lambda)a(1-u_\lambda)\|.$$

Thus $\lim\limits_{\lambda} \|a(1-u_\lambda)\| = 0$ and consequently, $a \in \mathcal{J}$.

(2) For each $\epsilon > 0$, define continuous function f_ϵ on $[0, +\infty)$ by

$$f_\epsilon(t) = \begin{cases} 1 & t \ge \epsilon \\ \frac{2}{\epsilon}(\frac{\epsilon}{2} - t) & \frac{\epsilon}{2} \le t < \epsilon \\ 0 & 0 \le t < \frac{\epsilon}{2} \end{cases}.$$

Put $d_n = f_{1/n}(a) \in \mathcal{A}_+$, $n \in \mathbb{N}$. Then $ad_n = d_n a$, $\|d_n\| \le 1$, $\forall\, n \in \mathbb{N}$ and

$$\|d_n a - a\| = \max\{|d_n(\lambda)\lambda - \lambda| \mid \lambda \in \sigma(a)\} \le \frac{1}{n}.$$

So $\lim\limits_{n\to\infty} \|d_n a - a\| = 0$.

Put $\mathcal{E} = \overline{a\mathcal{A}a}$. Since $a\mathcal{A}a$ is a $*$–algebra, $\mathcal{E}$ is a C^*–subalgebra of $\mathcal{A}$. Let $x \in \mathcal{E}_+$ and $y \in \mathcal{A}_+$ such that $y \leq x$. Then

$$\begin{aligned}&(1-d_n)y(1-d_n) \leq (1-d_n)x(1-d_n) \quad \text{and}\\ &\|y^{1/2}(1-d_n)\|^2 = \|(1-d_n)y(1-d_n)\| \leq \|(1-d_n)x(1-d_n)\|\\ &\qquad \leq \|1-d_n\|\|x(1-d_n)\| \leq \|x(1-d_n)\|. \end{aligned} \tag{1.5.7}$$

Choose $b_k \in \mathcal{A}$ such that $\|x - ab_ka\| < 1/k$, $\forall\, k \in \mathbb{N}$. Pick a subsequence $\{d_{n_k}\}$ of $\{d_n\}$ such that $\|ad_{n_k} - a\| < \dfrac{1}{k(\|ab_k\|+1)}$, $\forall\, k \in \mathbb{N}$. Then

$$\|x(1-d_{n_k})\| \leq \|1-d_{n_k}\|\|x - ad_ka\| + \|ad_k\|\|a(1-d_{n_k}\| < \frac{2}{k}$$

and consequently, $\lim\limits_{k\to\infty} \|x(1-d_{n_k})\| = 0$. Thus $\lim\limits_{k\to\infty} \|y^{1/2}(1-d_{n_k})\| = 0$ by (1.5.7) and hence $\lim\limits_{k\to\infty} \|y - d_{n_k}yd_{n_k}\| = 0$.

Note that $a^3 \in \mathcal{E}$. So $a = (a^3)^{1/3} \in \mathcal{E}$ by Theorem 1.5.10 and consequently, $d_{n_k} \in C^*(a) \subset \mathcal{E}$, $\forall\, k \in \mathbb{N}$. Choose $z_k \in \mathcal{A}$ such that $\|d_{n_k} - az_ka\| < 1/k$, $\forall\, k \in \mathbb{N}$. Then $\|y - a(z_kayaz_k)a\| \to 0$ as $k \to \infty$ and hence $y \in \mathcal{E}$.

(3) Let a, $b \in \mathcal{B}$ and $c \in \mathcal{A}$. Put $x = acb = y + iz$, where y, $z \in \mathcal{A}_{sa}$. Then $x^*x = b^*c^*a^*acb \leq \|ac\|^2b^*b$ and $xx^* \leq \|cb\|^2aa^*$. Since b^*b, $aa^* \in \mathcal{B}$ and $\mathcal{B}$ is hereditary, we have x^*x, $xx^* \in \mathcal{B}$.

From $2(y^2+z^2) = x^*x + xx^* \in \mathcal{B}$, we get that $y^2, z^2 \in \mathcal{B}$. Write $y = y_+ - y_-$, where y_+, $y_- \in \mathcal{B}_+$ and $y_+y_- = 0$. So $y^2 = (y_+)^2 + (y_-)^2$. This shows that $(y_+)^2, (y_-)^2 \in \mathcal{B}$ and hence $y_+ = ((y_+)^2)^{1/2} \in \mathcal{B}$, $y_- = ((y_-)^2)^{1/2} \in \mathcal{B}$. So $y \in \mathcal{B}$. Similarly, $z \in \mathcal{B}$. Therefore $x \in \mathcal{B}$. □

Definition 1.5.30. A linear functional ϕ on a C^*–algebra $\mathcal{A}$ is called to be positive if $\phi(a) \geq 0$ for every $a \in \mathcal{A}_+$.

Proposition 1.5.31. *Let ϕ be a positive linear functional on the C^*–algebra $\mathcal{A}$. Then*

(1) $\phi(x^*) = \overline{\phi(x)}$, $\forall\, x \in \mathcal{A}$.
(2) *ϕ is bounded.*
(3) $|\phi(y^*x)|^2 \leq \phi(x^*x)\phi(y^*y)$, $\forall\, x$, $y \in \mathcal{A}$.

Proof. (1) Let $x \in \mathcal{A}_{sa}$. Then $x = x_+ - x_-$, where x_+, $x_- \in \mathcal{A}_+$ and hence $\phi(x) \in \mathbb{R}$. Let $x \in \mathcal{A}$. Then $x = x_1 + ix_2$ with $x_1, x_2 \in \mathcal{A}_{sa}$ and consequently,

$$\phi(x^*) = \phi(x_1 - ix_2) = \phi(x_1) - i\phi(x_2) = \overline{\phi(x)}.$$

(2) If ϕ is not bounded, then for any $n \in \mathbb{N}$, there is $a_n \in D(\mathcal{A})$ such that $|\phi(a_n)| > \sqrt{2}n$. Since $a_n = a_1^{(n)} + ia_2^{(n)}$ with $a_j^{(n)} \in \mathcal{A}_{sa}$ and $\|a_j^{(n)}\| \leq 1$, $j = 1, 2$, we must have $|\phi(a_1^{(n)})| > n$ or $|\phi(a_2^{(n)})| > n$ for each n. So we obtain a sequence $\{c_n\} \subset \mathcal{A}_{sa} \cap D(\mathcal{A})$ such that $|\phi(c_n)| > n$.

Let $c_n = c_+^{(n)} - c_-^{(n)}$, where $c_+^{(n)}, c_-^{(n)} \in \mathcal{A}_+$ and $\|c_+^{(n)}\| \leq 1$, $\|c_-^{(n)}\| \leq 1$, $\forall\, n \in \mathbb{N}$. Then for each n, $|\phi(c_+^{(n)})| > n/2$ or $|\phi(c_-^{(n)})| > n/2$. Thus we get a sequence $\{b_n\} \subset \mathcal{A}_+ \cap D(\mathcal{A})$ such that $|\phi(b_n)| > n/2$, $\forall\, n \in \mathbb{N}$.

Set $a = \sum\limits_{n=1}^{\infty} \dfrac{a_n}{n^2} \in \mathcal{A}_+$. Then for each n, $\phi\Big(a - \sum\limits_{k=1}^{n} \dfrac{a_k}{k^2}\Big) \geq 0$, that is,

$$\phi(a) \geq \sum_{k=1}^{n} \frac{\phi(a_k)}{k^2} \geq \sum_{k=1}^{n} \frac{1}{2k}, \quad \forall\, n \in \mathbb{N}.$$

So $\sum\limits_{n=1}^{\infty} \dfrac{1}{n} \leq 2\phi(a)$. But this is impossible.

(3) From $\phi((x+ty)^*(x+ty)) \geq 0$, $\forall\, t \in \mathbb{C}$, we get that

$$\phi(x^*x) + t\overline{\phi(y^*x)} + \bar{t}\phi(y^*x) + |t|^2\phi(y^*y) \geq 0. \tag{1.5.8}$$

If $\phi(y^*y) \neq 0$, taking $t = -\dfrac{\phi(y^*x)}{\phi(y^*y)}$ in (1.5.8), we get that $|\phi(y^*x)|^2 \leq \phi(x^*x)\phi(y^*y)$; when $\phi(y^*y) = 0$, we take $t = \pm n$ and $\pm i\,n$ respectively in (1.5.8) and let $n \to \infty$. Then we finally get that $\phi(y^*x) = 0$. □

Theorem 1.5.32. *Let $\mathcal{A}$ be a C^*-algebra and $\phi \in \mathcal{A}^*$.*

(1) *If ϕ is positive and $\{u_\lambda\}$ is an approximate unite for $\mathcal{A}$, then $\lim\limits_\lambda \phi(u_\lambda) = \|\phi\|$.*
(2) *If there is some approximate unite $\{v_\lambda\}$ for $\mathcal{A}$ such that $\lim\limits_\lambda \phi(v_\lambda) = \|\phi\|$, then ϕ is positive.*

Proof. (1) By our assumptions, $\{\phi(u_\lambda)\}$ is an increasing net in $\mathbb{R}_+$, so

$$\lim_\lambda \phi(u_\lambda) = \sup_\lambda \phi(u_\lambda) = \alpha \leq \|\phi\|.$$

Let $x \in \mathcal{A}$. Then $|\phi(u_\lambda x)|^2 \leq \phi(x^*x)\phi(u_\lambda^2)$ by Proposition 1.5.31 (3). Since $\phi(x^*x) \leq \|\phi\|\|x\|^2$, $\phi(u_\lambda^2) \leq \phi(u_\lambda)$ and $\lim\limits_\lambda \|u_\lambda x - x\| = 0$, it follows that $|\phi(x)|^2 \leq \|\phi\|\|x\|^2\alpha$. Thus, $\alpha \geq \|\phi\|$ and hence $\alpha = \|\phi\|$.

(2) We may assume that $\|\phi\| = 1$. Let $x \in \mathcal{A}_{sa}$ with $\|x\| \leq 1$. Write $\phi(x) = a + ib$, where $a, b \in \mathbb{R}$. Fix $n \in \mathbb{N}$. From

$$\begin{aligned}\|x - i\,nv_\lambda\|^2 &= \|(x + i\,nv_\lambda)(x - i\,nv_\lambda)\| \\ &= \|x^2 + n^2v_\lambda^2 - i\,n(xv_\lambda - v_\lambda x)\| \\ &\leq 1 + n^2 + n\|xv_\lambda - v_\lambda x\|,\end{aligned}$$

we get that

$$|\phi(x) - in\phi(v_\lambda)|^2 = |\phi(x - inv_\lambda)|^2 \le 1 + n^2 + n\|xv_\lambda - v_\lambda x\|. \quad (1.5.9)$$

However, $\lim_\lambda \phi(v_\lambda) = 1$ and $\lim_\lambda \|xv_\lambda - v_\lambda x\| = 0$, so we have

$$a^2 + b^2 - 2nb + n^2 = |a + i(b - n)|^2 \le 1 + n^2 \quad (1.5.10)$$

from (1.5.9). Note that (1.5.10) holds for all $n \in \mathbb{N}$. So we deduce from (1.5.10) that $b \ge 0$.

Using above argument to the case $\phi(-x) = (-a) + i(-b)$, we get that $-b \ge 0$. Therefore, $\phi(x) \in \mathbb{R}$.

Let $x \in \mathcal{A}_+$ with $\|x\| \le 1$. Since $0 \le v_\lambda \le 1$ and $0 \le x \le 1$, it follows that $-1 \le x - v_\lambda \le 1$ in $\tilde{\mathcal{A}}$ and hence $\|x - v_\lambda\| \le 1$. Thus, $\phi(v_\lambda - x) \le \|\phi\|\|v_\lambda - x\| \le 1$. Finally, from $\lim_\lambda \phi(v_\lambda) = 1$ and $\phi(v_\lambda - x) \le 1$, we get that $\phi(x) \ge 0$. □

Definition 1.5.33. A representation of a C^*–algebra $\mathcal{A}$ is a pair (ρ, H), where H is a Hilbert space and $\rho\colon \mathcal{A} \to B(H)$ is a $*$–homomorphism.

We say that (ρ, H) is faithful if ρ is injective and that (ρ, H) is a cyclic representation if there is a vector $\xi \in H$ such that $\{\rho(a)\xi \mid a \in \mathcal{A}\}$ is dense in H and the vector ξ is called cyclic.

Theorem 1.5.34. *For each positive linear functional ϕ on a C^*–algebra $\mathcal{A}$, there is a cyclic representation (ρ_ϕ, H) of $\mathcal{A}$ with a cyclic vector ξ_ϕ such that $(\rho_\phi(a)\xi_\phi, \xi_\phi) = \phi(a)$ for all $a \in \mathcal{A}$.*

Proof. Put $L_\phi = \{a \in \mathcal{A} | \phi(a^*a) = 0\}$. Since ϕ is continuous, L_ϕ is closed in $\mathcal{A}$. It follows from Proposition 1.5.31 (3) that L_ϕ is a closed left ideal of $\mathcal{A}$. Put $\hat{H}_\phi = A/L_\phi$. Define

$$< [a], [b] >= \phi(b^*a), \quad \forall\, [a],\, [b] \in \hat{H}_\phi. \quad (1.5.11)$$

It is easy to see that (1.5.11) is well–defined functional on $\hat{H}_\phi \times \hat{H}_\phi$ since L_ϕ is a left ideal. By using Proposition 1.5.31 (3) again, (1.5.11) defines an inner product on $\hat{H}_\phi$.

Let H_ϕ be the completion about norm $\|[a]\| = (\phi(a^*a))^{1/2}$, $\forall\, a \in \mathcal{A}$. Define $\rho_\phi(a)[x] = [ax]$, $\forall\, a,\, x \in \mathcal{A}$. It is well-defined. In fact, if $x' \in \mathcal{A}$ such that $[x'] = [x]$, then $a(x' - x) \in L_\phi$ since L_ϕ is a left ideal. Thus $\rho_\phi(a)$ is a linear map on $\hat{H}_\phi$. Furthermore,

$$\|\rho_\phi(a)[x]\|^2 = \|[ax]\|^2 = \phi(x^*a^*ax) \le \|a\|^2\phi(x^*x) = \|a\|^2\|[x]\|^2.$$

So $\rho_\phi(a)$ can be uniquely extended to an operator (still denoted by $\rho_\phi(a)$) on H_ϕ such that $\|\rho_\phi(a)\| \le \|a\|$. Since for any a, b, $x \in \mathcal{A}$,

$$\rho_\phi(a+b)[x] = [ax+bx] = \rho_\phi(a)[x] + \rho_\phi(b)[x],$$
$$\rho_\phi(ab)[x] = [abx] = \rho_\phi(a)\rho_\phi(b)[x],$$

we get that $\rho_\phi\colon \mathcal{A} \to B(H_\phi)$ is a homomorphism. Also,

$$< \rho_\phi(a)[x], [y] >= \phi(y^*ax) =< [x], \rho_\phi(a^*)[y] >, \quad \forall\, a,\, b,\, x \in \mathcal{A},$$

which shows that $(\rho_\phi(a))^* = \rho_\phi(a^*)$. So $\rho_\phi\colon \mathcal{A} \to B(H_\phi)$ gives a representation.

Now let $\{u_\lambda\}$ be an approximate unit for $\mathcal{A}$. Then for $\lambda \le \mu$,

$$\|[u_\lambda] - [u_\mu]\|^2 = \phi((u_\lambda - u_\mu)^2) \le \phi(u_\mu - u_\lambda).$$

Since $\lim_\lambda \phi(u_\lambda) = \|\phi\|$ by Theorem 1.5.32 and $\|[u_\lambda]\| \le \|\phi\|$, the net $\{[u_\lambda]\}$ converges to a vector $\xi_\phi \in H_\phi$ with $\|\xi_\phi\| \le \|\phi\|$. Thus, for each $a \in \mathcal{A}$, we have

$$\rho_\phi(a)\xi_\phi = \lim_\lambda \rho_\phi(a)[u_\lambda] = \lim_\lambda [au_\lambda] = [a]$$

since $\|[au_\lambda] - [a]\|^2 = \phi((1-u_\lambda)a^*a(1-u_\lambda) \to 0$. Hence ξ_ϕ is cyclic. Moreover, for each $a \in \mathcal{A}$, we have

$$< \rho_\phi(a)\xi_\phi, \xi_\phi >= \lim_\lambda < [a], [u_\lambda] >= \lim_\lambda \phi(u_\lambda a) = \phi(a).$$

□

Let $\{(\rho_\lambda, H_\lambda) |\, \lambda \in \Lambda\}$ be a family of representations. Let $H = \bigoplus_\lambda H_\lambda$ be the direct sum of Hilbert spaces $\{H_\lambda\}$. Define $\rho(a)(\{\xi_\lambda\}) = \{\rho_\lambda(a)\xi_\lambda\}$. It is easy to verify that (ρ, H) gives a representation of $\mathcal{A}$. This representation is called the direct sum of $\{(\rho_\lambda, H_\lambda) |\, \lambda \in \Lambda\}$.

Definition 1.5.35. Let $\mathcal{A}$ be a C^*–algebra. A positive linear functional ψ on $\mathcal{A}$ is a state if $\|\psi\| = 1$. Let $S(\mathcal{A})$ denote the set of states on $\mathcal{A}$. $S(\mathcal{A})$ is called the state space of $\mathcal{A}$.

The direct sum representation (ρ_u, H_u) of $\{(\rho_\psi, H_\psi) |\, \psi \in S(\mathcal{A})\}$ is called the universal representation of $\mathcal{A}$, where (ρ_ψ, H_ψ) is constructed in Theorem 1.5.34.

Theorem 1.5.36 (Gelfand–Naimark–Segal). *If $\mathcal{A}$ is a C^*–algebra, then it has a faithful representation. In other words, every C^*–algebra is isometrically $*$–isomorphic to a C^*–subalgebra of $B(H)$ for some Hilbert space H.*

Proof. We will show that the universal representation (ρ_u, H_u) is faithful.

Let $a \in \mathcal{A}$ be nonzero. Consider the C^*–subalgebra $\mathcal{B}$ generated by 1 and a^*a in $\tilde{\mathcal{A}}$. Since $\|a^*a\| \in \sigma(a^*a)$, there is a multiplicative linear functional ϕ on $\mathcal{B}$ with $\phi(1) = 1$ such that $\phi(a^*a) = \|a^*a\|$ by Proposition 1.4.18. Note that $\mathcal{B} = \overline{\{p(a^*a) | p \in \mathfrak{P}(\mathbb{R})\}}$. So $\phi(x^*) = (\phi(x))^*$, $\forall x \in \mathcal{B}$ and consequently, ϕ is positive on $\mathcal{B}$. By Proposition 1.1.4, we can find $\psi \in \tilde{\mathcal{A}}^*$ such that

$$\psi(x) = \phi(x), \ \forall x \in \mathcal{B} \text{ and } \|\psi\| = \|\phi\|_{\mathcal{B}} = 1.$$

Since $\psi(1) = 1 = \|\psi\|$, it follows from Theorem 1.5.32 (2) that $\psi \in S(\mathcal{A})$. Thus, we get that

$$\begin{aligned}
\|\rho_u(a)\|^2 &\geq \|\rho_\psi(a)\|^2 \geq \|\xi_\phi\|^{-2} \|\rho_\psi(a)\xi_\psi\|^2 \\
&= \|\xi_\psi\|^{-2} < \rho_\psi(a^*a)\xi_\psi, \xi_\phi > \\
&= \|\xi_\psi\|^{-2} \psi(a^*a) = \|\xi_\psi\|^{-2} \|a^*a\| \\
&= \|a\|^2 \|\xi_\psi\|^{-2} \neq 0,
\end{aligned}$$

by Theorem 1.5.34. Therefore, ρ_u is injective. □

1.6 Notes

Section 1.1 is based on [Rudin (1973)], [Yosida (1980)].

Section 1.2 is based on [Goldberg (1985)].

Section 1.3 is based mainly on [Kato (1984)]. Definition 1.3.1, Proposition 1.3.2, Proposition 1.3.3, Proposition 1.3.4 and Lemma 1.3.7 all come from [Kato (1984)]. Lemma 1.3.5 comes from [Xue (2008a)], which is a generalization of corresponding results in [Markus (1959)]. (1.3.3) of Lemma 1.3.6 is a generalization of Theorem 3.1 of [Zhu *et al.* (2003)]. (1.3.4) of Lemma 1.3.6 is a generalization of Lemma 3.4 of [Ding and Huang (1996)] and Theorem 2.2 of [Christensen (1999)]. Theorem 1.3.8 and Corollary 1.3.9 come from [Kato (1984)] and [Xue and Chen (2004)].

Section 1.4 is based on [Blackadar (1998)], [Kaniuth (2009)], [Rickart (1960)] and [Rudin (1973)].

Section 1.5 is based on [Conway (2000)], [Dixmier (1977)], [Pedersen (1979)] and [Lin (2001)].

Chapter 2

Stable Perturbation Of Densely–Defined Closed Operators On Banach Spaces

The definitions and existences of various generalized inverses on infinite–dimensional spaces are fully discussed in [Nashed (1976b)]. In [Nashed (1976a)], Nashed proposed a condition that makes the perturbation of a bounded linear operator with a generalized inverse have a generalized inverse too. But it is not easy to verify this condition for a given operator and a perturbation. So, Chen and the author proposed a new notation so–called "stable perturbation" of a bounded linear operator with a generalized inverse in [Chen and Xue (1997)], which is equivalent to Nashed's condition and is also easy to be checked. Recently, the stable perturbation of the closed operator with a bounded generalized inverse is established in [Wang and Zhang (2007)] and [Huang (2011)].

This chapter is organized as follows. We first introduce the notation so–called the generalized inverse of a densely–defined closed operator on Banach spaces in Section 2.1 and then establish a stable perturbation theory of such generalized inverses in Section 2.2. In Section 2.3, we use the stable perturbation theory to establish the error estimations of the solutions of certain operator equations under stable perturbation. In Section 2.4, we discuss the generalized inverse and its stable perturbation in a Banach algebra. Finally, we will give a new characterization of the Supher–operator and existence of the generalized inverse of an element in the matrix space over a unital commutative Banach algebra in Section 2.5.

We will use X, Y to stand for Banach spaces throughout the chapter.

2.1 Generalized inverses of linear operators on Banach spaces

Definition 2.1.1. Let $A \in \mathcal{C}(X,Y)$. If there exists a $B \in B(Y,X)$ such that $\mathrm{Ran}\,(B) \subset \mathfrak{D}(A)$ and $ABA = A$ on $\mathfrak{D}(A)$, A is called to be regular or relatively invertible; In addition, if $BAB = B$, then A is called to be generalized invertible and B is called to be a generalized inverse of A, denoted by A^+.

Let $Gi(X,Y)$ denote the set of all generalized invertible operators in $\mathcal{C}(X,Y)$. Put $Gib(X,Y) = Gi(X,Y) \cap B(X,Y)$.

Remark 2.1.2. Suppose that $A \in \mathcal{C}(X,Y)$ is regular, that is, there is $C \in B(Y,X)$ such that $\mathrm{Ran}\,(C) \subset \mathfrak{D}(A)$ and $ACA = A$ on $\mathfrak{D}(A)$ by Definition 1.2.1. Then $B = CAC \in B(Y,X)$ satisfies conditions: $\mathrm{Ran}\,(B) \subset \mathfrak{D}(A)$, $ABA = A$ on $\mathfrak{D}(A)$ and $BAB = B$. So $A^+ = CAC$.

Proposition 2.1.3. *Let $A \in \mathcal{C}(X,Y)$ such that $ABA = A$ on $\mathfrak{D}(A)$ for some $B \in B(Y,X)$ with* $\mathrm{Ran}\,(B) \subset \mathfrak{D}(A)$.

(1) $\mathrm{Ran}\,(A)$ *is closed in Y.*
(2) $AB \in B(Y)$ *is an idempotent operator with* $\mathrm{Ran}\,(AB) = \mathrm{Ran}\,(A)$.
(3) BA *is idempotent on $\mathfrak{D}(A)$ with* $\mathrm{Ran}\,(I_X - BA) = \mathrm{Ker}\,A$.

In addition, if $BAB = B$, i.e., $B = A^+$, then $\mathrm{Ran}\,(I_Y - AA^+) = \mathrm{Ker}\,A^+$ *and* $\mathrm{Ran}\,(A^+A) = \mathrm{Ran}\,(A^+)$.

Proof. (1) Let $\{x_n\} \in \mathfrak{D}(A)$ such that $Ax_n \xrightarrow{\|\cdot\|} y_0$ as $n \to \infty$ for some $y_0 \in Y$. Since $AB \in B(Y)$ by Proposition 1.2.4 (4), we have $Ax_n = (AB)Ax_n \xrightarrow{\|\cdot\|} ABy_0$ and hence $y_0 = ABy_0 \in \mathrm{Ran}\,(A)$. Thus, $\mathrm{Ran}\,(A)$ is closed.

(2) We have known $AB \in B(Y)$ by Proposition 1.2.4 (4). From $A = ABA$, we get that $(AB)^2 = ABAB = AB$, i.e., AB is idempotent and $\mathrm{Ran}\,(A) \subset \mathrm{Ran}\,(AB)$. Note that $\mathrm{Ran}\,(AB) \subset \mathrm{Ran}\,(A)$. So $\mathrm{Ran}\,(AB) = \mathrm{Ran}\,(A)$.

(3) Since $\mathrm{Ran}\,(BA) \subset \mathrm{Ran}\,(B) \subset \mathfrak{D}(A)$, it follows from $A = ABA$ that BA is an operator from $\mathfrak{D}(A)$ to $\mathfrak{D}(A)$ with $(BA)^2 = BABA = BA$ on $\mathfrak{D}(A)$. On the other hand, $A = ABA$ on $\mathfrak{D}(A)$ indicates that $A(I_X - BA)x = 0$, $\forall\, x \in \mathfrak{D}(A)$ and so that $\mathrm{Ran}\,(I_X - BA) \subset \mathrm{Ker}\,A$. Now let $z \in \mathrm{Ker}\,A$. Then $z = (I_X - BA)z \in \mathrm{Ran}\,(I_X - BA)$. Therefore, $\mathrm{Ran}\,(I_X - BA) = \mathrm{Ker}\,A$.

When $B = A^+$, similar to (2) and (3), from $A^+ = A^+AA^+$, we get that $\mathrm{Ran}\,(A^+A) = \mathrm{Ran}\,(A^+)$ and $\mathrm{Ran}\,(I_Y - AA^+) = \mathrm{Ran}\,(A^+)$. □

Theorem 2.1.4. *Let $A \in \mathcal{C}(X,Y)$ with $ABA = A$ on $\mathfrak{D}(A)$ for some $B \in B(Y,X)$ with $\mathrm{Ran}\,(B) \subset \mathfrak{D}(A)$. Then $\gamma(A) \geq \|B\|^{-1}$. In addition, if BA is bounded on $\mathfrak{D}(A)$, then*

$$\frac{1}{\|B\|} \leq \gamma(A) \leq \frac{\|BA\|_{\mathfrak{D}(A)}\|AB\|}{\|BAB\|}, \tag{2.1.1}$$

where $\|BA\|_{\mathfrak{D}(A)} = \sup\{\|BAx\| \,|\, x \in \mathfrak{D}(A) \cap S(X)\}$. Moreover, if $A \in Gib(X,Y)$, then

$$\frac{1}{\|A^+\|} \leq \gamma(A) \leq \frac{\|A^+A\|\|AA^+\|}{\|A^+\|}. \tag{2.1.2}$$

Proof. Since $\mathrm{Ran}\,(I_X - BA) = \mathrm{Ker}\,A$ by Proposition 1.2.4, we have

$$\mathrm{dist}\,(x, \mathrm{Ker}\,A) \leq \|x - (I_X - BA)x\| = \|BAx\| \leq \|B\|\|Ax\|, \quad \forall\, x \in \mathfrak{D}(A)$$

and hence $\|Ax\| \geq \|B\|^{-1}\mathrm{dist}\,(x, \mathrm{Ker}\,A)$, $\forall\, x \in \mathfrak{D}(A)$ which means that $\gamma(A) \geq \|B\|^{-1}$ by the definition of the reduced minimum modulus.

Suppose that BA is bounded on $\mathfrak{D}(A)$. Then for any $x \in \mathfrak{D}(A)$ and $z \in \mathrm{Ker}\,A$, we have

$$\|BAx\| = \|BA(x - z)\| \leq \|BA\|_{\mathfrak{D}(A)}\|x - z\|.$$

Thus, $\|BAx\| \leq \|BA\|_{\mathfrak{D}(A)}\mathrm{dist}\,(x, \mathrm{Ker}\,A)$, $\forall\, x \in \mathfrak{D}(A)$ and consequently,

$$\|Ax\| \geq \gamma(A)\mathrm{dist}\,(x, \mathrm{Ker}\,A) \geq \frac{\gamma(A)}{\|BA\|_{\mathfrak{D}(A)}}\|BAx\|, \forall\, x \in \mathfrak{D}(A). \tag{2.1.3}$$

Let $y \in Y$ and put $x = By \in \mathfrak{D}(A)$. Then by (2.1.3),

$$\|y\|\|AB\| \geq \|ABy\| \geq \frac{\gamma(A)}{\|BA\|_{\mathfrak{D}(A)}}\|BABy\|, \quad \forall\, y \in Y$$

which implies that $\gamma(A) \leq \dfrac{\|BA\|_{\mathfrak{D}(A)}\|AB\|}{\|BAB\|}$.

Replacing B by A^+ in (2.1.1) when $A \in Gib(X,Y)$, we can obtain (2.1.2). □

Let $A \in Gi(X,Y)$ and put $P = I_X - A^+A$, $Q = AA^+$. By Proposition 1.2.4, P is idempotent on $\mathfrak{D}(A)$ with $\mathrm{Ran}\,(P) = \mathrm{Ker}\,A$ and $Q \in B(Y)$ is idempotent with $\mathrm{Ran}\,(Q) = \mathrm{Ran}\,(A)$. Moreover, Y and $\mathfrak{D}(A)$ can be decomposed as $Y = \mathrm{Ran}\,(A) \dotplus \mathrm{Ran}\,(I_Y - Q)$ and

$$\mathfrak{D}(A) = \mathrm{Ker}\,A + \mathrm{Ran}\,(I_X - P), \; \mathrm{Ker}\,A \cap \mathrm{Ran}\,(I_X - P) = \{0\}$$

respectively. Conversely, we have

Proposition 2.1.5. *Let* $A \in C(X,Y)$ *with* $\mathrm{Ran}\,(A)$ *closed. Suppose that there is a closed subspace* N *in* Y *and a closed subspace* M *in* X *satisfying following conditions:*

(1) $Y = \mathrm{Ran}\,(A) \dotplus N$.
(2) $\mathrm{Ker}\,A \cap M = \{0\}$ *and* $\mathfrak{D}(A) = \mathrm{Ker}\,A + M \cap \mathfrak{D}(A)$.

Then $A \in Gi(X,Y)$.

Proof. Define a linear operator $A_0 \colon M \cap \mathfrak{D}(A) \to \mathrm{Ran}\,(A)$ by $A_0 x = Ax$, $\forall x \in M \cap \mathfrak{D}(A)$. Then $\mathrm{Ker}\,A_0 = \{0\}$ for $M \cap \mathrm{Ker}\,A = \{0\}$ and $\mathrm{Ran}\,(A_0) = \mathrm{Ran}\,(A)$ is closed. Let $\{x_n\} \subset \mathfrak{D}(A_0)$ and $x_0 \in \mathfrak{D}(A)$, $y_0 \in Y$ such that $x_n \xrightarrow{\|\cdot\|} x_0$ and $A_0 x_n \xrightarrow{\|\cdot\|} y_0$. Since $G(A)$ is closed and M is closed, we get that $\{x_0, y_0\} \in G(A)$ and $x_0 \in M$. So $G(A_0)$ is closed and hence $A_0^{-1} \in B(\mathrm{Ran}\,(A), M)$ with $\mathrm{Ran}\,(A_0^{-1}) = M \cap \mathfrak{D}(A)$ by Lemma 1.2.2 (2). Set

$$By = \begin{cases} A_0^{-1} y & y \in \mathrm{Ran}\,(A) \\ 0 & y \in N \end{cases}$$

or $B = B_0 Q$, where $Q = P_{\mathrm{Ran}\,(A),N} \colon Y \to \mathrm{Ran}\,(A)$ is given by $Qy = y_1$ for $y = y_1 + y_2$, $y_1 \in \mathrm{Ran}\,(A)$, $y_2 \in N$. Since Q is bounded by Proposition 1.1.12, we have $B \in B(Y,X)$ with $\mathrm{Ran}\,(B) \subset \mathfrak{D}(A)$ and

$$\begin{aligned} ABAx &= AA_0 Ax = Ax, \quad \forall x \in M \cap \mathfrak{D}(A) \\ BABy &= BAA_0^{-1} y = By, \quad \forall y \in \mathrm{Ran}\,(A) \end{aligned}$$

so that $ABA = A$ on $\mathfrak{D}(A)$ and $BAB = B$, i.e., $A \in Gi(X,Y)$ with $A^+ = B$. □

Combining Proposition 1.1.11 with Proposition 2.2.11, we have

Corollary 2.1.6. *Let* $A \in C(X,Y)$ *with* $\mathrm{Ran}\,(A)$ *closed and* $\dim(\mathrm{Ker}\,A)$, $\dim(\mathrm{Coker}\,A)$ *finite, where* $\mathrm{Coker}\,A = X/\mathrm{Ran}\,(A)$. *Then* $A \in Gi(X,Y)$.

Corollary 2.1.7. *Let* $A \in C(X,Y)$ *with* $\mathrm{Ran}\,(A)$ *closed. Assume that there are idempotent operators* $P \in B(X)$ *and* $Q \in B(Y)$ *such that* $\mathrm{Ran}\,(P) = \mathrm{Ker}\,A$ *and* $\mathrm{Ran}\,(Q) = \mathrm{Ran}\,(A)$. *Then there is a unique operator* B *in* $B(Y,Y)$ *such that*

$$ABA = A, \; BA = I_X - P \text{ on } \mathfrak{D}(A), \; BAB = B, \; AB = Q. \tag{2.1.4}$$

Proof. It is easy to check that the B in the proof of Proposition 2.2.11 satisfies (2.1.4). Assume that there is $B_1 \in B(Y, X)$ such that

$$AB_1A = A, \ B_1A = I_X - P, \text{ on } \mathfrak{D}(A), \ B_1AB_1 = B_1, \quad AB_1 = Q.$$

Then

$$B_1 = B_1AB_1 = (I_X - P)B_1 = BAB_1 = BQ = BAB = B. \qquad \square$$

Remark 2.1.8. The generalized inverse B in Corollary 2.1.7 is denoted by $(A_{P,Q})^+$.

Corollary 2.1.9. *Let A, P, Q be as in Corollary* 2.1.7. Then $(A^+)^*$ is one of the generalized inverses of A^* and $((A_{P,Q})^+)^* = (A^*_{I_{Y^*}-Q^*, I_{X^*}-P^*})^+$.

Proof. For each $x^* \in X^*$ and $x \in \mathfrak{D}(\mathcal{A})$,

$$(I_{X^*} - P^*)x^*(x) = (I_X - P)^*x^*(x) = x^*(BAx) = B^*x^*(Ax).$$

So $B^*x^* \in \mathfrak{D}(A^*)$ and $A^*B^* = I_{X^*} - P^*$. Similarly, from $AB = Q$, we get that $Q^* = B^*A^*$ on $\mathfrak{D}(A^*)$. Thus

$$A^* = A^*(AB)^* = A^*B^*A^* \text{ on } \mathfrak{D}(A^*) \tag{2.1.5}$$

by Proposition 1.2.4 (4). Noting that $BQ = B = (I_X - P)B$, we have

$$B^* = B^*(I_{X^*} - P^*) = B^*A^*B^*. \tag{2.1.6}$$

Therefore $(A^+)^*$ is a generalized inverse of A^*.

Since $\operatorname{Ran}(I_{X^*} - P^*) = \operatorname{Ran}(A^*)$ and $\operatorname{Ran}(I_{Y^*} - Q^*) = \operatorname{Ker} Q^* = \operatorname{Ker} A^*$ by (2.1.5) and Proposition 2.1.3, it follows from (2.1.5) and (2.1.6) that $((A_{P,Q})^+)^* = (A^*_{I_{Y^*}-Q^*, I_{X^*}-P^*})^+$. $\square$

As an end of this section, we consider the reverse low of two generalized inverses as follows:

Proposition 2.1.10. *Let $A \in Gi(X, Y)$ and $C \in Gib(Y, Z)$. Then A^+C^+ is a generalized inverse of CA iff both AA^+C^+C and C^+CAA^+ are idempotent operators.*

Proof. Suppose that $CAA^+C^+CA = CA$ and $A^+C^+CAA^+C^+ = A^+C^+$. Then

$$C^+CAA^+C^+CAA^+ = C^+CAA^+, \quad AA^+C^+CAA^+C^+C = AA^+C^+C,$$

that is, AA^+C^+C and C^+CAA^+ are idempotent operators.

Conversely, if both AA^+C^+C and C^+CAA^+ are idempotent operators, then

$$\begin{aligned} CAA^+C^+CA &= CC^+CAA^+C^+CAA^+A = CC^+CAA^+A = CA \\ A^+C^+CAA^+C^+ &= A^+AA^+C^+CAA^+C^+CC^+ = A^+AA^+C^+CC^+ \\ &= A^+C^+. \end{aligned}$$

$\square$

2.2 Stable perturbation of operators

Let $A \in C(X,Y)$ and $\delta A \in B(X,Y)$. Put $\bar{A} = A + \delta A$. Then $\mathfrak{D}(\bar{A}) = \mathfrak{D}(A)$ and $\bar{A} \in C(X,Y)$. $\bar{A}$ is called the perturbation of A by δA or the perturbed operator.

Definition 2.2.1. Let $A \in Gi(X,Y)$ and $\delta A \in B(X,Y)$. We say $\bar{A} = A + \delta A$ is the stable perturbation of A if $\mathrm{Ran}\,(\bar{A}) \cap \mathrm{Ker}\, A^+ = \{0\}$.

Obviously, if $\mathrm{Ran}\,(\delta A) \subset \mathrm{Ran}\,(A)$, then $\mathrm{Ran}\,(\bar{A}) \subset \mathrm{Ran}\,(A)$ and $\mathrm{Ran}\,(\bar{A}) \cap \mathrm{Ker}\, A^+ = \{0\}$ by Preposition 2.1.3 (4). So $\mathrm{Ran}\,(\bar{A})$ is the stable perturbation of A.

Remark 2.2.2. Let $A \in C(X,Y)$ and $\delta A \in \mathcal{D}(X,Y)$ with $\mathfrak{D}(A) \subset \mathfrak{D}(\delta A)$. If there are constants a, $b > 0$ such that $\|\delta A x\| \leq a\|Ax\| + b\|x\|, \forall x \in \mathfrak{D}(A)$, we call δA is A–bounded (cf. [Kato (1984)]).

Define a norm $\|x\|_A = a\|Ax\| + b\|x\|, \forall x \in \mathfrak{D}(A)$ on $\mathfrak{D}(A)$. Since $G(A)$ is closed, it is easy to check that $\mathfrak{D}(A)$ is a Banach space with respect to this norm. So if δA is A–bounded, then $\delta A \in B(\mathfrak{D}(A), Y)$. This indicates that it is enough for us to consider the case that $\delta A \in B(X,Y)$.

Suppose that X, Y are finite dimensional Banach spaces and $A, \bar{A} = A + \delta A \in B(X,Y)$. If $\mathrm{rank}\, A = \mathrm{rank}\, \bar{A}$, we say that $\bar{A}$ is the rank–preserving perturbation of A.

The next theorem shows that the notation "stable perturbation of operators" is just the generalization of the notation "rank–preserving perturbation of matrices" in the set of infinite–dimensional cases.

Theorem 2.2.3. *Let X, Y be finite–dimensional Banach spaces and let $A, \bar{A} = A + \delta A \in B(X,Y)$ with $I_X + A^+\delta A$ invertible. Then* $\mathrm{Ran}\,(\bar{A}) \cap \mathrm{Ker}\, A^+ = \{0\}$ *iff* $\mathrm{rank}\,(\bar{A}) = \mathrm{rank}\,(A)$.

Proof. Assume that $X = \mathbb{C}^n$, $Y = \mathbb{C}^m$. Since $\mathrm{Ker}\, A^+ = \mathrm{Ran}\,(I_Y - AA^+)$ and $\mathrm{Ran}\,(I_Y - AA^+) + \mathrm{Ran}\,(A) = Y$ by Proposition 2.1.3, it follows that $\dim \mathrm{Ker}\, A^+ = m - \dim(\mathrm{Ran}\,(A))$. From $\mathrm{Ran}\,(\bar{A}) \cap \mathrm{Ker}\, A^+ = \{0\}$, we get that

$$\dim(\mathrm{Ran}\,(\bar{A})) + \dim(\mathrm{Ker}\, A^+) = \dim(\mathrm{Ran}\,(\bar{A}) + \mathrm{Ker}\, A^+) \leq m.$$

Thus, $\mathrm{rank}\,(\bar{A}) = \dim(\mathrm{Ran}\,(\bar{A})) \leq \dim(\mathrm{Ran}\,(A)) = \mathrm{rank}\,(A)$.

On the other hand, put $V = I_X + A^+\delta A$. Then

$$AV = A + AA^+\delta A = AA^+(A + \delta A) = AA^+\bar{A}.$$

Since V is invertible, we have

$$\text{rank}\,(A) = \text{rank}\,(AV) \le \min\{\text{rank}\,(AA^+), \text{rank}\,(\bar{A})\} \le \text{rank}\,(\bar{A}).$$

Thus, $\text{rank}\,(\bar{A}) = \text{rank}\,(A)$.

Conversely, set $W = (I_X + A^+\delta A)^{-1}(I_X - A^+A)$. Then $\text{rank}\,(W) = \text{rank}\,(I_X - A^+A) = \dim(\text{Ker}\,A)$. Let $x \in \text{Ker}\,\bar{A}$. Then

$$(I_X + A^+\delta A)x - (I_X - A^+A)x = 0.$$

Thus, $x = Wx$ and hence $\text{Ker}\,\bar{A} \subset \text{Ran}\,(W)$. Noting that

$$\text{rank}\,(A) + \dim(\text{Ker}\,A) = \text{rank}\,(\bar{A}) + \dim(\text{Ker}\,\bar{A}) = n, \ \text{rank}\,(\bar{A}) = \text{rank}\,(A),$$

we have $\dim(\text{Ker}\,\bar{A}) = \dim(\text{Ker}\,A)$. This implies that $\text{Ker}\,\bar{A} = \text{Ran}\,(W)$. Now let $y \in \text{Ker}\,A^+ \cap \text{Ran}\,(\bar{A})$. Then $A^+y = 0$ and $y = \bar{A}z$ for some $z \in X$. Thus, $A^+(A + \delta A)z = 0$ and so that $z = Wz$. Since $\text{Ker}\,\bar{A} = \text{Ran}\,(W)$, we get that $y = \bar{A}z = 0$, i.e., $\text{Ran}\,(\bar{A}) \cap \text{Ker}\,A^+ = \{0\}$. □

Lemma 2.2.4. *Let $A \in Gi(X, Y)$ and $\delta A \in B(X, Y)$. Suppose that $I_X + A^+\delta A$ is invertible in $B(X)$. Set $S = (I_X + A^+\delta A)^{-1}(I_X - A^+A)$. Then S is a linear operator from $\mathfrak{D}(A)$ into $\mathfrak{D}(A)$ such that $S^2 = S$ and* $\text{Ker}\,\bar{A} \subset \text{Ran}\,(S)$. *Furthermore,* $\text{Ker}\,\bar{A} = \text{Ran}\,(S)$ *iff* $\text{Ran}\,(\bar{A}) \cap \text{Ker}\,A^+ = \{0\}$.

Proof. Clearly, $\mathfrak{D}(S) = \mathfrak{D}(A)$. Note that $\text{Ran}\,(A^+) \subset \mathfrak{D}(A)$. So $(I_X + A^+\delta A)\mathfrak{D}(A) \subset \mathfrak{D}(A)$. Let $x \in \mathfrak{D}(A)$ and put $y = (I_X + A^+\delta A)^{-1}x$. then $y + A^+\delta Ay = x$ and hence $y = x - A^+\delta Ay \in \mathfrak{D}(A)$. Therefore, $\mathfrak{D}(A) = (I_X + A^+\delta A)^{-1}\mathfrak{D}(A)$. Since $\text{Ran}\,(I_X - A^+A) = \text{Ker}\,A \subset \mathfrak{D}(A)$, S is the linear operator from $\mathfrak{D}(A)$ into $\mathfrak{D}(A)$. Noting that

$$(I_X - A^+A)(I_X + A^+\delta A)x = (I_X - A^+A)x, \quad \forall\, x \in \mathfrak{D}(A)$$

and $\mathfrak{D}(A) = (I_X + A^+\delta A)^{-1}\mathfrak{D}(A)$, we have

$$(I_X - A^+A)y = (I_X - A^+A)(I_X + A^+\delta A)^{-1}y, \quad \forall\, y \in \mathfrak{D}(A).$$

Thus,

$$\begin{aligned} S^2 &= (I_X + A^+\delta A)^{-1}(I_X - A^+A)(I_X + A^+\delta A)^{-1}(I_X - A^+A) \\ &= (I_X + A^+\delta A)^{-1}(I_X - A^+A)(I_X - A^+A) \\ &= S. \end{aligned}$$

Let $x \in \mathfrak{D}(A)$. From $(I_X - A^+A)x = (I_X + A^+\delta A)x - A^+\bar{A}x$, we have

$$Sx = (I_X + A^+\delta A)^{-1}(I_X - A^+A)x = x - (I_X + A^+\delta A)^{-1}A^+\bar{A}x. \quad (2.2.1)$$

So if $x \in \text{Ker}\,\bar{A}$, then $x = Sx$ by (2.2.1). Thus, $\text{Ker}\,\bar{A} \subset \text{Ran}\,(S)$.

Now suppose that $\operatorname{Ker}\bar{A} = \operatorname{Ran}(S)$. Let $y \in \operatorname{Ran}(\bar{A}) \cap \operatorname{Ker} A^+$. Then there is $z \in \mathfrak{D}(A)$ such that $y = \bar{A}z$ and $A^+y = 0$. Thus, $A^+\bar{A}z = 0$ and consequently, $Sz = z$ by (2.2.1). From $\operatorname{Ker}\bar{A} = \operatorname{Ran}(S)$, we get that $y = \bar{A}z = 0$, i.e., $\operatorname{Ran}(\bar{A}) \cap \operatorname{Ker} A^+ = \{0\}$.

Conversely, assume that $\operatorname{Ran}(\bar{A}) \cap \operatorname{Ker} A^+ = \{0\}$. Let $x \in \operatorname{Ran}(S)$. Then $x = Sx$ (for $S^2 = S$ on $\mathfrak{D}(A)$). So $(I_X + A^+\delta A)^{-1}A^+\bar{A}x = 0$ by (2.2.1) and $A^+\bar{A}x = 0$. Put $y = \bar{A}x$. Then $y \in \operatorname{Ran}(\bar{A}) \cap \operatorname{Ker} A^+ = \{0\}$. Thus, $x \in \operatorname{Ker}\bar{A}$ and consequently, $\operatorname{Ker}\bar{A} = \operatorname{Ran}(S)$. □

Lemma 2.2.5. *Let $A \in Gi(X,Y)$ and $\delta A \in B(X,Y)$ with $I_X + A^+\delta A$ invertible in $B(X)$. Then $I_Y + \delta A A^+$ is invertible in $B(Y)$ and*

$$\delta A(I_X + A^+\delta A)^{-1} = (I_Y + \delta A A^+)^{-1}\delta A \tag{2.2.2}$$

$$(I_X + A^+\delta A)^{-1}A^+ = A^+(I_Y + \delta A A^+)^{-1}. \tag{2.2.3}$$

Proof. Since $\sigma(A^+\delta A)\backslash\{0\} = \sigma(\delta A A^+)\backslash\{0\}$ by Proposition 1.4.13, it follows that $I_X + A^+\delta A$ is invertible in $B(X)$ iff $I_Y + \delta A A^+$ is invertible in $B(Y)$. From the identities

$$\delta A(I_X + A^+\delta A) = (I_Y + \delta A A^+)\delta A, \ (I_X + A^+\delta A)A^+ = A^+(I_Y + \delta A A^+),$$

we can deduce (2.2.2) and (2.2.3). □

Theorem 2.2.6. *Let $A \in Gi(X,Y)$ and $\delta A \in B(X,Y)$ with $I_X + A^+\delta A$ invertible in $B(X)$. Then the following statements are equivalent:*

(1) $\bar{A} = A + \delta A \in Gi(X,Y)$ *with* $\bar{A}^+ = (I_X + A^+\delta A)^{-1}A^+$.
(2) $\bar{A}$ *is a stable perturbation of* A.
(3) $\bar{A}(I_X + A^+\delta A)^{-1}(I_X - A^+A) = 0$.
(4) $(I_Y - AA^+)(I_Y + \delta A A^+)^{-1}\bar{A} = 0$.
(5) $(I_Y - AA^+)\delta A(I_X - A^+A) = (I_Y - AA^+)\delta A(I_X + A^+\delta A)^{-1}A^+\delta A(I_X - A^+A)$.
(6) $(I_Y + \delta A A^+)^{-1}\bar{A}$ *maps* $\operatorname{Ker} A$ *into* $\operatorname{Ran}(A)$.

Proof. (1)⇒(2) Let $y \in \operatorname{Ran}(\bar{A}) \cap \operatorname{Ker} A^+$. Then $A^+y = 0$ and $y = \bar{A}x$ for some $x \in \mathfrak{D}(A)$. Since $(I_Y + \delta A A^+)y = y$ and

$$\begin{aligned}\bar{A}\bar{A}^+ &= (A + \delta A)A^+(I_Y + \delta A A^+)^{-1} \\ &= (AA^+ - I_Y + I_Y + \delta A A^+)(I_Y + \delta A A^+)^{-1} \\ &= I_Y - (I_Y - AA^+)(I_Y + \delta A A^+)^{-1}\end{aligned}$$

by Lemma 2.2.5, we have $\bar{A}\bar{A}^+y = 0$, i.e., $y = \bar{A}x = \bar{A}\bar{A}^+\bar{A}x = \bar{A}\bar{A}^+y = 0$.

(2)⇒(3) By Lemma 2.2.4, the range of $S = (I_X + A^+\delta A)^{-1}(I_X - A^+A)$ is $\operatorname{Ker}\bar{A}$. So $\bar{A}(I_X + A^+\delta A)^{-1}(I_X - A^+A) = 0$.

(3)⇔(4) In fact, on $\mathfrak{D}(A)$, we have

$$\begin{aligned}(I_Y - AA^+)(I_Y + \delta AA^+)^{-1}\bar{A}\\ &= (I_Y + \delta AA^+ - \delta AA^+ - AA^+)(I_Y + \delta AA^+)^{-1}\bar{A}\\ &= \bar{A} - \bar{A}A^+(I_Y + \delta AA^+)^{-1}\bar{A}\\ &= \bar{A} - \bar{A}(I_X + A^+\delta A)^{-1}(A^+A + A^+\delta A)\\ &= \bar{A} - \bar{A}(I_X + A^+\delta A)^{-1}(I_X + A^+\delta A + A^+A - I_X)\\ &= \bar{A}(I_X + A^+\delta A)^{-1}(I_X - A^+A).\end{aligned}$$

(3)⇒(5) We have

$$\begin{aligned}&(I_X - AA^+)\delta A(I_X + A^+\delta A)^{-1}A^+\delta A(I_X - A^+A)\\ &= (I_Y - AA^+)\delta A(I_X + A^+\delta A)^{-1}(I_X + A^+\delta A - I_X)(I_X - A^+A)\\ &= (I_Y - AA^+)\delta A(I_X - A^+A) - (I_Y - AA^+)\bar{A}(I_X + A^+\delta A)^{-1}(I_X - A^+A)\\ &= (I_Y - AA^+)\delta A(I_X - A^+A).\end{aligned}$$

(5)⇒(3) The computation above shows that if (5) holds, then

$$(I_Y - AA^+)\bar{A}(I_X + A^+\delta A)^{-1}(I_X - A^+A) = 0. \tag{2.2.4}$$

Since

$$\begin{aligned}AA^+\bar{A}(I_X + A^+\delta A)^{-1}(I_X - A^+A)\\ &= A(I_X + A^+\delta A)(I_X + A^+\delta A)^{-1}(I_X - A^+A)\\ &= A(I_X - A^+A) = 0,\end{aligned}$$

we have $\bar{A}(I_X + A^+\delta A)^{-1}(I_X - A^+A) = 0$ by (2.2.4).

(3)⇒(1) Put $B = (I_X + A^+\delta A)^{-1}A^+ \in B(Y, X)$. Noting that $\operatorname{Ran}(A^+) \subset \mathfrak{D}(A)$, $(I_X + A^+\delta A)^{-1}\mathfrak{D}(A) = \mathfrak{D}(A)$, we have $\operatorname{Ran}(B) \subset \mathfrak{D}(A) = D(\bar{A})$. Furthermore, by Lemma 2.2.5 and (3)⇔(4),

$$\begin{aligned}\bar{A}B\bar{A} &= (A + \delta A)(I_X + A^+\delta A)^{-1}A^+(A^+\delta A)\\ &= \bar{A}(I_X + A^+\delta A)^{-1}(A^+A - I_X + I_X + A^+\delta A)\\ &= \bar{A} - \bar{A}(I_X + A^+\delta A)^{-1}(I_X - A^+A)\\ &= \bar{A}\\ B\bar{A}B &= (I_X + A^+\delta A)^{-1}A^+(I_X + A^+\delta A + A^+A - I_X)(I_X + A^+\delta A)^{-1}A^+\\ &= (I_X + A^+\delta A)^{-1}A^+ - (I_X + A^+\delta A)^{-1}(I_X - A^+A)(I_Y + \delta AA^+)^{-1}\\ &= B.\end{aligned}$$

Therefore, $\bar{A} \in Gi(X, Y)$ with $\bar{A}^+ = (I_X + A^+\delta A)^{-1}A^+$.

(4)⇒(6) From $(I_Y - AA^+)(I_Y + \delta AA^+)^{-1}\bar{A} = 0$, we have

$$(I_Y + \delta AA^+)^{-1}\bar{A} = AA^+(I_Y + \delta AA^+)^{-1}\bar{A}.$$

Thus, for any $x \in \operatorname{Ker} A$, $(I_Y + \delta AA^+)^{-1}\bar{A}x \in \operatorname{Ran}(A)$.

(6)⇔(5) Note that $\operatorname{Ker} A = \operatorname{Ran}(I_X - A^+A)$ and $\operatorname{Ran}(A) = \operatorname{Ran}(AA^+)$ by Proposition 1.2.2. So the condition that $(I_Y + \delta AA^+)^{-1}\bar{A}$ maps $\operatorname{Ker} A$ into $\operatorname{Ran}(A)$ implies that $(I_Y - AA^+)(I_Y + \delta AA^+)^{-1}\bar{A}(I_X - A^+A) = 0$ on $\mathfrak{D}(A)$. Thus

$$(I_Y - AA^+)\delta A(I_X + A^+\delta A)^{-1}(I_X - A^+A) = 0$$

by Lemma 2.2.5 and hence

$$\begin{aligned}(I_Y - AA^+)&\delta A(I_X + A^+\delta A)^{-1}A^+\delta A(I_X - A^+A)\\ &= (I_Y - AA^+)\delta A(I_X - A^+A) - (I_Y - AA^+)\delta A\\ &\quad \times (I_X + A^+\delta A)^{-1}(I_X - A^+A)\\ &= (I_Y - AA^+)\delta A(I_X - A^+A).\end{aligned}$$

□

Corollary 2.2.7. *Let $A \in Gi(X,Y)$ and $\delta A \in B(X,Y)$ with $I_X + A^+\delta A$ invertible. Then the following conditions are equivalent:*

(1) $\operatorname{Ran}(\bar{A}) \cap \operatorname{Ker} A^+ = \{0\}$,
(2) $(I_X + A^+\delta A)^{-1}(\operatorname{Ker} A) = \operatorname{Ker} \bar{A}$,
(3) $(I_Y + \delta AA^+)^{-1}\operatorname{Ran}(\bar{A}) = \operatorname{Ran}(A)$.

Proof. (1)⇔(2) comes from Lemma 2.2.4 for $\operatorname{Ker} A = \operatorname{Ran}(I_X - A^+A)$.

(1)⇒(3) By Theorem 2.2.6 (4),

$$(I_Y + \delta AA^+)^{-1}\operatorname{Ran}(\bar{A}) \subset \operatorname{Ran}(AA^+) = \operatorname{Ran}(A).$$

On the other hand, from $(I_Y + \delta AA^+)A = \bar{A}A^+A$, we get that $(I_Y + \delta AA^+)\operatorname{Ran}(A) \subset \operatorname{Ran}(\bar{A})$, i.e., $(I_Y + \delta AA^+)^{-1}\operatorname{Ran}(\bar{A}) \subset \operatorname{Ran}(A)$. Thus, $(I_Y + \delta AA^+)^{-1}\operatorname{Ran}(\bar{A}) = \operatorname{Ran}(A)$.

(3)⇒(1) $(I_Y + \delta AA^+)^{-1}\operatorname{Ran}(\bar{A}) = \operatorname{Ran}(A)$ indicates that $(I_Y - AA^+)(I_Y + \delta AA^+)^{-1}\bar{A} = 0$ for AA^+ is the idempotent operator from Y to $\operatorname{Ran}(A)$. Thus, $\operatorname{Ran}(\bar{A}) \cap \operatorname{Ker} A^+ = \{0\}$ by Theorem 2.2.6. □

Corollary 2.2.8. *Let $A \in Gi(X,Y)$ and $\delta A \in B(X,Y)$ with $I_X + A^+\delta A$ invertible in $B(X)$. If one of the following conditions is satisfied, then $\bar{A} = A + \delta A$ is a stable perturbation of A.*

(1) $\dim \operatorname{Ker} A = \dim \operatorname{Ker} \bar{A} < \infty$.
(2) $\dim(X/\operatorname{Ker} A) = \dim(X/\operatorname{Ker} \bar{A}) < \infty$.

(3) $\dim(\mathrm{Ran}\,(\bar{A})) = \dim(\mathrm{Ran}\,(\bar{A})) < \infty$.
(4) $\mathrm{Ran}\,(\bar{A})$ *is closed and* $\dim(\mathrm{Coker}\,A) = \dim(\mathrm{Coker}\,\bar{A}) < \infty$.

Proof. We have by Lemma 2.2.4, $(I_X + A^+\delta A)^{-1}\mathrm{Ker}\,A \supset \mathrm{Ker}\,\bar{A}$. Thus, $\dim\mathrm{Ker}\,A = \dim\mathrm{Ker}\,A < \infty$ implies that $(I_X + A^+\delta A)^{-1}\mathrm{Ker}\,A = \mathrm{Ker}\,\bar{A}$. Consequently, $\mathrm{Ran}\,(\bar{A}) \cap \mathrm{Ker}\,A^+ = \{0\}$ by Corollary 2.2.7.

Now suppose that $\dim(X/\mathrm{Ker}\,A) = \dim(X/\mathrm{Ker}\,\bar{A}) < \infty$. Put $T = I_X + A^+\delta A$. Then $T(\mathrm{Ker}\,\bar{A}) \subset \mathrm{Ker}\,A$. Hence T induces a linear mapping $\hat{T}\colon X/\mathrm{Ker}\,\bar{A} \to X/\mathrm{Ker}\,A$ defined by $\hat{T}([x]) = [Tx]$. Clearly, $\hat{T}$ is surjective. So $\dim(X/\mathrm{Ker}\,A) = \dim(X/\mathrm{Ker}\,\bar{A}) < \infty$ indicates that $\hat{T}$ is injective. This means that $T(\mathrm{Ker}\,\bar{A}) = \mathrm{Ker}\,A$ and hence $\mathrm{Ran}\,(\bar{A}) \cap \mathrm{Ker}\,A^+ = \{0\}$ by Corollary 2.2.7.

Put $R = I_Y + \delta AA^+$. Since R is invertible in $B(Y)$ and $R(\mathrm{Ran}\,(A)) \subset \mathrm{Ran}\,(\bar{A})$, we get that $R(\mathrm{Ran}\,(A)) = \mathrm{Ran}\,(\bar{A})$ when $\dim(\mathrm{Ran}\,(A)) = \dim(\mathrm{Ran}\,(\bar{A})) < \infty$. In this case, $\mathrm{Ran}\,(\bar{A}) \cap \mathrm{Ker}\,A^+ = \{0\}$ by Corollary 2.2.7.

Assume that (4) holds. Let R be as above. Since $R(\mathrm{Ran}\,(A)) \subset \mathrm{Ran}\,(\bar{A})$, we can define a linear map $\hat{R}\colon Y/\mathrm{Ran}\,(A) \to Y/\mathrm{Ran}\,(\bar{A})$ by $\hat{R}([y]) = [R(y)]$, $\forall\, y \in Y$. Clearly, $\hat{R}$ is surjective and $\hat{R}$ is injective when $\dim(Y/\mathrm{Ran}\,(A)) = \dim(Y/\mathrm{Ran}\,(\bar{A})) < \infty$. Thus, $R(\mathrm{Ran}\,(A)) = \mathrm{Ran}\,(\bar{A})$ and the assertion follows from Corollary 2.2.7. □

Corollary 2.2.9. *let $A \in Gib(X,Y)$ and $\delta A \in B(X,Y)$ with $I_X + A^+\delta A$ invertible in $B(X)$. Then* $\mathrm{Ran}\,(\bar{A}) \cap \mathrm{Ker}\,A^+ = \{0\}$ *iff* $\mathrm{Ran}\,(\bar{A}^*) \cap \mathrm{Ker}\,(A^*)^+ = \{0\}$.

Proof. By Corollary 2.1.8, $(A^+)^*$ is one of the generalized inverses of A^*. Put $(A^*)^+ = (A^+)^*$. Note that $I_X + A^+\delta A$ is invertible if and only if $I_X + (\delta A)^*(A^*)^+$ is invertible in $B(X^*)$. Since $\mathrm{Ran}\,(\bar{A}) \cap \mathrm{Ker}\,A^+ = \{0\}$, it follows from Theorem 2.2.6 that

$$(I_{X^*} - A^*(A^+)^*)(I_{X^*} + (\delta A)^*(A^+)^*)^{-1}\bar{A}^* = 0.$$

Replacing A by A^* and $\bar{A}$ by $\bar{A}^*$ in Theorem 2.2.6, we get that $\mathrm{Ran}\,(\bar{A}^*) \cap \mathrm{Ker}\,(A^*)^+ = \{0\}$.

Assume that $\mathrm{Ran}\,(\bar{A}^*) \cap \mathrm{Ker}\,(A^*)^+ = \{0\}$. Since $(A^+)^{**}$ is one of the generalized inverses of A^{**} by Corollary 2.1.8, $\mathrm{Ran}\,(\bar{A}^{**}) \cap \mathrm{Ker}\,(A^+)^{**} = \{0\}$ by above argument. Let $y_0 \in \mathrm{Ran}\,(\bar{A}) \cap \mathrm{Ker}\,A^+$. Then $A^+y_0 = 0$ and $y_0 = \bar{A}x_0$ for some $x_0 \in X$. Define $f_0 \in X^{**}$ by $f_0(\phi) = \phi(x_0)$, $\forall\, \phi \in X^*$ and $g_0 \in Y^{**}$ by $g_0(\psi) = \psi(y_0)$, $\forall\, \psi \in Y^*$. Then

$$(\bar{A}^{**}f_0)(h) = f_0(\bar{A}^*h) = \bar{A}^*h(x_0) = h(Ax_0) = h(y_0) = g_0(h)$$

$$((A^+)^{**}(g_0))(\phi) = g_0((A^+)^*\phi) = (A^+)^*\phi(y_0) = \varphi(A^+y_0) = 0$$

for every h in Y and ϕ in X. So $g_0 \in \mathrm{Ran}\,(\bar{A}^{**} \cap \mathrm{Ker}\,(A^+)^{**} = \{0\}$. This means that $y_0 = 0$. □

Proposition 2.2.10. *Let $A \in Gi(X,Y)$ and $\delta A \in B(X,Y)$. Put $\bar{A} = A + \delta A$. If $\delta(\mathrm{Ran}\,(\bar{A}), \mathrm{Ran}\,(A)) < \|I_Y - AA^+\|^{-1}$, then $\mathrm{Ran}\,(\bar{A}) \cap \mathrm{Ker}\,A^+ = \{0\}$.*

Proof. If $\mathrm{Ran}\,(\bar{A}) \cap \mathrm{Ker}\,A^+ \neq \{0\}$, we can choose $y \in \mathrm{Ran}\,(\bar{A}) \cap \mathrm{Ker}\,A^+$ with $\|y\| = 1$. Then for any $z \in \mathfrak{D}(A)$,

$$1 = \|y\| = \|(I_Y - AA^+)(y - Az)\| \leq \|I_Y - AA^+\|\|y - Az\|$$

so that $\delta(\mathrm{Ran}\,(\bar{A}), \mathrm{Ran}\,(A)) \geq \|I_Y - AA^+\|^{-1}$. But this is in contradiction with the assumption that $\delta(\mathrm{Ran}\,(\bar{A}), \mathrm{Ran}\,(A)) < \|I_Y - AA^+\|^{-1}$. Thus, $\mathrm{Ran}\,(\bar{A}) \cap \mathrm{Ker}\,A^+ = \{0\}$. □

In Definition 2.2.1, we say $\bar{A}$ is a stable perturbation of A if $\mathrm{Ran}\,(\bar{A}) \cap \mathrm{Ker}\,A^+ = \{0\}$ for certain generalized inverse A^+ of A. One should ask: if $\mathrm{Ran}\,(\bar{A}) \cap \mathrm{Ker}\,A^+ = \{0\}$, is it true that $\mathrm{Ran}\,(\bar{A}) \cap \mathrm{Ker}\,B = \{0\}$ for any other generalized inverse B of A. The following proposition answers this question.

Proposition 2.2.11. *Let $A \in Gi(X,Y)$ and $\delta A \in B(X,Y)$. Let $B \in B(X,Y)$ with $\mathrm{Ran}\,(B) \subset \mathfrak{D}(A)$ and $ABA = A$. Suppose that $\mathrm{Ran}\,(\bar{A}) \cap \mathrm{Ker}\,A^+ = \{0\}$ and $\|A^+\|\|\delta A\| < (1 + \|I_Y - AB\|)^{-1}$. Then $\mathrm{Ran}\,(\bar{A}) \cap \mathrm{Ker}\,B = \{0\}$.*

Proof. By Theorem 2.2.6, $\bar{A}^+ = (I_X + A^+\delta A)^{-1}A^+ \in B(Y,X)$ and

$$\|\bar{A}^+\| \leq \|(I_X + A^+\delta A)^{-1}\|\|A^+\| \leq \frac{\|A^+\|}{1 - \|A^+\|\|\delta A\|}.$$

Therefore, by Theorem 2.1.4 and Lemma 1.3.5 (2),

$$\delta(\mathrm{Ran}\,(\bar{A}), \mathrm{Ran}\,(A)) \leq \frac{\|\delta A\|}{\gamma(\bar{A})} \leq \|\bar{A}^+\|\|\delta A\| < \frac{1}{\|I_Y - AB\|}.$$

If $\mathrm{Ran}\,(\bar{A}) \cap \mathrm{Ker}\,B \neq \{0\}$, we can pick $u \in \mathrm{Ran}\,(\bar{A}) \cap \mathrm{Ker}\,B$ with $\|u\| = 1$. Then for any $z \in Y$,

$$1 = \|u\| = \|(I_Y - AB)(u - Az)\| \leq \|I_Y - AB\|\|u - Az\|.$$

This shows that

$$\delta(\mathrm{Ran}\,(\bar{A}), \mathrm{Ran}\,(A)) \geq \mathrm{dist}\,(u, \mathrm{Ran}\,(A)) \geq \frac{1}{\|I_Y - AB\|}.$$

But it is in contradiction with the assumption that $\delta(\mathrm{Ran}\,(\bar{A}), \mathrm{Ran}\,(A)) < \|I_Y - AB\|^{-1}$. So $\mathrm{Ran}\,(\bar{A}) \cap \mathrm{Ker}\,B = \{0\}$. □

Theorem 2.2.6 and Corollary 2.2.7 give various equivalent conditions that make $\bar{A}$ because a stable perturbation of A. But one has not known how to choose δA such that $\mathrm{Ran}\,(\bar{A}) \cap \mathrm{Ker}\, A^+ = \{0\}$. This problem will be solved for bounded linear operators in the following theorem.

Theorem 2.2.12. *Let $A \in Gib(X, Y)$ and $\delta A \in B(X, Y)$. Then $I_X + A^+\delta A$ is invertible in $B(X)$ and $\mathrm{Ran}\,(\bar{A}) \cap \mathrm{Ker}\, A^+ = \{0\}$ iff there is a $\Delta A \in B(X, Y)$ such that $I_X + A^+\Delta A$ is invertible and*

$$(I_Y - AA^+)\Delta A(I_X - A^+A) = 0 \tag{2.2.5}$$

$$\delta A = \Delta A - (I_Y - AA^+)(I_Y + \Delta AA^+)^{-1}\Delta A(I_X - A^+A). \tag{2.2.6}$$

Proof. We first assume that there exists ΔA satisfying (2.2.5) and (2.2.6). Then $A^+\delta A = A^+\Delta A$ and hence $I_X + A^+\delta A$ is invertible. Let $y \in \mathrm{Ran}\,(\bar{A}) \cap \mathrm{Ker}\, A^+$. Then $A^+y = 0$ and $y = \bar{A}x$ for some $x \in X$. Put $x_1 = A^+Ax$, $x_2 = (I_X - A^+A)x$. Then

$$\begin{aligned} 0 &= A^+\bar{A}(x_1 + x_2) = x_1 + A^+\delta Ax_1 + A^+\delta Ax_2 \\ &= (I_X + A^+\Delta A)x_1 + A^+\Delta Ax_2 \end{aligned}$$

and consequently, $x_1 = -(I_X + A^+\Delta A)^{-1}A^+\Delta Ax_2$. By (2.2.5) and (2.2.6),

$$\begin{aligned} y &= (I_Y - AA^+)y = (I_Y - AA^+)(A + \delta A)(x_1 + x_2) \\ &= (I_Y - AA^+)\delta Ax_1 + (I_Y - AA^+)\delta Ax_2 \\ &= (I_Y - AA^+)\Delta Ax_1 - (I_Y - AA^+)(I_Y + \Delta AA^+)^{-1}\Delta Ax_2 \\ &= -(I_Y - AA^+)\Delta A(I_X + A^+\Delta A)^{-1}A^+\Delta Ax_2 \\ &\quad - (I_Y - AA^+)(I_Y + \Delta AA^+)^{-1}\Delta Ax_2 \\ &= -(I_Y - AA^+)(I_Y + \Delta AA^+)^{-1}[(I_Y + \Delta AA^+) - I_Y]\Delta Ax_2 \\ &\quad - (I_Y - AA^+)(I_Y + \Delta AA^+)^{-1}\Delta Ax_2 \\ &= -(I_Y - AA^+)\Delta A(I_X - A^+A)x \\ &= 0. \end{aligned}$$

Now assume that $I_X + A^+\delta A$ is invertible and $\mathrm{Ran}\,(\bar{A}) \cap \mathrm{Ker}\, A^+ = \{0\}$. Set

$$\Delta A = \delta A - (I_Y - AA^+)\delta A(I_X - A^+A).$$

Then $(I_Y - AA^+)\Delta A(I_X - A^+A) = 0$ and $I_X + A^+\Delta A = I_X + A^+\delta A$ is invertible in $B(X)$. Using the formula

$$I_Y - (I_Y + \delta AA^+)^{-1}\delta AA^+ = (I_Y + \delta AA^+)^{-1},$$

we get that

$$\begin{aligned}
&(I_Y - AA^+)(I_Y + \Delta AA^+)^{-1}\Delta A(I_X - A^+A) \\
&= (I_Y - AA^+)[I_Y - (I_Y + \delta AA^+)^{-1}\delta AA^+]\Delta A(I_X - A^+A) \\
&= -(I_Y - AA^+)(I_Y + \delta AA^+)^{-1}\delta AA^+\delta A(I_X - A^+A) \\
&= -(I_Y - AA^+)(I_Y + \delta AA^+)^{-1}(I_Y + \delta AA^+ - I_Y)\delta A(I_X - A^+A) \\
&= -(I_Y - AA^+)\delta A(I_X - A^+A) \\
&\quad + (I_Y - AA^+)(I_Y + \delta AA^+)^{-1}\delta A(I_X - A^+A) \\
&= \Delta A - \delta A
\end{aligned}$$

for $(I_Y - AA^+)(I_Y + \delta AA^+)^{-1}\delta A(I_X - A^+A) = 0$ by Theorem 2.2.6 when $\mathrm{Ran}\,(\bar{A}) \cap \mathrm{Ker}\,A^+ = \{0\}$. □

In Theorem 2.2.6 and its corollaries, the hypothesis that $I_X + A^+\delta A$ is invertible in $B(X)$ is necessary. The simplest fact is that $I_X + A^+\delta A$ is always invertible in $B(X)$ when $\|A^+\|\|\delta A\| < 1$. Besides, we can find another conditions that make $I_X + A^+\delta A$ invertible as follows.

Theorem 2.2.13. *Let $T \in Gi(X,Y)$ and $\delta T \in B(X,Y)$. Put $\bar{T} = T+\delta T \in \mathcal{C}(X,Y)$. Then $I_X + T^+\delta T \in GL(B(X))$ and* $\mathrm{Ran}\,(\bar{T}) \cap \mathrm{Ker}\,T^+ = \{0\}$ *iff*

$$\begin{aligned}
&\mathrm{Ran}\,(\bar{T}) \cap \mathrm{Ker}\,T^+ = \{0\}, &\qquad &\mathrm{Ker}\,\bar{T} \cap \mathrm{Ran}\,(T^+) = \{0\}, \\
&Y = \mathrm{Ran}\,(\bar{T}) + \mathrm{Ker}\,T^+, &\qquad &\mathfrak{D}(T) = \mathrm{Ker}\,\bar{T} + \mathrm{Ran}\,(T^+).
\end{aligned}$$

Proof. Suppose that $\mathrm{Ran}\,(\bar{T})\cap\mathrm{Ker}\,T^+ = \{0\}$ and $I_X + T^+\delta T$ is invertible in $B(X)$. Then $S = (I_X + T^+\delta T)^{-1}(I_X - T^+T)$ is an idempotent operator on $\mathfrak{D}(T)$ with $\mathrm{Ran}\,(S) = \mathrm{Ker}\,\bar{T}$ by Lemma 2.2.4 and $\bar{T}^+$ exists by Theorem 2.2.6. Put $V = I_X + T^+\delta T$ and $V' = I_Y + \delta TT^+$. Then V' is also invertible in $B(X)$ by Proposition 1.4.13. Since

$$V = I_X - T^+T + T^+\bar{T}, \;\; V' = I_Y - TT^+ + \bar{T}T^+,$$

it follows that

$$\begin{aligned}
I_X &= (I_X + T^+\delta T)^{-1}(I_X - T^+T) + (I_X + T^+\delta T)^{-1}T^+\bar{T} \\
&= S + T^+(I_X + \delta TT^+)^{-1}\bar{T}, \;\text{ on } \mathfrak{D}(T), \\
Y &= \mathrm{Ran}\,(V') \subset \mathrm{Ran}\,(\bar{T}) + \mathrm{Ker}\,(T^+).
\end{aligned}$$

Thus $\mathfrak{D}(T) = \mathrm{Ker}\,\bar{T} + \mathrm{Ran}\,(T^+)$ and $Y = \mathrm{Ran}\,(\bar{T}) + \mathrm{Ker}\,T^+$.

Now let $\xi \in \mathrm{Ker}\,\bar{T} \cap \mathrm{Ran}\,(T^+)$. Then $\bar{T}\xi = 0$ and $\xi = T^+\eta$ for some $\eta \in Y$. So

$$(I_X + T^+\delta T)\xi = \xi - T^+T\xi = T^+\eta - T^+TT^+\xi = 0.$$

and hence $\xi = 0$, that is, $\operatorname{Ker} \bar{T} \cap \operatorname{Ran}(\bar{T}^+) = \{0\}$.

Conversely, to prove $I_X + T^+\delta T$ is invertible in $B(X)$, we need only show that $\operatorname{Ker}(I_Y + \delta T T^+) = \{0\}$ and $\operatorname{Ran}(I_Y + \delta T T^+) = Y$ by Theorem 1.1.7 and Proposition 1.4.13.

Let $\xi \in \operatorname{Ker}(I_Y + \delta T T^+)$. Then

$$(I_Y - TT^+)\xi + \bar{T}T^+\xi = 0. \tag{2.2.7}$$

Since $(I_Y - TT^+)\xi \in \operatorname{Ker} T^+$ and $\operatorname{Ran}(\bar{T}) \cap \operatorname{Ker} T^+ = \{0\}$, we have

$$(I_Y - TT^+)\xi = \bar{T}T^+\xi = 0$$

by (2.2.7). From $\bar{T}T^+\xi = 0$, we get that $T^+\xi \in \operatorname{Ker} \bar{T} \cap \operatorname{Ran}(T^+) = \{0\}$. Thus $\xi = (I_Y - TT^+)\xi + TT^+\xi = 0$, i.e., $\operatorname{Ker}(I_Y + \delta T T^+) = \{0\}$. Let $y \in Y$. Then from $Y = \operatorname{Ran}(\bar{T}) + \operatorname{Ker} T^+$, we can choose $y_1 \in \operatorname{Ran}(\bar{T})$ and $y_2 \in \operatorname{Ker} T^+$ such that $y = y_1 + y_2$. From $\mathfrak{D}(T) = \operatorname{Ker} \bar{T} + \operatorname{Ran}(T^+)$, we have $\operatorname{Ran}(\bar{T}) = \bar{T}(\operatorname{Ran}(T^+))$. So we can pick $y_3 \in Y$ such that $y_1 = \bar{T}T^+y_3$. Put $z = TT^+y_3 + y_2$. Then

$$V'z = (I_Y - TT^+ + \bar{T}T^+)z = y_1 + y_2 = y,$$

that is, $\operatorname{Ran}(V') = Y$. □

2.3 Perturbation analysis of the operator equation $Ax = b$

In this section, we assume that $A \in Gib(X, Y)$ and $\bar{A} = A + \delta A \in B(X, Y)$. Take $b \in \operatorname{Ran}(A) \setminus \{0\}$ and $\bar{b} = b + \delta b \in \operatorname{Ran}(\bar{A})$. Put $\epsilon_A = \|A\|^{-1}\|\delta A\|$, $\epsilon_b = \|b\|^{-1}\|\delta b\|$ and $\kappa_A = \|A\|\|A^+\|$ (the condition number of A). Set $S(A, b) = \{x \in X \,|\, Ax = b\}$, $S(\bar{A}, \bar{b}) = \{x \in X \,|\, \bar{A}x = \bar{b}\}$.

Lemma 2.3.1. *Set $m = \inf\{\|x\| \,|\, x \in S(A, b)\}$. Then*

$$\|b\|\|A\|^{-1} \le m \le \|A^+\|\|b\|.$$

Proof. For any $x \in S(A, b)$, $\|b\| = \|Ax\| \le \|A\|\|x\|$. So $m \ge \|A\|^{-1}\|b\|$. Since $AA^+b = b$, it follows that $S(A, b) = A^+b + \operatorname{Ker} A$. Thus,

$$m = \operatorname{dist}(A^+b, \operatorname{Ker} A) \le \|A^+b\| \le \|A^+\|\|b\|.$$

□

Lemma 2.3.2. *For any $\bar{x} \in S(\bar{A}, \bar{b})$, we have*

$$\operatorname{dist}(\bar{x}, S(A, b)) \le \|A^+\|\|\delta b\| + \kappa_A \epsilon_A \|\bar{x}\|. \tag{2.3.1}$$

Proof. By the definition of $\gamma(A)$,

$$\operatorname{dist}(\bar{x}, S(A,b)) = \operatorname{dist}(\bar{x} - A^+b, \operatorname{Ker} A) \leq \gamma(A)^{-1}\|A(\bar{x} - A^+b)\|. \quad (2.3.2)$$

Note that $A(\bar{x} - A^+b) = (\bar{A} - \delta A)\bar{x} - AA^+b = \delta b - \delta A\bar{x}$. Thus, from (2.3.2), we get that

$$\operatorname{dist}(\bar{x}, S(A,b)) \leq \|A^+\|\|\delta b - \delta A\bar{x}\| \leq \|A^+\|\|\delta b\| + \kappa_A \epsilon_A \|\bar{x}\|.$$

□

Lemma 2.3.3. *Set* $\bar{m} = \inf\{\|\bar{x}\| \mid \bar{x} \in S(\bar{A}, \bar{b}\}$. *Assume that* $\|A^+\|\|\delta A\| < 1$ *and* $\operatorname{Ran}(\bar{A}) \cap \operatorname{Ker} A^+ = \{0\}$. *Then*

$$\bar{m} \leq \frac{m}{1 - \kappa_A\epsilon_A} + \frac{\|A^+\|}{1 - \kappa_A\epsilon_A}(\|\delta b\| + 2\kappa_A\epsilon_A\|b\|). \quad (2.3.3)$$

Proof. By Theorem 2.2.6, $\bar{A} \in Gib(X,Y)$ with $\bar{A}^+ = (I_X + A^+\delta A)^{-1}A^+$. So $S(\bar{A}, \bar{b}) = \bar{A}^+\bar{b} + \operatorname{Ker}\bar{A}$. Given $\epsilon > 0$, there is $x_\epsilon \in S(A,b)$ such that $m > \|x_\epsilon\| - \epsilon$. Put $y_\epsilon = x_\epsilon - A^+b \in \operatorname{Ker} A$. Then $\|y_\epsilon\| \leq \|x_\epsilon\| + \|A^+b\| < m + \|A^+b\| + \epsilon$ and for any $\bar{y} \in \operatorname{Ker}\bar{A}$,

$$\begin{aligned}\|\bar{A}^+\bar{b} + \bar{y}\| &= \|\bar{A}^+\bar{b} - A^+b + A^+b - y_\epsilon + y_\epsilon + \bar{y}\| \\ &\leq \|\bar{A}^+\delta b\| + \|\bar{A}^+ - A^+\|\|b\| + \|x_\epsilon\| + \|y_\epsilon + \bar{y}\|.\end{aligned}$$

Therefore,

$$\begin{aligned}\bar{m} &\leq m + \epsilon + \frac{\|A^+\|\|\delta b\|}{1 - \kappa_A\epsilon_A} + \frac{\|A^+\|\|\delta A\|}{1 - \kappa_A\epsilon_A}\|A^+b\|\|b\| + \operatorname{dist}(y_\epsilon, \operatorname{Ker}\bar{A}) \\ &\leq m + \epsilon + \frac{\|A^+\|\|\delta b\|}{1 - \kappa_A\epsilon_A} + \frac{\kappa_A\epsilon A}{1 - \kappa_A\epsilon_A}\|A^+\|\|b\| + \|y_\epsilon\|\delta(\operatorname{Ker} A, \operatorname{Ker}\bar{A}).\end{aligned}$$

Since

$$\begin{aligned}\delta(\operatorname{Ker} A, \operatorname{Ker}\bar{A}) &\leq \gamma(\bar{A})^{-1}\|\delta A\| \leq \|\bar{A}^+\|\|\delta A\| \\ &\leq \frac{\|A^+\|\|\delta\|}{1 - \|A^+\|\|\delta A\|} = \frac{\kappa_A\epsilon_A}{1 - \kappa_A\epsilon_A},\end{aligned}$$

by Lemma 1.3.5, we have

$$\begin{aligned}\bar{m} &\leq m + \epsilon + \frac{\|A^+\|\|b\|}{1 - \kappa_A\epsilon_A} + \frac{\|A^+\|\|\delta b\|}{1 - \kappa_A\epsilon_A} + \frac{\kappa_A\epsilon_A}{1 - \kappa_A\epsilon_A}\|A^+\|\|b\| \\ &\quad + (m + \epsilon + \|A^+\|\|b\|)\frac{\kappa_A\epsilon_A}{1 - \kappa_A\epsilon_A} \\ &= \frac{m}{1 - \kappa_A\epsilon_A} + \frac{\|A^+\|\|\delta b\|}{1 - \kappa_A\epsilon_A} + \frac{2\kappa_A\epsilon_A}{1 - \kappa_A\epsilon_A}\|A^+\|\|b\| + (1 + \frac{\kappa_A\epsilon_A}{1 - \kappa_A\epsilon_A})\epsilon.\end{aligned}$$

Finally, let $\epsilon \to 0^+$, we obtain (2.3.3). □

Theorem 2.3.4. *Let $A \in Gib(X,Y)$ and $\bar{A} = A + \delta A \in B(X,Y)$ with $\|A^+\|\|\delta A\| < 1$ and $\operatorname{Ran}(\bar{A}) \cap \operatorname{Ker} A^+ = \{0\}$. Assume that there is $\bar{x}_m \in S(\bar{x}, \bar{b})$ such that $\|\bar{x}_m\| = \inf\{\|\bar{x}\| \,|\, \bar{x} \in S(\bar{A}, \bar{b})\} = \bar{m}$, i.e., $\bar{x}_m$ is the minimum norm solution of $\bar{A}x = \bar{b}$. Then,*

$$\frac{\operatorname{dist}(\bar{x}_m, S(A,b))}{m} \leq \frac{\kappa_A}{1 - \kappa_A \epsilon_A}(\epsilon_A + \epsilon_b + 2(\kappa_A \epsilon_A)^2).$$

Proof. By (2.3.1) and (2.3.3),

$$\begin{aligned}\operatorname{dist}(\bar{x}_m, S(A,b)) &\leq \|A^+\|\|\delta b\| + \kappa_A \epsilon_A \|\bar{x}_m\| \\ &\leq \frac{\|A^+\|\|\delta b\|}{1-\kappa_A\epsilon_A} + \frac{2(\kappa_A\epsilon_A)^2\|A^+\|\|b\|}{1-\kappa_A\epsilon_A} + \frac{\kappa_A\epsilon_A m}{1-\kappa_A\epsilon_A}.\end{aligned}$$

Thus, by Lemma 2.3.1,

$$\begin{aligned}\frac{\operatorname{dist}(\bar{x}_m, S(A,b))}{m} &\leq \frac{\kappa_A}{1-\kappa_A\epsilon_A}\Big(\epsilon_A + \frac{\|b\|}{m\|A\|}(\epsilon_b + 2(\kappa_A\epsilon_A)^2)\Big) \\ &\leq \frac{\kappa_A}{1-\kappa_A\epsilon_A}(\epsilon_A + \epsilon_b + 2(\kappa_A\epsilon_A)^2).\end{aligned}$$

□

Remark 2.3.5. Consider an example. Let A, $\bar{A}$, b and $\bar{b}$ be

$$A = \begin{bmatrix} 1 & 1 \\ 2 & 2 \end{bmatrix}, \; \bar{A} = \begin{bmatrix} 1+\epsilon & 1+\epsilon \\ 2 & 2 \end{bmatrix}, \; b = \begin{bmatrix} 3 \\ 6 \end{bmatrix}, \; \bar{b} = \begin{bmatrix} 3+3\epsilon \\ 6 \end{bmatrix}, \; 0 < \epsilon < \frac{5}{2}.$$

We have $\operatorname{rank}(\bar{A}) = \operatorname{rank}(A)$. Take $\|x\| = |x_1| + |x_2|$ for $x = \begin{bmatrix} x_1 \\ x_2 \end{bmatrix} \in \mathbb{C}^2$. Then $\|A^+\|\|\bar{A} - A\| = \frac{2}{5}\epsilon < 1$ and it is easy to check that

$$S(A,b) = S(\bar{A},\bar{b}) = \left\{ \begin{bmatrix} \frac{3}{2} \\ \frac{3}{2} \end{bmatrix} + s\begin{bmatrix} -1 \\ 1 \end{bmatrix} \,\middle|\, s \in \mathbb{C} \right\}.$$

Furthermore, when $|s| \leq \frac{2}{3}$, $x(s) = \begin{bmatrix} \frac{3}{2} \\ \frac{3}{2} \end{bmatrix} + s\begin{bmatrix} -1 \\ 1 \end{bmatrix}$ is the minimum norm solution of two equations $Ax = b$, $\bar{A}x = \bar{b}$. Thus $\|x(\frac{1}{2}) - x(\frac{1}{4})\| = \frac{1}{2}$ and $\operatorname{dist}(x(\frac{1}{2}), S(A,b)) = 0$. From such a viewpoint, we should estimate the $\operatorname{dist}(\bar{x}_m, S(A,b))$ instead of $\|\bar{x}_m - x_m\|$, where x_m is the minimum norm solution of $Ax = b$.

Remark 2.3.6. The discussion above is the special case of the perturbation analysis of the least problem

$$\min \|x\| \quad \text{subject to} \quad \|Ax - b\| = \inf_{z \in X} \|Az - b\|. \tag{2.3.4}$$

for $A \in B(X,Y)$ and $b \in Y$.

The following proposition shows the existence of the solution of (2.3.4).

Proposition 2.3.7. *Let $A \in C(X,Y)$ with* Ran (A) *closed and $b \in Y\backslash\{0\}$. If one of following conditions*

(A) *X is reflexive and $\mathfrak{D}(A) = X$, i.e., $A \in B(X,Y)$.*
(B) *X and Y are all reflexive.*

are satisfied, then there is at least one $x_m \in \mathfrak{D}(A)$ satisfying (2.3.4).

Moreover, x_m is unique when X is strictly convex.

Proof. We first assume that Condition (A) holds.

Let $\{x_n\} \subset X$ such that $\|Ax_n - b\| \to \inf_{z\in X} \|Az - b\| = M \geq 0$ as $n \to \infty$. Then there is $M_0 > 0$ such that $\|Ax_n\| \leq M_0, \forall n \geq 1$. Thus

$$\operatorname{dist}(x_n, \operatorname{Ker} A) \leq \gamma(A)^{-1}\|Ax_n\| \leq \gamma(A)^{-1}M_0.$$

Choose $z_n \in \operatorname{Ker} A$ such that $\|x_n - z_n\| \leq \operatorname{dist}(x_n, \operatorname{Ker} A) + \dfrac{1}{n}$, $n \geq 1$ and put $y_n = x_n - z_n, n \geq 1$. Then $\{y_n\}$ is a bounded sequence in X. Since X is reflexive, there is $x \in X$ and a subsequence $\{y_{n_k}\}$ of $\{y_n\}$ such that $y_{n_k} \xrightarrow{w} x$ by Theorem 1.1.20. Thus $Ax_{n_k} = Ay_{n_k} \xrightarrow{w} Ax$ and consequently,

$$\|Ax - b\| \leq \liminf_{k\to\infty} \|Ax_{n_k} - b\| = M \leq \|Ax - b\|$$

by Proposition 1.1.17.

Set $\hat{S}(A,b) = \{x \in X | \, \|Ax - b\| = M\}$. It is easy to verify that $\hat{S}(A,b)$ is a closed convex subset of X. So by Proposition 1.1.21, we can find $x_m \in X$ such that $\inf\{\|x\| | \, x \in \hat{S}(A,b)\} = \operatorname{dist}(0, \hat{S}(A,b)) = \|x_m\|$ and moreover x_m is unique when X is strictly convex.

Now assume that Condition (B) holds. Then there is $b_0 \in \operatorname{Ran}(A)$ such that $\|b_0 - b\| = M$ by Proposition 1.1.21. So

$$\hat{S}(A,b) = \{x \in \mathfrak{D}(A) | \, \|Ax - b\| = M\} \neq \emptyset.$$

Clearly, $\hat{S}(A,b)$ is convex. We have to show that $\hat{S}(A,b)$ is closed. Let $\{x_n\} \subset \hat{S}(A,b)$ and $x_0 \in X$ such that $\|x_n - x_0\| \to 0 \ (n \to \infty)$. Since $\{Ax_n\}$ is bounded in Y and Y is reflexive, it follows from Theorem 1.1.20 that there is a subsequence $\{Ax_{n_k}\}$ of $\{Ax_n\}$ and $y_0 \in Y$ such that $Ax_{n_k} \xrightarrow{w} y_0$.

We claim that $\{x_0, y_0\} \in G(A)$. If it is false, then there is $h \in (X \times Y)^*$ such that

$$h(\{x_0, y_0\}) = \operatorname{dist}(\{x_0, y_0\}, G(T)) > 0, \quad h(\{x, Ax\}) = 0, \quad \forall x \in \mathfrak{D}(A).$$

Let $h_1(x) = h(\{x, 0\})$ and $h_2(y) = h(\{0, y\})$, $\forall\, x \in X$ and $y \in Y$. Then

$$h(\{x, y\}) = h_1(x) + h_2(y), \quad \forall\, \{x, y\} \in X \times Y.$$

Since $h_1 \in X^*$ and $h_2 \in Y^*$, we have $h_1(x_{n_k}) \to h_1(x_0)$ and $h_2(Ax_{n_k}) \to h_2(y_0)$ as $n \to \infty$ and hence $h(\{x_{n_k}, Ax_{n_k}\}) \to h(\{x_0, y_0\})$. Note that $h(\{x_{n_k}, Ax_{n_k}\}) = 0$. So $h(\{x_0, y_0\}) = 0$, a contradiction. Therefore, $x_0 \in \mathfrak{D}(A)$ and $Ax_0 = y_0$.

Finally, by Proposition 1.1.21, we can find $x_m \in X$ such that

$$\inf\{\|x\| \,|\, x \in \hat{S}(A, b)\} = \operatorname{dist}(0, \hat{S}(A, b)) = \|x_m\|$$

and moreover x_m is unique when X is strictly convex. $\square$

Remark 2.3.8. Let $A \in B(X, Y)$ with $\operatorname{Ran}(A)$ closed and $\bar{A} = A + \delta A \in B(X, Y)$. Suppose that

$$\gamma(A)^{-1}\|\delta A\| < 1 \text{ and } \delta(\operatorname{Ker} A, Ker\bar{A}) < \frac{1}{2}(1 - \gamma(A)^{-1}\|\delta A\|).$$

Then by Lemma 1.3.6, $\operatorname{Ran}(\bar{A})$ is closed in Y. Let $b \in Y\backslash\{0\}$ and $\bar{b} = b + \delta b \in Y$. Assume that X is a reflexive and strictly convex Banach space. Then by Proposition 2.3.7,

$$\min \|x\| \quad \text{subject to} \quad \|\bar{A}x - \bar{b}\| = \inf_{z \in x} \|\bar{A} - \bar{b}\| \tag{2.3.5}$$

has a unique solution $\bar{x}_m$. it seems to be a problem how to estimate the upper bound of $\dfrac{\|\bar{x}_m - x_m\|}{\|x_m\|}$. But if X, Y are all Hilbert spaces, we can give an answer.

If we do not consider the perturbation of the minimum norm solution of the equation $Ax = b$, we have the following more general result.

Proposition 2.3.9. *Let $A \in Gib(X, Y)$ and $\bar{A} = A + \delta A \in B(X, Y)$ with $\|A^+\|\|\delta A\| < 1$. Let $b \in \operatorname{Ran}(A)\backslash\{0\}$ and $\bar{b} = b + \delta b \in \operatorname{Ran}(\bar{A})$. Then for any $\bar{x} \in S(\bar{A}, \bar{b})$, there is an $x \in S(A, b)$ such that*

$$\frac{\|\bar{x} - x\|}{\|x\|} \le \frac{\kappa_A}{1 - \kappa_A \epsilon_A}(\epsilon_b + \epsilon_A).$$

Proof. Put $x = A^+b + (I_X - A^+A)\bar{x}$. Then $Ax = b$ and

$$\bar{x} - x = -A^+b + A^+(\bar{A} - \delta A)\bar{x} = -A^+b - A^+\delta A\bar{x} + A^+\bar{b} = A^+\delta b - A^+\delta A\bar{x}.$$

Thus, $\bar{x} = (I_X + A^+\delta A)^{-1}(A^+\delta b + x)$ for $\|A^+\|\|\delta A\| < 1$ and hence

$$\bar{x} - x = (I_X + A^+\delta A)^{-1}A^+\delta b + ((I_X + A^+\delta A)^{-1} - I_X)x. \tag{2.3.6}$$

By (2.3.6) and Lemma 2.3.1, we have

$$\begin{aligned}\frac{\|\bar{x}-x\|}{\|x\|} &\leq \frac{1}{\|x\|}\frac{\|A^+\|\|\delta b\|}{1-\|A^+\|\|\delta A\|}+\frac{\|A^+\|\|\delta A\|}{1-\|A^+\|\|\delta A\|}\\ &\leq \frac{\|A\|}{\|b\|}\frac{\|A^+\|\|\delta b\|}{1-\|A^+\|\|\delta A\|}+\frac{\|A^+\|\|deltaA\|}{1-\|A^+\|\|\delta A\|}\\ &= \frac{\kappa_A}{1-\kappa_A\epsilon_A}(\epsilon_b+\epsilon_A).\end{aligned}$$

□

Remark 2.3.10. Let A and $\bar{A}=A+\delta A$ be in $B(X,Y)$. Suppose that A has the bounded inverse A^{-1} and $\|A^{-1}\|\|\delta A\|<1$. Then $\bar{A}=A(I_X+A^{-1}\delta A)$ has the bounded inverse $\bar{A}^{-1}\colon Y\to X$. Let b and $\bar{b}=b+\delta b\in Y\backslash\{0\}$. Then we have the classical result:

$$\frac{\|\bar{A}^{-1}\bar{b}-A^{-1}b\|}{\|A^{-1}b\|}\leq\frac{\|A^{-1}\|\|A\|}{1-\|A^{-1}\|\|\delta A\|}\Big(\frac{\|\delta b\|}{\|b\|}+\frac{\|\delta A\|}{\|A\|}\Big).$$

2.4 Generalized inverses in Banach algebras

In this section, we assume that $\mathcal{A}$ is a Banach algebra.

Definition 2.4.1. A nonzero element a in $\mathcal{A}$ is called to be generalized invertible if there is $b\in\mathcal{A}$ such that $a=aba$.

In addition, if b is invertible (i.e., $b\in GL(\tilde{\mathcal{A}})$), then a is called to be decomposably generalized invertible.

Denote by $Gi(\mathcal{A})$ and $G_d(\mathcal{A})$ the set of all generalized invertible and decomposably generalized invertible elements in $\mathcal{A}$ respectively.

Remark 2.4.2. Let $a\in Gi(\mathcal{A})$. Then there is $b\in\mathcal{A}$ such that $aba=a$. Put $c=bab$. Then we have

$$aca=ababa=aba=a,\quad cac=bababab=babab=c.$$

In this case, we call c is the generalized inverse of a and denote it by a^+.

Obviously, the notation "generalized inverse" in Banach algebra is a generalization of the corresponding notation in $B(X)$.

Let $a\in\tilde{\mathcal{A}}$ and let V be a subspace of $\mathcal{A}$. Set

$$\begin{aligned}K_r(a)&=\{x\in\mathcal{A}\,|\,ax=0,\}, & R_r(a)&=\{ax\,|\,x\in\mathcal{A}\}\\ K_l(a)&=\{x\in\mathcal{A}\,|\,xa=0\}, & R_l(a)&=\{xa\,|\,x\in\mathcal{A}\}\\ V\mathcal{A}&=\{va\,|\,v\in V,\ a\in\mathcal{A}\}, & \mathcal{A}V&=\{av\,|\,a\in\mathcal{A},\ v\in V\}.\end{aligned}$$

It is routine to check the following:

Lemma 2.4.3. *Let $p \in \mathcal{A}$ be an idempotent element ($p^2 = p$). Then $K_r(p)$, $K_l(p)$ are closed and*

$$K_r(p) = R_r(1-p),\ K_l(p) = R_l(1-p),\ K_r(p)\mathcal{A} \subset K_r(p),\ \mathcal{A}K_l(p) \subset K_l(p).$$

Corresponding to Proposition 2.1.3, we have

Theorem 2.4.4. *Let a be a nonzero element in a unital Banach algebra $\mathcal{A}$. Then the following statements are equivalent.*

(1) $a \in Gi(\mathcal{A})$.
(2) *$R_r(a)$ is closed in $\mathcal{A}$ and there are closed subspaces M, N in $\mathcal{A}$ such that $M\mathcal{A} \subset M, N\mathcal{A} \subset N$ and*
$$\mathcal{A} = K_r(a) \dotplus M = N \dotplus R_r(a).$$
(3) *$R_l(a)$ is closed in $\mathcal{A}$ and there are closed spaces M', N' in $\mathcal{A}$ such that $\mathcal{A}M' \subset M'$, $\mathcal{A}N' \subset N'$ and*
$$\mathcal{A} = K_l(a) \dotplus M' = N' \dotplus R_l(a).$$

Proof. Suppose (1) holds. Put $p = a^+a$, $q = aa^+$. It is easy to check that $R_r(a) = R_r(q)$, $R_l(a) = R_l(q)$ and $K_r(a) = R_r(1-p)$, $K_l(a) = R_l(1-q)$. Thus, $R_r(a)$ and $R_l(a)$ are closed in $\mathcal{A}$. Moreover, we have

$$\begin{aligned}\mathcal{A} &= K_r(a) \dotplus R_r(p) = R_r(1-q) \dotplus R_r(a)\\ &= K_l(a) \dotplus R_l(q) = R_l(a) \dotplus R_r(p).\end{aligned}$$

Therefore (2) and (3) hold.

Now suppose that (2) is true. Define a linear map $L_a \colon \mathcal{A} \to \mathcal{A}$ by $L_a(x) = ax$, $\forall\, x \in \mathcal{A}$. Then $\|L_a\| \leq \|a\|$. Let φ be the restriction of L_a on M. Then $\varphi \in B(M, R_r(a))$ with $\operatorname{Ker}\varphi = \{0\}$ and $\operatorname{Ran}(\varphi) = R_r(a)$. Thus $\varphi^{-1} \colon R_r(a) \to M$ is bounded. Noting that $\varphi(xc) = \varphi(x)c$, $\forall\, x \in M$ and $c \in \mathcal{A}$, we have $\varphi^{-1}(yc) = \varphi^{-1}(y)c$, $\forall\, y \in R_r(a)$ and $c \in \mathcal{A}$.

Let $Q \colon \mathcal{A} \to R_r(a)$ be the bounded idempotent mapping. Since

$$\mathcal{A} = N \dotplus R_r(a), \quad N\mathcal{A} \subset N \text{ and } R_r(a)\mathcal{A} \subset R_r(a),$$

$Q(xc) = Q(x)c$, $\forall\, x,\ c \in \mathcal{A}$. Put $W = \varphi^{-1} \circ Q$. Then

$$W(xc) = W(x)c,\ (L_aWL_a)(x) = L_a(x),\ (WL_aW)(x) = W(x),\ \forall\, c,\ x \in \mathcal{A}.$$

Put $b = W(1)$. Then from above argument, we get that $aba = a$, $bab = a$, i.e., $a \in Gi(A)$.

Similarly, if (3) holds, then (1) is true. $\square$

For the relation between $Gi(\mathcal{A})$ and $G_d(\mathcal{A})$, we have following:

Proposition 2.4.5. *Let $\mathcal{A}$ be a untial Banach algebra*

(1) $G_d(\mathcal{A}) = \overline{GL(\mathcal{A})} \cap Gi(\mathcal{A})$.
(2) *Let $a \in Gi(\mathcal{A})$. If $\{\lambda \in \mathbb{C} | \, |\lambda| < \|a\|^{-1}\} \cap (\mathbb{C} \backslash \sigma(a^+)) \neq \emptyset$, then $a \in G_d(\mathcal{A})$.*

Proof. (1) Let $x \in G_d(\mathcal{A})$. Then there is $y \in GL(\mathcal{A})$ such that $x = xyx$. Put $p = yx$. Then $p^2 = p$ and $x = y^{-1}p$. Given $\epsilon > 0$ and set $x_\epsilon = y^{-1}(p + \epsilon(1-p))$. Since $(p + \epsilon(1-p)) \in GL(\mathcal{A})$ with $(p + \epsilon(1-p))^{-1} = p + \epsilon^{-1}(1-p)$ and $\|x - x_\epsilon\| \leq \epsilon \|y^{-1}\| \|1-p\|$, it follows that $x \in \overline{GL(\mathcal{A})}$. Thus, $G_d(\mathcal{A}) \subset \overline{GL(\mathcal{A})} \cap Gi(\mathcal{A})$.

On the other hand, let $x \in \overline{GL(\mathcal{A})} \cap Gi(\mathcal{A})$. Then we can find $z \in GL(\mathcal{A})$ such that $\|x - z\| < \|x^+\|^{-1}$. Hence $\|xx^+ - zx^+\| < 1$. Put $c = zx^+ + 1 - xx^+$. Then $cx = zx^+x$ and

$$\|1 - c\| = \|xx^+ + (1 - xx^+) - (zx^+ + 1 - xx^+)\| < 1.$$

and hence $c \in GL(\mathcal{A})$. Therefore, $xz^{-1}cx = xz^{-1}(zx^+x) = x$ and $z^{-1}c \in GL(A)$.

(2) Choose $\lambda_0 \in \mathbb{C}$ such that $|\lambda_0| < \|a\|^{-1}$ and $\lambda_0 \notin \sigma(a^+)$. Since $\|\lambda_0 a\| < 1$, $1 - \lambda_0 a \in GL(\mathcal{A})$. Put $d = -(1 - \lambda_0 a)^{-1}(\lambda_0 - a^+) \in GL(\mathcal{A})$. Noting that

$$(1 - \lambda_0 a)a^+a = a^+a - \lambda_0 a = -(\lambda_0 - a^+)a,$$

we have

$$a = aa^+a = -a(1 - \lambda_0 a)^{-1}(\lambda_0 - a^+)a = ada \in G_d(\mathcal{A}).$$

□

Let $a \in Gi(\mathcal{A})$ and $\bar{a} = a + \delta a \in \mathcal{A}$. Then $\operatorname{Ran}(L_{\bar{a}}) = R_r(\bar{a})$ and $\operatorname{Ker}(L_{a^+}) = K_r(a^+) = R_r(1 - aa^+)$. This leads to the following definition.

Definition 2.4.6. Let $\mathcal{A}$ be a unital Banach algebra and $a \in Gi(\mathcal{A})$, $\bar{a} = a + \delta a \in \mathcal{A}$. We say that $\bar{a}$ is a stable perturbation of a (with respect to a^+) if $R_r(\bar{a}) \cap R_r(1 - aa^+) = \{0\}$ (or $\bar{a}\mathcal{A} \cap (1 - aa^+)\mathcal{A} = \{0\}$).

Similar to Theorem 2.2.6, we have following conditions that provide means for efficient handling of stable perturbation of elements of $Gi(\mathcal{A})$.

Theorem 2.4.7. *Let $\mathcal{A}$ be a unital Banach algebra. Let $a \in Gi(\mathcal{A})$ and $\bar{a} = a + \delta a \in \mathcal{A}$ with $1 + a^+\delta a \in GL(\mathcal{A})$. Then the following conditions are equivalent.*

(1) $\bar{a} \in Gi(\mathcal{A})$ *with* $\bar{a}^+ = (1 + a^+\delta a)^{-1}a^+$.
(2) $\bar{a}$ *is a stable perturbation of* a.
(3) $\bar{a}(1 + a^+\delta a)^{-1}(1 - a^+a) = 0$.
(4) $(1 - aa^+)(1 + \delta aa^+)^{-1}\bar{a} = 0$.
(5) $(1 - aa^+)\delta a(1 - a^+a) = (1 - aa^+)\delta a(1 + a^+\delta a)^{-1}a^+\delta a(1 - a^+a)$.
(6) $\mathcal{A}\bar{a} \cap \mathcal{A}(1 - a^+a) = \{0\}$.

Proof. (1)⇒(2) Let $x \in \bar{a}\mathcal{A} \cap (1 - aa^+)\mathcal{A}$. Then $a^+x = 0$ and $x = \bar{a}y$ for some $y \in \mathcal{A}$. Put $z = (1+\delta aa^+)x$. Using $a^+(1+\delta aa^+)^{-1} = (1+a^+\delta a)^{-1}a^+$, we have

$$\bar{a}\bar{a}^+ = (a + \delta a)a^+(1 + \delta aa^+)^{-1} = 1 - (1 - aa^+)(1 + \delta aa^+)^{-1}.$$

Thus, $(1 - \bar{a}\bar{a}^+)z = x = \bar{a}y$ and hence $x = 0$.

(2)⇒(3) Set $z = \bar{a}(1 + a^+\delta a)^{-1}(1 - a^+a) \in \bar{a}\mathcal{A}$. Since

$$aa^+z = a(1 + a^+\delta a)(1 + a^+\delta a)^{-1}(1 - a^+a) = 0,$$

it follows that $z \in \bar{a}\mathcal{A} \cap (1 - aa^+)\mathcal{A} = \{0\}$, i.e., $z = 0$.

(3)⇔(4) In fact:

$$\begin{aligned}(1 - aa^+)(1 + \delta aa^+)^{-1}\bar{a} =&(1 + \delta aa^+ - \bar{a}a^+)(1 + \delta aa^+)^{-1}\bar{a}\\ =&\bar{a} - \bar{a}(1 + a^+\delta a)^{-1}(a^+\delta a + 1 + a^+a - 1)\\ =&\bar{a}(1 + a^+\delta a)^{-1}(1 - a^+a).\end{aligned}$$

(3)⇒(5) We have

$$\begin{aligned}&(1 - aa^+)\delta a(1 + a^+\delta a)^{-1}a^+\delta a(1 - a^+a)\\ &\quad=(1 - aa^+)\delta a(1 - a^+a) - (1 - aa^+)\delta a(1 + a^+\delta a)^{-1}(1 - a^+a)\\ &\quad=(1 - aa^+)\delta a(1 - a^+a).\end{aligned}$$

(5)⇒(1) Above computation shows that if (5) is true, then

$$(1 - aa^+)\delta a(1 + a^+\delta a)^{-1}(1 - a^+a) = 0.$$

Since $aa^+\bar{a}(1+a^+\delta a)^{-1}(1-a^+a) = 0$, we have $\bar{a}(1+a^+\delta a)^{-1}(1-a^+a) = 0$. Put $b = (1 + a^+\delta a)^{-1}a^+$. Then $\bar{a}b\bar{a} = \bar{a}$ and $b\bar{a}b = b$, that is, $\bar{a} \in Gi(\mathcal{A})$ with $\bar{a}^+ = (1 + a^+\delta a)^{-1}a^+$.

(4)⇒(6) Let $z \in \mathcal{A}\bar{a} \cap \mathcal{A}(1 - a^+a)$. Then $z = x\bar{a} = y(1 - a^+a)$ for some $x, y \in \mathcal{A}$ and consequently,

$$0 = za^+ = x(a + \delta a)a^+ = x(1 + \delta aa^+ + aa^+ - 1). \tag{2.4.1}$$

So, $x = x(1-aa^+)(1+\delta aa^+)^{-1}$ by (2.4.1). Since $(1-aa^+)(1+\delta aa^+)^{-1}\bar{a} = 0$, we have $z = x\bar{a} = 0$.

(6)⇒(3) Put $c = (1 - aa^+)(1 + \delta aa^+)^{-1}\bar{a}$. Then $c \in \mathcal{A}\bar{a}$. By (3)⇔(4),

$$c = \bar{a}(1 + a^+\delta a)^{-1}(1 - a^+a) \in \mathcal{A}(1 - a^+a).$$

Thus, $c \in \mathcal{A}\bar{a} \cap \mathcal{A}(1 - a^+a) = \{0\}$. □

Corollary 2.4.8. *Let B be a unital Banach subalgebra of the unital Banach algebra $\mathcal{A}$. Let $a \in Gi(B)$ and $\bar{a} = a + \delta a \in B$. with $1 + a^+\delta a \in GL(B)$. Then $\bar{a}\,B \cap (1 - aa^+)B = \{0\}$ iff $R_r(\bar{a}) \cap R_r(1 - aa^+) = \{0\}$.*

Proof. The "f" part is obvious. We now prove the "only if" part. By Theorem 2.4.7, $\bar{a}B \cap (1 - aa^+)B = \{0\}$ implies that $\bar{a}^+ = (1 + a^+\delta a)^{-1}a^+$ for $\bar{a} \in Gi(B)$. Since $Gi(B) \subset Gi(\mathcal{A})$, we have

$$\bar{a}\mathcal{A} \cap (1 - aa^+)\mathcal{A} = R_r(\bar{a}) \cap R_r(1 - aa^+) = \{0\}$$

by using Theorem 2.4.7 again. □

Remark 2.4.9. Let $\mathcal{A} = B(X)$ and $A \in Gi(\mathcal{A})$, $\bar{A} = A + \delta A \in B(X)$. Then we have two notations about the stable perturbation of A given by Definition 2.2.1 and Definition 2.4.1, respectively. But, in fact,

$$\operatorname{Ran} \bar{A} \cap \operatorname{Ker} A^+ = \{0\} \Leftrightarrow R_r(\bar{A}) \cap R_r(I - AA^+) = \{0\}.$$

We prove this. Let $T \in R_r(\bar{A}) \cap R_r(I - AA^+)$. Then $A^+T = 0$ and $T = \bar{A}S$ for some $S \in B(X)$ and hence for any $x \in X$, $Tx \in \operatorname{Ker} A^+ \cap \operatorname{Ran}(\bar{A})$. So, if $\operatorname{Ran}(\bar{A}) \cap \operatorname{Ker} A^+ = \{0\}$, then $T = 0$, i.e., $R_r(\bar{A}) \cap R_r(I - AA^+) = \{0\}$.

Conversely, assume that $R_r(\bar{A}) \cap R_r(I - AA^+) = \{0\}$. Let $\xi \in \operatorname{Ran}(\bar{A}) \cap \operatorname{Ker} A^+$. Then $A^+\xi = 0$ and $\xi = \bar{A}\eta$ for some $\eta \in X$. Let f be a nonzero continuous functional on X and define operators C_1, $C_2 \in B(X)$ by

$$C_1x = f(x)\xi \text{ and } C_2x = f(x)\eta, \quad \forall x \in X.$$

Then $\bar{A}C_2 = C_1$. So $C_1 \in R_r(\bar{A}) \cap R_r(I - AA^+) = \{0\}$ and consequently, $\xi = 0$, that is, $\operatorname{Ran}(\bar{A}) \cap \operatorname{Ker} A^+ = \{0\}$.

In the rest of this section, we will investigate further conditions for stable perturbation.

Lemma 2.4.10. *For idempotents p_1 and p_2 in $\mathcal{A}$, $\hat{\delta}(p_1\mathcal{A}, p_2\mathcal{A}) \le \|p_1 - p_2\|$.*

Proof. The assertion is trivial when $p_1 = 0$. If $p_1 \neq 0$, then for any $z \in p_1\mathcal{A}$ with $\|z\| = 1$, $\operatorname{dist}(z, p_2\mathcal{A}) \le \|p_1z - p_2z\| \le \|p_1 - p_2\|$. Thus, $\delta(p_1\mathcal{A}, p_2\mathcal{A}) \le \|p_1 - p_2\|$ and hence $\hat{\delta}(p_1\mathcal{A}, p_2\mathcal{A}) \le \|p_1 - p_2\|$. □

Proposition 2.4.11. *Let $a \in Gi(\mathcal{A})$ and $\bar{a} = a + \delta a \in \mathcal{A}$ with $\|a^+\|\|\delta a\| < 1$.*

(1) *If $aa^+ \neq 1$ and $\delta(\bar{a}\mathcal{A}, a\mathcal{A}) < \dfrac{1}{\|1 - aa^+\|}$, then $\bar{a}\mathcal{A} \cap (1 - aa^+)\mathcal{A} = \{0\}$.*

(2) *If $\bar{a}\mathcal{A} \cap (1 - aa^+)\mathcal{A} = \{0\}$, then $\hat{\delta}(\bar{a}\mathcal{A}, a\mathcal{A}) \le \dfrac{\|1 - aa^+\|\|a^+\|\|\delta a\|}{1 - \|a^+\|\|\delta a\|}$.*

Proof. (1) If $\bar{a}\mathcal{A} \cap (1 - aa^+)\mathcal{A} \neq \{0\}$, we can find $x \in \bar{a}\mathcal{A} \cap (1 - aa^+)\mathcal{A}$ with $\|x\| = 1$. Then for any $y \in \mathcal{A}$, $(1 - aa^+)(x - ay) = x$ and it follows that $1 \leq \|1 - aa^+\|\|x - ay\|$. Thus $\delta(\bar{a}\mathcal{A}, a\mathcal{A}) \geq \|1 - aa^+\|^{-1}$ which contradicts the assumption.

(2) In this case, $\bar{a} \in Gi(A)$ with $\bar{a}^+ = (1 + a^+\delta a)^{-1}a^+$ by Theorem 2.4.7. Noting that $R_r(a) = R_r(aa^+)$ and $R_r(\bar{a}) = R_r(\bar{a}\bar{a}^+)$, we have by Lemma 2.4.10,

$$\begin{aligned}\hat{\delta}(\bar{a}\mathcal{A}, a\mathcal{A}) \leq \|\bar{a}\bar{a}^+ - aa^+\| &= \|(a + \delta a)a^+(1 + \delta aa^+)^{-1} - aa^+\| \\ &= \|(1 - aa^+)(1 - (1 + \delta aa^+)^{-1})\| \\ &\leq \frac{\|1 - aa^+\|\|a^+\|\|\delta\|}{1 - \|a^+\|\|\delta a\|}.\end{aligned}$$

□

By Proposition 2.4.11, we have.

Corollary 2.4.12. *Let $a \in Gi(\mathcal{A})$ and $\bar{a} = a + \delta a \in \mathcal{A}$ with $\|a^+\|\|\delta a\| < \dfrac{1}{1 + \|1 - aa^+\|^2}$. Then $\bar{a}\mathcal{A} \cap (1 - aa^+)\mathcal{A} = \{0\}$ iff $\hat{\delta}(\bar{a}\mathcal{A}, a\mathcal{A}) < \dfrac{1}{\|1 - aa^+\|}$.*

Definition 2.4.13. Let p, q be idempotent in $\mathcal{A}$. p and q are equivalent in $\mathcal{A}$ (in symbol, $p \sim q$) if there are x, $y \in \mathcal{A}$ such that $p = xy$, $q = yx$. p and q are said to similar (in symbol, $p \overset{s}{\sim} q$), if there is $x \in GL(\tilde{\mathcal{A}})$ such that $p = x^{-1}qx$.

Clearly, if $p \overset{s}{\sim} q$, then $p \sim q$.

Definition 2.4.14. Let $\mathcal{A}$ be a unital Banach algebra. $\mathcal{A}$ is called to be finite if for any idempotent element p in $\mathcal{A}$ with $p \sim 1$, we have $p = 1$.

If $\mathcal{A}$ is nonunital and $\tilde{\mathcal{A}}$ is finite, we will say $\mathcal{A}$ is finite.

A nonzero idempotent element p in $\mathcal{A}$ is called to be finite, if $p\mathcal{A}p = \{pxp|\, x \in \mathcal{A}\}$ is finite.

Lemma 2.4.15. *Let p, q, r be nonzero idempotent elements in $\mathcal{A}$.*

(1) *If $p \sim q$ and $q \sim r$, then $p \sim r$.*
(2) *If p is finite and $pq = qp = q$, $p \sim q$, then $p = q$.*
(3) *If p is finite and $p \sim q$, then q is finite.*

Proof. (1) Let x_1, y_1, x_2, $y_2 \in \mathcal{A}$ such that $p = x_1y_1$, $q = y_1x_1$, $q = x_2y_2$, $r = y_2x_2$. Put $z_1 = x_1x_2$, $z_2 = y_2y_1$. Then

$$z_1z_2 = x_1x_2y_2y_1 = x_1qy_1 = x_1y_1x_1y_1 = p^2 = p,$$
$$z_2z_1 = y_2y_1x_1x_2 = y_2qx_2 = y_2x_2y_2x_2 = r^2 = r.$$

(2) Let $x,\ y \in \mathcal{A}$ such that $p = xy$ and $q = yx$. Put $x_1 = pxq,\ y_1 = qyp$. Since $pq = qp = q$, it follows $px_1p = x_1,\ y_1p = y_1$. Thus, $x_1,\ y_1 \in p\mathcal{A}p$ and $p = x_1y_1,\ q = y_1x_1$ in $p\mathcal{A}p$. Noting that p is the unit of $p\mathcal{A}p$ and $p\mathcal{A}p$ is finite, we have $p = q$.

(3) Let e be idempotent in $q\mathcal{A}q$ with $q \sim e$. Let $a,\ b \in \mathcal{A}$ such that

$$p = ab,\ q = ba,\ pa = aq = a,\ qb = bp = b. \tag{2.4.2}$$

Put $f = aeb$. Then $fp = pf = f$ by (2.4.2). From $eq = qe = e$, we get that $f^2 = f$ and $ebae = e$. Since $f \sim e \sim q \sim p$ by (1) and p is finite, we have $f = p$ by (2) and hence $q = bpa = b(aeb)a = qeq = e$ by (2.4.2). □

Theorem 2.4.16. *Let $\mathcal{A}$ be a unital Banach algebra and let $a,\ \bar{a} = a+\delta a \in Gi(\mathcal{A})$ with $1 + a^+\delta a \in GL(\mathcal{A})$. If $1 - a^+a$ (or $1 - aa^+$) is finite and $1 - a^+a \sim 1 - \bar{a}^+\bar{a}$ (or $1 - aa^+ \sim 1 - \bar{a}\bar{a}^+$), then $\bar{a}$ is a stable perturbation of a.*

Proof. Put $p_a = (1 + a^+\delta a)^{-1}(1 - a^+a),\ q_a = (1 - aa^+)(1 + \delta aa^+)^{-1}$. Since,

$$p_a = (1+a^+\delta a)^{-1}(1-a^+a)(1+a^+\delta a),\ q_a = (1+\delta aa^+)(1-aa^+)(1+\delta aa^+)^{-1},$$

it follows that $p_a,\ q_a$ are idempotent and $p_a \overset{s}{\sim} 1 - a^+a, q_a \overset{s}{\sim} 1 - aa^+$.

Let $x \in K_r(\bar{a}) = R_r(1 - \bar{a}^+\bar{a})$ and $y \in K_l(\bar{a}) = R_l(1 - \bar{a}\bar{a}^+)$. Then $(a + \delta a)x = y(a + \delta a) = 0$ and hence

$$(1 + a^+\delta a + a^+a - 1)x = 0 = y(1 + \delta aa^+ + aa^+ - 1).$$

So $p_ax = x,\ yq_a = y$, that is,

$$p_a(1 - \bar{a}^+\bar{a}) = 1 - \bar{a}^+\bar{a} = \bar{p},\ (1 - \bar{a}\bar{a}^+)q_a = 1 - \bar{a}\bar{a}^+ = \bar{q}. \tag{2.4.3}$$

Set $w_1 = 1+\bar{p}p_a(1-\bar{p}),\ w_2 = 1+(1-\bar{q})q_a$. Then $w_1, w_2 \in GL(\mathcal{A})$ with

$$w_1^{-1} = 1 - \bar{p}p_a(1 - \bar{p}),\ w_2^{-1} = 1 - (1 - \bar{q})q_a\bar{q}.$$

Set $\bar{p_1} = w_1p_aw_1^{-1},\ \bar{q_1} = w_2^{-1}q_aw_2$. Using (2.4.3), we obtain

$$\bar{p_1} = p_a - \bar{p}p_a(1 - \bar{p}),\quad \bar{p_1}\bar{p} = \bar{p}\bar{p_1} = \bar{p}, \tag{2.4.4}$$

$$\bar{q_1} = q_a - (1 - \bar{q})q_a\bar{q},\quad \bar{q_1}\bar{q} = \bar{q}\bar{q_1} = \bar{q}. \tag{2.4.5}$$

Since $1 - a^+a \sim \bar{p}$ (or $1 - aa^+ \sim \bar{q}$),

$$\bar{p_1} \overset{s}{\sim} p_a \overset{s}{\sim} 1 - a^+a,\quad \bar{q_1} \overset{s}{\sim} q_a \overset{s}{\sim} 1 - aa^+$$

and $1 - a^+a$ (or $1 - aa^+$) is finite, it follows from (2.4.4) (or (2.4.5)) and Lemma 2.4.15 that $\bar{p_1} = \bar{p}$ (or $\bar{q_1} = \bar{q}$), that is, $p_a = \bar{p}p_a$ (or $q_a = q_a\bar{q}$) and hence $\bar{a}(1 + a^+\delta a)^{-1}(1 - a^+a) = 0$ (or $(1 - aa^+)(1 + \delta aa^+)^{-1}\bar{a} = 0$). Therefore, $R_r(\bar{a}) \cap R_r(1 - aa^+) = \{0\}$ by Theorem 2.4.7. □

Finally, similar to Proposition 2.2.11, we have

Proposition 2.4.17. *Let $\mathcal{A}$ be a unital Banach algebra. Suppose that $a \in Gi(\mathcal{A})$ and $\bar{a} = a + \delta a \in \mathcal{A}$. Let $b \in \mathcal{A}$ such that $aba = a$. Suppose that $\|a^+\|\|\delta a\| < (1 + \|1 - ab\|\|1 - aa^+\|)^{-1}$ and $R_r(\bar{a}) \cap R_r(1 - aa^+) = \{0\}$. Then $R_r(\bar{a}) \cap K_r(b) = \{0\}$.*

Proof. By Theorem 2.4.7, $\bar{a}^+ = (1 + a^+\delta a)^{-1}a^+ \in \mathcal{A}$. Consequently,

$$\hat{\delta}(R_r(\bar{a}), R_r(a)) \leq \frac{\|1 - aa^+\|\|a^+\|\|\delta a\|}{1 - \|a^+\|\|\delta a\|} < \frac{1}{\|1 - ab\|} \tag{2.4.6}$$

by Proposition 2.4.11 (2). If $R_r(\bar{a}) \cap K_r(b) \neq \{0\}$, we can find $c \in R_r(\bar{a}) \cap K_r(b)$ with $\|c\| = 1$. Thus, for any $z \in \mathcal{A}$,

$$1 = \|c\| = \|(1 - ab)(c - az)\| \leq \|1 - ab\|\|c - az\|$$

and hence $\delta(R_r(\bar{a}), R_r(a)) \geq \|1 - ab\|^{-1}$, which contradicts (2.4.6). So $R_r(\bar{a}) \cap K_r(b) = \{0\}$. □

2.5 Some applications of stable perturbations

Definition 2.5.1. An operator $T \in B(X)$ is called Supher if $T \in Gib(X)$ and $\operatorname{Ker} T \subset \bigcap_{n=1}^{\infty} \operatorname{Ran}(T^n)$.

Lemma 2.5.2. *Let $T \in B(X)$ and $k, n \in \mathbb{N}$ with $1 \leq k \leq n - 2$. Suppose that $\operatorname{Ker} T^k \subset \operatorname{Ran}(T^{n-k})$. Then $\operatorname{Ker} T^{k+1} \subset \operatorname{Ran}(T^{n-k-1})$.*

Proof. Let $x \in \operatorname{Ker} T^{k+1}$. Then $Tx \in \operatorname{Ker} T^k \subset \operatorname{Ran}(T^{n-k})$. So $Tx = T^{n-k}y$ for some $y \in X$ Then

$$x - T^{n-k-1}y \in \operatorname{Ker} T \subset \operatorname{Ker} T^k \subset \operatorname{Ran}(T^{n-k}) \subset \operatorname{Ran}(T^{n-k-1}).$$

Therefore, $x \in \operatorname{Ran}(T^{n-k-1})$. □

Theorem 2.5.3. *Let $T \in B(X)$. Then the following conditions are equivalent.*

(1) *T is a supher operator.*
(2) *$T \in Gib(X)$ and $\bigvee_{k=1}^{\infty} \operatorname{Ker} T^k \subset \operatorname{Ran}(T)$.*
(3) *$T \in Gib(X)$ and $\lim_{\lambda \to 0} \gamma(T + \lambda I) = \gamma(T)$.*
(4) *$T \in Gib(X)$ and $\operatorname{Ran}(T + \lambda I) \cap \operatorname{Ker} T^+ = \{0\}$, $\forall \lambda \in \mathbb{C}$ with $|\lambda| < \delta$ for some $\delta > 0$.*

Proof. (1)⇒(2) By Definition 2.5.1, we need only to show that $\operatorname{Ker} T \subset \bigcap_{n=1}^{\infty} \operatorname{Ran}(T^n)$ implies that $\bigvee_{k=1}^{\infty} \operatorname{Ker} T^k \subset \operatorname{Ran}(T)$. Since $\operatorname{Ker} T \subset \operatorname{Ran}(T^n)$, $\forall\, n \geq 1$, by using Lemma 2.5.2 repeatedly, we get that $\operatorname{Ker} T^k \subset \operatorname{Ran}(T)$, $\forall\, k \geq 1$. So $\bigvee_{k=1}^{\infty} \operatorname{Ker} T^k \subset \operatorname{Ran}(T)$ for $\operatorname{Ran}(T)$ is closed in X.

(2)⇒(3) If $\operatorname{Ker} T = \{0\}$, $\delta(\operatorname{Ker} T, \operatorname{Ker}(T+\lambda I)) = 0$ and $\gamma(T+\lambda I) \geq \gamma(T) - |\lambda|$ by Lemma 1.3.6 since

$$\|(T+\lambda I)x\| \geq \|Tx\| - |\lambda|\|x\|, \quad \forall\, x \in X.$$

From $\operatorname{Ker} T = \{0\}$ and $TT^+T = T$, we get that $T^+T = I$ and consequently, $\lambda I + T = (\lambda T^+ + I)T$. Let λ be in $\{\lambda \in \mathbb{C} \,|\, |\lambda| < \|T^+\|^{-1}\}$. Then $\lambda T^+ + I$ is invertible in $B(X)$ and hence $\operatorname{Ker}(T+\lambda I) = \{0\}$. Thus,

$$\|x\|\gamma(T+\lambda I) \leq \|(T+\lambda I)x\| \leq (|\lambda|\|T^+\| + 1)\|Tx\|, \quad \forall\, x \in X.$$

and so that

$$-|\lambda| \leq \gamma(T+\lambda I) - \gamma(T) \leq |\lambda|\|T^+\|\gamma(T), \quad |\lambda| < \|T^+\|^{-1}.$$

Therefore, $\lim\limits_{\lambda \to 0} \gamma(T+\lambda I) = \gamma(T)$.

Now suppose that $\operatorname{Ker} T \neq \{0\}$. Given $x \in \operatorname{Ker} T$ with $\|x\| = 1$ and set $x_0 = x$. Then there is $x_0' \in X$ such that $x_0 = Tx_0'$ for $\operatorname{Ker} T^k \subset \operatorname{Ran}(T)$, $\forall\, k \geq 1$. Hence

$$1 = \|x_0\| = \|Tx_0'\| \geq \gamma(T)\operatorname{dist}(x_0', \operatorname{Ker} T).$$

Choose $x_0'' \in \operatorname{Ker} T$ such that $\|x_0' - x_0''\| < \operatorname{dist}(x_0', \operatorname{Ker} T) + \gamma^{-1}(T)$ and put $x_1 = x_0' - x_0''$. Then $Tx_1 = Tx_0' = x_0$ and $\|x_1\| < 2\gamma^{-1}(T)$. Since $T^2 x_1 = 0$, it follows that $x_1 = Tx_1'$ for some $x_1' \in X$. From

$$\operatorname{dist}(x_1', \operatorname{Ker} T) \leq \frac{1}{\gamma(T)}\|Tx_1'\| = \frac{1}{\gamma(T)}\|x_1\|,$$

we find $x_1'' \in \operatorname{Ker} T$ such that

$$\|x_1' - x_1''\| < \operatorname{dist}(x_1', \operatorname{Ker} T) + \frac{\|x_1\|}{\gamma(T)} < \frac{2}{\gamma(T)}\|x_1\|.$$

Set $x_2 = x_1' - x_1''$, Then $Tx_2 = x_1$ and $\|x_2\| < 2\gamma^{-1}(T))\|x_1\|$. In this way, we can find a sequence $\{x_n\} \subset X$ such that $Tx_{n+1} = x_n$ and $\|x_{n+1}\| < 2\gamma^{-1}(T)\|x_n\|$, $n \geq 0$. So $\|x_n\| < (2\gamma^{-1}(T))^n$, $n \geq 1$.

Define a holomorphic function $f\colon D(0, 2^{-1}\gamma(T)) \to X$ by $f(\lambda) = \sum_{n=0}^{\infty} x_n(-\lambda)^n$. Then $(T+\lambda I)f(\lambda) = 0$, $\forall\, \lambda \in D(0, 2^{-1}\gamma(T))$. Therefore, for any $\lambda \in D(0, 2^{-1}\gamma(T))$,

$$\operatorname{dist}(x, \operatorname{Ker}(T+\lambda I)) \leq \|x_0 - f(\lambda)\| \leq \sum_{n=1}^{\infty} \|x_n\||\lambda|^n < \frac{2|\lambda|}{\gamma(T) - 2|\lambda|}$$

and hence $\lim_{\lambda\to 0} \delta(\operatorname{Ker} T, \operatorname{Ker}(T+\lambda I)) = 0$. Finally, by Lemma 1.3.5, $\delta(\operatorname{Ker}(T+\lambda I), \operatorname{Ker} T) \le \gamma(T)^{-1}|\lambda|$ and by Lemma 1.3.6,

$$\gamma(T+\lambda I) \ge \gamma(T)(1 - 2\delta(\operatorname{Ker} T, \operatorname{Ker}(T+\lambda I))) - |\lambda|$$
$$\gamma(T) \ge \gamma(T+\lambda I)(1 - 2\delta(\operatorname{Ker}(T+\lambda I), \operatorname{Ker}(T))) - |\lambda|$$

(for $\|(T+\lambda I)x\| \ge \|Tx\| - \lambda\|x\|$, $\|Tx\| \ge \|(T+\lambda I)x\| - \lambda\|x\|$, $\forall\, x \in X$), we get that $\lim_{\lambda\to 0} \gamma(T+\lambda I) = \gamma(T)$.

(3)$\Rightarrow$(4) $\lim_{\lambda\to 0} \gamma(T+\lambda) = \gamma(T) > 0$ means that there is a $\delta_0 > 0$ such that $\gamma(T+\lambda I) > 2^{-1}\gamma(T)$, $\forall\, \lambda \in D(0,\delta_0)$. Then

$$\delta(\operatorname{Ran}(T+\lambda I), \operatorname{Ran}(T)) \le \gamma^{-1}(T+\lambda I)|\lambda| < \frac{2}{\gamma(T)}|\lambda|, \ \forall\, \lambda \in D(0,\delta_0)$$

by Lemma 1.3.5. Choose $\delta > 0$ such that $2\gamma^{-1}(T)|\lambda| < \|I - TT^+\|^{-1}$, $\forall\, \lambda \in D(0,\delta)$. Then $\operatorname{Ran}(T+\lambda I) \cap \operatorname{Ker} T^+ = \{0\}$, $\forall\, \lambda \in D(0,\delta)$ by Proposition 2.2.10.

(4)$\Rightarrow$(1) Set $\gamma = \min\{\delta, \|T^+\|^{-1}\}$. Then $I + \lambda T^+$ is invertible in $B(X)$ and $(I+\lambda T^+)^{-1}\operatorname{Ker} T = \operatorname{Ker}(T+\lambda I)$, $\forall\, \lambda \in D(0,\gamma)$ by Corollary 2.2.7, So for any $x \in \operatorname{Ker} T$, $f(\lambda) = (I+\lambda T^+)^{-1}x \in \operatorname{Ker}(T+\lambda I)$, that is, $Tf(\lambda) = -\lambda f(\lambda), \forall\, \lambda \in D(0,\gamma)$. Write $f(\lambda)$ as

$$f(\lambda) = \sum_{n=0}^{\infty} (-1)^n x_n \lambda^n, \quad |\lambda| < \gamma,$$

where $x_0 = x$, $x_n = (T^+)^n x$, $n \ge 1$. It follows that $Tx_0 = 0$, $Tx_{n+1} = x_n$, $n \ge 0$ and hence $x = x_0 = T^n x_n$, $n \ge 1$, i.e., $x \in \bigcap_{n=1}^{\infty} \operatorname{Ran}(T^n)$. □

Furthermore, we can prove the following by using the method in the proof of the implication (4)$\Rightarrow$(1) of Theorem 2.5.3.

Proposition 2.5.4. *Let $T \in B(X)$ be a Supher operator. Suppose that $\operatorname{Ran}(T+\lambda I) \cap \operatorname{Ker} T^+ = \{0\}$ for any $\lambda \in D(0, \|T^+\|^{-1})$. Then for any $\lambda \in D(0,\|T^+\|^{-1}), T+\lambda I$ is Supher.*

Proof. Given $\lambda \in D(0, \|T^+\|^{-1})$. Then there is $\rho > 0$ such that $D(\lambda,\rho) \subset D(0,\|T^+\|^{-1})$. Noting that for any $\mu \in D(\lambda,\rho)$,

$$(I+\lambda T^+)^{-1}\operatorname{Ker} T = \operatorname{Ker}(T+\lambda I), \ (I+\mu T^+)^{-1}\operatorname{Ker} T = \operatorname{Ker}(T+\mu I),$$

by Corollary 2.2.7, we get that for any $y \in \operatorname{Ker}(T+\lambda I)$, $\mu \in D(\lambda,\rho)$,

$$g_y(\mu) = (I+\mu T^+)^{-1}(I+\lambda T^+)y \in \operatorname{Ker}(T+\mu I),$$

i.e., $(T+\lambda I)g_y(\mu) = (\lambda-\mu)g_y(\mu)$, $\forall \mu \in D(\lambda,\rho)$. Express $g_y(\mu)$ as

$$g_y(\mu) = \sum_{n=0}^{\infty} y_n(\lambda-\mu)^n, \quad \forall\, \mu \in D(\lambda,\rho),$$

where $\{y_n\} \subset X$ and $g_y(\lambda) = y_0 = y$. Then we have $(T+\lambda I)y_0 = 0$, $(T+\lambda I)y_{n+1} = y_n$, $n \in \mathbb{N}$. Hence $y \in \bigcap_{n=1}^{\infty} \mathrm{Ran}\,((T+\lambda I)^n)$, that is, $\mathrm{Ker}\,(T+\lambda I) \subset \bigcap_{n=1}^{\infty} \mathrm{Ran}\,((T+\lambda I)^n)$. □

We have known that the map $Gi\colon Gib(X,Y) \to Gib(X,Y)$ given by $Gi(A) = A^+$ is a set-valued mapping. In terms of the notation "stable perturbation", we can characterize the lower semi–continuity of Gi. First we give the definition of the lower semi–continuity of set–valued mapping, which comes from Definition 2.1.2 of [Aubin and Cellina (1984)].

Definition 2.5.5. Let D be a subset of X, a set–valued map $F\colon D \to Y$ is called lower semi–continuous at $x \in D$, if for any $y \in F(x)$ and any sequence $\{x_n\} \subset D$ converging to x, there exists a sequence $\{y_n\} \subset F(x_n)$ converging to y (with respect to the norm of Y)

Then we give the equivalent conditions of the lower semi–continuity of Gi as follows.

Proposition 2.5.6. *Let $A_0 \in Gib(X,Y)$. Then the following statements are equivalent:*

(1) *$Gi(A) = A^+$ is lower semi–continuous at A_0.*
(2) *For any generalized inverse A_0^+ of A_0, there is a $\delta > 0$ such that* $\mathrm{Ran}\,(A) \cap \mathrm{Ker}\, A_0^+ = \{0\}$ *for any $A \in Gib(X,Y)$ with $\|A - A_0\| < \delta$.*
(3) *There is a generalized inverse A_0^+ of A_0 and a $\delta > 0$ such that* $\mathrm{Ran}\,(A) \cap \mathrm{Ker}\, A_0^+ = \{0\}$ *for any $A \in Gib(X,Y)$ with $\|A - A_0\| < \delta$.*

Proof. The implication (2)⇒(3) is obvious.

(3)⇒(1) Let $\{A_n\}$ be a sequence in $Gib(X,Y)$ such that $\|A_n - A_0\| \to 0$ $(n \to \infty)$. Let $B \in Gi(A_0)$, i.e., $ABA = A, BAB = B$. Choose $N \in \mathbb{N}$ such that $\|A_n - A_0\| < \min\{\delta, \|B\|^{-1}, \|A_0^+\|(1+\|I_Y - AB\|)^{-1}\}, \forall\, n > N$. Thus, $\mathrm{Ran}\,(A_n) \cap \mathrm{Ker}\, A_0^+ = \{0\}$ by Condition (3) and hence $\mathrm{Ran}\,(A_n) \cap \mathrm{Ker}\, B = \{0\}, \forall\, n > N$ by Proposition 2.2.11. Thus A_n has the generalized inverse $A_n^+ = (I_X + B(A_n - A_0))^{-1}B \in Gi(A_n)$, $n > N$, by Theorem 2.2.6. Clearly, $\|A_n^+ - B\| \to 0$ as $n \to \infty$ by Corollary 1.4.8.

(1)⇒(2) Suppose that there is a generalized inverse A_0^+ and a sequence $\{A_n\} \subset Gib(X,Y)$ such that $\operatorname{Ran}(A_n) \cap \operatorname{Ker} A_0^+ \neq \{0\}$ and $\|A_n - A_0\| \to 0$ $(n \to \infty)$. Since Gi is lower semi-continuous at A_0, we can find a sequence $\{C_n\} \subset Gi(A_n)$ such that $C_n \xrightarrow{\|\cdot\|} A_0^+$ $(n \to \infty)$. Since

$$\delta(\operatorname{Ran}(A_n), \operatorname{Ran}(A)) \leq \gamma(A_n)^{-1}\|A_n - A\| \leq \|C_n\|\|A_n - A_0\|, \ \forall\, n \geq 1,$$

by Lemma 1.3.5 and Theorem 2.1.4, we can pick $N \in \mathbb{N}$ such that $\delta(\operatorname{Ran}(A_n), \operatorname{Ran}(A)) < \|I_Y - A_0 A_0^+\|^{-1}$, $\forall\, n > N$. So $\operatorname{Ran}(A_n) \cap \operatorname{Ker} A_0^+ = \{0\}$, $\forall\, n > N$ by Proposition 2.2.10. But it contradicts the assumption that $\operatorname{Ran}(A_n) \cap \operatorname{Ker} A_0 \neq \{0\}$, $n \geq 1$. □

Similar to Proposition 2.5.6, for the set-valued map $Gi\colon Gi(\mathcal{A}) \to Gi(\mathcal{A})$ given by $Gi(a) = a^+$ for the united Banach algebra $\mathcal{A}$, we have

Proposition 2.5.7. *let $\mathcal{A}$ be a unital Banach algebra and $a_0 \in Gi(A)$. Then the following statement are equivalent:*

(1) *$Gi(a) = a^+$ is lower semi-continuous at a_0.*
(2) *For any generalized inverse a_0^+ of a_0, there is a $\delta > 0$ such that $R_r(a) \cap K_r(a_0) = \{0\}, \forall\, a \in Gi(A)$ with $\|a - a_0\| < \delta$.*
(3) *There is a generalized inverse a_0^+ of a_0 and a $\delta > 0$ such that $R_r(a) \cap K_r(a_0) = \{0\}, \forall\, a \in Gi(A)$ with $\|a - a_0\| < \delta$.*

Theorem 2.4.4 gives the existence of the generalized inverse of an element in a unital Banach algebra. But the verification of the conditions given in Theorem 2.4.4 is quite not easy. So we want to search for some easy checked conditions that make the generalized inverse of a given element exist in a Banach algebra.

Let $\mathcal{A}$ be a unital commutative Banach algebra and let $M_{nm}(\mathcal{A})$ be the set of all $n \times m$ matrices over $\mathcal{A}$ with the norm $\|A\| = \sum\limits_{i=1}^{n}\sum\limits_{j=1}^{m}\|a_{ij}\|$ for $A = (a_{ij})_{n\times m} \in M_{nm}(A)$. Let $M_{\mathcal{A}}$ be the set of all non-zero characters on $\mathcal{A}$ and let $A = (a_{ij})_{n\times m}$. Put $\hat{A}(\phi) = (\hat{a}_{ij}(\phi))_{n\times m}$, $\forall\, \phi \in M_{\mathcal{A}}$.

Theorem 2.5.8. *Let $\mathcal{A}$ be a unital commutative Banach algebra. Suppose that $\mathcal{A}$ is semi-simple and $M_{\mathcal{A}}$ is connected. Let $A = (a_{ij})_{n\times m} \in M_{nm}(\mathcal{A})$. Then A is generalized invertible (i.e., there is $B \in M_{nm}(\mathcal{A})$) such that $ABA = A$ iff* $\operatorname{rank}(\hat{A}(\phi))$ *is a constant for every $\phi \in M_{\mathcal{A}}$.*

Proof. "Only if" part. Suppose that there is a $B \in M_{mn}(\mathcal{A})$ such that $ABA = A$. Then $\hat{A}(\phi)\hat{B}(\phi)\hat{A}(\phi) = \hat{A}(\phi)$, $\forall\, \phi \in M_{\mathcal{A}}$. Since $M_{\mathcal{A}}$ is compact and $\phi \mapsto \hat{B}(\phi)$ is continuous, we have $c = \sup_{\phi \in M_{\mathcal{A}}} \|\hat{B}(\phi)\| < +\infty$. Choose $\phi_0 \in M_{\mathcal{A}}$ such that $\operatorname{rank}(\hat{A}(\phi_0)) = k \in \mathbb{N}$ and put

$$Y(k) = \{\phi \in M_{\mathcal{A}} | \operatorname{rank}(\hat{A}(\phi)) = k\}.$$

Let $\phi_1 \in Y(k)$. Since

$$\begin{aligned}\delta(\operatorname{Ran}(\hat{A}(\phi)), \operatorname{Ran}(\hat{A}(\phi_1))) &\leq \gamma^{-1}(\hat{A}(\phi))\|\hat{A}(\phi) - \hat{A}(\phi_1)\| \\ &\leq \|\hat{B}(\phi)\|\|\hat{A}(\phi) - \hat{A}(\phi_1)\| \\ &\leq c\|\hat{A}(\phi) - \hat{A}(\phi_1)\|\end{aligned}$$

and $\phi \mapsto \hat{A}(\phi)$ is continuous on $M_{\mathcal{A}}$, it follows that

$$\lim_{\phi \to \phi_1} \delta(\operatorname{Ran}(\hat{A}(\phi)), \operatorname{Ran}(\hat{A}(\phi_1))) = 0, \ \lim_{\phi \to \phi_1} \|\hat{B}(\phi_1)\|\|\hat{A}(\phi) - \hat{A}(\phi_1)\| = 0.$$

So there is a neighborhood $U(\phi_1)$ of ϕ_1 such that

$$\|\hat{B}(\phi_1)\|\|\hat{A}(\phi) - \hat{A}(\phi_1)\| < 1, \ \operatorname{Ran}(\hat{A}(\phi)) \cap \operatorname{Ker}\hat{B}(\phi_1) = \{0\}, \ \forall\, \phi \in U(\phi_1)$$

by Proposition 2.2.10 and consequently, $\operatorname{rank}(\hat{A}(\phi)) = \operatorname{rank}(\hat{A}(\phi_1)) = k$, $\forall\, \phi \in U(\varphi_1)$ by Theorem 2.2.3, which means that $Y(k)$ is open in $M_{\mathcal{A}}$.

To prove $Y(k)$ is also closed, let $\{\psi_n\} \subset Y(k)$ and $\psi \in M_{\mathcal{A}}$ such that $\psi_n \xrightarrow{w^*} \psi$. Then from $\lim_{n\to\infty} \|B(\psi)\|\|\hat{A}(\psi_n) - \hat{A}(\psi)\| = 0$ and

$$\lim_{n\to\infty} \delta(\operatorname{Ran}(\hat{A}(\psi_n)), \operatorname{Ran}(\hat{A}(\psi))) \leq \lim_{n\to\infty} \|B(\psi)\|\|\hat{A}(\psi_n) - \hat{A}(\psi)\| = 0,$$

we get that

$$\operatorname{Ran}(\hat{A}(\psi_n)) \cap \operatorname{Ker}(\hat{B}(\psi)) = \{0\} \text{ and } \|\hat{B}(\psi)\|\|\hat{A}(\psi_n) - \hat{A}(\psi)\| < 1$$

for n large enough. Thus, $\operatorname{rank}(\hat{A}(\psi)) = \operatorname{rank}(\hat{A}(\psi_n)) = k$ by Theorem 2.2.3 and hence $Y(k)$ is closed.

Since $M_{\mathcal{A}}$ is connected and $Y(k) \neq \emptyset$, it follows that $Y(k) = M_{\mathcal{A}}$, that is, $\operatorname{rank}(\hat{A}(\phi)) = k$, $\forall\, \phi \in M_{\mathcal{A}}$.

"If" part. Assume that $\operatorname{rank}(\hat{A}(\phi)) = k \geq 1$, $\forall\, \phi \in M_{\mathcal{A}}$. Then there exist $\binom{n}{k} \times \binom{m}{k}$ $k \times k$ sub–matrix $A_{\alpha,\beta}$ of A determined by α rows and β columns, where

$$\begin{aligned}\alpha &= \{i_1 < i_2 < \cdots < i_k\} \subset \{1, \cdots, n\}, \\ \beta &= \{j_1 < j_2 < \cdots < j_k\} \subset \{1, \cdots, m\}.\end{aligned}$$

Let $|A_{\alpha,\beta}|$ denote the determinant of $A_{\alpha,\beta}$.

Consider the closed ideal I generated by $\{A_{\alpha,\beta}\}$ in $\mathcal{A}$. If $I \neq \mathcal{A}$, there is a maximal ideal $\operatorname{Ker}\phi_0$ ($\phi_0 \in M_{\mathcal{A}}$) such that $I \subset \operatorname{Ker}\phi_0$. Thus,

$|\hat{A}_{\alpha,\beta}(\phi_0)| = 0$ for all $k \times k$ minors $|A_{\alpha,\beta}|$, which contradicts the condition that $\text{rank}\,(\hat{A}(\phi)) = k \geq 1$, $\forall\,\phi \in M_{\mathcal{A}}$. Therefore, $1 \in I$, that is, there are $g_{\alpha,\beta} \in \mathcal{A}$ such that $\| \sum\limits_{\alpha,\beta} g_{\alpha,\beta}|A_{\alpha,\beta}| - 1\| < 1$ and consequently, $g = \sum_{\alpha,\beta} g_{\alpha,\beta}|A_{\alpha,\beta}| \in \mathcal{A}$ is invertible. Put $c_{\alpha,\beta} = g^{-1}g_{\alpha,\beta} \in \mathcal{A}$. Then

$$\sum_{\alpha,\beta} c_{\alpha,\beta}|A_{\alpha,\beta}| = 1. \tag{2.5.1}$$

For any $\alpha = \{i_1 < \cdots < i_k\}, \beta = \{j_1 < \cdots < j_k\}$, let $\text{adj}(A_{\alpha,\beta})$ denote the adjoint matrix of $A_{\alpha,\beta}$. Then

$$A_{\alpha,\beta}(\text{adj}(A_{\alpha,\beta})) = (\text{adj}(A_{\alpha,\beta}))A_{\alpha,\beta} = |A_{\alpha,\beta}|I_k. \tag{2.5.2}$$

Let $P_{\alpha,\beta}$ be the $m \times n$ matrix, in which $\text{adj}(A_{\alpha,\beta})$ is distributed in the intersection of β rows and α columns of $P_{\alpha,\beta}$ and other entries of $P_{\alpha,\beta}$ are zeros. By permutations of rows and columns, we may assume, without loss of the generality, that $A_{\alpha,\beta} = A_{11}$ is at the upper left corner of $A = \begin{bmatrix} A_{11} & A_{12} \\ A_{21} & A_{22} \end{bmatrix}$. Thus, by (2.5.1),

$$\begin{bmatrix} A_{11} & A_{12} \\ A_{21} & A_{22} \end{bmatrix} \begin{bmatrix} \text{adj}(A_{11}) & 0 \\ 0 & 0 \end{bmatrix} \begin{bmatrix} A_{11} & A_{12} \\ A_{21} & A_{22} \end{bmatrix} = \begin{bmatrix} |A_{11}|A_{11} & |A_{11}|A_{12} \\ |A_{11}|A_{21} & C \end{bmatrix}, \tag{2.5.3}$$

where $C = A_{21}(\text{adj}(A_{11}))A_{12}$. Let a_{ij} be any element in A_{22}. Since $\text{rank}\,(\hat{A}(\phi)) = k$, $\forall\,\phi \in M_{\mathcal{A}}$ and A_{11} is a $k \times k$ sub-matrix of A, we have

$$\begin{aligned} 0 = \det &\left[\begin{array}{c|c} \hat{A}_{11}(\phi) & \begin{matrix} \hat{a}_{1j}(\phi) \\ \vdots \\ \hat{a}_{kj}(\phi) \end{matrix} \\ \hline \hat{a}_{i1}(\phi) \cdots \hat{a}_{ik}(\phi) & \hat{a}_{ij}(\phi) \end{array}\right] \\ &= \hat{a}_{ij}(\phi)|\hat{A}_{11}(\phi)| - \sum_{l=1}^{k}\sum_{\gamma=1}^{k} \hat{a}_{il}(\phi)\hat{A}_{l\gamma}(\phi)\hat{a}_{ij}(\phi), \end{aligned}$$

$\forall\,\phi \in M_{\mathcal{A}}$, where $A_{l\gamma}$ is the cofactor of $a_{\gamma l}$ in $|A_{11}|$. The above identity means that $|A_{11}|A_{22} = A_{21}(\text{adj}(A_{11}))A_{12}$ and so $AP_{\alpha,\beta}A = |A_{\alpha,\beta}|A$ by (2.5.3). Put $B = \sum\limits_{\alpha,\beta} c_{\alpha,\beta}P_{\alpha,\beta}$, where $\{c_{\alpha,\beta}\}$ is given by (2.5.1). Then

$$A = A(\sum_{\alpha,\beta} c_{\alpha,\beta}|A_{\alpha,\beta|}) = \sum_{\alpha,\beta} c_{\alpha,\beta}(AP_{\alpha,\beta}A) = ABA.$$

$\square$

2.6 Notes

Section 2.1 is based on [Nashed (1976b)] and [Chen *et al.* (2004)]. Theorem 2.1.4 is a generalization of Lemma 2.1 of [Chen and Xue (1997)].

Section 2.2 is based mainly on [Chen and Xue (1997)]. Theorem 2.2.3 is Corollary 3.2 of [Chen and Xue (1997)]. Lemma 2.2.4 is a generalization of Proposition 3.1 in [Chen and Xue (1997)]. Theorem 2.2.6 is a generalization of Proposition 2.2 of [Xue (2007c)]. Corollary 2.2.7 is a generalization of Theorem 2.2 in [Huang and Zhai (2011)]. Proposition 2.2.11 is a generalization of Theorem 2.2 in [Huang and Ma (2005)]. Theorem 2.2.12 generalizes Theorem 3.2 of [Chen and Xue (1997)].

Section 2.3 is based on §4 of [Chen and Xue (1997)]. Proposition 2.3.9 is Proposition 4.2 of [Chen and Xue (1997)]. But its proof comes from [Ding and Huang (1999)].

Section 2.4 is based on [Xue (2007c)]. Proposition 2.4.5 (1) is Theorem 1.1 of [Harte (1987)].

Section 2.5. The equivalent condition (1), (2) and (3) of Theorem 2.5.3 come from Theorem 20 (P124) and Theorem 9 (P133) respectively in [Muller (2007)]. The equivalent condition (4) is new. Proposition 2.5.6 is Theorem 3.1 of [Huang and Ma (2005)]. Theorem 2.5.8 is due to Huang (cf. [Huang (1992)]). But its proof is not same as the original one.

Chapter 3

The Moore–Penrose Inverses And Its Stable Perturbation

Tseng first introduced generalized inverses of unbounded operators on Hilbert spaces in his paper [Tseng (1949)]. Then a very useful theory of generalized inverses of operators on Hilbert spaces was established by means of Arghiriade's paper [Arghiriade (1968)], Hestenes' paper [Hestenes (1961)] and Petryshyn's papers [Petryshyn (1967)]. But there are two important problems remaining unsolved at that time. One is the continuity of the generalized inverse and the other is the error–estimations of the perturbation of the generalized inverses. The first problem is solved by Izumino in 1983 (cf. [Izumino (1983)]). The second problem is solved by Chen, Wei and the author in 1996 (cf. [Chen *et al.* (1996)]). The results in this paper generalize corresponding results in [Stewart (1977)].

This chapter is organized as follows. We first introduce the Moore–Penrose inverse of a densely–defined closed operator on Hilbert spaces and discuss its basic properties in Section 3.1 and then investigate the reverse order law of the Moore–Penrose inverses in Section 2.2. In Section 3.3, we establish the stable perturbation theory of Moore–Penrose inverses and give the perturbation analysis for the least square solution. In Section 3.4, we discuss the expressions of Moore–Penrose inverses of some operators and block operator matrices. We will discuss the Moore–Penrose inverses in a C^*–algebra in Section 3.5.

3.1 The Moore–Penrose inverse

In this section, H, L, K are always Hilbert spaces.

Lemma 3.1.1. *Let* $A \in \mathcal{C}(H,K)$ *and* $y \in \mathrm{Ran}\,(A)$. *Set*

$$S(A,y) = \{x \in \mathfrak{D}(A) |\, Ax = y\}.$$

Then there is a unique $x_y \in S(A,y)$ *such that*

$$x_y \in (\operatorname{Ker} A)^{\perp} \text{ and } \|x_y\| = \inf\{\|x\| \,|\, x \in S(A,y)\}.$$

Proof. Let $m = \inf\{\|x\| \,|\, x \in S(A,y)\}$. Then there is $\{x_n\} \subset S(A,y)$ such that $\|x_n\| \to m$ as $n \to \infty$. Noting that $\frac{1}{2}(x_n + x_m) \in S(A,y)$ and

$$\Big\|\frac{1}{2}(x_n - x_m)\Big\|^2 + \Big\|\frac{1}{2}(x_n + x_m)\Big\|^2 = 2\Big(\Big\|\frac{x_n}{2}\Big\|^2 + \Big\|\frac{x_m}{2}\Big\|^2\Big), \ \forall\, m, n \geq 1,$$

we obtain that $\{x_n\}$ is a Cauchy sequence in H and hence there exists an $x_y \in H$ such that $\|x_n - x_y\| \to 0$ as $n \to \infty$. Thus, $m = \|x_y\|$. Since $\{x_n, y\} \in G(A)$, $n \geq 1$ and $G(A)$ is closed, $\{x_y, y\} \in G(A)$, i.e., $Ax_y = y$.

If there is another $x'_y \in \mathfrak{D}(A)$ such that $\|x'_y\| = m$ and $Ax'_y = y$. Then from $\|\frac{1}{2}(x_y + x'_y)\| \geq m$ and

$$\Big\|\frac{1}{2}(x_y - x'_y)\Big\|^2 + \Big\|\frac{1}{2}(x_y + x'_y)\Big\|^2 = 2\Big(\Big\|\frac{x_y}{2}\Big\|^2 + \Big\|\frac{x'_y}{2}\Big\|^2\Big) = m^2,$$

we get that $x_y = x'_y$.

Now choose $x_1 \in \operatorname{Ker} A$ and $x_2 \in (\operatorname{Ker} A)^{\perp}$ such that $x_y = x_1 + x_2$. So $Ax_y = Ax_2 = y$ and $\|x_2\| \geq \|x_y\|$. Since $\|x_y\|^2 = \|x_2\|^2 + \|x_1\|^2$, we have $x_1 = 0$, i.e., $x_y = x_2 \in (\operatorname{Ker} A)^{\perp}$. □

By Lemma 3.1.1, we can define a linear operator $B_0\colon \operatorname{Ran}(A) \to (\operatorname{Ker} A)^{\perp} \cap \mathfrak{D}(A)$ by $B_0 y = x_y$, $\forall\, y \in \operatorname{Ran}(A)$. In fact, for any $y_1, y_2 \in \operatorname{Ran}(A)$, $\lambda_1, \lambda_2 \in \mathbb{C}$,

$$Ax_{\lambda_1 y_1 + \lambda_2 y_2} = \lambda_1 y_1 + \lambda_2 y_2 = A(\lambda_1 x_{y_1} + \lambda_2 x_{y_2}).$$

Since $x_{\lambda_1 y_1 + \lambda_2 y_2} - (\lambda_1 x_{y_1} + \lambda_2 x_{y_2}) \in \operatorname{Ker} A$ and $x_{\lambda_1 y_1 + \lambda_2 y_2}, x_{y_1}, x_{y_2} \in (\operatorname{Ker} A)^{\perp} \cap \mathfrak{D}(A)$, it follows that $x_{\lambda_1 y_1 + \lambda_2 y_2} = \lambda_1 x_{y_1} + \lambda_2 x_{y_2}$, i.e.,

$$B_0(\lambda_1 y_1 + \lambda_2 y_2) = \lambda_1 B_0 y_1 + \lambda_2 B_0 y_2.$$

Note that for any $z \in (\operatorname{Ker} A)^{\perp} \cap \mathfrak{D}(A)$, the equation $Ax = Az = u$, has the unique solution $x_u \in \mathfrak{D}(A)$ by Lemma 3.1.1. Since $z - x_u \in \operatorname{Ker} A$ and $x_u, z \in (\operatorname{Ker} A)^{\perp} \cap \mathfrak{D}(A)$, we have $z = x_u$, which means that $\operatorname{Ran}(B_0) = (\operatorname{Ker} T)^{\perp} \cap \mathfrak{D}(A)$.

Now we extend B_0 to B on $\mathfrak{D}(B) = \operatorname{Ran}(A) + (\operatorname{Ran}(A))^{\perp}$ by

$$By = \begin{cases} B_0 y & y \in \operatorname{Ran}(A) \\ 0 & y \in (\operatorname{Ran}(A))^{\perp} \end{cases}. \tag{3.1.1}$$

Proposition 3.1.2. *The operator B given in* (3.1.1) satisfies following conditions:

(1) $\mathfrak{D}(B) = \text{Ran}\,(A) + \text{Ran}\,(A)^\perp$ is dense in K and $\text{Ran}\,(B) = (\text{Ker}\,A)^\perp \cap \mathfrak{D}(A)$.

(2) $B \in \mathcal{C}(K, H)$ and for any $x \in \mathfrak{D}(A)$, $y \in \mathfrak{D}(B)$,

$$ABA = A \text{ on } \mathfrak{D}(A) \quad BAB = B \text{ on } \mathfrak{D}(B)$$

$$ABy = P_{\overline{\text{Ran}\,(A)}}\, y \qquad BAx = P_{(\text{Ker}\,A)^\perp}\, x.$$

Moreover, if there is an another linear operator B' satisfying (1) and (2), then $B' = B$.

Proof. (1) Since $K = \overline{\text{Ran}\,(A)} \oplus (\text{Ran}\,(A))^\perp$, it follows that $\mathfrak{D}(B)$ is dense in K. $\text{Ran}\,(B) = (\text{Ker}\,A)^\perp \cap \mathfrak{D}(A)$ is obvious.

(2) By Lemma 3.1.1 and (3.1.1), we have

$$ABAx = AB_0(Ax) = Ax,\ \forall\, x \in \mathfrak{D}(A), \quad ABy = P_{\overline{\text{Ran}\,(A)}} y, \quad \forall\, y \in \mathfrak{D}(B).$$

Since $A_1 = A|_{(\text{Ker}\,A)^\perp} \colon (\text{Ker}\,A)^\perp \cap \mathfrak{D}(A) \to \text{Ran}\,(A)$ is injective, $B_1 = A_1^{-1} \colon \text{Ran}\,(A) \to (\text{Ker}\,A)^\perp \cap \mathfrak{D}(A)$ exists and $AB_1 y = y$, $\forall\, y \in \text{Ran}\,(A)$, $B_1 Ax = x$, $\forall\, x \in (\text{Ker}\,A)^\perp \cap \mathfrak{D}(A)$. Thus, $B_0 y = B_1 y$, $\forall\, y \in \text{Ran}\,(A)$ (for $B_0 y, B_1 y \in (\text{Ker}\,A)^\perp, B_0 y - B_1 y \in \text{Ker}\,A$) and consequently, $BAB = B$ on $\mathfrak{D}(B)$, $BAx = P_{(\text{Ker}\,A)^\perp} x$, $\forall\, x \in \mathfrak{D}(A)$.

Now if $B' \in \mathcal{C}(K, H)$ satisfies $\mathfrak{D}(B') = \mathfrak{D}(B)$, $\text{Ran}\,(B') = \text{Ran}\,(B)$ and

$$AB'A = A \text{ on } \mathfrak{D}(A) \quad B'AB' = B' \text{ on } \mathfrak{D}(B')$$

$$AB'y = P_{\overline{\text{Ran}\,(A)}}\, y \qquad B'Ax = P_{(\text{Ker}\,A)^\perp}\, x,$$

$\forall\, x \in \mathfrak{D}(A)$ and $y \in \mathfrak{D}(B')$, then

$$B'y = B'(AB')y = B'(AB)y = (B'A)By = BABy = By, \quad \forall\, y \in \mathfrak{D}(B').$$

To prove $B \in \mathcal{C}(K, H)$, let $\{\{y_n, x_n\}\} \subset G(B)$ with $\|y_n - y_0\| \to 0$, $\|x_n - x_0\| \to 0$ as $n \to \infty$. Since $By_n = x_n \in (\text{Ker}\,A)^\perp \cap \mathfrak{D}(A)$, $G(A)$ is closed and $ABy_n = P_{\overline{\text{Ran}\,(A)}} y_n$, we have $\{x_0, P_{\overline{\text{Ran}\,(A)}} y_0\} \in G(A)$ and $x_0 \in (\text{Ker}\,A)^\perp$. So, $x_0 \in (\text{Ker}\,A)^\perp \cap \mathfrak{D}(A)$ and $Ax_0 = P_{\overline{\text{Ran}\,(A)}} y_0$. Write y_0 as $y_0 = y_1 + y_2$, where $y_1 \in \overline{\text{Ran}\,(A)}$ and $y_2 \in \overline{\text{Ran}\,(A)}^\perp$. Then $y_1 = Ax_0$ and hence, $By_0 = B_0 y_1 = x_0$. □

Definition 3.1.3. Let $A \in \mathcal{C}(H, K)$. The operator B in Proposition 3.1.2 is called the Moore–Penrose inverse of A, denoted by $A^\dagger$.

Remark 3.1.4. (1) According to (3.1.1) and the proof of (2) in Proposition 3.1.2, $A^\dagger = B$ is bounded when $\text{Ran}\,(A)$ is closed.

(2) When $A \in B(H, K)$ with $\text{Ran}\,(A)$ closed, $A^\dagger$ is the unique solution of the operator equations:

$$AXA = A,\ XAX = X,\ (AX)^* = AX,\ (XA)^* = XA$$

by Proposition 3.1.2.

Proposition 3.1.5. *Let $A \in \mathcal{C}(H,K)$. Then the following conditions are equivalent:*

(1) $A^\dagger \in B(K,H)$.
(2) $\mathrm{Ran}\,(A)$ *is closed in* K.
(3) $(A^*)^\dagger \in B(H,K)$ *with* $(A^*)^\dagger = (A^\dagger)^*$.
(4) $\mathrm{Ran}\,(A^*A)$ *is closed in* H.

Proof. The implication of (1)⇒(2) comes from Proposition 3.1.2.

(2)⇒(3) That $\mathrm{Ran}\,(A)$ is closed implies that $\mathrm{Ran}\,(A^*)$ is closed too by Theorem 1.2.17. Since $A^* \in \mathcal{C}(K,H)$, we have $(A^*)^\dagger \in B(H,K)$ and $A^\dagger \in B(K,H)$ by Remark 3.1.4 (1). From $A^\dagger Ax = P_{(\mathrm{Ker}\,A)^\perp}x$, $\forall x \in \mathfrak{D}(A)$, we have, for any $z \in H$,

$$(Ax, (A^\dagger)^*z) = (A^\dagger Ax, z) = (P_{(\mathrm{Ker}\,A)^\perp}x, z) = (x, P_{(\mathrm{Ker}\,A)^\perp}z).$$

This means that $(A^\dagger)^*z \in \mathfrak{D}(A^*)$ and $A^*(A^\dagger)^* = P_{(\mathrm{Ker}\,A)^\perp} = P_{\mathrm{Ran}\,(A^*)}$. From $AA^\dagger = P_{\mathrm{Ran}\,(A)}$, we get that for any $z \in K$ and $y \in \mathfrak{D}(A^*)$,

$$(z, (A^\dagger)^*A^*y) = (A^\dagger z, A^*y) = (AA^\dagger z, y) = (P_{\mathrm{Ran}\,(A)}z, y) = (z, P_{\mathrm{Ran}\,(A)}y)$$

for $\mathrm{Ran}\,(A^\dagger) \subset \mathfrak{D}(A)$. Thus, $(A^\dagger)^*A^*y = P_{\mathrm{Ran}\,(A)}y = P_{(\mathrm{Ker}\,A^*)^\perp}y$, $\forall y \in \mathfrak{D}(A^*)$. Therefore,

$$A^*(A^\dagger)^*A^* = P_{\mathrm{Ran}\,(A^*)}A^* = A^* \text{ on } \mathfrak{D}(A^*) \text{ and}$$
$$(A^\dagger)^*A^*(A^\dagger)^* = P_{(\mathrm{Ker}\,A^*)^\perp}(A^\dagger)^* = (A^\dagger)^*$$

for $\overline{\mathrm{Ran}\,(A^\dagger)^*} = (\mathrm{Ker}\,A^\dagger)^\perp = (\mathrm{Ran}\,(A)^\perp)^\perp = \mathrm{Ran}\,(A) = (\mathrm{Ker}\,A^*)^\perp$ and consequently, $(A^*)^\dagger = (A^\dagger)^*$.

(3)⇒(4) $(A^*)^\dagger \in B(H,K)$ indicates that $\mathrm{Ran}\,(A^*)$ is closed and so is the $\mathrm{Ran}\,(A)$. Clearly, $\mathrm{Ran}\,(A^*A) \subset \mathrm{Ran}\,(A^*)$. On the hand, let $x \in \mathrm{Ran}\,(A^*)$. Then $x = A^*y$ for some $y \in \mathfrak{D}(A^*)$. Write y as the form $y = y_1 + y_2$, where $y_1 \in \mathrm{Ran}\,(A)$ and $y_2 \in (\mathrm{Ran}\,(A))^\perp = \mathrm{Ker}\,A^*$. So $y_1 \in \mathfrak{D}(A^*)$ and $x = A^*y_1 = A^*Ax_1$ for some $x_1 \in \mathfrak{D}(A)$ with $Ax_1 = y_1$. Therefore, $\mathrm{Ran}\,(A^*A) = \mathrm{Ran}\,(A^*)$ is closed in H.

(4)⇒(1) Let $x \in \mathrm{Ker}\,(A^*A)$. Then $x \in \mathfrak{D}(A)$ and $Ax \in \mathfrak{D}(A^*)$. Consequently, $\|Ax\|^2 = (Ax, Ax) = (A^*Ax, x) = 0$, i.e., $x \in \mathrm{Ker}\,A$. Thus, $\mathrm{Ker}\,(A^*A) = \mathrm{Ker}\,A$ and so that

$$\overline{\mathrm{Ran}\,(A^*)} \subset (\mathrm{Ker}\,A)^\perp = (\mathrm{Ker}\,(A^*A))^\perp = \mathrm{Ran}\,(A^*A) \subset \mathrm{Ran}\,(A^*)$$

by Theorem 1.2.15. This means that $\mathrm{Ran}\,(A^*)$ is closed and so is the $\mathrm{Ran}\,(A)$. Thus, $A^\dagger \in B(K,H)$. □

Corollary 3.1.6. *Let $A \in B(H)$ with* $\mathrm{Ran}\,(A)$ *closed. Suppose that* $A^* = A$. *Then* $(A^\dagger)^* = A^\dagger$ *and* $(A^2)^\dagger = (A^\dagger)^2$.

Proof. We have $(A^\dagger)^* = A^\dagger$ and $\mathrm{Ran}\,(A^2)$ is closed by Proposition 3.1.5. Noting that $A^\dagger A = (A^\dagger A)^* = A^*(A^\dagger)^* = AA^\dagger$, we have

$$A^2(A^\dagger)^2A^2 = A^2,\ (A^\dagger)^2A^2(A^\dagger)^2 = (A^\dagger)^2,\ A^2(A^\dagger)^2 = AA^\dagger = (A^\dagger)^2A^2.$$

Thus, $(A^2)^\dagger = (A^\dagger)^2$ by Remark 3.1.4 (2). □

Proposition 3.1.7. *Let $A \in B(H,K)$. Then the following conditions are equivalent:*

(1) $A^\dagger \in B(K,H)$.
(2) $\mathrm{Ran}\,(A)$ *is closed in* K.
(3) $\mathrm{Ran}\,(|A|)$ *is closed in* H.
(4) $0 \notin \sigma(|A|)$ *or* $0 \in \sigma(|A|)$ *is an isolated point.*

Proof. The equivalence of (1) and (2) comes from Proposition 3.1.5.

(2)⇒(3) $\mathrm{Ran}\,(A^*A)$ is closed in H by Proposition 3.1.5. Noting that $\mathrm{Ker}\,(A^*A) = \mathrm{Ker}\,(|A|)$, we have

$$\overline{\mathrm{Ran}\,(|A|)} = \mathrm{Ran}\,(A^*A) = \mathrm{Ran}\,(|A|^2) \subset \mathrm{Ran}\,(|A|),$$

that is, $\mathrm{Ran}\,(|A|)$ is closed.

(3)⇒(4) Assume that $0 \in \sigma(|A|)$. By Condition (3), $|A|^\dagger \in B(H)$ and

$$|A|^\dagger|A| = P_{(\mathrm{Ker}\,|A|)^\perp} = P_{\mathrm{Ran}\,(|A|)} = |A||A|^\dagger. \tag{3.1.2}$$

Put $A_0 = |A|\big|_{(\mathrm{Ker}\,|A|)^\perp}$, $B_0 = |A|^\dagger\big|_{(\mathrm{Ker}\,|A|)^\perp}$. Then (3.1.2) shows that B_0 is the inverse of A_0 in $B((\mathrm{Ker}\,|A|)^\perp)$. So $0 \notin \sigma(A_0)$. Finally, from $|A| = A_0 \oplus |A|\big|_{\mathrm{Ker}\,|A|} = A_0 \oplus 0$, we get that $\sigma(|A|) = \sigma(A_0) \cup \{0\}$ and 0 is the isolated point of $\sigma(|A|)$.

(4)⇒(2) If $0 \notin \sigma(|A|)$, then $|A|$ and hence A^*A is invertible in $B(H)$. Set $B = (A^*A)^{-1}A^* \in B(K,H)$. It is easy to check that $A^\dagger = B$.

Assume that $0 \in \sigma(|A|)$ is an isolated point. Set

$$f(t) = \begin{cases} 0 & t = 0 \\ 1 & t \in \sigma(|A|)\backslash\{0\} \end{cases}, \quad g(t) = \begin{cases} 0 & t = 0 \\ t^{-1} & t \in \sigma(|A|)\backslash\{0\} \end{cases}.$$

Then $f,\, g \in C_0(\sigma(|A|)\backslash\{0\})$ and $tg(t) = f(t)$, $tf(t) = t$, $f(t)g(t) = g(t)$, $\forall t \in \sigma(|A|)$. Let $\mathcal{A}$ be the C^*–algebra generated by $|A|$ in $B(H)$. Put $P = f(|A|)$ and $C = g(|A|)$. Then $P, C \in \mathcal{A}$ and

$$P^2 = P = P^*,\ C|A| = |A|C = P,\ PC = CP = C,\ P|A| = |A|P = |A|.$$

This shows that $\mathrm{Ker}\,|A| = \mathrm{Ran}\,(I_H - P)$ and $\mathrm{Ran}\,(|A|) = \mathrm{Ran}\,(P)$. So $\mathrm{Ran}\,(|A|)$ is closed. Since $\|Ax\|^2 = \||A|x\|^2$, $\forall\, x \in H$ and $\mathrm{Ker}\,|A| = \mathrm{Ker}\,A$, it follows that $\gamma(A) = \inf\{\|Ax\| \,|\, \mathrm{dist}\,(x, \mathrm{Ker}\,A) = 1\} = \gamma(|A|) > 0$. □

Proposition 3.1.8. *Let* $A \in B(H,K)\backslash\{0\}$ *with* $\mathrm{Ran}\,(A)$ *closed. Then*

(1) $(A^*A)^\dagger = A^\dagger(A^*)^\dagger$.
(2) $A^\dagger = (A^*A)^\dagger A^* = A^*(AA^*)^\dagger$.
(3) $\gamma(A) = \|A^\dagger\|^{-1} = \inf\{\lambda |\, \lambda \in \sigma(|A|)\backslash\{0\}\}$.

Proof. (1) Since $A^\dagger A = P_{(\mathrm{Ker}\,A)^\perp}$, $AA^\dagger = P_{\mathrm{Ran}\,(A)}$ and $\mathrm{Ran}\,(A^\dagger) = (\mathrm{Ker}\,A)^\perp$, it follows from Proposition 3.1.5 and Proposition 3.1.7 that

$$\begin{aligned} A^*AA^\dagger(A^*)^\dagger A^*A &= A^*(AA^\dagger)(AA^\dagger)^*A = A^*AA^\dagger A = (AA^\dagger A)^*A = A^*A \\ A^\dagger(A^*)^\dagger A^*AA^\dagger(A^*)^\dagger &= A^\dagger(AA^\dagger)^*(AA^\dagger)(A^\dagger)^* = A^\dagger AA^\dagger(A^\dagger)^* = A^\dagger(A^\dagger)^* \\ A^*AA^\dagger(A^*)^\dagger &= (AA^\dagger A)^*(A^*)^\dagger = P_{(\mathrm{Ker}\,A)^\perp} = P_{\mathrm{Ran}\,(A^*A)} \\ A^\dagger(A^*)^\dagger A^*A &= A^\dagger(AA^\dagger)^*A = P_{(\mathrm{Ker}\,A)^\perp} = P_{(\mathrm{Ker}\,A^*A)^\perp}. \end{aligned}$$

Thus, $(A^*A)^\dagger = A^\dagger(A^*)^\dagger$.

(2) We can check that by using (1):

$$(A^*A)^\dagger A^* = A^\dagger(A^*)^\dagger A^* = A^\dagger(AA^\dagger)^* = A^\dagger AA^\dagger = A^\dagger.$$

So $(A^*)^\dagger = (AA^*)^\dagger A$ and that $A^\dagger = ((A^*)^\dagger)^* = A^*(AA^*)^\dagger$.

(3) We have $\gamma(A) = \|A^\dagger\|^{-1}$ by Theorem 2.1.4 since $\|AA^\dagger\| = \|A^\dagger A\| = 1$. Note that, by Corollary 3.1.6,

$$\begin{aligned} \|A^\dagger\|^2 &= \|A^\dagger(A^\dagger)^*\| = \|A^\dagger(A^*)^\dagger\| = \|(A^*A)^\dagger\| \\ &= \|(|A||A|)^\dagger\| = \|(|A|^\dagger)(|A|^\dagger)\| = \||A|^\dagger\|^2. \end{aligned}$$

So we need to show that $\||A|^\dagger\|^{-1} = \inf\{\lambda |\, \lambda \in \sigma(|A|)\backslash\{0\}\}$.

If $0 \notin \sigma(|A|)$, then $|A|^\dagger = |A|^{-1}$ and

$$\||A|^{-1}\| = \max\{\lambda |\, \lambda \in \sigma(|A|^{-1})\} = \max\{\lambda^{-1} |\, \lambda \in \sigma(|A|)\}.$$

So $\||A|^\dagger\|^{-1} = \inf\{\lambda |\, \lambda \in \sigma(|A|)\backslash\{0\}\}$.

If $0 \in \sigma(|A|)$ is an isolated point, then by the proof of Proposition 3.1.7, $A_0 = |A||_{(\mathrm{Ker}\,|A|)^\perp}$ is invertible in $B((\mathrm{Ker}\,|A|)^\perp)$ and $|A|^\dagger = (A_0)^{-1}P_{(\mathrm{Ker}\,|A|)^\perp}$. So we have

$$\||A|^\dagger\| = \|(A_0)^{-1}\| = \max\{\lambda^{-1} |\, \lambda \in \sigma(A_0)\} = \max\{\lambda^{-1} |\, \lambda \in \sigma(|A|)\backslash\{0\}\}$$

for $|A| = A_0 \oplus 0$. Consequently, $\||A|^\dagger\|^{-1} = \inf\{\lambda |\, \lambda \in \sigma(|A|)\backslash\{0\}\}$. □

Proposition 3.1.9. *Let* $A \in C(H,K)\backslash\{0\}$ *with* $\mathrm{Ran}\,(A)$ *closed. Then* $\gamma(A) = \|A^\dagger\|^{-1}$.

Proof. Since $AA^\dagger A = A$, we have $\gamma(A) \geq \|A^\dagger\|^{-1}$ by Theorem 2.1.4. On the other hand, from $\|Ax\| \geq \gamma(A)\text{dist}\,(x, \text{Ker}\, A)$, $\forall\, x \in \mathfrak{D}(A)$, we get that

$$\|AA^\dagger y\| \geq \gamma(A)\text{dist}\,(A^\dagger y, \text{Ker}\, A) = \gamma(A)\|P_{(\text{Ker}\, A)^\perp} A^\dagger y\|, \forall\, y \in K. \quad (3.1.3)$$

Since $AA^\dagger = P_{\text{Ran}\,(A)}$, $A^\dagger Ax = P_{(\text{Ker}\, A)^\perp} x$, $\forall\, x \in \mathfrak{D}(A)$ and $\text{Ran}\,(A^\dagger) = (\text{Ker}\, A)^\perp \cap \mathfrak{D}(A)$, it follows from (3.1.3) that $\gamma(A)\|A^\dagger y\| \leq \|AA^\dagger y\| \leq \|y\|$, $\forall\, y \in K$. Therefore, $\gamma(A) = \|A^\dagger\|^{-1}$. □

Proposition 3.1.10. *Let A be in $C(H,K)$ such that $ACA = A$ for some $C \in B(K,H)$ with* $\text{Ran}\,(C) \subset \mathfrak{D}(A)$*. Then $A^\dagger = P_{(\text{Ker}\, A)^\perp} CAO(AC)$, where $O(Q) = -Q(I_H - Q - Q^*)^{-1}$ is the projection of H onto the range of the idempotent $Q \in B(H)$ given in Proposition* 1.1.30. In addition, if A is bounded, then

$$A^\dagger = (CA + A^*C^* - I_H)^{-1} CAC(AC + C^*A^* - I_K)^{-1}.$$

Proof. Note that $\text{Ker}\, A \subset \mathfrak{D}(A)$ and $\text{Ker}\, A$ is closed. So $P_{(\text{Ker}\, A)^\perp} x \in \mathfrak{D}(A)$ for any $x \in \mathfrak{D}(A)$. Put $B = P_{(\text{Ker}\, A)^\perp} CO(AC)$. Using the identities:

$$O(AC)AC = AC,\ O(AC) = P_{\text{Ran}\,(A)},\ ACO(AC) = O(AC),$$
$$P_{(\text{Ker}\, A)^\perp} CAx = P_{(\text{Ker}\, A)^\perp} x, \qquad AP_{(\text{Ker}\, A)^\perp} x = Ax,\ x \in \mathfrak{D}(A),$$

we can obtain that $A^\dagger = P_{(\text{Ker}\, A)^\perp} CAO(AC)$ easily.

When A is bounded,

$$\begin{aligned} P_{(\text{Ker}\, A)^\perp} &= I_H - O(I_H - CA) = I_H - (CA - I_H)(CA + A^*C^* - I_H)^{-1} \\ &= A^*C^*(CA + A^*C^* - I_H)^{-1} = (CA + A^*C^* - I_H)^{-1} CA. \end{aligned}$$

Therefore,

$$\begin{aligned} A^\dagger &= (CA + A^*C^* - I_H)^{-1} CAC(AC)(AC + C^*A^* - I_K)^{-1} \\ &= (CA + A^*C^* - I_H)^{-1} CAC(AC + C^*A^* - I_K)^{-1}. \end{aligned}$$
□

3.2 The reverse order law of the Moore–Penrose inverses

Let $A \in B(K, L)$ and $B \in C(H,K)$ with $\text{Ran}\,(A)$ and $\text{Ran}\,(B)$ closed. We will investigate the representation $(AB)^\dagger$ by means of $A^\dagger$ and $B^\dagger$ and give some conditions that make $(AB)^\dagger = B^\dagger A^\dagger$.

Definition 3.2.1. Let P, Q be projections in $B(H)$. Then the projection of H onto $\overline{\text{Ran}\,(P + Q)}$ is denoted by $P \vee Q$.

Proposition 3.2.2. *Let P, Q be projections in $B(H)$. Then*

(1) $P \vee Q \geq P$, $P \vee Q \geq Q$.
(2) $(I-Q)(P \vee Q) = (P \vee Q)(I-Q) = P \vee Q - Q$.
(3) $Q(P \vee (I-Q))$ *is the projection of H onto* $\overline{\mathrm{Ran}\,(QP)}$.

Proof. (1) Since $\mathrm{Ran}\,(I - P \vee Q) = \mathrm{Ker}\,(P+Q) = \mathrm{Ker}\,Q \cap \mathrm{Ker}\,P$, we have $P(I - P \vee Q) = 0 = Q(I - P \vee Q)$. Thus, $P \vee Q \geq P$, $P \vee Q \geq Q$.

(2) By (1), $(I-Q)(P \vee Q) = P \vee Q - Q = (P \vee Q)(I-Q)$.

(3) Put $S = (P \vee (I-Q))Q$. Then $S = P \vee (I-Q) - (I-Q)$ is a projection. Moreover, from

$$\begin{aligned}\mathrm{Ran}\,(QP) &= Q(\mathrm{Ran}\,(P)) \subset Q(\mathrm{Ran}\,(P \vee (I-Q))) \\ &= \mathrm{Ran}\,(S) \subset Q(\overline{\mathrm{Ran}\,(P+I-Q)}) \subset \overline{\mathrm{Ran}\,(QP)},\end{aligned}$$

we get that $\mathrm{Ran}\,(S) = \overline{\mathrm{Ran}\,(QP)}$, that is, $Q(P \vee (I-P))$ is the projection of H onto $\overline{\mathrm{Ran}\,(QP)}$. □

Lemma 3.2.3. *Let P, Q be projections in $B(H)$. Suppose that $PQ \neq 0$ and* $\mathrm{Ran}\,(PQ)$ *is closed. Then* $\gamma(PQ) = (1 - \|(I-P)Q(P \vee (I-Q))\|^2)^{1/2}$.

Proof. Let $x \in (\mathrm{Ker}\,(PQ))^\perp$ with $\|x\| = 1$. Since

$$(\mathrm{Ker}\,(PQ))^\perp = \overline{\mathrm{Ran}\,(PQ)^*} = \overline{\mathrm{Ran}\,(QP)}$$

and $\mathrm{Ran}\,(PQ)$ is closed, it follows from Proposition 3.2.2 that $R = Q(P \vee (I-Q))$ is the projection of H onto $(\mathrm{Ker}\,(PQ))^\perp$. Thus, $Rx = x = Qx$ and so that

$$\begin{aligned}\|(I-P)R\| &\geq \|(I-P)Rx\| = \|(I-P)Qx\| \\ \|PQx\|^2 + \|(I-P)R\|^2 &\geq \|PQx\|^2 + \|(I-P)Qx\|^2 = \|Qx\|^2 = 1.\end{aligned}$$

Note that $\gamma(PQ) = \inf\{\|PQx\| \,|\, x \in (\mathrm{Ker}\,(PQ))^\perp, \|x\| = 1\}$. So we have $\gamma(PQ) \geq (1 - \|(I-P)Q(P \vee (I-Q))\|^2)^{1/2}$. Since $I - R$ is the projection of H onto $\mathrm{Ker}\,(PQ)$, we get that

$$\begin{aligned}&\sup\{\|(I-P)Qx\| \,|\, x \in (\mathrm{Ker}\,(PQ))^\perp, \|x\| = 1\} \\ &= \sup\{\|(I-P)Qx\| \,|\, x \in \mathrm{Ran}\,(R), \|x\| = 1\} \\ &= \|(I-P)R\|.\end{aligned}$$

Thus, from

$$(\gamma(PQ))^2 + \|(I-P)Qx\|^2 \leq \|PQx\|^2 + \|(I-P)Qx\|^2 = \|Qx\|^2 = 1,$$

$\forall x \in (\mathrm{Ker}\,(PQ))^\perp$, $\|x\| = 1$, we get that $(\gamma(PQ))^2 + \|(I-P)R\|^2 \leq 1$. The assertion follows. □

Proposition 3.2.4. *Let P, Q be projections in $B(H)$ with $PQ \neq 0$ and* $\mathrm{Ran}\,(PQ)$ *closed. Then $R_{P,Q} = I - (P \vee (I - Q))Q + PQ$ is invertible in $B(H)$ and $(PQ)^\dagger = R_{P,Q}^{-1} P((I - P) \vee Q)$.*

Proof. Put $R = Q(P \vee (I - Q))$ and $S = P(Q \vee (I - P))$. By Proposition 3.2.2, R and S are the projections of H onto $\overline{\mathrm{Ran}\,(QP)}$ and $\overline{\mathrm{Ran}\,(PQ)}$, respectively. Since $\mathrm{Ran}\,(QP)$ and $\mathrm{Ran}\,(PQ)$ are closed, it follows that

$$(PQ)^\dagger (PQ) = P_{(\mathrm{Ker}\,(PQ))^\perp} = P_{\mathrm{Ran}\,(QP)} = R, \ (PQ)(PQ)^\dagger = S.$$

So $R_{P,Q}(PQ)^\dagger = (PQ)^\dagger - R(PQ)^\dagger + (PQ)(PQ)^\dagger = S$.

Since $\mathrm{Ran}\,(PQ)$ is closed, we have $(1 - \|(I - P)R\|^2)^{1/2} = \gamma(PQ) > 0$, i.e., $\|(I - P)R\| < 1$ by Lemma 3.2.3. Thus,

$$R_{P,Q} = I - (P \vee (I - Q) - P)Q = I - (I - P)R$$

is invertible in $B(H)$ and $(PQ)^\dagger = R_{P,Q}^{-1} S$. □

Lemma 3.2.5. *Let $A \in B(K, L)$ and $B \in C(H, K)$ with* $\mathrm{Ran}\,(A)$ *and* $\mathrm{Ran}\,(B)$ *closed. Then the following conditions are equivalent*

(1) $\mathrm{Ran}\,(A^\dagger ABB^\dagger)$ *is closed.*
(2) $\mathrm{Ran}\,(BB^\dagger A^\dagger A)$ *is closed.*
(3) $\mathrm{Ran}\,(AB)$ *is closed.*
(4) $\mathrm{Ran}\,(I_K - A^\dagger A + BB^\dagger)$ *is closed.*
(5) $\mathrm{Ker}\,A + \mathrm{Ran}\,(B)$ *is closed.*

Proof. Since $(A^\dagger ABB^\dagger)^* = BB^\dagger A^\dagger A$, it follows from Theorem 1.2.17 that (1) and (2) are equivalent.

(1)⇒(3) By Proposition 3.1.7, $C = (A^\dagger ABB^\dagger)^\dagger \in B(K)$ with $A^\dagger ABB^\dagger = A^\dagger ABB^\dagger CA^\dagger ABB^\dagger$. Put $D = B^\dagger CA^\dagger \in B(L, K)$. Then $\mathrm{Ran}\,(D) \subset \mathrm{Ran}\,(B^\dagger) = (\mathrm{Ker}\,B)^\perp \cap \mathfrak{D}(B)$ and

$$ABDAB = A(A^\dagger ABB^\dagger CA^\dagger ABB^\dagger)B = AA^\dagger ABB^\dagger B = AB.$$

So $\mathrm{Ran}\,(AB)$ is closed by Proposition 2.1.3.

(3)⇒(1) In this case, we can find $E \in B(L, H)$ such that $ABEAB = AB$ on $\mathfrak{D}(B)$ and $\mathrm{Ran}\,(E) \subset (\mathrm{Ker}\,(AB))^\perp \cap \mathfrak{D}(B)$. Put $F = BEA \in B(K)$. Then from

$$A^\dagger ABB^\dagger FA^\dagger ABB^\dagger = A^\dagger ABEABB^\dagger = A^\dagger ABB^\dagger,$$

we obtain that $\mathrm{Ran}\,(A^\dagger ABB^\dagger)$ is closed.

(1)⇒(4) Set $D = I_K - A^\dagger A + BB^\dagger = (I_K - AA^\dagger)(I_K + BB^\dagger) + A^\dagger ABB^\dagger$. Let $\{x_n\} \subset K$ and $y_0 \in K$ such that $Dx_n \xrightarrow{\|\cdot\|} y_0$. Then $A^\dagger ADx_n =$

$A^\dagger ABB^\dagger x_n \xrightarrow{\|\cdot\|} A^\dagger Ay_0$ and hence $(I_K - AA^\dagger)(I_K + BB^\dagger)x_n \xrightarrow{\|\cdot\|} y_0 - A^\dagger Ay_0$. Put $z_n = (I_K + BB^\dagger)x_n$, $n \in \mathbb{N}$. Then

$$x_n = (I_K + BB^\dagger)^{-1}z_n = (I_K - 2^{-1}BB^\dagger)z_n, \; n \in \mathbb{N}$$

$$A^\dagger ABB^\dagger z_n \xrightarrow{\|\cdot\|} 2A^\dagger Ay_0, \; (I_K - A^\dagger A)z_n \xrightarrow{\|\cdot\|} (I_K - A^\dagger A)y_0.$$

Thus,

$$\begin{aligned} A^\dagger ABB^\dagger A^\dagger Az_n &= A^\dagger ABB^\dagger z_n - A^\dagger ABB^\dagger(I_K - A^\dagger A)z_n \\ &\xrightarrow{\|\cdot\|} 2A^\dagger Ay_0 - A^\dagger ABB^\dagger(I_K - A^\dagger A)y_0. \end{aligned}$$

Noting that $\operatorname{Ran}(A^\dagger ABB^\dagger A^\dagger A)$ is closed by Proposition 3.1.5, we can find $z_0 \in K$ such that $A^\dagger ABB^\dagger A^\dagger Az_0 = 2A^\dagger Ay_0 - A^\dagger ABB^\dagger(I_K - A^\dagger A)y_0$. Set $z = (I_K + BB^\dagger)^{-1}(A^\dagger Az_0 + (I_K - A^\dagger A)y_0)$. Then

$$\begin{aligned} (I_K - A^\dagger A)(I_K + BB^\dagger)z &= (I_K - A^\dagger A)y_0 \\ A^\dagger ABB^\dagger(I_K - A^\dagger A)z &= 2^{-1}(A^\dagger ABB^\dagger A^\dagger Az_0 + A^\dagger ABB^\dagger(I_K - A^\dagger A)y_0) \\ &= A^\dagger Ay_0 \end{aligned}$$

and consequently, $Dz = y_0$, i.e., $\operatorname{Ran}(D)$ is closed.

(4)⇒(5) Note that $\operatorname{Ran}(I_K - A^\dagger A + BB^\dagger) \subset \operatorname{Ker} A + \operatorname{Ran}(B)$ and

$$\begin{aligned} \operatorname{Ker} A + \operatorname{Ran}(B) &= ((I_K - A^\dagger A) \vee BB^\dagger)(\operatorname{Ker} A + \operatorname{Ran}(B)) \\ &\subset \overline{\operatorname{Ran}(I_K - A^\dagger A + BB^\dagger)}. \end{aligned}$$

So $\operatorname{Ker} A + \operatorname{Ran}(B)$ is closed when $\operatorname{Ran}(I_K - A^\dagger A + BB^\dagger)$ is closed.

(5)⇒(1) Let G be the projection of K onto $\operatorname{Ker} A + \operatorname{Ran}(B)$. Then $I_K - G$ is the projection of K onto $(\operatorname{Ker} A + \operatorname{Ran}(B))^\perp = (\operatorname{Ker} A)^\perp \cap (\operatorname{Ran}(B))^\perp$. So $A^\dagger A(I_K - G) = I_K - G$ and hence $A^\dagger A = GA^\dagger AG + I_K - G$. This indicates that $GA^\dagger AG$ is closed and so is the $\operatorname{Ran}(A^\dagger AG)$ by Proposition 3.1.5. But

$$\begin{aligned} \operatorname{Ran}(A^\dagger AG) &= A^\dagger A(\operatorname{Ran}(G)) = A^\dagger A(\operatorname{Ker} A + \operatorname{Ran}(B)) \\ &= A^\dagger A(\operatorname{Ran}(B)) = A^\dagger A(\operatorname{Ran}(BB^\dagger)) = \operatorname{Ran}(A^\dagger ABB^\dagger). \end{aligned}$$

Thus, $\operatorname{Ran}(A^\dagger ABB^\dagger)$ is closed. □

Theorem 3.2.6. *Let $A \in B(K, L)$ and $B \in C(H, K)$ with* $\operatorname{Ran}(A)$ *and* $\operatorname{Ran}(B)$ *and* $\operatorname{Ran}(AB)$ *is closed. Suppose that $AB \in C(H, L)$. Then*

$$\begin{aligned} (AB)^\dagger = P_{(\operatorname{Ker}(AB))^\perp} B^\dagger [I_K - (A^\dagger A \vee (I_K - BB^\dagger))BB^\dagger + A^\dagger ABB^\dagger]^{-1} A^\dagger \\ \times (A(BB^\dagger \vee (I_K - A^\dagger A))A^\dagger + (A^*)^\dagger(BB^\dagger \vee (I_K - A^\dagger A))A^* - I_K). \end{aligned}$$

In addition, if B is bounded, then

$$\begin{aligned}(AB)^\dagger &= (B^\dagger(A^\dagger A \vee (I_K - BB^\dagger))B + B^*(A^\dagger A \vee (I_K - BB^\dagger))(B^*)^\dagger - I_H)^{-1} \\ &\quad \times B^\dagger[I_K - (A^\dagger A \vee (I_K - BB^\dagger))BB^\dagger + A^\dagger ABB^\dagger]^{-1}A^\dagger \\ &\quad \times (A(BB^\dagger \vee (I_K - A^\dagger A))A^\dagger + (A^*)^\dagger(BB^\dagger \vee (I_K - A^\dagger A))A^* - I_K)^{-1}.\end{aligned}$$

Proof. We have $\operatorname{Ran}(A^\dagger ABB^\dagger)$ and $\operatorname{Ran}(BB^\dagger A^\dagger A)$ are closed by Lemma 3.2.5 and

$$(A^\dagger ABB^\dagger)(A^\dagger ABB^\dagger)^\dagger = P_{\operatorname{Ran}(A^\dagger ABB^\dagger)} = A^\dagger A(BB^\dagger \vee (I_K - A^\dagger A)) \tag{3.2.1}$$

$$(A^\dagger ABB^\dagger)^\dagger(A^\dagger ABB^\dagger) = P_{(\operatorname{Ker}(A^\dagger ABB^\dagger))^\perp} = BB^\dagger((I_K - BB^\dagger) \vee A^\dagger A) \tag{3.2.2}$$

by Proposition 3.2.2. Put $C = B^\dagger(A^\dagger ABB^\dagger)^\dagger A^\dagger$. It is easy to verify that $ABCAB = AB$, $CABC = C$ and $\operatorname{Ran}(C) \subset \mathfrak{D}(B)$. Now by Proposition 3.2.4, $C = B^\dagger R^{-1}_{A^\dagger A, BB^\dagger} A^\dagger A(BB^\dagger \vee (I_K - A^\dagger A))A^\dagger$ and

$$\begin{aligned}B^\dagger R^{-1}_{A^\dagger A, BB^\dagger} &= B^\dagger BB^\dagger[I_K - (I_K - A^\dagger A)(A^\dagger A \vee (I_K - BB^\dagger))BB^\dagger]^{-1} \\ &= B^\dagger[I_K - BB^\dagger(I_K - A^\dagger A)(A^\dagger A \vee (I_K - BB^\dagger))]^{-1}BB^\dagger.\end{aligned}$$

So $C = B^\dagger R^{-1}_{A^\dagger A, BB^\dagger} A^\dagger$. On the other hand, by (3.2.1) and (3.2.2),

$$\begin{aligned}ABC &= AA^\dagger ABB^\dagger(A^\dagger ABB^\dagger)^\dagger A^\dagger = A(BB^\dagger \vee (I_K - A^\dagger A))A^\dagger \\ CAB &= B^\dagger(A^\dagger ABB^\dagger)^\dagger A^\dagger ABB^\dagger B = B^\dagger(A^\dagger A \vee (I_K - BB^\dagger))B.\end{aligned}$$

Therefore, we can obtain the result in terms of Proposition 3.1.10. □

Corollary 3.2.7. *Let $A \in B(K, L)$ and $B \in C(H, K)$ such that* $\operatorname{Ran}(A)$, $\operatorname{Ran}(B)$ *and* $\operatorname{Ran}(AB)$ *are all closed. If*

$$BB^\dagger \vee (I_K - A^\dagger A) = I_K, \quad A^\dagger A \vee (I_K - BB^\dagger) = I_K,$$

then $(AB)^\dagger = B^\dagger(I_K - BB^\dagger + A^\dagger ABB^\dagger)^{-1}A^\dagger$.

Theorem 3.2.8. *Let $A \in B(K, L)$ and $B \in C(H, K)$ with* $\operatorname{Ran}(A)$ *and* $\operatorname{Ran}(B)$ *and* $\operatorname{Ran}(AB)$ *closed. Suppose that $AB \in C(H, L)$. Then $(AB)^\dagger = B^\dagger A^\dagger$ iff following conditions are satisfied:*

(1) $A^\dagger ABB^\dagger = BB^\dagger A^\dagger A$.
(2) $\operatorname{Ran}(B^\dagger A^*) = \operatorname{Ran}(B^* A^*) \cap \mathfrak{D}(B)$ *and* $\operatorname{Ran}((A^*)^\dagger B) = \operatorname{Ran}(AB)$.

Proof. Since $(AB)^\dagger = B^\dagger A^\dagger$, we have $AB = ABB^\dagger A^\dagger AB$ and hence $(A^\dagger ABB^\dagger)^2 = A^\dagger ABB^\dagger$. Since $\|A^\dagger ABB^\dagger\| \le 1$, it follows from Corollary 1.5.21 that $A^\dagger ABB^\dagger$ is a projection in $B(K)$, that is, $A^\dagger ABB^\dagger = BB^\dagger A^\dagger A$.

Now $\mathrm{Ran}\,(B^\dagger A^*) = B^\dagger(\mathrm{Ran}\,(A^\dagger)) = \mathrm{Ran}\,(B^\dagger A^\dagger)$. On the other hand, from

$$\begin{aligned}\mathrm{Ran}\,(B^\dagger A^\dagger) &= \mathrm{Ran}\,((AB)^\dagger) = (\mathrm{Ker}\,(AB))^\perp \cap \mathfrak{D}(B)\\ &= \mathrm{Ran}\,(B^*A^*) \cap \mathfrak{D}(B),\end{aligned}$$

we get that $\mathrm{Ran}\,(B^\dagger A^*) = \mathrm{Ran}\,(B^*A^*) \cap \mathfrak{D}(B)$. Noting that

$$\begin{aligned}(A^\dagger)^*B(B^\dagger A^*)(A^\dagger)^*B &= (A^\dagger)^*BB^\dagger A^\dagger AB\\ &= (A^\dagger)^*A^\dagger ABB^\dagger B = (A^\dagger)^*B,\end{aligned}$$

we have $\mathrm{Ran}\,((A^\dagger)^*B) = \mathrm{Ran}\,((A^\dagger)^*B(B^\dagger A^*))$. Since

$$AB(AB)^\dagger = ABB^\dagger A^\dagger = P_{\mathrm{Ran}\,(AB)} = (A^\dagger)^*BB^\dagger A^*,$$

it follows that $\mathrm{Ran}\,((A^\dagger)^*B) = \mathrm{Ran}\,((A^\dagger)^*B(B^\dagger A^*)) = \mathrm{Ran}\,(AB)$.

Conversely, suppose that A and B satisfy (1) and (2). Put

$$X = B^\dagger A^\dagger,\ Q = (AB)X = ABB^\dagger A^\dagger,\ P = X(AB) = B^\dagger A^\dagger AB.$$

Since $ABXAB = AB$ and $XABX = X$, we have $\mathrm{Ran}\,(AB) = \mathrm{Ran}\,(Q)$ and $\mathrm{Ker}\,(AB) = \mathrm{Ran}\,(I_H - P)$.

From $Q^* = (A^\dagger)^*BB^\dagger A^*$ and $(A^\dagger)^*BB^\dagger A^*(A^\dagger)^*B = (A^\dagger)^*B$, we obtain that $\mathrm{Ran}\,(Q^*) = \mathrm{Ran}\,((A^\dagger)^*B) = \mathrm{Ran}\,(AB) = \mathrm{Ran}\,(Q)$ and consequently, Q is a projection by Proposition 1.1.30.

Note that

$$\begin{aligned}\mathrm{Ran}\,(P) &= B^\dagger A^\dagger A\mathrm{Ran}\,(B) = \mathrm{Ran}\,(B^\dagger A^\dagger ABB^\dagger)\\ &= \mathrm{Ran}\,(B^\dagger BB^\dagger A^\dagger A) = \mathrm{Ran}\,(B^\dagger A^\dagger A)\\ &= B^\dagger\mathrm{Ran}\,(A^\dagger A) = B^\dagger\mathrm{Ran}\,(A^\dagger) = \mathrm{Ran}\,(B^\dagger A^*)\end{aligned}$$

and $B^*A^* = (AB)^*$ on $\mathfrak{D}(B^*A^*) = \mathfrak{D}((AB)^*)$ by Proposition 1.2.4. So

$$\begin{aligned}\mathrm{Ran}\,((AB)^\dagger) &= (\mathrm{Ker}\,(AB))^\perp \cap \mathfrak{D}(AB) = \mathrm{Ran}\,((AB)^*) \cap \mathfrak{D}(B)\\ &= \mathrm{Ran}\,(B^*A^*) \cap \mathfrak{D}(B) = \mathrm{Ran}\,(B^\dagger A^*)\\ &= \mathrm{Ran}\,(P).\end{aligned}$$

and hence $(AB)^\dagger(AB)P = P$ on $\mathfrak{D}(B)$. Therefore,

$$\begin{aligned}P &= (AB)^\dagger(AB)P = (AB)^\dagger(AB)B^\dagger A^\dagger AB\\ &= (AB)^\dagger AA^\dagger ABB^\dagger B = (AB)^\dagger(AB)\end{aligned}$$

is a projection. □

Corollary 3.2.9. *Let* $A \in B(K, L)$, $B \in B(H, K)$ *with* $\mathrm{Ran}\,(A)$, $\mathrm{Ran}\,(B)$ *and* $\mathrm{Ran}\,(AB)$ *closed. Then* $(AB)^\dagger = B^\dagger A^\dagger$ *iff* $A^\dagger ABB^* = BB^* A^\dagger A$ *and* $A^* ABB^\dagger = BB^\dagger A^* A$.

Proof. First assume that $(AB)^\dagger = B^\dagger A^\dagger$. Then

$$(AB)^\dagger AB = B^\dagger A^\dagger AB = (B^\dagger A^\dagger AB)^* = B^* A^\dagger A (B^\dagger)^* \tag{3.2.3}$$

$$AB(AB)^\dagger = ABB^\dagger A^\dagger = (ABB^\dagger A^\dagger)^* = (A^\dagger)^* BB^\dagger A^*. \tag{3.2.4}$$

From (3.2.3) and (3.2.4), we have

$$BB^\dagger A^\dagger ABB^* = BB^* A^\dagger A (B^\dagger)^* B^* = BB^* A^\dagger ABB^\dagger \tag{3.2.5}$$

$$A^* ABB^\dagger A^\dagger A = A^* (A^\dagger)^* BB^\dagger A^* A = A^\dagger ABB^\dagger A^* A. \tag{3.2.6}$$

By Theorem 3.2.8, $A^\dagger ABB^\dagger = BB^\dagger A^\dagger A$. So (3.2.5) and (3.2.6) imply that

$$A^\dagger ABB^* = BB^\dagger A^\dagger ABB^* = BB^* BB^\dagger A^\dagger A = BB^* A^\dagger A$$

$$A^* ABB^\dagger = A^* ABB^\dagger A^\dagger A = BB^\dagger A^\dagger AA^* A = BB^\dagger A^* A.$$

Conversely, we assume that $A^\dagger ABB^* = BB^* A^\dagger A$ and $A^* ABB^\dagger = BB^\dagger A^* A$. Then $BB^\dagger (I_K - A^\dagger A) K \subset \mathrm{Ker}\, A$ and hence

$$(I_K - A^\dagger A) BB^\dagger (I_K - A^\dagger A) = BB^\dagger (I_K - A^\dagger A). \tag{3.2.7}$$

From (3.2.7), we get that $A^\dagger AB^\dagger B = A^\dagger ABB^\dagger A^\dagger A = BB^\dagger A^\dagger A$. Thus,

$$B^\dagger A^\dagger ABB^\dagger A^\dagger = B^\dagger A^\dagger, \; (A^\dagger)^* (B^\dagger)^* = (A^\dagger)^* (B^\dagger)^* B^* A^* (A^\dagger)^* (B^\dagger)^*$$

and consequently,

$$\mathrm{Ran}\,(B^\dagger A^\dagger) = \mathrm{Ran}\,(B^\dagger A^\dagger AB), \; \mathrm{Ran}\,((A^\dagger)^* (B^\dagger)^*) = \mathrm{Ran}\,((A^\dagger)^* (B^\dagger)^* B^* A^*).$$

Using the relations

$$\mathrm{Ran}\,(A^\dagger) = \mathrm{Ran}\,(A^\dagger A) = \mathrm{Ran}\,(A^*) = \mathrm{Ran}\,(A^* A)$$

$$\mathrm{Ran}\,(B) = \mathrm{Ran}\,(BB^\dagger) = \mathrm{Ran}\,(BB^*) = \mathrm{Ran}\,((B^*)^\dagger),$$

we obtain that

$$\begin{aligned}
\mathrm{Ran}\,(B^\dagger A^*) &= \mathrm{Ran}\,(B^\dagger A^\dagger) = \mathrm{Ran}\,(B^\dagger A^\dagger AB) \\
&= \mathrm{Ran}\,(B^\dagger A^\dagger ABB^*) = \mathrm{Ran}\,(B^\dagger BB^* A^\dagger A) \\
&= \mathrm{Ran}\,(B^* A^*), \\
\mathrm{Ran}\,((A^*)^\dagger B) &= \mathrm{Ran}\,((A^*)^\dagger (B^*)^\dagger) = \mathrm{Ran}\,((A^*)^\dagger (B^*)^\dagger B^* A^*) \\
&= \mathrm{Ran}\,((A^*)^\dagger BB^\dagger A^*) = \mathrm{Ran}\,((A^*)^\dagger BB^\dagger A^* A) \\
&= \mathrm{Ran}\,((A^*)^\dagger A^* ABB^\dagger) = \mathrm{Ran}\,(AA^\dagger ABB^\dagger) \\
&= \mathrm{Ran}\,(AB).
\end{aligned}$$

Thus, $(AB)^\dagger = B^\dagger A^\dagger$ by Theorem 3.2.8. □

3.3 Stable perturbation of closed operators in Hilbert spaces

We have investigated the stable perturbation of closed operators in Banach spaces in §2.2 of Chapter 2. In this section, we will consider the stable perturbation of closed operators in Hilbert spaces.

Let $A \in \mathcal{C}(H,K)$ with $\operatorname{Ran}(A)$ closed and $\delta A \in B(H,K)$. Put $\bar{A} = A + \delta A \in \mathcal{C}(H,K)$. Since $\operatorname{Ker} A^\dagger = \operatorname{Ker}(AA^\dagger) = \operatorname{Ran}(I_K - AA^\dagger) = \operatorname{Ran}(A)^\perp$, it follows from Definition 2.2.1 that $\bar{A}$ is the stable perturbation of A iff $\operatorname{Ran}(\bar{A}) \cap \operatorname{Ran}(A)^\perp = \{0\}$.

Lemma 3.3.1. *Let $A \in \mathcal{C}(H,K)$ with $\operatorname{Ran}(A)$ closed and $\delta A \in B(H,K)$. Put $\bar{A} = A + \delta A \in \mathcal{C}(H,K)$.*

(1) *If $\operatorname{Ker}(I_H + A^\dagger \delta A) = \{0\}$ and $\overline{\operatorname{Ran}(I_K + \delta A A^\dagger)} = K$, then $\operatorname{Ker} \bar{A} \cap (\operatorname{Ker} A)^\perp = \{0\}$ and $\operatorname{Ran}(\bar{A})^\perp \cap \operatorname{Ran}(A) = \{0\}$.*
(2) *If $\operatorname{Ker} \bar{A} \cap (\operatorname{Ker} A)^\perp = \{0\}$, $(\operatorname{Ker} \bar{A})^\perp \cap \operatorname{Ker} A = \{0\}$ and $\operatorname{Ran}(\bar{A})^\perp \cap \operatorname{Ran}(A) = \{0\}$, $\operatorname{Ran}(\bar{A}) \cap \operatorname{Ran}(A)^\perp = \{0\}$, then $\operatorname{Ker}(I_K + \delta A A^\dagger) = \{0\}$, $\operatorname{Ker}(I_H + A^\dagger \delta A) = \{0\}$, $\overline{\operatorname{Ran}(I_K + \delta A A^\dagger)} = K$, $\overline{\operatorname{Ran}(I_H + A^\dagger \delta A)} = H$.*

Proof. (1) Let $x \in \operatorname{Ker} \bar{A} \cap (\operatorname{Ker} A)^\perp$. Then $\bar{A}x = 0$ and $A^\dagger A x = x$, $x \in \mathfrak{D}(A)$. Since

$$0 = A^+ \bar{A} x = (A^\dagger A - I_H)x + (I_H + A^\dagger \delta A)x = (I_H + A^\dagger \delta A)x$$

and $\operatorname{Ker}(I_H + A^\dagger \delta A) = \{0\}$, we have $x = 0$.

Note that $\overline{\operatorname{Ran}(I_K + \delta A A^\dagger)} = K$ implies that $\operatorname{Ker}(I_K + (A^*)^\dagger (\delta A)^*) = \{0\}$ for $(A^*)^\dagger = (A^\dagger)^*$ by Proposition 3.1.5. So, $\operatorname{Ker}(\bar{A}^*) \cap (\operatorname{Ker} A^*)^\perp = \{0\}$ by above argument. But $(\operatorname{Ker} A^*)^\perp = \operatorname{Ran}(A)$ by Theorem 1.2.17 and $\operatorname{Ker} \bar{A}^* = (\operatorname{Ran}(\bar{A}))^\perp$ by Theorem 1.2.15. So $(\operatorname{Ran}(\bar{A}))^\perp \cap \operatorname{Ran}(A) = \{0\}$.

(2) Let $x \in \operatorname{Ker}(I_K + \delta A A^\dagger)$ and put

$$x_1 = AA^\dagger x, \quad x_2 = (I_K - AA^\dagger)x \in (\operatorname{Ran}(A))^\perp.$$

Then $x_1 + x_2 + \delta A A^\dagger x = 0$ and hence $(A + \delta A)A^\dagger x_1 = -x_2$. So $x_2 \in \operatorname{Ran}(\bar{A}) \cap (\operatorname{Ran}(A))^\perp = \{0\}$ and $\bar{A}A^\dagger x_1 = 0$, that is, $A^\dagger x_1 \in \operatorname{Ker} \bar{A} \cap (\operatorname{Ker} A)^\perp = \{0\}$ for $\operatorname{Ran}(A^\dagger) = (\operatorname{Ker} A)^\perp \cap \mathfrak{D}(A)$. Thus, $AA^\dagger x_1 = 0$ and consequently, $x = 0$.

To prove that $\overline{\operatorname{Ran}(I_H + A^\dagger \delta A)} = H$, it is enough to show that $\operatorname{Ker}(I_H + (\delta A)^*(A^*)^\dagger) = \{0\}$. By above argument, we need to check that $\operatorname{Ran}(\bar{A}^*) \cap (\operatorname{Ran}(A^*))^\perp = \{0\}$ and $\operatorname{Ker} \bar{A}^* \cap (\operatorname{Ker} A^*)^\perp = \{0\}$. But it follows from the following relations

$$\operatorname{Ran}(\bar{A}^*) \subset (\operatorname{Ker} \bar{A})^\perp, \qquad \operatorname{Ker} \bar{A}^* = \operatorname{Ran}(\bar{A})^\perp$$
$$\operatorname{Ran}(A^*)^\perp = \operatorname{Ker} A, \qquad (\operatorname{Ker} A^*)^\perp = \operatorname{Ran}(A).$$

Let $x \in \operatorname{Ker}(I_H + A^\dagger \delta A)$. Then $A^\dagger \delta A x = -x \in \mathfrak{D}(A)$ for $\operatorname{Ran}(A^\dagger) \subset \mathfrak{D}(A)$. So $(I_H - A^\dagger A)x = -A^\dagger(A + \delta A)x$ and hence $(I_H - A^\dagger A)x = 0$, $A^\dagger(A + \delta A)x = 0$, that is, $x \in (\operatorname{Ker} A)^\perp$ and $\bar{A}x \in \operatorname{Ran}(A)^\perp$. From $\bar{A}x \in \operatorname{Ran}(\bar{A}) \cap \operatorname{Ran}(A)^\perp = \{0\}$, we obtain that $x \in \operatorname{Ker} \bar{A}$. Thus, $x \in \operatorname{Ker} \bar{A} \cap (\operatorname{Ker} A)^\perp = \{0\}$, i.e., $\operatorname{Ker}(I_H + A^\dagger \delta A) = 0$.

Similarly, we can obtain $\overline{\operatorname{Ran}(I_K + \delta A A^\dagger)} = K$. □

Let $T \in \mathcal{C}(H, K)$ and $\delta T \in B(H, K)$. Put $\bar{T} = T + \delta T$ and set

$$K(T, \bar{T}) = I_H - P_{\operatorname{Ker} \bar{T}} - P_{\operatorname{Ker} T}, \quad R(T, \bar{T}) = I_K - P_{\overline{\operatorname{Ran}(\bar{T})}} - P_{\overline{\operatorname{Ran}(T)}}.$$

Lemma 3.3.2. *Let $T \in \mathcal{C}(H, K)$ and $\delta T \in B(H, K)$. Put $\bar{T} = T + \delta T$.*

(1) $\operatorname{Ker}(K(T, \bar{T})) = \{0\}$ *if and only if* $\operatorname{Ker} \bar{T} \cap (\operatorname{Ker} T)^\perp = \{0\}$ *and* $\operatorname{Ker} T \cap (\operatorname{Ker} \bar{T})^\perp = \{0\}$.

(2) $\operatorname{Ker}(R(T, \bar{T})) = \{0\}$ *if and only if* $\overline{\operatorname{Ran}(\bar{T})} \cap \operatorname{Ran}(T)^\perp = \{0\}$ *and* $\operatorname{Ran}(\bar{T})^\perp \cap \overline{\operatorname{Ran}(T)} = \{0\}$.

Proof. We only prove (1) since the proof of (2) is similar.

Suppose $\operatorname{Ker}(K(T, \bar{T})) = \{0\}$. Let $x \in \operatorname{Ker} \bar{T} \cap (\operatorname{Ker} T)^\perp$. Then $P_{\operatorname{Ker} \bar{T}} x = x$ and $P_{\operatorname{Ker} T} x = 0$. So $x \in \operatorname{Ker}(K(T, \bar{T})) = \{0\}$. Similarly, we have $(\operatorname{Ker} \bar{T})^\perp \cap \operatorname{Ker} T = \{0\}$.

Conversely, let $x \in \operatorname{Ker}(K(T, \bar{T}))$. Then we have

$$P_{\operatorname{Ker} T} x = (I_H - P_{\operatorname{Ker} \bar{T}}) x \in \operatorname{Ker} T \cap (\operatorname{Ker} \bar{T})^\perp = \{0\}$$

and $P_{\operatorname{Ker} T} x = 0$ and $P_{\operatorname{Ker} \bar{T}} x = x$. So $x \in \operatorname{Ker} \bar{T} \cap (\operatorname{Ker} T)^\perp = \{0\}$. □

Theorem 3.3.3. *Let $A \in \mathcal{C}(H, K)$ with* $\operatorname{Ran}(A)$ *closed and $\delta A \in B(H, K)$. Put $\bar{A} = A + \delta A$. Then the following conditions are equivalent:*

(1) *$I_H + A^\dagger \delta A$ is invertible in $B(H)$ and* $\operatorname{Ran}(\bar{A}) \cap \operatorname{Ran}(A)^\perp = \{0\}$.

(2) *$I_H + A^\dagger \delta A$ is invertible in $B(H)$ and* $\operatorname{Ran}(\bar{A})$ *is closed with $\bar{A}^+ = (I_H + A^\dagger \delta A)^{-1} A^\dagger$.*

(3) *$I_H + A^\dagger \delta A$ is invertible in $B(H)$ and* $\operatorname{Ker} A \cap (\operatorname{Ker} \bar{A})^\perp = \{0\}$.

(4) $\operatorname{Ran}(I_H + A^\dagger \delta A)$ *is closed and* $\operatorname{Ker} \bar{A} \cap (\operatorname{Ker} A)^\perp = \{0\}$, $(\operatorname{Ker} \bar{A})^\perp \cap \operatorname{Ker} A = \{0\}$, $\operatorname{Ran}(\bar{A})^\perp \cap \operatorname{Ran}(A) = \{0\}$, $\operatorname{Ran}(\bar{A}) \cap \operatorname{Ran}(A)^\perp = \{0\}$.

(5) $\operatorname{Ran}(I_H + A^\dagger \delta A)$ *is closed and $K(A, \bar{A})$, $R(A, \bar{A})$ are all injective.*

Proof. The implication (1)⇒(2) comes from Theorem 2.2.6 and the implication (5)⇒(4) comes from Lemma 3.3.2.

(2)⇒(3) By Theorem 2.2.6, $(I_K - AA^\dagger)(I_K + \delta A A^\dagger)^{-1} \bar{A} = 0$ on $\mathfrak{D}(A)$. Note that $(I_K + (A^*)^\dagger (\delta A)^*)^{-1}(I_K - (A^*)^\dagger A^*)$ is an idempotent operator from $\mathfrak{D}(A^*)$ to $\mathfrak{D}(A^*)$ (cf. Lemma 2.2.4). We obtain that

$$\bar{A}^*(I_K + (A^*)^\dagger (\delta A)^*)^{-1}(I_K - (A^*)^\dagger A^*) = 0 \text{ on } \mathfrak{D}(A^*)$$

from $(I_K - AA^\dagger)(I_K + \delta AA^\dagger)^{-1}\bar{A} = 0$ and $(AA^\dagger) = (AA^\dagger)^* = (A^*)^\dagger A^*$ on $\mathfrak{D}(A^*)$. So $\operatorname{Ran}(\bar{A}^*)$ is closed and $\operatorname{Ran}(\bar{A}^*) \cap \operatorname{Ran}(A^*)^\perp = \{0\}$ by using Theorem 2.2.6. Since $\operatorname{Ran}(\bar{A}^*) = (\operatorname{Ker}\bar{A})^\perp$ and $\operatorname{Ran}(A^*)^\perp = \operatorname{Ker} A$, we have $(\operatorname{Ker}\bar{A})^\perp \cap \operatorname{Ker} A = \{0\}$.

(3)⇒(1) Noting that $\operatorname{Ran}(\bar{A}^*) \subset (\operatorname{Ker}\bar{A})^\perp$ and $\operatorname{Ran}(A^*)^\perp = \operatorname{Ker} A$, we have $\operatorname{Ran}(\bar{A}^*) \cap \operatorname{Ran}(A^*)^\perp = \{0\}$. So

$$(I_H - A^*(A^*)^\dagger)(I_H + (\delta A)^*(A^*)^\dagger)^{-1}\bar{A}^* = 0 \text{ on } \mathfrak{D}(A^*)$$

by Theorem 2.2.6. Since $(\bar{A}^*)^* = \bar{A}$, $(A^*)^* = A$ on $\mathfrak{D}(A)$ and $((A^*)^\dagger)^* = A^\dagger$ by Theorem 1.2.8, it follows that $\bar{A}(I_H + A^\dagger\delta A)^{-1}(I_H - A^\dagger A) = 0$ on $\mathfrak{D}(A)$ ($A^\dagger A = A^*(A^*)^\dagger$ on $\mathfrak{D}(A)$). Thus, $\operatorname{Ran}(\bar{A}) \cap \operatorname{Ran}(A)^\perp = \{0\}$ by Theorem 2.2.6.

(3)⇒(4) By Lemma 3.3.1, $\operatorname{Ker}\bar{A} \cap (\operatorname{Ker} A)^\perp = \{0\}$ and $\operatorname{Ran}(\bar{A})^\perp \cap \operatorname{Ran}(A) = \{0\}$. Furthermore, $\operatorname{Ran}(\bar{A}) \cap \operatorname{Ran}(A)^\perp = \{0\}$ by the equivalence of (1) and (3).

(4)⇒(1) By Lemma 3.3.1 and the Inverse Operator Theorem, $I_H + A^\dagger\delta A$ is invertible in $B(H)$.

(4)⇒(5) If (4) holds, then by Lemma 3.3.1, the equivalence of (1) and (4) and by Lemma 3.3.2, we can obtain (5). □

It is well-known that $\operatorname{Ran}(I_H + T)$ is closed when $T \in B(H)$ is a compact operator. So we have by Theorem 3.3.3:

Corollary 3.3.4. *Let $A \in C(H,K)$ with $\operatorname{Ran}(A)$ closed and $\delta A \in B(H,K)$ be a compact operator. Put $\bar{A} = A + \delta A$. If both $K(A,\bar{A})$ and $R(A,\bar{A})$ are all injective, then $I_H + A^\dagger\delta A$ is invertible in $B(H)$ and $\bar{A}^+ = (I_H + A^\dagger\delta A)^{-1}A^\dagger$.*

Theorem 3.3.5. *Let $A \in C(H,K)$ with $\operatorname{Ran}(A)$ closed and $\delta A \in B(H,K)$ with $I_H + A^\dagger\delta A$ invertible in $B(H)$. Put $\bar{A} = A + \delta A$. Assume that $\bar{A}$ is the stable perturbation of A. Then $\bar{A}^\dagger \in B(K,H)$ and*

$$\|\bar{A}^\dagger\| \le \|(I_H + A^\dagger\delta A)^{-1}\|\|A^\dagger\|,$$

$$\|\bar{A}^\dagger - A^\dagger\| \le \frac{1+\sqrt{5}}{2}\|(I_H + A^\dagger\delta A)^{-1}\|\|A^\dagger\|^2\|\delta A\|.$$

Proof. By Theorem 3.3.3, $\bar{A}^+ = (I_H + A^\dagger\delta A)^{-1}A^\dagger$. Therefore, $\bar{A}^\dagger = P_{(\operatorname{Ker}\bar{A})^\perp}\bar{A}^+O(\bar{A}\bar{A}^+)$ by Proposition 3.1.10 and hence

$$\|\bar{A}^\dagger\| \le \|\bar{A}^+\| \le \|(I_H + A^\dagger\delta A)^{-1}\|\|A^\dagger\|.$$

From the identity

$$\bar{A}^\dagger - A^\dagger = -\bar{A}^\dagger\delta AA^\dagger + \bar{A}^\dagger(I_K - AA^\dagger) - (I_H - \bar{A}^\dagger\bar{A})A^\dagger, \tag{3.3.1}$$

we get that by applying the orthogonality of the operators on the right side of (3.3.1), i.e., $\operatorname{Ran}(\bar{A}^\dagger) = (\operatorname{Ker}\bar{A})^\perp \cap \mathfrak{D}(A)$ and $\operatorname{Ran}(I_H - \bar{A}^\dagger\bar{A}) = \operatorname{Ker}\bar{A}$,

$$\|(\bar{A}^\dagger - A^\dagger)x\|^2 = \|\bar{A}^\dagger(-\delta AA^\dagger + I_K - AA^\dagger)x\|^2 + \|(I_H - \bar{A}^\dagger\bar{A})A^\dagger x\|^2 \quad (3.3.2)$$

where $x \in K$ with $\|x\| = 1$. Set $\sin\phi = \|AA^\dagger x\|$, $\cos\phi = \|(I_K - AA^\dagger)x\|$. We have

$$\begin{aligned}
&\|\bar{A}^\dagger(-\delta AA^\dagger + I_K - AA^\dagger)x\| \\
&\quad = \|-\bar{A}^\dagger\delta AA^\dagger(AA^\dagger x) + \bar{A}^\dagger(\bar{A}\bar{A}^\dagger - AA^\dagger)(I_K - AA^\dagger)x\| \\
&\quad \le \|\bar{A}^\dagger\|\|\delta A\|\|A^\dagger\|\sin\phi + \|\bar{A}^\dagger\|\|P_{\operatorname{Ran}(\bar{A})} - P_{\operatorname{Ran}(A)}\|\cos\phi \\
\|(I_H - \bar{A}^\dagger\bar{A})A^\dagger x\| &= \|(A^\dagger A - \bar{A}^\dagger\bar{A})A^\dagger(AA^\dagger x)\| \\
&= \|(P_{\operatorname{Ker}\bar{A}} - P_{\operatorname{Ker}A})A^\dagger(AA^\dagger x)\| \\
&\le \|P_{\operatorname{Ker}\bar{A}} - P_{\operatorname{Ker}A}\|\|A^\dagger\|\sin\phi.
\end{aligned}$$

We also have, by Theorem 3.3.3, Corollary 1.3.9 and Proposition 3.1.9,

$$\|P_{\operatorname{Ran}(\bar{A})} - P_{\operatorname{Ran}(A)}\| \le \|A^\dagger\|\|\delta A\|, \ \|P_{\operatorname{Ker}\bar{A}} - P_{\operatorname{Ker}A}\| \le \|\bar{A}^\dagger\|\|\delta A\|.$$

Thus, from (3.3.2), we get that

$$\begin{aligned}
\|(\bar{A}^\dagger - A^\dagger)x\|^2 &\le (\|\bar{A}^\dagger\|\|A^\dagger\|\|\delta A\|)^2((\sin\phi + \cos\phi)^2 + \sin^2\phi) \\
&= \frac{1}{2}\|\bar{A}^\dagger\|^2\|A^\dagger\|^2\|\delta A\|^2(3 + 2\sin 2\phi - \cos 2\phi) \\
&\le \frac{3+\sqrt{5}}{2}\|\bar{A}^\dagger\|^2\|A^\dagger\|^2\|\delta A\|^2
\end{aligned}$$

and hence $\|\bar{A}^\dagger - A^\dagger\| \le \dfrac{1+\sqrt{5}}{2}\|(I_H + A^\dagger\delta A)^{-1}\|\|A^\dagger\|^2\|\delta A\|$. □

We now characterize the continuity of the Moore–Penrose inverse of the bounded linear operator on Hilbert spaces as follows:

Corollary 3.3.6. *Let* $\{T_n\} \subset B(H,K)\backslash\{0\}$ *and* $T \in B(H,K)\backslash\{0\}$ *with* $\operatorname{Ran}(T_n)$ *and* $\operatorname{Ran}(T)$ *closed for any* n. *Suppose that* $T_n \xrightarrow{\|\cdot\|} T$ *as* $n \to \infty$. *Then the following statements are equivalent:*

(1) $T_n^\dagger \xrightarrow{\|\cdot\|} T^\dagger$ *as* $n \to \infty$.
(2) $T_n^\dagger T_n \xrightarrow{\|\cdot\|} T^\dagger T$ *as* $n \to \infty$.
(3) $T_n T_n^\dagger \xrightarrow{\|\cdot\|} TT^\dagger$ *as* $n \to \infty$.
(4) $\|T_n^\dagger\| \xrightarrow{\|\cdot\|} \|T^\dagger\|$ *as* $n \to \infty$.
(5) $\sup\limits_{n\ge 1}\|T_n^\dagger\| < +\infty$.

(6) $(\operatorname{Ker} T_n)^\perp \cap \operatorname{Ker} T = \{0\}$ *for n large enough.*
(7) $\operatorname{Ran}(T_n) \cap \operatorname{Ran}(T)^\perp = \{0\}$ *for n large enough.*

Proof. The implications (1)$\Rightarrow$(2), (1)$\Rightarrow$(3), (1)$\Rightarrow$(4) and (4)$\Rightarrow$(5) are obvious.

(2)$\Rightarrow$(6) Choose a nature number N such that $\|T_n^\dagger T_n - T^\dagger T\| < 1$, $\forall\, n > N$. Let $x \in (\operatorname{Ker} T_n)^\perp \cap \operatorname{Ker} T = \{0\}$. Then $Tx = 0$ and $T_n^\dagger T_n x = x$, $\forall\, n > N$. If $x \neq 0$, then

$$\|x\| = \|(T_n^\dagger T_n - T^\dagger T)x\| \leq \|T_n^\dagger T_n - T^\dagger T\|\|x\| < \|x\|.$$

But this is impossible. So $(\operatorname{Ker} T_n)^\perp \cap \operatorname{Ker} T = \{0\}$, $\forall\, n > N$.

Similarly, we can obtain (3)$\Rightarrow$(7).

(6)$\Rightarrow$(7) Pick N such that $\|T^\dagger\|\|T_n - T\| < 1$ for any $n > N$. Thus, by Theorem 3.3.3. $\operatorname{Ran}(T_n) \cap \operatorname{Ran}(T)^\perp = \{0\}$, $n > N$.

Similarly, we can obtain (7)$\Rightarrow$(6) by using Theorem 3.3.3.

(6)$\Rightarrow$(1) For n large enough, we have $\operatorname{Ran}(T_n) \cap \operatorname{Ran}(T)^\perp = \{0\}$ and $\|T^\dagger\|\|T_n - T\| < 1$. Then by Theorem 3.3.5, $T_n^\dagger \xrightarrow{\|\cdot\|} T^\dagger$.

(5)$\Rightarrow$(6) By Lemma 1.3.5 and Proposition 3.1.8 (3),

$$\delta(\operatorname{Ran}(T_n), \operatorname{Ran}(T)) \leq \|T_n^\dagger\|\|T_n - T\|, \ n \in \mathbb{N}.$$

Since $M = \sup\limits_{n \geq 1} \|T_n^\dagger\| < +\infty$, we have $M\|T_n - T\| < 1$ for n large enough. Thus, $\operatorname{Ran}(T_n) \cap \operatorname{Ran}(T)^\perp = \{0\}$ by Proposition 2.2.10. $\square$

Let $A \in \mathcal{C}(H, K)$ with $\operatorname{Ran}(A)$ closed. Let $b \in K\backslash\{0\}$. Consider the least square problem

$$\min \|x\| \text{ subject to } \|Ax - b\| = \inf_{z \in \mathfrak{D}(A)} \|Az - b\|, \ x \in \mathfrak{D}(A). \tag{3.3.3}$$

Since $\operatorname{Ran}(A)$ is closed, $\inf\limits_{z \in \mathfrak{D}(A)} \|Az - b\| = \|b - P_{\operatorname{Ran}(A)} b\|$. So the solution of (3.3.3) is $x_m \in S(A, P_{\operatorname{Ran}(A)} b) = \{x \in \mathfrak{D}(A) |\, Ax = P_{\operatorname{Ran}(A)} b\}$ such that $\|x_m\| = \inf\{\|x\| |\, x \in S(A, P_{\operatorname{Ran}(A)} b)\}$. By means of Parallogma law, we can deduce that x_m is the unique solution of (3.3.3). Note that $A^\dagger b \in S(A, P_{\operatorname{Ran}(A)} b)$. So for any $x \in S(A, P_{\operatorname{Ran}(A)} b)$, $AA^\dagger b = P_{\operatorname{Ran}(A)} b = Ax$ and $A^\dagger Ax = A^\dagger AA^\dagger b = A^\dagger b$. Since

$$x = A^\dagger Ax + (I_K - A^\dagger A)x = P_{(\operatorname{Ker} A)^\perp} x + P_{\operatorname{Ker} A} x,$$

we have $\|A^\dagger b\| = \|A^\dagger Ax\| \leq \|x\|$, which means that $x_m = A^\dagger b$.

Suppose that the problem (3.3.3) is perturbed to

$$\min \|x\| \text{ subject to } \|\bar{A}x - \bar{b}\| = \inf_{z \in \mathfrak{D}(A)} \|\bar{A}z - \bar{b}\|, \ x \in \mathfrak{D}(A), \tag{3.3.4}$$

where $\bar{b} = b + \delta b \in K$ and $\bar{A} = A + \delta A \in \mathcal{C}(H, K)$ with $\delta A \in B(H, K)$. If $\mathrm{Ran}\,(\bar{A})$ is closed, (3.3.4) has the unique solution $\bar{x}_m = \bar{A}^\dagger \bar{b}$.

Theorem 3.3.7. *Let $A \in \mathcal{C}(H, K)$ with $\mathrm{Ran}\,(A)$ closed and $\delta A \in B(H, K)$ with $\|A^\dagger\|\|\delta A\| < 1$. Let $b \in K \backslash \{0\}$ and put $\bar{A} = A + \delta A$, $\bar{b} = b + \delta b \in K$. Let x_m and $\bar{x}_m$ be as above. Suppose that $\mathrm{Ran}\,(\bar{A}) \cap \mathrm{Ran}\,(A)^\perp = \{0\}$. Then*

$$\frac{\|\bar{x}_m - x_m\|}{\|x_m\|} \le \frac{\|A^\dagger\|}{1 - \|A^\dagger\|\|\delta A\|} \Big(\frac{\|\delta b\|}{\|x_m\|} + \|\delta A\| + \|A^\dagger\|\|\delta A\| \frac{\|b - Ax_m\|}{\|x_m\|} \Big) + \|A^\dagger\|\|\delta A\|.$$

Proof. According to the assumption, we have $\mathrm{Ran}\,(\bar{A})$ is closed and $\|\bar{A}^\dagger\| \le \dfrac{\|A^\dagger\|}{1 - \|A^\dagger\|\|\delta A\|}$ by using Theorem 3.3.3.

Since $\bar{x}_m - x_m = \bar{A}^\dagger \delta b + (\bar{A}^\dagger - A^\dagger) b$, it follows from (3.3.1) that

$$\|\bar{x}_m - x_m\| \le \|\bar{A}^\dagger\|\|\delta b\| + \|\bar{A}^\dagger\|\|\delta A\|\|x_m\| + \|\bar{A}^\dagger\|\|\bar{A}\bar{A}^\dagger - AA^\dagger\|\|b - AA^\dagger b\| + \|P_{\mathrm{Ker}\,\bar{A}} - P_{\mathrm{Ker}\,A}\|\|A^\dagger b\|.$$

Now by Theorem 3.3.3, Corollary 1.3.9 and Proposition 3.1.9,

$$\|P_{\mathrm{Ran}\,(\bar{A})} - P_{\mathrm{Ran}\,(A)}\| \le \|A^\dagger\|\|\delta A\|, \ \|P_{\mathrm{Ker}\,\bar{A}} - P_{\mathrm{Ker}\,A}\| \le \|A^\dagger\|\|\delta A\|.$$

Therefore the assertion follows. □

3.4 The representations of certain Moore–Penrose inverses

In this section, we give some representations of the Moore–Penrose inverses, such as the perturbed operator under the stable perturbation, $A - XY^*$ and $\begin{bmatrix} A & B \\ C & D \end{bmatrix}$ etc..

Proposition 3.4.1. *Let $T \in \mathcal{C}(H, K)$ with $\mathrm{Ran}\,(T)$ closed and $\delta A \in B(H, K)$ with $I_H + T^\dagger \delta T$ invertible in $B(H)$. Put $\bar{T} = T + \delta T$. Then the following statements are equivalent:*

(1) $\bar{T}^\dagger = (I_H + T^\dagger \delta T)^{-1} T^\dagger = T^\dagger (I_K + \delta T T^\dagger)^{-1}$.
(2) $\mathrm{Ran}\,(\bar{T}) = \mathrm{Ran}\,(T)$ *and* $\mathrm{Ker}\,\bar{T} = \mathrm{Ker}\,T$.
(3) $\mathrm{Ran}\,(\delta T) \subset \mathrm{Ran}\,(T)$ *and* $\mathrm{Ker}\,T \subset \mathrm{Ker}\,\delta T$.

Proof. (1)⇒(2) We have $\mathrm{Ran}\,(\bar{T}^\dagger) = \mathrm{Ran}\,(T^\dagger)$ and $\mathrm{Ker}\,\bar{T}^\dagger = \mathrm{Ker}\,T^\dagger$. Since $\mathrm{Ran}\,(\bar{T})$ is closed by Theorem 3.3.3 and $\mathrm{Ran}\,(T)$ is closed and $\mathrm{Ker}\,\bar{T}^\dagger = \mathrm{Ran}\,(\bar{T})^\perp = \mathrm{Ker}\,T^\dagger = \mathrm{Ran}\,(T)^\perp$, we get that $\mathrm{Ran}\,(\bar{T}) = \mathrm{Ran}\,(T)$.

Let $x \in \operatorname{Ker}\bar{T}$. Decompose x as $x = x_1 + x_2$, where $x_1 \in \operatorname{Ker} T$ and $x_2 \in (\operatorname{Ker} T)^{\perp}$. Then $x_2 \in \mathfrak{D}(A)$. Note that

$$\operatorname{Ran}(\bar{T}^{\dagger}) = (\operatorname{Ker}\bar{T})^{\perp} \cap \mathfrak{D}(T) = \operatorname{Ran}(T^{\dagger}) = (\operatorname{Ker} T)^{\perp} \cap \mathfrak{D}(T).$$

So $x_2 \in (\operatorname{Ker}\bar{T})^{\perp}$ and hence $0 = (x, x_2) = (x_1, x_2) + (x_2, x_2) = (x_2, x_2)$. Thus, $x = x_1$, that is, $\operatorname{Ker}\bar{T} \subset \operatorname{Ker} T$. Similarly, $\operatorname{Ker} T \subset \operatorname{Ker}\bar{T}$.

(2)⇒(3) Let $x \in \operatorname{Ker} T = \operatorname{Ker}\bar{T}$. Then $Tx = 0$ and $Tx + \delta T x = 0$. So $\delta T x = 0$, that is, $\operatorname{Ker} T \subset \operatorname{Ker}\delta T$.

Let $x \in \mathfrak{D}(T)$. Then there exists $x_0 \in \mathfrak{D}(T)$ such that $Tx_0 = \bar{T}x = Tx + \delta T x$. Thus, $\delta T x = T(x_0 - x) \in \operatorname{Ran}(T)$. Since δT is bounded, $\operatorname{Ran}(T)$ is closed and $\overline{\mathfrak{D}(T)} = H$, we have $\operatorname{Ran}(\delta T) \subset \operatorname{Ran}(T)$.

(3)⇒(1) $\operatorname{Ran}(\delta T) \subset \operatorname{Ran}(T)$ implies that $\operatorname{Ran}(\bar{T}) \subset \operatorname{Ran}(T)$. Thus, $\operatorname{Ran}(\bar{T}) \cap \operatorname{Ran}(T)^{\perp} = \{0\}$. So $\operatorname{Ran}(\bar{T})$ is closed and

$$\bar{T}^{+} = T^{+}(I_K + \delta T T^{\dagger})^{-1} = (I_H + T^{\dagger}\delta T)^{-1} T^{\dagger}$$

by Theorem 3.3.3. Using $TT^{\dagger}\delta T = \delta T$ and $\delta T = \delta T T^{\dagger} T$ on $\mathfrak{D}(T)$, we have

$$\bar{T}\bar{T}^{+} = (T + \delta T)T^{\dagger}(I_K + \delta T T^{\dagger})^{-1} = TT^{+}(I_K + \delta T T^{\dagger})(I_K + \delta T T^{\dagger})^{-1} = TT^{\dagger}$$
$$\bar{T}^{+}\bar{T} = (I_H + T^{\dagger}\delta T)^{-1} T^{\dagger}(T + \delta T) = (I_H + T^{\dagger}\delta T)^{-1}(I_H + T^{\dagger}\delta T)T^{\dagger}T = T^{\dagger}T.$$

This shows that $\operatorname{Ran}(\bar{T}) = \operatorname{Ran}(T)$ and $\operatorname{Ker}\bar{T} = \operatorname{Ker} T$ since $\bar{T}\bar{T}^{+}\bar{T} = \bar{T}$ and $\bar{T}^{+}\bar{T}\bar{T}^{+} = \bar{T}^{+}$. Thus, $\bar{T}\bar{T}^{+} = P_{\operatorname{Ran}(\bar{T})}$ and $\bar{T}^{+}\bar{T}x = P_{(\operatorname{Ker}\bar{T})^{\perp}}x$ on $\mathfrak{D}(T)$. Consequently, $\bar{T}^{\dagger} = \bar{T}^{+}$. □

Let $A \in B(H)$, $B \in B(K, H)$, $C \in B(H, K)$ and $D \in B(K)$. Then $\begin{bmatrix} A & B \\ C & D \end{bmatrix}$ is the bounded linear operator from $H \oplus K$ to $H \oplus K$.

Proposition 3.4.2. *$A \in B(H)$ with $\operatorname{Ran}(A)$ closed, $B \in B(K, H)$, $C \in B(H, K)$ and $D \in B(K)$ with $\operatorname{Ran}(D)$ closed.*

(1) *Suppose that $M = I_K - D^{\dagger}CA^{\dagger}B$ is invertible in $B(K)$, then*

$$\begin{bmatrix} A & B \\ C & D \end{bmatrix}^{\dagger} = \begin{bmatrix} A^{\dagger} + A^{\dagger}BM^{-1}D^{\dagger}CA^{\dagger} & -A^{\dagger}BM^{-1}D^{\dagger} \\ -M^{-1}D^{\dagger}CA^{\dagger} & M^{-1}D^{\dagger} \end{bmatrix}$$

iff $\operatorname{Ran}(B) \subset \operatorname{Ran}(A)$, $\operatorname{Ran}(C) \subset \operatorname{Ran}(D)$, $\operatorname{Ran}(C^*) \subset \operatorname{Ran}(A^*)$ *and* $\operatorname{Ran}(B^*) \subset \operatorname{Ran}(D^*)$.

Especially, When $C = 0$, then $\begin{bmatrix} A & B \\ 0 & D \end{bmatrix}^{\dagger} = \begin{bmatrix} A^{\dagger} & -A^{\dagger}BD^{\dagger} \\ 0 & D^{\dagger} \end{bmatrix}$ *iff* $\operatorname{Ran}(B) \subset \operatorname{Ran}(A)$ *and* $\operatorname{Ran}(B^*) \subset \operatorname{Ran}(D^*)$.

(2) *If* $(I_H - AA^\dagger)B(I_K - D^\dagger D) = 0$, *then* $\begin{bmatrix} A & B \\ 0 & D \end{bmatrix}^\dagger = \begin{bmatrix} X_1 & X_2 \\ X_3 & X_4 \end{bmatrix}$, *where*

$$\begin{aligned}
&E = A^\dagger B(I_K - D^\dagger D), \quad F = (I_H - AA^\dagger)BD^\dagger \\
&X_1 = (I_H + EE^*)^{-1}A^\dagger - (I_H + EE^*)^{-1}A^\dagger BD^\dagger(I_H + FF^*)^{-1}F^*, \\
&X_2 = -(I_H + EE^*)^{-1}A^\dagger BD^\dagger(I_K + F^*F)^{-1}, \\
&X_3 = (I_K + E^*E)^{-1}E^*A^\dagger + D^\dagger(I_K + F^*F)^{-1}F^* \\
&\qquad - (I_K + E^*E)^{-1}E^*A^\dagger BD^\dagger(I_K + F^*F)^{-1}F^* \\
&X_4 = D^\dagger(I_K + F^*F)^{-1} - (I_K + E^*E)^{-1}E^*A^\dagger BD^\dagger(I_K + F^*F)^{-1}.
\end{aligned}$$

Proof. (1) Set $T = \begin{bmatrix} A & \\ & D \end{bmatrix}$ and $\delta T = \begin{bmatrix} 0 & B \\ C & 0 \end{bmatrix}$. Set $\bar{T} = T + \delta T = \begin{bmatrix} A & B \\ C & D \end{bmatrix}$. It is easy to verify that $\mathrm{Ran}\,(\delta T) \subset \mathrm{Ran}\,(T)$ iff $\mathrm{Ran}\,(B) \subset \mathrm{Ran}\,(A)$, $\mathrm{Ran}\,(C) \subset \mathrm{Ran}\,(D)$ and $\mathrm{Ker}\,T \subset \mathrm{Ker}\,\delta T$ iff $\mathrm{Ran}\,(C^*) \subset \mathrm{Ran}\,(A^*)$, $\mathrm{Ran}\,(B^*) \subset \mathrm{Ran}\,(D^*)$. Noting that $I_{H\oplus K} + T^\dagger \delta T = \begin{bmatrix} I_H & A^\dagger B \\ D^\dagger C & I_K \end{bmatrix}$ is invertible in $B(H \oplus K)$ when M is invertible in $B(K)$. Thus, by Proposition 3.4.1, $\bar{T}^\dagger$ has the form $\bar{T}^\dagger = (I_{H\oplus K} + T^\dagger \delta T)^{-1}T^\dagger$ iff $\mathrm{Ker}\,T \subset \mathrm{Ker}\,\delta T$ and $\mathrm{Ran}\,(\delta T) \subset \mathrm{Ran}\,(T)$. Since

$$(I_{H\oplus K} + T^\dagger \delta T)^{-1} = \begin{bmatrix} I_H & A^\dagger B \\ D^\dagger C & I_K \end{bmatrix}^{-1} = \begin{bmatrix} I_H + A^\dagger BM^{-1}D^\dagger C & -A^\dagger BM^{-1} \\ -M^{-1}D^\dagger C & M^{-1} \end{bmatrix},$$

we get that

$$(I_{H\oplus K} + T^\dagger \delta T)^{-1}T^\dagger = \begin{bmatrix} A^\dagger + A^\dagger BM^{-1}D^\dagger CA^\dagger & -A^\dagger BM^{-1}D^\dagger \\ -M^{-1}D^\dagger CA^\dagger & M^{-1}D^\dagger \end{bmatrix}. \tag{3.4.1}$$

So by Proposition 3.4.1, $\begin{bmatrix} A & B \\ C & D \end{bmatrix}^\dagger$ has the form (3.4.1) iff $\mathrm{Ran}\,(B) \subset \mathrm{Ran}\,(A)$, $\mathrm{Ran}\,(C) \subset \mathrm{Ran}\,(D)$, $\mathrm{Ran}\,(C^*) \subset \mathrm{Ran}\,(A^*)$ and $\mathrm{Ran}\,(B^*) \subset \mathrm{Ran}\,(D^*)$.

When $C = 0$, $M = I_K$. In this case, we have $\begin{bmatrix} A & B \\ 0 & D \end{bmatrix}^\dagger = \begin{bmatrix} A^\dagger & -A^\dagger BD^\dagger \\ 0 & D^\dagger \end{bmatrix}$ iff $\mathrm{Ran}\,(B) \subset \mathrm{Ran}\,(A)$ and $\mathrm{Ran}\,(B^*) \subset \mathrm{Ran}\,(D^*)$.

(2) Put $T = \begin{bmatrix} A & \\ & D \end{bmatrix}$ and $\delta T = \begin{bmatrix} 0 & B \\ 0 & 0 \end{bmatrix}$. Then $\bar{T} = T + \delta T = \begin{bmatrix} A & B \\ 0 & D \end{bmatrix}$. Let $y = \begin{pmatrix} y_1 \\ y_2 \end{pmatrix} \in \mathrm{Ran}\,(\bar{T}) \cap \mathrm{Ran}\,(T)^\perp = \mathrm{Ran}\,(\bar{T}) \cap \mathrm{Ker}\,(T^\dagger)$. Then

$$A^\dagger y_1 = 0, \; D^\dagger y_2 = 0, \; y_1 = Ax_1 + Bx_2, \; y_2 = Dx_2$$

for some $x_1 \in H$ and $x_2 \in K$. We have $x_2 \in \operatorname{Ker} D$, $y_2 = 0$ and $y_1 = (I_H - AA^\dagger)Bx_2$. Consequently, $(I_H - AA^\dagger)B(I_K - D^\dagger D) = 0$ implies that $y_1 = 0$ and hence $\bar{T}$ is the stable perturbation of T. Therefore, $\operatorname{Ran}(\bar{T})$ is closed and $\bar{T}^+ = (I_{H\oplus K} + T^\dagger \delta T)^{-1} T^\dagger = \begin{bmatrix} A^\dagger & -A^\dagger B D^\dagger \\ 0 & D^\dagger \end{bmatrix}$ by Theorem 3.3.3. Simple computation shows that

$$\bar{T}^+\bar{T} = \begin{bmatrix} A^\dagger A & A^\dagger B(I_K - D^\dagger D) \\ 0 & D^\dagger D \end{bmatrix}, \quad \bar{T}\bar{T}^+ = \begin{bmatrix} AA^\dagger & (I_H - AA^\dagger)BD^\dagger \\ 0 & DD^\dagger \end{bmatrix}.$$

Noting that $(2AA^\dagger - I_H)(I_H - AA^\dagger) = AA^\dagger - I_H$, $(2D^\dagger D - I_K)(I_K - D^\dagger D) = D^\dagger D - I_K$, we have

$$(\bar{T}^+\bar{T} + (\bar{T}^+\bar{T})^* - I_{H\oplus K})^{-1} = \begin{bmatrix} I_H & E \\ -E^* & I_K \end{bmatrix}^{-1} \begin{bmatrix} 2AA^\dagger - I_H & \\ & 2D^\dagger D - I_K \end{bmatrix}$$

$$= \begin{bmatrix} (I_H + EE^*)^{-1} & -E(I_K + E^*E)^{-1} \\ (I_K + E^*E)^{-1}E^* & (I_K + E^*E)^{-1} \end{bmatrix} \begin{bmatrix} 2A^\dagger A - I_H & \\ & 2D^\dagger D - I_K \end{bmatrix}$$

$$(\bar{T}\bar{T}^+ + (\bar{T}\bar{T}^+)^* - I_{H\oplus K})^{-1} = \begin{bmatrix} 2AA^\dagger - I_H & \\ & 2DD^\dagger - I_K \end{bmatrix} \begin{bmatrix} I_H & F \\ -F^* & I_K \end{bmatrix}^{-1}$$

$$= \begin{bmatrix} 2AA^\dagger - I_H & \\ & 2DD^\dagger - I_K \end{bmatrix} \begin{bmatrix} (I_H + FF^*)^{-1} & -F(I_K + F^*F)^{-1} \\ (I_K + F^*F)^{-1}F^* & (I_K + F^*F)^{-1} \end{bmatrix}.$$

Finally, applying the above to Proposition 3.1.10 and using identities

$$ED^\dagger = 0, \ (I_K + E^*E)^{-1}D^\dagger = D^\dagger, \ A^\dagger F = 0, \ A^\dagger(I_H + FF^*)^{-1} = A^\dagger,$$

we can obtain the assertion. □

Let $A \in B(H)$ and $X, Y \in B(K, H)$. If A is invertible in $B(H)$, then from the identities

$$\begin{bmatrix} A - XY^* & 0 \\ 0 & I_K \end{bmatrix} = \begin{bmatrix} I_H & -X \\ 0 & I_K \end{bmatrix} \begin{bmatrix} A & X \\ Y^* & I_K \end{bmatrix} \begin{bmatrix} I_H & 0 \\ -Y^* & I_K \end{bmatrix}$$

$$\begin{bmatrix} A & X \\ Y^* & I_K \end{bmatrix} = \begin{bmatrix} I_H & 0 \\ Y^*A^{-1} & I_K \end{bmatrix} \begin{bmatrix} A & 0 \\ 0 & I_K - Y^*A^{-1}X \end{bmatrix} \begin{bmatrix} I_H & A^{-1}X \\ 0 & I_K \end{bmatrix},$$

we see that $A - XY^*$ is invertible in $B(H)$ iff $I_K - Y^*A^{-1}X$ is invertible in $B(K)$ and in this case, the well–known Shermen–Morrison–Woodbury (SMW) formula can be expressed as

$$(A - XY^*)^{-1} = A^{-1} + A^{-1}X(I_K - Y^*A^{-1}X)^{-1}Y^*A^{-1}$$

(cf. [Kurt and Riedel (1991)]). This formula can be generalized to the case of Moore–Penrose inverses as follows:

Proposition 3.4.3. *Let $A \in B(H)$ with* $\operatorname{Ran}(A)$ *closed and $X, Y \in B(K, H)$ with* $\operatorname{Ran}(X) \subset \operatorname{Ran}(A)$, $\operatorname{Ran}(Y) \subset \operatorname{Ran}(A^*)$.

(1) *If* $I_K - Y^*A^\dagger X$ *is invertible in* $B(K)$, *then*

$$(A - XY^*)^\dagger = A^\dagger + A^\dagger X(I_K - Y^*A^\dagger X)^{-1}Y^*A^\dagger. \tag{3.4.2}$$

(2) *If* $XY^*A^\dagger XY^* = XY^*$, *then*

$$(A-XY^*)^\dagger = [I_H-(A^\dagger XY^*)(A^\dagger XY^*)^\dagger]A^\dagger[I_H-(XY^*A^\dagger)^\dagger(XY^*A^\dagger)]. \tag{3.4.3}$$

Especially, if $XY^*A^\dagger X = X$ *and* $Y^*A^\dagger XY^* = Y^*$, *then*

$$(A - XY^*)^\dagger = [I_H - (A^\dagger X)(A^\dagger X)^\dagger]A^\dagger[I_H - (Y^*A^\dagger)^\dagger(Y^*A^\dagger)]. \tag{3.4.4}$$

(3) *Assume that* $\mathrm{Ran}\,(A - XY^*)$ *is closed. If* $\mathrm{Ran}\,(XY^*)$ *is closed, then* (3.4.3) *implies that* $YX^*A^\dagger YX^* = XY^*$; *if* $\mathrm{Ran}\,(X)$ *and* $\mathrm{Ran}\,(Y)$ *are closed, then* (3.4.4) *implies that* $XY^*A^\dagger X = X$ *and* $Y^*A^\dagger XY^* = Y^*$.

Proof. (1) Since $\mathrm{Ran}\,(X) \subset \mathrm{Ran}\,(A)$ and $\mathrm{Ran}\,(Y) \subset \mathrm{Ran}\,(A^*)$ implies that $\mathrm{Ran}\,(XY^*) \subset \mathrm{Ran}\,(A)$, $\mathrm{Ker}\,A = \mathrm{Ran}\,(A^*)^\perp \subset \mathrm{Ran}\,(Y)^\perp \subset \mathrm{Ker}\,(XY^*)$ and the invertiblity of $I_K - Y^*A^\dagger X$ in $B(K)$ indicates that $I_K - A^\dagger XY^*$ is invertible in $B(K)$, it follows from Proposition 3.4.1 that

$$\begin{aligned}(A - XY^*)^\dagger &= (I_K - A^\dagger XY^*)^{-1}A^\dagger \\ &= A^\dagger + (I_K - A^\dagger XY^*)^{-1}[A^\dagger - (I_K - A^\dagger XY^*)A^\dagger] \\ &= A^\dagger + (I_K - A^\dagger XY^*)^{-1}A^\dagger XY^*A^\dagger \\ &= A^\dagger + A^\dagger X(I_K - Y^*A^\dagger X)^{-1}Y^*A^\dagger\end{aligned}$$

(2) Set $W = A - XY^*$. Since $XY^*A^\dagger XY^* = XY^*$ and $AA^\dagger X = X$, $A^\dagger AY = Y$, we have

$$\begin{aligned}WA^\dagger W &= (AA^\dagger - XY^*A^\dagger)(A - XY^*) \\ &= A - XY^*A^\dagger A - AA^\dagger XY^* + XY^*A^\dagger XY^* \\ &= W.\end{aligned}$$

Thus, $\mathrm{Ran}\,(W)$ is closed and

$$W^\dagger = (A^\dagger W + (A^\dagger W)^* - I_H)^{-1}A^\dagger WA^\dagger(WA^\dagger + (WA^\dagger)^* - I_H)^{-1}$$

by Proposition 3.1.10. Since

$$\begin{aligned}A^\dagger WA^\dagger &= A^\dagger WA^\dagger WA^\dagger = (I_H - A^\dagger XY^*)A^\dagger(I_H - XY^*A^\dagger), \\ (WA^\dagger + (WA^\dagger)^* - I_H)^{-1} &= (2AA^\dagger - I_H - XY^*A^\dagger - (XY^*A^\dagger)^*) \\ &= (2AA^\dagger - I_H)(I_H - XY^*A^\dagger - (XY^*A^\dagger)^*)^{-1}, \\ (A^\dagger W + (A^\dagger W)^* - I_H)^{-1} &= (2A^\dagger A - I_H - A^\dagger XY^* - (A^\dagger XY^*)^*)^{-1} \\ &= (I_H - A^\dagger XY^* - (A^\dagger XY^*)^*)^{-1}(2A^\dagger A - I_H),\end{aligned}$$

it follows that

$$\begin{aligned}(A-XY^*)^\dagger &= (I_H - A^\dagger XY^* - (A^\dagger XY^*)^*)^{-1}(I_H - A^\dagger XY^*)A^\dagger \\ &\quad \times (I_H - XY^*A^\dagger)(I_H - XY^*A^\dagger - (XY^*A^\dagger)^*)^{-1}.\end{aligned}$$

From $XY^*A^\dagger XY^* = XY^*$, we get that $A^\dagger XY^*$ and $XY^*A^\dagger$ are idempotent. Thus, by Proposition 1.1.30,

$$\begin{aligned}&(I_H - XY^*A^\dagger)(I_H - XY^*A^\dagger - (XY^*A^\dagger)^*)^{-1} = O(I_H - XY^*A^\dagger),\\ &(I_H - A^\dagger XY^* - (A^\dagger XY^*)^*)^{-1}(I_H - A^\dagger XY^*)\\ &\qquad = I_H + (I_H - A^\dagger XY^* - (A^\dagger XY^*)^*)^{-1}(A^\dagger XY^*)^*\\ &\qquad = I_H + (A^\dagger XY^*)(I_H - A^\dagger XY^* - (A^\dagger XY^*)^*)^{-1}\\ &\qquad = I_H - O(A^\dagger XY^*).\end{aligned}$$

Note that $\mathrm{Ran}\,(I_H - XY^*A^\dagger) = \mathrm{Ker}\,(XY^*A^\dagger)$. So $O(I_H - XY^*A^\dagger) = I_H - (XY^*A^\dagger)^\dagger(XY^*A^\dagger)$ and

$$\begin{aligned}(A-XY^*)^\dagger &= (I_H - O(A^\dagger XY^*))A^\dagger O(I_H - XY^*A^\dagger)\\ &= [I_H - (A^\dagger XY^*)(A^\dagger XY^*)^\dagger]A^\dagger[I_H - (XY^*A^\dagger)^\dagger(XY^*A^\dagger)].\end{aligned}$$

If $XY^*A^\dagger X = X$ and $Y^*A^\dagger XY^* = Y^*$, then $(A^\dagger X)^+ = Y^*$, $(Y^*A^\dagger)^+ = X$. Hence $\mathrm{Ran}\,(A^\dagger X) = \mathrm{Ran}\,(A^\dagger XY^*)$, $\mathrm{Ker}\,(Y^*A^\dagger) = \mathrm{Ker}\,(XY^*A^\dagger)$ and so that $(A^\dagger X)(A^\dagger X)^\dagger = O(A^\dagger XY^*)$, $I_H - (Y^*A^\dagger)^\dagger(Y^*A^\dagger) = O(I_H - XY^*A^\dagger)$. Therefore,

$$\begin{aligned}(A-XY^*)^\dagger &= (I_H - O(A^\dagger XY^*))A^\dagger O(I_H - XY^*A^\dagger)\\ &= [I_H - (A^\dagger X)(A^\dagger X)^\dagger]A^\dagger[I_H - (Y^*A^\dagger)^\dagger(Y^*A^\dagger)].\end{aligned}$$

(3) Since

$$\mathrm{Ker}\,A^\dagger = \mathrm{Ran}\,(A)^\perp,\quad \mathrm{Ran}\,(XY^*) \subset \mathrm{Ran}\,(X) \subset \mathrm{Ran}\,(A) \tag{3.4.5}$$

$$\mathrm{Ker}\,(A^\dagger)^* = \mathrm{Ran}\,(A^*)^\perp,\ \mathrm{Ran}\,(YX^*) \subset \mathrm{Ran}\,(Y) \subset \mathrm{Ran}\,(A^*), \tag{3.4.6}$$

we get that $\mathrm{Ker}\,A^\dagger + \mathrm{Ran}\,(XY^*)$ and $\mathrm{Ker}\,(A^\dagger)^* + \mathrm{Ran}\,(YX^*)$ are closed when $\mathrm{Ran}\,(XY^*)$ (or equivalently, $\mathrm{Ran}\,(YX^*)$ is closed. It follows from Lemma 3.2.5 that $\mathrm{Ran}\,(A^\dagger XY^*)$ and $\mathrm{Ran}\,((A^\dagger)^*YX^*)$ (or $\mathrm{Ran}\,(XY^*A^\dagger)$) are closed. When (3.4.3) holds, we have

$$\mathrm{Ran}\,((XY^*A^\dagger)^*) = \mathrm{Ran}\,((XY^*A^\dagger)^\dagger) \subset \mathrm{Ker}\,(A-XY^*)^\dagger = \mathrm{Ran}\,(A-XY^*)^\perp$$

and hence $\mathrm{Ran}\,(A - XY^*) \subset \mathrm{Ran}\,((XY^*A^\dagger)^*)^\perp = \mathrm{Ker}\,(XY^*A^\dagger)$, that is, $XY^*A^\dagger XY^* = XY^*$.

When $\mathrm{Ran}\,(X)$ and $\mathrm{Ran}\,(Y)$ are closed, $\mathrm{Ran}\,(A^\dagger X)$ and $\mathrm{Ran}\,((A^\dagger)^* Y)$ (or $\mathrm{Ran}\,(Y^* A^\dagger)$) are all closed by (3.4.5), (3.4.6) and Lemma 3.2.5. When (3.4.4) holds, we have

$$\mathrm{Ran}\,((Y^*A^\dagger)^*) = \mathrm{Ran}\,((Y^*A^\dagger)^\dagger) \subset \mathrm{Ker}\,(A - XY^*)^\dagger = \mathrm{Ker}\,(A - XY^*)^*$$
$$\mathrm{Ran}\,((A - XY^*)^*) = \mathrm{Ran}\,((A - XY^*)^\dagger) \subset \mathrm{Ker}\,(A^\dagger X)^\dagger) = \mathrm{Ker}\,(A^\dagger X)^*.$$

Thus, $(A - XY^*)^*(Y^*A^\dagger)^* = 0$ and $(A^\dagger X)^*(A - XY^*)^* = 0$ and hence $Y^*A^\dagger XY^* = Y^*$, $XY^*A^\dagger X = X$. □

Lemma 3.4.4. *Let* $A \in B(K, H)$ *and* $B \in B(L, H)$ *with* $\mathrm{Ran}\,(A)$ *and* $\mathrm{Ran}\,(B)$ *closed. Put* $T = \begin{bmatrix} A & B \end{bmatrix} \in B(K \oplus L, H)$.

(1) $\mathrm{Ran}\,(T)$ *is closed in* H *iff* $\mathrm{Ran}\,(A) + \mathrm{Ran}\,(B)$ *is closed in* H *and iff* $\mathrm{Ran}\,((I_H - AA^\dagger)B)$ *is closed.*
(2) *When* $\mathrm{Ran}\,(T)$ *is closed,* $T^\dagger$ *has the form*

$$T^\dagger = \begin{bmatrix} (I_H + D_{A,B}D_{A,B}^*)^{-1}(A^\dagger - A^\dagger BC_{A,B}^\dagger) \\ D_{A,B}^*(I_H + D_{A,B}D_{A,B}^*)^{-1}(A^\dagger - A^\dagger BC_{A,B}^\dagger) + C_{A,B}^\dagger \end{bmatrix}, \quad (3.4.7)$$

where $C_{A,B} = (I_H - AA^\dagger)B$ *and* $D_{A,B} = A^\dagger B(I_L - C_{A,B}^\dagger C_{A,B})$.

Proof. (1) Noting that

$$\mathrm{Ran}\,(T) = \{Ax_1 + Bx_2 \,|\, x_1 \in K,\ x_2 \in L\} = \mathrm{Ran}\,(A) + \mathrm{Ran}\,(B),$$

we have $\mathrm{Ran}\,(T)$ is closed iff $\mathrm{Ran}\,(A)+\mathrm{Ran}\,(B)$ is closed. On the other hand, since $\mathrm{Ran}\,(A) = \mathrm{Ran}\,(AA^\dagger) = \mathrm{Ker}\,(I_H - AA^\dagger)$, it follows from Lemma 3.2.5 that $\mathrm{Ran}\,(A) + \mathrm{Ran}\,(B) = \mathrm{Ker}\,(I_H - AA^\dagger) + \mathrm{Ran}\,(B)$ is closed iff $\mathrm{Ran}\,((I_H - AA^\dagger)B)$ is closed.

(2) Let R denote the right side of (3.4.7). Then

$$TR = (A + BD_{A,B}^*)(I_H + D_{A,B}D_{A,B}^*)^{-1}(A^\dagger - A^\dagger BC_{A,B}^\dagger) + BC_{A,B}^\dagger.$$

Note that

$$\begin{aligned} A + BD_{A,B}^* &= A + AA^\dagger B(I_L - C_{A,B}^\dagger C_{A,B})B^*(A^\dagger)^* \\ &\quad + (I_H - AA^\dagger)B(I_L - C_{A,B}^\dagger C_{A,B})B^*(A^\dagger)^* \\ &= A(I_H + D_{A,B}D_{A,B}^*). \end{aligned}$$

We have

$$TR = A(A^\dagger - A^\dagger BC_{A,B}^\dagger) + BC_{A,B}^\dagger = AA^\dagger + C_{A,B}C_{A,B}^\dagger = (TR)^*.$$

Now put $RT = \begin{bmatrix} T_1 & T_2 \\ T_3 & T_4 \end{bmatrix}$, where

$$
\begin{aligned}
T_1 &= (I_H + D_{A,B}D^*_{A,B})^{-1}(A^\dagger - A^\dagger BC^\dagger_{A,B})A \\
T_2 &= (I_H + D_{A,B}D^*_{A,B})^{-1}(A^\dagger - A^\dagger BC^\dagger_{A,B})B \\
T_3 &= [D^*_{A,B}(I_H + D_{A,B}D^*_{A,B})^{-1}(A^\dagger - A^\dagger BC^\dagger_{A,B}) + C^\dagger_{A,B}]A \\
T_4 &= [D^*_{A,B}(I_H + D_{A,B}D^*_{A,B})^{-1}(A^\dagger - A^\dagger BC^\dagger_{A,B}) + C^\dagger_{A,B}]B.
\end{aligned}
$$

Note that $\operatorname{Ker} C^\dagger_{A,B} = \operatorname{Ker} C^*_{A,B}$ and $A^\dagger AD_{A,B}D^*_{A,B} = D_{A,B}D^*_{A,B}$. We have $C^\dagger_{A,B}A = 0$, $C^\dagger_{A,B}(I_H - AA^\dagger) = C^\dagger_{A,B}$ and

$$
\begin{aligned}
T_1 &= (I_H + D_{A,B}D^*_{A,B})^{-1}A^\dagger A = T_1^* \\
T_3^* &= [D^*_{A,B}(I_H + D_{A,B}D^*_{A,B})^{-1}A^\dagger A]^* = (I_H + D_{A,B}D^*_{A,B})^{-1}A^\dagger AD_{A,B} \\
&= (I_H + D_{A,B}D^*_{A,B})^{-1}(A^\dagger B - A^\dagger BC^\dagger_{A,B}(I_H - AA^\dagger)B) = T_2 \\
T_4 &= (I_H + D^*_{A,B}D_{A,B})^{-1}D^*_{A,B}(A^\dagger B - A^\dagger BC^\dagger_{A,B}C_{A,B}) + C^\dagger_{A,B}B \\
&= (I_H + D^*_{A,B}D_{A,B})^{-1}D^*_{A,B}D_{A,B} + C^\dagger_{A,B}C_{A,B} = T_4^*.
\end{aligned}
$$

Thus, $(RT)^* = RT$.

Finally, we have $RTR = R$ since $C^\dagger_{A,B}(AA^\dagger + C_{A,B}C^\dagger_{A,B}) = C^\dagger_{A,B}$, $(A^\dagger - A^\dagger BC^\dagger_{A,B})(AA^\dagger + C_{A,B}C^\dagger_{A,B}) = A^\dagger - A^\dagger BC^\dagger_{A,B}$ and also

$$
\begin{aligned}
TRT &= (AA^\dagger + C_{A,B}C^\dagger_{A,B})\begin{bmatrix} A & B \end{bmatrix} = \begin{bmatrix} A & AA^\dagger B + C_{A,B}C^\dagger_{A,B}B \end{bmatrix} \\
&= \begin{bmatrix} A & AA^\dagger B + C_{A,B} \end{bmatrix} = T.
\end{aligned}
$$

The above proves that $T^\dagger = R$. □

Corollary 3.4.5. *Let* $A \in B(K,H)$ *and* $B \in B(L,H)$ *with* $\operatorname{Ran}(A)$, $\operatorname{Ran}(B)$ *and* $\operatorname{Ran}(A) + \operatorname{Ran}(B)$ *closed. Then* $\operatorname{Ran}(AA^* + BB^*)$ *is closed and*

$$
(AA^* + BB^*)^\dagger = (A^\dagger - A^\dagger BC^\dagger)^*(I_H + DD^*)^{-1}(A^\dagger - A^\dagger BC^\dagger) + (CC^*)^\dagger,
$$

where $C = (I_H - AA^\dagger)B$ *and* $D = A^\dagger B(I_L - C^\dagger C)$.

Proof. Put $T = \begin{bmatrix} A & B \end{bmatrix}$. Then $\operatorname{Ran}(T)$ and $\operatorname{Ran}(C)$ are closed by Lemma 3.4.4(1) and so are the $\operatorname{Ran}(T^*)$ and $\operatorname{Ran}(TT^*)$ by Proposition 3.1.5.

Now by Proposition 3.1.8 and Lemma 3.4.4 (2),

$$\begin{aligned}(AA^* + BB^*)^\dagger &= (TT^*)^\dagger = (T^\dagger)^* T^\dagger \\ &= (A^\dagger - A^\dagger BC^\dagger)^*(I_H + DD^*)^{-2}(A^\dagger - A^\dagger BC^\dagger) \\ &\quad + [C^\dagger + D^*(I_H + DD^*)^{-1}(A^\dagger - A^\dagger BC^\dagger)]^* \\ &\quad \times [C^\dagger + D^*(I_H + DD^*)^{-1}(A^\dagger - A^\dagger BC^\dagger)] \\ &= (A^\dagger - A^\dagger BC^\dagger)^*[(I_H + DD^*)^{-2} + (I_H + DD^*)^{-2}DD^*] \\ &\quad \times (A^\dagger - A^\dagger BC^\dagger) + (C^\dagger)^* C^\dagger \\ &= (A^\dagger - A^\dagger BC^\dagger)^*(I_H + DD^*)^{-1}(A^\dagger - A^\dagger BC^\dagger) + (CC^*)^\dagger,\end{aligned}$$

here we use the relation $DC^\dagger = 0$. □

Corollary 3.4.6. *Let* $A, B \in B(H, K)$ *with* $\operatorname{Ran}(A)$, $\operatorname{Ran}(B)$ *and* $\operatorname{Ran}(A) + \operatorname{Ran}(B)$ *closed. If* $AB^* = 0$, *then* $\operatorname{Ran}(A + B)$ *is closed and*

$$(A + B)^\dagger = (I_H + D^*)(I_H + DD^*)^{-1}(A^\dagger - A^\dagger BC^\dagger) + C^\dagger,$$

where $C = (I_K - AA^\dagger)B$ *and* $D = A^\dagger B(I_H - C^\dagger C)$.

Proof. $AB^* = 0$ implies that $(A + B)(A + B)^* = AA^* + BB^*$. Thus, $\operatorname{Ran}(A + B)$ is closed by Corollary 3.4.5 and Proposition 3.1.5. Now by Proposition 3.1.8 (2),

$$\begin{aligned}(A + B)^\dagger &= (A + B)^*[(A + B)(A + B)^*]^\dagger = (A^* + B^*)(AA^* + BB^*)^\dagger \\ &= (A^* + B^*)(A^\dagger - A^\dagger BC^\dagger)^*(I_H + DD^*)^{-1}(A^\dagger - A^\dagger BC^\dagger) \\ &\quad + (A^* + B^*)(CC^*)^\dagger.\end{aligned}$$

Since $A^*CC^\dagger = ((I_K - AA^\dagger)A)^* BC^\dagger = 0$, $C^\dagger(I_K - AA^\dagger) = C^\dagger$, we have $C^\dagger A = 0$ and $C^\dagger B = C^\dagger C$. Thus,

$$\begin{aligned}(A^* + B^*)(CC^*)^\dagger &= (C^\dagger A)^* C^\dagger + (C^\dagger B)^* C^\dagger \\ &= C^\dagger CC^\dagger = C^\dagger \\ (A^* + B^*)(A^\dagger - A^\dagger BC^\dagger)^* &= (A^\dagger A + A^\dagger B - A^\dagger BC^\dagger B)^* \\ &= A^\dagger A + D^*.\end{aligned}$$

Finally, from $A^\dagger ADD^* = DD^*A^\dagger A$ and above two identities, we get that

$$\begin{aligned}(A + B)^\dagger &= (A^\dagger A + D^*)(I_H + DD^*)^{-1}(A^\dagger - A^\dagger BC^\dagger) + C^\dagger \\ &= (I_H + D^*)(I_H + DD^*)^{-1}(A^\dagger - A^\dagger BC^\dagger) + C^\dagger.\end{aligned}$$

□

Let $A \in B(H)$, $B \in B(K,H)$, $C \in B(H,K)$ and $D \in B(K)$. We now deduce an expression of $\begin{bmatrix} A & B \\ C & D \end{bmatrix}^\dagger$ as follows.

Put $S = \begin{bmatrix} A & 0 \\ C & 0 \end{bmatrix}$ and $T = \begin{bmatrix} 0 & B \\ 0 & D \end{bmatrix}$. Then $ST^* = 0$ and $\mathrm{Ran}\,(S) + \mathrm{Ran}\,(T) = \mathrm{Ran}\left(\begin{bmatrix} A & B \\ C & D \end{bmatrix} \right)$. Suppose that $\mathrm{Ran}\,(A^*A + C^*C)$ and $\mathrm{Ran}\,(B^*B + D^*D)$ are closed. Then both $S^*S = \begin{bmatrix} A^*A + C^*C & 0 \\ 0 & 0 \end{bmatrix}$ and $T^*T = \begin{bmatrix} 0 & 0 \\ B^*B + D^*D & 0 \end{bmatrix}$ all have closed range. By Proposition 3.1.8 (2),

$$S^\dagger = (S^*S)^\dagger S^* = \begin{bmatrix} (A^*A + C^*C)^\dagger A^* & (A^*A + C^*C)^\dagger C^* \\ 0 & 0 \end{bmatrix}.$$

Put $X = (A^*A + C^*C)^\dagger (A^*B + C^*D)$. Then

$$C_{S,T} = (I - SS^\dagger)T = \begin{bmatrix} 0 & B - AX \\ 0 & D - CX \end{bmatrix}.$$

Put $Y = (B - AX)^*(B - AX) + (D - CX)^*(D - CX)$ and assume that $\mathrm{Ran}\,(Y)$ is closed. Then

$$C_{S,T}^\dagger = (C_{S,T}^* C_{S,T})^\dagger C_{S,T}^* = \begin{bmatrix} 0 & 0 \\ Y^\dagger (B - AX)^* & Y^\dagger (D - CX)^* \end{bmatrix}$$

$$\begin{aligned} D_{S,T} = S^\dagger T (I - C_{S,T}^\dagger C_{S,T}) &= \begin{bmatrix} 0 & X \\ 0 & 0 \end{bmatrix} \begin{bmatrix} I_H & 0 \\ 0 & I_K - Y^\dagger Y \end{bmatrix} \\ &= \begin{bmatrix} 0 & X(I_K - Y^\dagger Y) \\ 0 & 0 \end{bmatrix} \end{aligned}$$

$$\begin{aligned} S^\dagger T C_{S,T}^\dagger &= \begin{bmatrix} 0 & X \\ 0 & 0 \end{bmatrix} \begin{bmatrix} 0 & 0 \\ Y^\dagger (B - AX)^* & Y^\dagger (B - AX)^* \end{bmatrix} \\ &= \begin{bmatrix} XY^\dagger (B - AX)^* & XY^\dagger (D - CX)^* \\ 0 & 0 \end{bmatrix}. \end{aligned}$$

Now assume that $\mathrm{Ran}\left(\begin{bmatrix} A & B \\ C & D \end{bmatrix} \right)$ is closed. Then by Corollary 3.4.6,

$$\begin{aligned} \begin{bmatrix} A & B \\ C & D \end{bmatrix}^\dagger &= (S + T)^\dagger \\ &= (I + D_{S,T}^*)(I + D_{S,T} D_{S,T}^*)^{-1} (S^\dagger - S^\dagger T C_{S,T}^\dagger) + C_{S,T}^\dagger \\ &= \begin{bmatrix} Z_1 & Z_2 \\ Z_3 & Z_4 \end{bmatrix}, \end{aligned}$$

where,

$$Z_1 = [I_H + X(I_K - Y^\dagger Y)X^*]^{-1}[(A^*A + C^*C)^\dagger A^* - XY^\dagger(B - AX)^*]$$
$$Z_2 = [I_H + X(I_K - Y^\dagger Y)X^*]^{-1}[(A^*A + C^*C)^\dagger C^* - XY^\dagger(D - CX)^*]$$
$$Z_3 = (I_K - Y^\dagger Y)X^* Z_1 - Y^\dagger(B - AX)^*$$
$$Z_4 = (I_K - Y^\dagger Y)X^* Z_2 - Y^\dagger(D - CX)^*$$

3.5 Moore–Penrose inverses in C^*–algebras

In this section, we will define the Moore–Penrose inverse of an element in a $*$–algebra and generalize some properties of the Moore–Penrose inverse in previous sections of this chapter to set of C^*–algebras.

Lemma 3.5.1. *Let $\mathcal{A}$ be a $*$–algebra over field $\mathbb{C}$ and let a be a non–zero element in $\mathcal{A}$. Consider the equations:*

$$axa = a, \ xax = x, \ (ax)^* = ax, \ (xa)^* = xa. \tag{3.5.1}$$

If (3.5.1) has at least one solution, then the solution of (3.5.1) is unique.

Proof. Let b and c be two solutions of (3.5.1). Then

$$\begin{aligned} b &= bab = b(ab)^* = bb^*a^* = bb^*(aca)^* = bb^*a^*(ac)^* = bb^*a^*ac \\ &= bb^*a^*(aca)c = bb^*a^*a(ca)^*c = bb^*a^*aa^*c^*c \\ c &= cac = (ca)^*c = a^*c^*c = (aba)^*c^*c = (ba)^*a^*c^*c = baa^*c^*c \\ &= b(aba)a^*c^*c = b(ab)^*aa^*c^*c = bb^*a^*aa^*c^*c. \end{aligned}$$

Thus, $b = c$. □

By Lemma 3.5.1, we can give following definition:

Definition 3.5.2. Let a be a non–zero element in a $*$–algebra $\mathcal{A}$. If there is a $b \in \mathcal{A}$ such that

$$aba = a, \ bab = b, \ (ab)^* = ab, \ (ba)^* = ba,$$

we call a is Moore–Penrose invertible and b is the Moore–Penrose inverse of a, denoted by $a^\dagger$.

Let $G_{mp}(\mathcal{A})$ denote the set of all non–zero Moore–Penrose invertible elements in the $*$–algebra $\mathcal{A}$. According to Definition 3.5.2, we can verify the following proposition easily.

Proposition 3.5.3. *Let $\mathcal{A}$ be a $*$–algebra and $a \in G_{mp}(\mathcal{A})$. Then*

(1) $a^\dagger a$ *and* $aa^\dagger$ *are projections in* $\mathcal{A}$.
(2) $a^\dagger \in G_{mp}(\mathcal{A})$ *and* $(a^\dagger)^\dagger = a$.
(3) $a^* \in G_{mp}(\mathcal{A})$ *and* $(a^*)^\dagger = (a^\dagger)^*$.
(4) $a^*a \in G_{mp}(\mathcal{A})$ *and* $(a^*a)^\dagger = a^\dagger(a^\dagger)^*$.
(5) $a^\dagger = (a^*a)^\dagger a^* = a^*(aa^*)^\dagger$.

Definition 3.5.4. Let $\mathcal{A}$ be a unital $*$–algebra. $\mathcal{A}$ is called to be Hermitian if $1 + x^*x$ is invertible in $\mathcal{A}$ for every $x \in \mathcal{A}$.

Lemma 3.5.5. *Let* $\mathcal{A}$ *be a unital Hermitian* $*$*–algebra. Let* p *be an idempotent in* $\mathcal{A}$. *Then* $1 - p - p^*$ *is invertible in* $\mathcal{A}$ *and* $o(p) = p(p + p^* - 1)^{-1}$ *is a projection (i.e.,* $(o(p))^2 = o(p) = (o(p))^*$*) in* $\mathcal{A}$ *satisfying* $p\,o(p) = o(p)$ *and* $o(p)p = p$.

Proof. By assumption, $s = 1 + (p - p^*)^*(p - p^*)$ is invertible in $\mathcal{A}$. Since $p^2 = p$, $(p^*)^2 = p^*$, we have $s = (1 - p - p^*)^2$ is invertible and so that $1 - p - p^*$ is invertible in $\mathcal{A}$. Noting that

$$(p + p^* - 1)p = p^*p = p^*(p + p^* - 1),\ (p + p^* - 1)p^* = pp^* = p(p + p^* - 1),$$

we obtain that $(o(p))^2 = o(p) = (o(p))^*$ and moreover, $p\,o(p) = o(p)$,

$$o(p)p = p(p + p^* - 1)^{-1}p = pp^*(p + p^* - 1)^{-1} = p.$$

□

Proposition 3.5.6. *Let* $\mathcal{A}$ *be a unital Hermitian* $*$*–algebra and* $a \in \mathcal{A}\backslash\{0\}$. *Suppose that there is a* $b \in \mathcal{A}$ *such that* $aba = a$. *Then* $a \in G_{mp}(\mathcal{A})$ *with* $a^\dagger = [1 - o(1 - ba)]bo(ab)$.

Proof. $aba = a$ implies that ab and ba are all idempotent. So

$$o(ab)(ab) = ab, \qquad (1 - ba)o(1 - ba) = o(1 - ba)$$
$$(ab)o(ab) = o(ab), \quad o(1 - ba)(1 - ba) = 1 - ba$$

by Lemma 3.5.5. Put $x = [1 - o(1 - ba)]bo(ab)$. Then

$$ax = ab\,o(ab) - a(1 - ba)\,o(1 - ba)bo(ab) = o(ab) = (ax)^*$$
$$xa = (1 - o(1 - ba))bo(ab)(aba) = (1 - o(1 - ba))ba$$
$$= (1 - o(1 - ba))(ba - 1 + 1) = 1 - o(1 - ba) = (xa)^*$$

and $axa = o(ab)aba = aba = a$, $xax = x$. Thus, $a^\dagger = x$. □

In the rest of the section, we assume that $\mathcal{A}$ is a C^*–algebra.

Definition 3.5.7. Let $a \in \mathcal{A}\backslash\{0\}$. a is called to be well–supported if there is a projection $p \in \mathcal{A}$ such that $a = ap$ and $|a|$ is invertible in $p\mathcal{A}p$.

Proposition 3.5.8. *Let $a \in \mathcal{A}\backslash\{0\}$. Then the following statements are equivalent:*

(1) $a \in Gi(\mathcal{A}) = \{a \in \mathcal{A}|\, a = aba \text{ for some } b \in \mathcal{A}\}$.
(2) $a \in G_{mp}(\mathcal{A})$.
(3) *a is well–supported.*
(4) $a^*a \in G_{mp}(\mathcal{A})$.
(5) $a^*a \in Gi(\mathcal{A})$.
(6) $0 \notin \sigma(|a|)$ *or* $0 \in \sigma(|a|)$ *is an isolated point.*

Proof. (1)⇒(2) Let $b \in A$ such that $aba = a$. Since $1 + x^*x$ is invertible in $\tilde{\mathcal{A}}$, it follows from Lemma 3.5.5 that $o(1 - ba)$ and $o(ab)$ are projections in $\tilde{\mathcal{A}}$ and $a^\dagger = (1 - o(1 - ba))bo(ab) \in \mathcal{A}$ for $b \in \mathcal{A}$.

(2)⇒(3) Put $p = a^\dagger a$. Then p is a projection in $\mathcal{A}$ and $ap = aa^\dagger a = a$. Since

$$a^*a(a^\dagger(a^\dagger)^*) = (aa^\dagger a)^*(a^\dagger)^* = p,\ (a^\dagger(a^\dagger)^*)a^*a = a^\dagger(aa^\dagger)^*a = a^\dagger aa^\dagger a = p$$
$$p(a^*a)p = (aa^\dagger a)^*a = a^*a, \quad p(a^\dagger(a^\dagger)^*)p = a^\dagger(a^\dagger aa^\dagger)^* = a^\dagger(a^\dagger)^*,$$

it follows that a^*a and $a^\dagger(a^\dagger)^* \in p\mathcal{A}p$ and a^*a is invertible in $p\mathcal{A}p$ with $(a^*a)^{-1} = a^\dagger(a^\dagger)^*$. Consequently, $|a|$ is invertible in $p\mathcal{A}p$.

(3)⇒(4) By Definition 3.5.7, there is a projection $p \in \mathcal{A}$ and $c \in (p\mathcal{A}p)_{sa}$ such that $a = ap$ and $|a|c = c|a| = p$. Noting that $ap = a$ and $cp = c = pc$, we have $c^2(a^*a) = (a^*a)c^2 = p$ and

$$(a^*a)c^2(a^*a) = p(a^*a) = a^*a, \quad c^2(a^*a)c^2 = c^2p = c^2,$$

that is, $a^*a \in G_{mp}(\mathcal{A})$.

Now by Theorem 1.5.36, there is a $*$–isomorphism $\pi\colon \tilde{A} \to B(H_\pi)$ for some Hilbert space H_π. Put $B = \pi(\tilde{\mathcal{A}}) \subset B(H_\pi)$, $A = \pi(a)$.

(4)⇒(6) $a^*a \in G_{mp}(\tilde{\mathcal{A}})$ means that $A^*A \in G_{mp}(B) \subset B(H_\pi)$. So by Proposition 3.1.7, $0 \notin \sigma(|A|)$ or $0 \in \sigma(|A|)$ is an isolated point. Note that $\sigma(|a|) = \sigma_B(|A|) = \sigma_{B(H_\pi)}(|A|) = \sigma(|A|)$ by Corollay 1.5.8. Thus, $0 \notin \sigma(|a|)$ or $0 \in \sigma(|a|)$ is an isolated.

(6)⇒(1) If $0 \notin \sigma(|a|)$, then a^*a is invertible in $\tilde{\mathcal{A}}$, In this case, it is easy to check that $a^\dagger = (a^*a)^{-1}a^*$. If $0 \in \sigma(|a|)$ is an isolated point, then using the same method in the proof of the implication (4)⇒(2) of Proposition 3.1.7, we can find a projection p and a self-adjoint element c, which are contained in the C^*–subalgebra $C^*(|a|)$ of $\mathcal{A}$ generated by $|a|$, such that

$$|a|c = p = c|a|,\ |a|p = |a|,\ pc = cp = c.$$

Put $b = c^2 a^* \in \mathcal{A}$. From $|a|p = p|a| = |a|$, we get that $pa^*a = a^*a$, and consequently, $(1-p)a^*a(1-p) = 0$, that is, $ap = a$. Therefore, we can verify that $a^\dagger = b = c^2 a^*$, i.e., $a \in Gi(A)$.

The implication (4)⇒(5) is clear and (5)⇒(4) comes from the implication (1)⇒(2) by replacing a by a^*a. □

Corollary 3.5.9. *Let B be a C^*-subalgebra of $\mathcal{A}$ and let $a \in G_{mp}(\mathcal{A}) \cap B$. Then $a^\dagger \in B$.*

Proof. By Proposition 3.5.8, $0 \notin \sigma_{\tilde{\mathcal{A}}}(|a|)$ or $0 \in \sigma_{\tilde{\mathcal{A}}}(|a|)$ is an isolated point. Since $\sigma_{\tilde{\mathcal{A}}}(|a|) = \sigma_{\tilde{B}}(|a|)$ and $|a| = (a^*a)^{1/2} \in B$, it follows from Proposition 3.5.8 that $a \in G_{mp}(B)$, i.e., $a^\dagger \in B$. □

Now for $a \in \mathcal{A}\backslash\{0\}$, define linear operators L_a, R_a from $\mathcal{A}$ to $\mathcal{A}$, respectively, by $L_a(x) = ax$, $R_a(x) = xa$, $\forall x \in \mathcal{A}$. Then $\|L_a(x)\| \le \|a\|\|x\|$, $\forall x \in \mathcal{A}$, i.e.,

$$\|L_a\| \le \|a\|, \quad \|L_a(a^*)\| = \|aa^*\| = \|a\|^2 \le \|L_a\|\|a^*\|.$$

Thus, $\|L_a\| = \|a\|$. Similarly, $\|R_a\| = \|a\|$. Set $\gamma^l(a) = \gamma(L_a)$ and $\gamma^r(a) = \gamma(R_a)$, $\gamma^l(a)$ and $\gamma^r(a)$ are called to be the left and the right conorm of a respectively, defined in [Harte and Mbekhta (1992)].

Definition 3.5.10. Let $a \in A\backslash\{0\}$.

$$\gamma_{\mathcal{A}}(a) = \inf\{\|b-a\| \mid K_r(a) \subsetneqq K_r(b), b \in \mathcal{A}\}$$

is called to be the reduced minimum modulus of a (with respect to $\mathcal{A}$), where $K_r(b) = \{x \in \mathcal{A} \mid bx = 0\}$, $b \in \mathcal{A}$.

Proposition 3.5.11. *Let a be a non-zero element in $\mathcal{A}$, then*

(1) $\gamma_{\mathcal{A}}(a) \ge \gamma^l(a)$.
(2) $\gamma^l(a) = \gamma^r(a) = \|a^\dagger\|^{-1}$, *when $a \in Gi(\mathcal{A})$.*

Proof. (1) For any $b \in \mathcal{A}$ with $K_r(b) \supsetneqq K_r(a)$ and $\forall \epsilon > 0$, there is $c \in K_r(a)$ with $\|c\| = 1$ and $\operatorname{dist}(c, K_r(b)) > 1 - \epsilon$ by Proposition 1.3.3. Then

$$\begin{aligned}\|a-b\| &\ge \|(a-b)c\| = \|ac\| = \|L_a(c)\| \\ &\ge \gamma(L_a)\operatorname{dist}(c, K_r(a)) > (1-\epsilon)\gamma^l(a).\end{aligned}$$

Thus, $\gamma_{\mathcal{A}}(a) \ge \gamma^l(a)$.

(2) When $a \in Gi(A)$,

$$L_a L_{a^\dagger} L_a = L_a, \; L_{a^\dagger} L_a L_{a^\dagger} = L_{a^\dagger}, \; R_a R_{a^\dagger} R_a = R_a, \; R_{a^\dagger} R_a R_{a^\dagger} = R_{a^\dagger}.$$

Therefore, $(L_a)^+ = L_{a^\dagger}$ and $(R_a)^+ = R_{a^\dagger}$ and hence by Theorem 2.1.3,

$$\frac{1}{\|L_{a^\dagger}\|} \le \gamma(L_a) \le \frac{\|L_{a^\dagger}L_a\|\|L_aL_{a^\dagger}\|}{\|L_{a^\dagger}\|}, \tag{3.5.2}$$

$$\frac{1}{\|R_{a^\dagger}\|} \le \gamma(R_a) \le \frac{\|R_{a^\dagger}R_a\|\|R_aR_{a^\dagger}\|}{\|R_{a^\dagger}\|}. \tag{3.5.3}$$

Since $\|L_{a^\dagger}\| = \|a^\dagger\| = \|R_{a^\dagger}\|$ and

$$\|L_{a^\dagger}L_a\| = \|L_{a^\dagger a}\| = \|a^\dagger a\| = 1 = \|L_aL_{a^\dagger}\| = \|L_{aa^\dagger}\| = \|aa^\dagger\|$$
$$\|R_{a^\dagger}R_a\| = \|R_{aa^\dagger}\| = \|aa^\dagger\| = 1 = \|R_aR_{a^\dagger}\| = \|R_{a^\dagger a}\| = \|a^\dagger a\|,$$

it follows from (3.5.2) and (3.5.3) that $\gamma^l(a) = \gamma^r(a) = \|a^\dagger\|^{-1}$. □

The following theorem characterizes $\gamma_{\mathcal{A}}(a)$.

Theorem 3.5.12. *Let a be a non-zero element in $\mathcal{A}$. Then*

$$\gamma_{\mathcal{A}}(a) = \inf\{\lambda \in \mathbb{R} |\, \lambda \in \sigma(|a|)\backslash\{0\}\}.$$

In order to prove Theorem 3.5.12, we need following lemma.

Lemma 3.5.13. *Let $a \in \mathcal{A}\backslash\{0\}$.*

(1) *If $a \in G_{mp}(\mathcal{A})$, then* $\inf\{\lambda \in \mathbb{R} |\, \lambda \in \sigma(|a|)\backslash\{0\}\} = \|a^\dagger\|^{-1}$.
(2) *If $0 \in \sigma(|a|)$ is not an isolated point, then $\gamma_{\mathcal{A}}(a) = \gamma_{\mathcal{A}}(|a|) = 0$.*
(3) $\gamma_{\mathcal{A}}(a) = \gamma_{\mathcal{A}}(|a|)$.

Proof. We may assume that $\mathcal{A}$ is a C^*–subalgebra in $B(H)$ for some Hilbert space H. By Proposition 3.1.8 and Corollary 3.5.9, we obtain (1).

(2) Let $a = u|a|$ be the polar decomposition of a in $B(H)$. Since for any $P \in \mathfrak{P}_0(\mathbb{R})$, $uP(|a|) \in \mathcal{A}$, it follows that for any $f \in C_0((0, \|a\|])$, $uf(|a|) \in \mathcal{A}$.

Given $\epsilon \in (0, \|a\|)$, define continuous functions f_ϵ and g_ε on $[0, \|a\|]$ by

$$f_\epsilon(t) = \begin{cases} 0 & 0 \le t < \frac{\epsilon}{2} \\ 2(t - \frac{\epsilon}{2}) & \frac{\epsilon}{2} \le t < \epsilon \\ t & \epsilon \le t \le \|a\| \end{cases}, \quad g_\epsilon(t) = \begin{cases} 2(\frac{\epsilon}{2} - t) & 0 \le t \le \frac{\epsilon}{2} \\ 0 & \frac{\epsilon}{2} \le t \le \|a\| \end{cases}$$

and set $b_\epsilon = f_\epsilon(|a|)$, $y = g_\epsilon(|a|)$. Noting that $\max\limits_{0 \le t \le \|a\|} |t - f_\epsilon(t)| \le \epsilon/2$, we have $\||a| - f_\epsilon(|a|)\| \le \epsilon/2$. From $|a|u^*u = u^*u|a| = |a|$, we get that $K_r(a) = K_r(|a|)$. Noting that $f_\epsilon(t)g_\epsilon(t) = 0$, $\forall\, t \in [0, \|a\|]$ implies that $y \in K_r(|b_\epsilon|)$ and $tg_\epsilon(t) \ne 0$ for some $t \in [0, \|a\|]$ indicates that $y \notin K_r(|a|)$, we have

$$K_r(|a|) \subsetneqq K_r(|b_\epsilon|), \quad \gamma_{\mathcal{A}}(|a|) \le \||a| - b_\epsilon\| \le \epsilon/2$$

and $\gamma_{\mathcal{A}}(|a|) = 0$. Note that $ub_\epsilon, uy \in \mathcal{A}$, $K_r(a) \subset K_r(ub_\epsilon)$, $y \in K_r(ub_\epsilon)$ and $y \notin K_r(a)$. So

$$\gamma_{\mathcal{A}}(a) \le \|a - ub_\epsilon\| = \|u(|a| - b_\epsilon)\| \le \||a| - b_\epsilon\| \le \frac{\epsilon}{2}$$

and hence $\gamma_{\mathcal{A}}(a) = \gamma_{\mathcal{A}}(|a|) = 0$.

(3) When $0 \in \sigma(|a|)$ is not an isolated point, then $\gamma_{\mathcal{A}}(a) = \gamma_{\mathcal{A}}(|a|) = 0$ by (1). Suppose that $0 \notin \sigma(|a|)$ or $0 \in \sigma(|a|)$ is an isolated point. Put $u = a|a|^\dagger$. Then $u \in \mathcal{A}$ is unitary or partial isometry and $u|a| = a$. Since $|a||a|^\dagger|a| = |a|$ implies that

$$(1 - |a|^\dagger|a|)a^*a(1 - |a|^\dagger|a|) = 0 \text{ and } a(1 - |a|^\dagger|a|) = 0,$$

it follows that for any $c \in \mathcal{A}$ with $K_r(a) = K_r(|a|) \subsetneqq K_r(c)$, we have $K_r(|a|) \subsetneqq K_r(u^*c)$ and $K_r(a) \subsetneqq K_r(uc)$. Thus,

$$\gamma_{\mathcal{A}}(|a|) \le \||a| - u^*c\| = \|u^*(a - c)\| \le \|a - c\|$$

$$\gamma_{\mathcal{A}}(a) \le \|a - uc\| = \|u(|a| - c)\| \le \||a| - c\|$$

and consequently, $\gamma_{\mathcal{A}}(|a|) \le \gamma_{\mathcal{A}}(a) \le \gamma_{\mathcal{A}}(|a|)$. □

Proof of Theorem 3.5.12. If 0 is not an isolated point of $\sigma(|a|)$, then $\inf\{\lambda \in \mathbb{R} | \lambda \in \sigma(|a|)\backslash\{0\}\} = 0$ and $\gamma_{\mathcal{A}}(a) = 0$ by Lemma 3.5.13.

In the following, we assume that 0 is an isolated point or $0 \notin \sigma(|a|)$. By Proposition 3.5.11 and Lemma 3.5.13, we have

$$\gamma_{\mathcal{A}}(a) \ge \|a^\dagger\|^{-1} = \inf\{\lambda \in \mathbb{R} | \lambda \in \sigma(|a|)\backslash\{0\}\}.$$

So we need to prove $\gamma_{\mathcal{A}}(a) \le \inf\{\lambda \in \mathbb{R} | \lambda \in \sigma(|a|)\backslash\{0\}\}$.

Put $\mu = \min\{\lambda \in \mathbb{R} | \lambda \in \sigma(|a|)\backslash\{0\}\}$. If $\mu = \|a\|$, then $\sigma(|a|) = \{\mu\}$ or $\{0, \mu\}$ and so $|a| = \mu p$ for some projection p in $\mathcal{A}$. Since $p \notin K_r(|a|)$ and $p \in K_r(0)$,

$$\gamma_{\mathcal{A}}(a) = \gamma_{\mathcal{A}}(|a|) \le \|\mu p - 0\| = \mu.$$

Suppose that $\mu < \|a\|$. If μ is an isolated point of $\sigma(|a|)$, then $|a|$ can be written as $|a| = a_1 + \mu p_1$, where $p_1 = \chi_{\{\mu\}}(|a|)$ is a projection in $\mathcal{A}$ and a_1 is a positive element in $(1 - p_1)\mathcal{A}(1 - p_1)$. It follows that $K_r(|a|) \subsetneqq K_r(a_1)$ ($p_1 \in K_r(a_1)$) and $p_1 \notin K_r(a)$). Consequently,

$$\gamma_{\mathcal{A}}(a) = \gamma_{\mathcal{A}}(|a|) \le \||a| - a_1\| = \mu.$$

Now assume that μ is an accumulation point of $\sigma(|a|)$. Let $\epsilon \in (0, \|a\| - \mu)$ and define continuous functions h_ϵ and k on $[0, \|a\|]$ by

$$h_\epsilon(t) = \begin{cases} t & 0 \le t \le \frac{\mu}{2} \\ \mu - t & \frac{\epsilon}{2} < t \le \mu \\ 0 & \mu < t \le \mu + \frac{\epsilon}{2} \\ \frac{2(\mu+\epsilon)}{\epsilon}(t - \mu - \frac{\epsilon}{2}) & \mu + \frac{\epsilon}{2} < t \le u + \epsilon \\ t & u + \epsilon < t \le \|a\| \end{cases},$$

$$k(t) = \begin{cases} 0 & 0 \le t \le \mu \\ \frac{4}{\epsilon}(\mu + \frac{\epsilon}{4})(t - \mu) & \mu < t \le \mu + \frac{\epsilon}{4} \\ -\frac{4}{\epsilon}(\mu + \frac{\epsilon}{4})(t - \mu - \frac{\epsilon}{2}) & \mu + \frac{\epsilon}{4} < t \le \mu + \frac{\epsilon}{2} \\ 0 & \mu + \frac{\epsilon}{2} < t \le \|a\| \end{cases}$$

and put $a_\epsilon = h_\epsilon(|a|)$, $d = k(|a|)$. It is easy to verify that $K_r(|a|) \subset K_r(a_\epsilon)$, $d \in K_r(a_\epsilon)$ and $d \notin K_r(|a|)$. Noting that $|t - h_\epsilon(t)| \le \mu + \epsilon$, $\forall t \in [0, \|a\|]$, we get that

$$\gamma_{\mathcal{A}}(a) = \gamma_{\mathcal{A}}(|a|) \le \||a| - a_\epsilon\| \le \mu + \epsilon$$

and hence $\gamma_{\mathcal{A}}(a) \le \mu$. This completes the proof.

Corollary 3.5.14. *Let a be a non-zero element in a C^*-algebra $\mathcal{A}$. Then $\gamma_{\mathcal{A}}(a^*) = \gamma_{\mathcal{A}}(a)$.*

Proof. Since $\sigma(a^*a)\backslash\{0\} = \sigma(aa^*)\backslash\{0\}$ by Proposition 1.4.14, we have $\sigma(|a|)\backslash\{0\} = \sigma(|a^*|)\backslash\{0\}$. So $\gamma_{\mathcal{A}}(a^*) = \gamma_{\mathcal{A}}(a)$ by Theorem 3.5.12. □

Corollary 3.5.15. *Let $\mathcal{A}$ be a unital C^*-algebra and $a \in GL(\mathcal{A})$. Then* $\mathrm{dist}\,(a, \mathcal{A}\backslash GL(\mathcal{A})) = \|a^{-1}\|^{-1}$.

Proof. Let $b \in \mathcal{A}$ such that $\|a - b\| < \|a^{-1}\|^{-1}$. Then $b \in GL(\mathcal{A})$ by Corollary 1.4.8. So for any $c \in \mathcal{A}\backslash GL(\mathcal{A})$, $\|a - c\| \ge \|a^{-1}\|^{-1}$ and hence $\mathrm{dist}\,(a, \mathcal{A}\backslash GL(\mathcal{A})) \ge \|a^{-1}\|^{-1}$.

Now for any $\epsilon > 0$, there exists $b_0 \in \mathcal{A}$ with $K_r(b_0) \supsetneqq K_r(a) = \{0\}$ such that $\gamma_{\mathcal{A}}(a) > \|a - b_0\| - \epsilon$. Since $b_0 \in \mathcal{A}\backslash GL(\mathcal{A})$, it follows that $\gamma_{\mathcal{A}}(a) \ge \mathrm{dist}\,(a, \mathcal{A}\backslash GL(\mathcal{A})) \ge \|a^{-1}\|^{-1}$. Note that by Theorem 3.5.12 and Lemma 3.5.13, $\gamma_{\mathcal{A}}(a) = \|a^{-1}\|^{-1}$. So $\mathrm{dist}\,(a, \mathcal{A}\backslash GL(\mathcal{A})) = \|a^{-1}\|^{-1}$. □

Some further properties of $\gamma_{\mathcal{A}}(\cdot)$ will be mentioned in Chapter 6. In the following, we will discuss the Moore–Penrose inverses of the product, sum and difference of two projections in a C^*-algebra.

First by Theorem 1.5.36, Corollary 3.2.9 and Proposition 3.5.8, we have

Proposition 3.5.16. *Let $a, b \in G_{mp}(\mathcal{A})$ such that $ab \in G_{mp}(\mathcal{A})$. Then $(ab)^\dagger = b^\dagger a^\dagger$ iff $a^\dagger abb^* = bb^*a^\dagger a$ and $a^*abb^\dagger = bb^\dagger a^*a$.*

Now let p, q be projections in $\mathcal{A}$ such that $p + q \in G_{mp}(\mathcal{A})$. Set

$$q \vee p = p \vee q = (p+q)^\dagger(p+q) = (p+q)(p+q)^\dagger \in \mathcal{A}.$$

Then

$$(p+q)(1 - p \vee q) = 0 = (1 - p \vee q)(p+q)(1 - p \vee q)$$

and so that

$$(1 - p \vee q)p(1 - p \vee q) = 0, \quad (1 - p \vee q)q(1 - p \vee q) = 0,$$

i.e., $p \leq p \vee q$, $q \leq p \vee q$. Furthermore, by Theorem 1.5.36, Proposition 3.2.2, Lemma 3.2.3 and Proposition 3.2.4, Lemma 3.2.5, we have

Proposition 3.5.17. *Let p, q be projections in the unital C^*-algebra $\mathcal{A}$. Then $pq \in G_{mp}(\mathcal{A})$ iff $1 - p + q \in G_{mp}(\mathcal{A})$ and iff $1 - q + p \in G_{mp}(\mathcal{A})$. In this case, $r(p, q) = 1 - (p \vee (1 - q))q + pq \in GL(\mathcal{A})$ and*

$$(pq)^\dagger = (r(p, q))^{-1} p(q \vee (1 - p)).$$

Proposition 3.5.18. *Let $\mathcal{A}$ be a unital C^*-algebra and p, q be projections in $\mathcal{A}$. Then following statements are equivalent:*

(1) $p + q \in G_{mp}(\mathcal{A})$.
(2) $(1 - p)q \in G_{mp}(\mathcal{A})$.
(3) $q(1 - p) \in G_{mp}(\mathcal{A})$.
(4) $2 - p - q \in G_{mp}(\mathcal{A})$.
(5) $p - q \in G_{mp}(\mathcal{A})$.

Proof. By using Proposition 3.5.17, we obtain that (1), (2), (3) and (4) are equivalent.

Now suppose that $2 - p - q \in G_{mp}(\mathcal{A})$. Then $p + q \in G_{mp}(A)$ by the equivalence of (1) and (4). Thus, 0 (resp. 2) $\notin \sigma(p + q)$ or 0 (resp. 2) $\in \sigma(p + q)$ is an isolated point. Since

$$(p - q)^2 = 2(p + q) - (p + q)^2 \text{ and} \tag{3.5.4}$$

$$\sigma((p - q)^2) = \{2\lambda - \lambda^2 \in \mathbb{R} | \, \lambda \in \sigma(p + q)\}, \tag{3.5.5}$$

it follows that $0 \notin \sigma(p - q)$ or $0 \in \sigma(p - q)$ is an isolated point and hence $p - q \in G_{mp}(\mathcal{A})$.

Conversely, from (3.5.4) and (3.5.5), we get that $0 \notin \sigma(p - q)$ implies that 0 or 2 doses not contains in $\sigma(p + q)$ and that $0 \in \sigma(p - q)$ is an isolated point indicates that 0 or 2 is an isolated point of $\sigma(p + q)$. Therefore, $p + q \in G_{mp}(\mathcal{A})$ or $2 - (p + q) \in G_{mp}(\mathcal{A})$. □

Corollary 3.5.19. *Let p, q be projections in a unital C^*-algebra $\mathcal{A}$ with $p + q \in G_{mp}(\mathcal{A})$. Then*

(1) $\rho(p, q) = 1 + p + q - p \vee q \in GL(\mathcal{A})$ *and* $(p + q)^\dagger = (\rho(p, q))^{-1}(p \vee q)$.
(2) $(p - q)^\dagger = (2 - p - q)^\dagger (p + q)^\dagger (p - q)$.

Proof. (1) Put $s = 1 - p \vee q + (p+q)^\dagger$. Using

$$(p \vee q)p = p, \ (p \vee q)q = q, \ (p \vee q)(p+q)^\dagger = (p+q)^\dagger,$$

we have $s\rho(p,q) = \rho(p,q)s = 1$, that is, $\rho(p,q) \in GL(\mathcal{A})$. Since

$$\rho(p,q)(p+q)^\dagger = (p+q)^\dagger + p \vee q - (p+q)^\dagger = p \vee q,$$

we get that $(p+q)^\dagger = (\rho(p,q))^{-1}(p \vee q)$.

(2) By Proposition 3.5.18, $p + q \in G_{mp}(\mathcal{A})$ implies that $p - q$ and $2 - p - q \in G_{mp}(\mathcal{A})$. Since $(p-q)^2 = (p+q)(2-p-q)$, it follows from Proposition 3.5.3 that

$$(p-q)^\dagger = ((p-q)^2)^\dagger(p-q) = ((p+q)(2-p-q))^\dagger(p-q).$$

Note that

$$((1-p) \vee (1-q))(p+q) = (p+q)((1-p) \vee (1-q)),$$
$$(2-p-q)(p \vee q) = (p \vee q)(2-p-q).$$

Thus $((p+q)(2-p-q))^\dagger = (2-p-q)^\dagger(p+q)^\dagger$ by Proposition 3.5.16 and hence $(p-q)^\dagger = (2-p-q)^\dagger(p+q)^\dagger(p-q)$. □

Let $\mathcal{A}$ be a C^*–algebra and $a \in G_{mp}(\mathcal{A})$. Let $\delta a \in \mathcal{A}$ and put $\bar{a} = a + \delta a \in \mathcal{A}$. According to Definition 2.4.6, we say $\bar{a}$ is the stable perturbation of a if $\bar{a}\tilde{\mathcal{A}} \cap (1 - aa^\dagger)\tilde{\mathcal{A}} = \{0\}$, or equivalently, $\bar{a}\mathcal{A} \cap (1 - aa^\dagger)\mathcal{A} = \{0\}$.

Proposition 3.5.20. *Let $\mathcal{A}$ be a C^*–algebra and $a \in G_{mp}(\mathcal{A})$. Let $\delta a \in \mathcal{A}$ and put $\bar{a} = a + \delta a$. Suppose that $\|a^\dagger\|\|\delta a\| < 1$ and $\bar{a}\mathcal{A} \cap (1 - aa^\dagger)\mathcal{A} = \{0\}$. Then $\bar{a} \in G_{mp}(\mathcal{A})$ and*

$$\|\bar{a}^\dagger\| \le \frac{\|a^\dagger\|}{1 - \|a^\dagger\|\|\delta a\|}, \qquad \frac{\|\bar{a}^\dagger - a^\dagger\|}{\|a^\dagger\|} \le \frac{1+\sqrt{5}}{2}\,\frac{\|a^\dagger\|\|\delta a\|}{1 - \|a^\dagger\|\|\delta a\|}.$$

Proof. We may assume that $\mathcal{A}$ is a C^*–subalgebra of $B(H)$ for some Hilbert space H. Then $\bar{a}\tilde{\mathcal{A}} \cap (1 - aa^\dagger)\tilde{\mathcal{A}} = \{0\}$ and $\|a^\dagger\|\|\delta a\| < 1$ in $\mathcal{A}$ implies that $\bar{a}(B(H)) \cap (1 - aa^\dagger)(B(H)) = \{0\}$ by Corollary 2.4.8 and also means that $\mathrm{Ran}\,(\bar{a}) \cap \mathrm{Ker}\, a^\dagger = \{0\}$ by Remark 2.4.2. Thus, the assertions follows from Theorem 3.3.5 and Corollary 3.5.9. □

Combining Proposition 3.5.20 with Corollary 3.3.6, we have

Corollary 3.5.21. *Let $\{a_n\} \subset G_{mp}(\mathcal{A})$ and $a \in G_{mp}(\mathcal{A})$. Suppose that $a_n \xrightarrow{\|\cdot\|} a$ as $n \to \infty$. Then the following statements are equivalent.*

(1) *$a_n^\dagger \xrightarrow{\|\cdot\|} a^\dagger$ as $n \to \infty$.*

(2) $a_n^\dagger a_n \xrightarrow{\|\cdot\|} a^\dagger a$ *as* $n \to \infty$.
(3) $a_n a_n^\dagger \xrightarrow{\|\cdot\|} aa^\dagger$ *as* $n \to \infty$.
(4) $\|a_n^\dagger\| \to \|a^\dagger\|$ *as* $n \to \infty$.
(5) $\sup_{n\geq 1} \|a_n^\dagger\| < +\infty$.
(6) $a_n\mathcal{A} \cap (1 - aa^\dagger)\mathcal{A} = \{0\}$ *for* n *large enough.*

3.6 Notes

Section 3.1 is based on Chapter 9 of [Ben-Israel and Greville (2003)].

Section 3.2 is based on [Izumino (1982)]. Theorem 3.2.8 and Corollary 3.2.9 tell us when the Moore–Penrose inverse has the reverse order law. In fact, there are many equivalent conditions that make the the reverse order law for Moore–Penrose inverses hold. The interesting readers may refer to [Djordjević (2001)], [Djordjević (2007)] and [Djordjević and Dinčić (2010)].

Section 3.3 is based on [Chen *et al.* (1996)], [Chen and Xue (1998)] and [Xue and Chen (2004)]. Now most part of results in this section is extended to the set of C^*–module operators in [Xu *et al.* (2010b)].

Section 3.4. Proposition 3.4.1 generalizes Theorem 1 of [Wei and Ding (2001)]. Proposition 3.4.2 (1) generalizes Theorem 3.4.1 of [Campbell and Meyer (1979)]. Proposition 3.4.2 (2) is a special case of Theorem 11 in [Deng and Du (2010)]. But it does not give the explicit expression of $\begin{bmatrix} A & B \\ 0 & D \end{bmatrix}^\dagger$. More results about the representation of $\begin{bmatrix} A & B \\ C & D \end{bmatrix}^\dagger$ can be found in [Deng and Du (2009)] and [Deng and Du (2010)]. Proposition 3.4.3 comes from [Du and Xue (2010)], which generalizes Theorem 3.2 of [Chen *et al.* (2009)], Theorem 3.3 (1) and Theorem 3.5 of [Deng (2009)]. Lemma 3.4.4 is a generalization of Theorem 3.3.3 of [Campbell and Meyer (1979)]. There are several expressions about $\begin{bmatrix} A & B \end{bmatrix}^\dagger$. Please see [Deng and Du (2009)], [Xu and Hu (2008)] for details. Corollary 3.4.5 (resp. Corollary 3.4.6) may be regarded as the generalization of Theorem 1 (resp. Theorem 2) in [Cline (1965)]. But the expression of $(AA^* + BB^*)^\dagger$ (resp. $(A + B)^\dagger$) in both statements are different.

Section 3.5. Definition 3.5.7 is Definition 6.5.3 in [Blackadar (1998)]. Definition 3.5.10 and Theorem 3.5.12 come from [Xue (2007b)]. Some of results about the Moore–Penrose inverses of the sum, difference and product of two projections in a C^*–algebra are contained in [Li (2008)]. But the approaches to deal with them are different from these in [Li (2008)].

Chapter 4

Drazin Inverses, T–S Generalized Inverses And Their Perturbations

The Drazin inverse of an element was originally defined by Drazin in [Drazin (1958)] in the context of semigroups and rings. It was later generalized to bounded linear operators whose resolvent had a pole, and more generally, any singularity at zero by [Koliha (1996)]. Applications to semigroups of operators require the Drazin inverse of a closed linear operator; a first account was give in 1992 by Nashed and Zhao [Nashed and Zhao (1992)] without proofs for the case of a resolvent pole. The fully general case is studied in [Koliha and Tran (2001)].

The continuity and perturbation analysis for the Drazin inverse play important roles in its applications. Around 2000, many equivalent conditions for the continuity of the Drazin inverse on a Banach space or in a Banach algebra are obtained (cf. [Koliha and Rakočević (1998)], [Rakočević (1999)]). Meanwhile, many results about the perturbation analysis of the Drazin inverse on a Banach Space are presented in [Castro-González *et al.* (2002b)], [Castro-González *et al.* (2002a)], etc. Chen and the author give up–bounds of the perturbations of the group inverse and the Drazin inverse under the stable perturbation on a Banach space or in a Banach algebra in [Xue and Chen (2007)] and [Xue (2007c)].

The outer generalized inverse with prescribed range and kernel is very useful in solving some restricted operator equations. It unifies the Moore–Penrose inverse, Drazin inverse, generalized Drazin inverse into one.

This chapter is organized as follows. We first introduce the definition of Drazin inverse of a bounded operator and the definition of Drazin inverse of a densely–defined closed operator on a Banach space in Section 2.1 and then present some useful properties of them. We introduce the outer generalized inverse with prescribed range and kernel and give its various expressions and their reverse order law in Section 4.2. In Section 4.3, we establish a stable perturbation theory of group inverses and Drazin inverses. In Section 4.4, we present the perturbation analysis for $A_{T,S}^{(2)}$. In Section 4.5, the Drazin inverse and the generalized Drazin inverse in a Banach algebra are discussed. In Section 4.6, the perturbation analysis for group inverses, Drazin inverses and generalized Drazin inverses are given. We will compute the Drazin inverse and generalized Drazin inverse of sums of two certain elements and some matrices over a unital Banach algebra in the final section.

Throughout the chapter, X is a Banach space, H is a Hilbert space and $\mathcal{A}$ is a unital Banach algebra.

4.1 Drazin inverses on Banach spaces

Let $T \in B(X)$. Define T^n by induction, with $T^0 = I$, $T^1 = T$.

Lemma 4.1.1. *Let $T \in B(X)$.*

(1) $\operatorname{Ker} T^n \subset \ker T^{n+1}$, $n = 0, 1, 2, \cdots$. *If* $\operatorname{Ker} T^k = \operatorname{Ker} T^{k+1}$ *for some* $k \geq 0$, *then* $\operatorname{Ker} T^n = \operatorname{Ker} T^k$, $\forall\, n \geq k$.
(2) $\operatorname{Ran}(T^{n+1}) \subset \operatorname{Ran}(T^n)$, $n = 0, 1, 2, \cdots$. *If* $\operatorname{Ran}(T^{i+1}) = \operatorname{Ran}(T^i)$ *for some* $i \geq 0$, *then* $\operatorname{Ran}(T^n) = \operatorname{Ran}(T^i)$, $\forall\, n \geq i$.

Proof. (1) $\operatorname{Ker} T^n \subset \operatorname{Ker} T^{n+1}$, $n = 0, 1, 2, \cdots$ is evident.

Suppose that there is k such that $\operatorname{Ker} T^k = \operatorname{Ker} T^{k+1}$. Let $\xi \in \operatorname{Ker} T^{k+2}$. Then $T\xi \in \operatorname{Ker} T^{k+1} = \operatorname{Ker} T^k$, that is, $T^{k+1}\xi = 0$. Thus, $\operatorname{Ker} T^{k+2} \subset \operatorname{Ker} T^{k+1}$ and consequently, $\operatorname{Ker} T^{k+2} = \operatorname{Ker} T^{k+1} = \operatorname{Ker} T^k$. The assertion now follows by induction.

(2) It is clear that $\operatorname{Ran}(T^{n+1}) \subset \operatorname{Ran}(T^n)$, $\forall\, n \geq 0$. Assume that $\operatorname{Ran}(T^i) = \operatorname{Ran}(T^{i+1})$ for some i. Let $\eta \in \operatorname{Ran}(T^{i+1})$. Then $\eta = T(T^i\xi)$ for some $\xi \in X$ and $T^i\xi = T^{i+1}z$ for some $z \in X$. Thus, $\eta = T^{i+2}z \in \operatorname{Ran}(T^{i+2}\eta)$ and hence $\operatorname{Ran}(T^{i+1}) \subset \operatorname{Ran}(T^{i+2})$. Finally, by induction, we can obtain the assertion. □

According to Lemma 4.1.1, we have following definition.

Definition 4.1.2. Let $T \in B(X)$. If there is an integer $n \geq 0$ such that $\operatorname{Ker} T^n = \operatorname{Ker} T^{n+1}$ (resp. $\operatorname{Ran}(T^n) = \operatorname{Ran}(T^{n+1})$), the smallest such integer is called the ascent (resp. descent) of T and denoted by $\alpha(T)$ (resp. $\delta(T)$). If no such integer, we say that $\alpha(T) = \infty$ (resp. $\delta(T) = \infty$).

Clearly, $\alpha(T) = 0$ iff $\operatorname{Ker} T = \{0\}$ and $\delta(T) = 0$ iff $\operatorname{Ran}(T) = X$.

Proposition 4.1.3. *Let $T \in B(X)$ with $\alpha(T)$ and $\delta(T)$ finite. Then*

(1) $\alpha(T) = \delta(T)$.
(2) $X = \operatorname{Ran}(T^j) + \operatorname{Ker} T^j$, $\operatorname{Ker} T^j \cap \operatorname{Ran}(T^j) = \{0\}$, $\forall j \geq \alpha(T)$.
(3) $\operatorname{Ran}(T^j)$ *is closed,* $\forall j \geq \alpha(T)$.

Proof. (1) Assume that $\delta(T) = 0$, i.e., $\operatorname{Ran}(T) = X$. If $\alpha(T) > 0$, then there is $x_0 \in X \backslash \{0\}$ such that $Tx_0 = 0$. By induction, we can find $\{x_n\} \subset X$ such that $Tx_{n+1} = x_n, n \geq 0$. Then $x_0 = T^n x_n \neq 0$ and $T^{n+1}x_n = 0$. Thus $\operatorname{Ker} T^n \subsetneqq \operatorname{Ker} T^{n+1}$, $n \geq 0$ and hence $\alpha(T) = \infty$.

Assume that $p = \delta(T) \geq 1$. Then there is $y \in \operatorname{Ran}(T^{p-1}) \backslash \operatorname{Ran}(T^p)$. Write $y = T^{p-1}x$ for some $x \in X$. Then

$$Ty = T^p x \in \operatorname{Ran}(T^p) = \operatorname{Ran}(T^{2p}) = T^p(\operatorname{Ran}(T^p))$$

by Lemma 4.1.1. So $Ty = T^p u$ for some $u \in \operatorname{Ran}(T^p)$. Put $v = x - u$. Then

$$T^p v = T^p x - Ty = Ty - Ty = 0$$
$$T^{p-1}v = T^{p-1}x - T^{p-1}u = y - T^{p-1}u.$$

Since $T^{p-1}u \in \operatorname{Ran}(T^{2p-1}) = \operatorname{Ran}(T^p)$ and $y \notin \operatorname{Ran}(T^p)$, it follows that $T^{p-1}v \neq 0$ and hence $\operatorname{Ker} T^{p-1} \subsetneqq \operatorname{Ker} T^p$. Thus, $\alpha(T) \geq p$.

On the other hand, let $x \in \operatorname{Ker} T^{p+1}$ and put $k = \alpha(T)$. From $T^p x \in \operatorname{Ran}(T^p) = \operatorname{Ran}(T^{p+k})$, we get that $T^p x = T^{p+k}w$ for some $w \in X$. Thus, $x - T^k w \in \operatorname{Ker} T^p \subset \operatorname{Ker} T^{p+1}$ and $T^k w \in \operatorname{Ker} T^{p+1}$ for $x \in \operatorname{Ker} T^{p+1}$, that is, $w \in \operatorname{Ker} T^{p+k+1} = \operatorname{Ker} T^k$ and consequently, $x \in \operatorname{Ker} T^p$. Therefore, $\operatorname{Ker} T^p = \operatorname{Ker} T^{p+1}$ and $\alpha(T) \leq p$.

(2) Let $x \in \operatorname{Ran}(T^j) \cap \operatorname{Ker} T^j$, $j \geq \alpha(T) = p = \delta(T)$. Then $T^j x = 0$ and $x = T^j y$ for some $y \in X$. So $y \in \operatorname{Ker} T^{2j} = \operatorname{Ker} T^j$ and hence $\operatorname{Ran}(T^j) \cap \operatorname{Ker} T^j = \{0\}$.

Now for any $z \in X$, $T^j z \in \operatorname{Ran}(T^j) = \operatorname{Ran}(T^{2j})$, $j \geq \delta(T)$. Thus there is $y \in X$ such that $T^j z = T^{2j}y$ and furthermore, $z - T^j y \in \operatorname{Ker} T^j$. The assertion follows.

(3) Applying Corollary 1.2.13 to (2), we get the assertion. □

Proposition 4.1.3 leads to the following definition.

Definition 4.1.4. Let $T \in B(X)$ with $\alpha(T)$ and $\delta(T)$ finite. $k = \alpha(T) = \delta(T)$ is called the index of T and is denoted by $\operatorname{ind}(T) = k$.

Theorem 4.1.5. *Let $T \in B(X)$ with* $\operatorname{ind}(T) = k$. *Then there is $A \in B(X)$ with* $\operatorname{Ran}(A) = \operatorname{Ran}(T^k)$ *and* $\operatorname{Ker} A = \operatorname{Ker}(T^k)$ *such that*

$$TAT^k = T^k, \quad ATA = A, \quad AT = TA.$$

Proof. By Proposition 4.1.3 (2) and (3), $X = \text{Ran}\,(T^k) \dot{+} \text{Ker}\,(T^k)$. Since

$$\text{Ran}\,(T^k) = \text{Ran}\,(T^{k+1}) = T(\text{Ran}\,(T^k)),\ \ \text{Ran}\,(T^k) \cap \text{Ker}\,(T^k) = \{0\},$$

it follows that $A_0 = T|_{\text{Ran}\,(T^k)}$ is invertible in $B(\text{Ran}\,(T^k))$. Define a bounded linear operator $A\colon X \to X$ by

$$Ax = \begin{cases} A_0^{-1}x & x \in \text{Ran}\,(T^k) \\ 0 & x \in \text{Ker}\,(T^k) \end{cases}. \tag{4.1.1}$$

Clearly, $\text{Ran}\,(A) = \text{Ran}\,(T^k)$ and $\text{Ker}\,A = \text{Ker}\,(T^k)$ by (4.1.1). Note that $T(\text{Ker}\,(T^k)) \subset \text{Ker}\,(T^k)$. It follows from (4.1.1) that $TAT^k = T^k$, $ATA = A$ and $TA = AT$. □

From Theorem 4.1.5, we can give a new kind generalized inverse of a bounded linear operator on a Banach space as follows.

Definition 4.1.6. Let $T \in B(X)$. If there is a positive integer k and an $S \in B(X)$ such that

$$TST^k = T^k,\ \ STS = S,\ \ TS = TS, \tag{4.1.2}$$

T is called to be Drazin invertible and S is called the Drazin inverse of T, denoted by T^D.

Especially, when $k = 1$, T^D is called the group inverse of T and denoted by $T^{\#}$; if $k = 0$, $T^D = T^{-1}$.

Let $Dr(X)$ denote the set of all Drazin invertible operators in $B(X)$.

Remark 4.1.7. (1) The T^D in Definition 4.1.6 is uniquely determined by (4.1.2). Suppose that there is another $R \in B(X)$ satisfying

$$TRT^n = T^n,\ \ RTR = R,\ \ RT = TR$$

for some integer $n \geq 0$. Set $p = \max\{k, n\}$. Since TT^D and TR are idempotent operators and $TT^D = T^DT$, $TR = RT$, we have

$$T^jR^j = (TR)^j = TR,\quad TT^D = (TT^D)^j = T^j(T^D)^j,\ j \geq 1.$$

Thus,

$$\begin{aligned} T^D &= TT^DT^D = T^n(T^D)^nT^D = TRT^n(T^D)^{n+1} \\ &= RTT^D = R^{p+1}T^{p+1}T^D = R^{p+1}T^{p-k}T^{k+1}T^D \\ &= R^{p+1}T^{p-k}T^k = RR^pT^p = RTR \\ &= R. \end{aligned}$$

(2) The least positive integer k satisfying (4.1.2) is equal to $\mathrm{ind}(T)$. In fact, from (4.1.2), we have

$$TST^m = T^m, \quad STS = S, \quad TS = TS, \quad \forall m \geq k. \tag{4.1.3}$$

By (4.1.3), we have $\mathrm{Ran}\,(T^m) = \mathrm{Ran}\,(T^{m+1})$ and $\mathrm{Ker}\,(T^m) = \mathrm{Ker}\,(T^{m+1})$. Thus $\alpha(T) = \delta(T) \leq k$ by Proposition 4.1.3 (1) and consequently, $\mathrm{ind}(T) = k$ by Theorem 4.1.5.

According to Definition 4.1.6 and Remark 4.1.7, we have following properties about the Drazin inverses.

Proposition 4.1.8. *Let $T \in Dr(X)$ with* $\mathrm{ind}(T) = k$. *Then*

(1) $T^* \in Dr(X^*)$ *with* $(T^*)^D = (T^D)^*$.
(2) $(T^l)^D = (T^D)^l$, $\forall\, l \in \mathbb{N}$.
(3) $(T^D)^D = T$ *iff* $\mathrm{ind}(T) = 1$.
(4) T^l *has the group inverse* $(T^l)^\#$ *and* $T^D = (T^l)^\# T^{l-1}$, $\forall\, l \geq k$.
(5) $\mathrm{Ker}\,(T^D) = \mathrm{Ker}\,(T^k)$ *and* $\mathrm{Ran}\,(T^D) = \mathrm{Ran}\,(T^k)$.
(6) $T^{l+1}T^D = T^l$, $\forall\, l \geq k$.
(7) $VTV^{-1} \in Dr(X)$ *and* $(VTV^{-1})^D = VT^DV^{-1}$, $\forall\, V \in GL(B(X))$.

Now we characterize the existence of the Drazin invertible operator on a Banach space.

Theorem 4.1.9. *Let $T \in B(X)$. Then the following statements are equivalent.*

(1) *T is Drazin invertible.*
(2) *$\alpha(T) < \infty$ and $\delta(T) < \infty$.*
(3) *There are closed subspaces M and N in X such that $X = M \dotplus N$, $TM \subset M$, $TN \subset N$ and moreover, $T\big|_M$ is invertible in $B(M)$ and $T\big|_N$ is nilpotent in $B(N)$.*
(4) *$0 \in \sigma(T)$ is an isolated point and $\lambda = 0$ is a pole of $(\lambda I - T)^{-1}$, $\lambda \in \mathbb{C}\backslash\sigma(T)$.*
(5) *There is $S \in B(X)$ such that $TS = ST$, $B = BTB$ and $T(I - TS)$ is nilpotent.*

Proof. The implication (1)⇒(2) comes from Remark 4.1.7 (2).

(2)⇒(3) By Proposition 4.1.3, $\alpha(T) = \delta(T) = k$ and

$$X = \mathrm{Ran}\,(T^k) \dotplus \mathrm{Ker}\, T^k. \tag{4.1.4}$$

Put $M = \mathrm{Ran}\,(T^k)$ and $N = \mathrm{Ker}\, T^k$. Then

$$TM = \mathrm{Ran}\,(T^{k+1}) = \mathrm{Ran}\,(T^k) = M, \quad TN \subset \mathrm{Ker}\, T^k = N. \tag{4.1.5}$$

Put $T_1 = T\big|_M$, $T_2 = T\big|_N$. Then $T_2^k = T^k\big|_N = 0$ and T_1 is invertible in $B(M)$ by (4.1.4) and (4.1.5).

(3)$\Rightarrow$(4) Write T as $T = \begin{bmatrix} T_M & \\ & T_N \end{bmatrix}$ with respect to the decomposition $X = M \dotplus N$, where $T_M = T\big|_M$ is invertible in $B(M)$ and $T_N = T\big|_N$ is nilpotent in $B(N)$. From $\sigma(T) = \sigma(T_M) \cup \sigma(T_N)$ and $0 \notin \sigma(T_M)$, $\sigma(T_N) = \{0\}$, we get that $0 \in \sigma(T)$ is an isolated point. Let $T_N^k = 0$, for some $k \geq 0$. Then for any $\lambda \in \mathbb{C}$ with $0 < |\lambda| < \|T_M^{-1}\|^{-1}$,

$$(\lambda I_M - T_M)^{-1} = T_M^{-1}(\lambda T_M^{-1} - I_M)^{-1} = -\sum_{n=0}^{\infty} T_M^{-(n+1)} \lambda^n,$$

$$(\lambda I_N - T_N)^{-1} = \lambda^{-1}(I_N - \lambda^{-1} T_N)^{-1} = \sum_{n=0}^{k-1} T_N^n \lambda^{-(n+1)}$$

by Lemma 1.4.7 and hence

$$(\lambda I - T)^{-1} = -\sum_{n=0}^{\infty} \begin{bmatrix} T_M^{-1} & \\ & 0 \end{bmatrix}^{n+1} \lambda^n + \sum_{n=0}^{k-1} \begin{bmatrix} 0 & \\ & T_N \end{bmatrix}^n \lambda^{-(n+1)}.$$

This means that $\lambda = 0$ is a pole of order k.

(4)$\Rightarrow$(5) Assume that $\lambda = 0$ is a pole of $(\lambda I - T)^{-1}$, $\lambda \in \mathbb{C}\backslash\sigma(T)$ with order $k \geq 1$. Then in the annulus $0 < |\lambda| < \delta$,

$$(\lambda I - T)^{-1} = \sum_{n=0}^{\infty} A_n \lambda^n + \sum_{n=0}^{k} B_n \lambda^{-n},$$

where $\{A_n\} \subset B(X)$ and $\{B_1, \cdots, B_k\} \subset B(X)$.

Choose open subsets U_1, U_2 in $\mathbb{C}$ such that $U_1 \subset \{\lambda \in \mathbb{C} \,|\, |\lambda| < \delta\}$, $U_1 \cap U_2 = \phi$ and $0 \in U_1$, $\sigma(T)\backslash\{0\} \subset U_2$. Let Γ_1 (resp. Γ_2) be any contour surrounding $\{0\}$ (resp.$\sigma(T)/\{0\}$) in U_1 (resp. U_2). Put $h(z) = z$,

$$g(z) = \begin{cases} 0 & z \in U_1 \\ z^{-1} & z \in U_2 \end{cases}, \quad k(z) = h(z)(1 - h(z)g(z)), \ \forall\, z \in U_1 \cup U_2.$$

Then $g(z)$ and $k(z)$ are holomorphic on $U_1 \cup U_2$. Put $S = g(T)$. Then $TS = ST$ and $STS = (ghg)(T) = g(T) = S$. Moreover,

$$\begin{aligned} (T(I - TS))^k = k^k(T) &= \frac{1}{2\pi i} \oint_{\Gamma_1 + \Gamma_2} k^k(z)(\lambda I - T)^{-1} \mathrm{d}\,\lambda \\ &= \frac{1}{2\pi i} \oint_{\Gamma_1} \lambda^k \Big(\sum_{n=0}^{\infty} A_n \lambda^n + \sum_{n=0}^{k} B_n \lambda^{-n} \Big) \mathrm{d}\,\lambda = 0. \end{aligned}$$

(5)⇒(1) $S = STS$ implies that TS and $I - TS$ are idempotent in $B(X)$. Let k be the least natural number such that $(T(I - TS))^k = 0$. Since $TS = ST$ and $(I - TS)^k = I - TS$, we get that $T^k(I - TS) = 0$. So $S \in B(X)$ satisfies conditions (4.1.2), that is, $S = T^D$ and $\mathrm{ind}(T) = k$. □

Corollary 4.1.10. *Let $T \in Dr(X)$ and $A \in B(X)$. If $AT = TA$, then $AT^D = T^D A$.*

Proof. By Theorem 4.1.9, there are closed subspaces M and N in X such that $X = M \dotplus N$, $T\big|_M: M \to M$ is invertible in $B(M)$ and $T\big|_N: N \to N$ is nilpotent in $B(N)$. Write A and T as $A = \begin{bmatrix} A_1 & A_2 \\ A_3 & A_4 \end{bmatrix}$ and $T = \begin{bmatrix} T_1 & \\ & T_2 \end{bmatrix}$ with respect to the decomposition $X = M \dotplus N$, where $T_1 = T\big|_M$, $T_2 = T\big|_N$ and $A_1: M \to M$, $A_2: N \to M$, $A_3: M \to N$, $A_4: N \to N$. Since T_1 is invertible and $AT = TA$, it follows that

$$A_1 T_1 = T_1 A_1, \ A_2 T_2 = T_1 A_2, \ A_3 T_1 = T_2 A_3.$$

Let $k \in N$ such that $T_2^k = 0$. Then

$$T_1^k A_2 = T_2^k A_2 = 0, \ A_3 T_1^k = T_2^k A_3 = 0$$

and consequently, $A_2 = 0$, $A_3 = 0$ for T_1 is invertible. Therefore,

$$AT^D = \begin{bmatrix} A_1 & 0 \\ 0 & A_4 \end{bmatrix} \begin{bmatrix} T_1^{-1} & 0 \\ 0 & 0 \end{bmatrix} = \begin{bmatrix} A_1 T_1^{-1} & 0 \\ 0 & 0 \end{bmatrix} = T^D A.$$

□

Now we turn to discuss the Drazin invertibility of unbounded linear operators on a Banach space.

Let $T \in \mathcal{C}(X)$. By induction, if $n \geq 1$, $\mathfrak{D}(T^n)$ is the subspace of all x in $\mathfrak{D}(T^{n-1})$ such that $T^{n-1}x \in \mathfrak{D}(T)$.

Lemma 4.1.11. *Let $T \in \mathcal{C}(X)$ and $B \in B(X)$ with $\mathrm{Ran}\,(B) \subset \mathfrak{D}(T)$. If $TB = BT$ on $\mathfrak{D}(T)$ and $B = BTB$, $\mathrm{Ran}\,(I - TB) \subset \mathfrak{D}(T)$. Then*

(1) $\mathrm{Ran}\,(I - TB) \subset \mathfrak{D}(T^n)$, $\forall\, n \in \mathbb{N}$.
(2) $T(\mathrm{Ran}\,(I - TB)) \subset \mathrm{Ran}\,(I - TB)$.
(3) $T\big|_{\mathrm{Ran}\,(I-TB)}$ *is bounded.*

Proof. (1) By induction, $TB = BT$ on $\mathfrak{D}(T)$ implies that $T^n B = BT^n$ on $\mathfrak{D}(T^n)$, $\forall n \in \mathbb{N}$. Let $\xi \in \mathrm{Ran}\,(I - TB) \subset \mathfrak{D}(T^{n-1})$. Since $B = BTB$, it follows from the proof of Proposition 2.1.3 (2) that TB and $I - TB$ are all bounded and are all idempotent. Thus $\xi = (I - TB)\xi \in \mathfrak{D}(T)$ and hence $0 = TB\xi = BT\xi$. Moreover,

$$T^{n-1}\xi = T^{n-1}\xi - TBT^{n-1}\xi = (I - TB)T^{n-1}\xi \in \mathfrak{D}(T).$$

Consequently, $\xi \in \mathfrak{D}(T^n)$ and $\operatorname{Ran}(I-TB) \subset \mathfrak{D}(T^n)$.

(2) Let $y \in \operatorname{Ran}(I-TB)$. Then $TBy = 0$ and $y \in \mathfrak{D}(T)$. Thus, $TBTy = TTBy = 0$ and hence

$$Ty = Ty - T(TBy) = Ty - T(BTy) = (I - TB)Ty,$$

that is, $T(\operatorname{Ran}(I-TB)) \subset \operatorname{Ran}(I-TB)$.

(3) The closeness of $G(T)$ and $\operatorname{Ran}(I–TB)$ implies that $G(T|_{\operatorname{Ran}(I–TB)})$ is closed. So $T|_{\operatorname{Ran}(I-TB)}$ is bounded by Theorem 1.1.8. □

Lemma 4.1.12. *Let $T \in \mathcal{C}(X)$ and $B \in B(X)$ with* $\operatorname{Ran}(B) \subset \mathfrak{D}(T)$. *Suppose that $B = BTB$ and $TB = BT$ on $\mathfrak{D}(T)$.*

(1) *If $\xi \in \mathfrak{D}(T^n)$, then $B\xi \in \mathfrak{D}(T^{n+1})$ and $T^{n+1}B\xi = TBT^n\xi$, $n \geq 1$.*
(2) *If $\xi \in \mathfrak{D}(T^n)$, then $T^nB\xi = BT^n\xi$, $n \geq 1$.*
(3) $\operatorname{Ran}(B^n) \subset \mathfrak{D}(T^n)$ *and $TB = T^nB^n$, $n \geq 1$.*

Proof. (1) Let $\xi \in \mathfrak{D}(T)$, then $TB\xi = BT\xi$. This implies that $TB\xi \in \mathfrak{D}(T)$ and $B\xi \in \mathfrak{D}(T^2)$ and moreover, $T^2B\xi = T(TB)\xi = TBT\xi$.

Now suppose that $B\xi \in \mathfrak{D}(T^{k+1})$ and $T^{k+1}B\xi = TBT^k\xi$ when $\xi \in \mathfrak{D}(T^k)$. Then for any $\xi \in \mathfrak{D}(T^{k+1})$, $T\xi \in \mathfrak{D}(T^k)$ and hence

$$BT\xi \in \mathfrak{D}(T^{k+1}),\ T^{k+1}BT\xi = TBT^kT\xi.$$

Noting that $\xi \in \mathfrak{D}(T)$, we have $TB\xi = BT\xi \in \mathfrak{D}(T^{k+1})$, i.e., $B\xi \in \mathfrak{D}(T^{k+2})$ and $T^{k+2}B\xi = TBT^{k+1}\xi$.

(2) Let $\xi \in \mathfrak{D}(T)$. Then $TB\xi = BT\xi$. Suppose that $T^kB\xi = BT^k\xi$, $\forall \xi \in \mathfrak{D}(T^k)$. Now let $\xi \in \mathfrak{D}(T^{k+1})$. Then

$$T^{k+1}B\xi = T(T^kB\xi) = TBT^k\xi = BT^{k+1}\xi$$

for $T^k\xi \in \mathfrak{D}(T)$ and $T^kB\xi \in \mathfrak{D}(T)$. The assertion follows.

(3) Let $\xi \in X$. Then $B\xi \in \mathfrak{D}(T)$ and $B^2\xi \in \mathfrak{D}(T^2)$ and furthermore $B^n\xi \in \mathfrak{D}(T^n)$ by (1). Note that $TB \in B(X)$ is an idempotent operator. So $TB = (TB)^n$. Let $\xi \in X$. Then

$$(TB)^2\xi = TBTB\xi = T(TB)B\xi = T^2B^2\xi = TB\xi,$$

By induction, if $TB = T^nB^n$, then

$$TB = TBTB = TBT^nB^n = T^{n+1}B^{n+1}.$$

□

Proposition 4.1.13. *Let $T \in \mathcal{C}(X)$. Suppose that there is $B \in B(X)$ with* $\operatorname{Ran}(B) \subset \mathfrak{D}(T)$ *satisfying conditions*

(1) *$B = BTB$ and $TB = BT$ on $\mathfrak{D}(T)$.*

(2) $\mathrm{Ran}\,(I - TB) = \{(I - TB)x \,|\, x \in X\} \subset \mathfrak{D}(T)$.
(3) $T\big|_{\mathrm{Ran}\,(I-TB)}$ *is quasi–nilpotent.*

Then such B is unique.

Proof. Put $P = I - TB$. Then $P \in B(X)$ is an idempotent operator. Put $T_1 = T(I - P)$ on $\mathfrak{D}(T)$ and $T_2 = TP$. Then $T = T_1 + T_2$. By Lemma 4.1.12 and Lemma 4.1.11, $T_1^n = T^n(I - P)$ on $\mathfrak{D}(T^n)$ and $T_2^n = T^nP$, $\forall\, n \in \mathbb{N}$ for $T^nP = PT^n$ on $\mathfrak{D}(T^n)$. Consequently,

$$T^n = T^n(I - P) + T^nP \text{ on } \mathfrak{D}(T^n).$$

Let $C \in B(X)$ with $\mathrm{Ran}\,(C) \subset \mathfrak{D}(T)$ satisfy conditions

(1) $C = CTC$ and $TC = CT$ on $\mathfrak{D}(T)$.
(2) $\mathrm{Ran}\,(I - TC) \subset \mathfrak{D}(T)$.
(3) $T\big|_{\mathrm{Ran}\,(I-TC)}$ is quasi–nilpotent.

Put $P_1 = I - TC$ and $T_1' = T(I - P_1)$, $T_2' = TP_1$. Then $T^n = (T_1')^n + (T_2')^n$, $\forall\, n \in \mathbb{N}$. So by Lemma 4.1.12,

$$\begin{aligned} C = CTC = TCC = T^nC^nC &= (T_1^n + T_2^n)C^nC \\ &= T_1^nC^nC + T_2^nC^{n+1} = T^nTBC^nC + T_2^nC^{n+1} \\ &= T_2^nC^{n+1} + T^nT^nB^nC^{n+1}. \end{aligned}$$

Since $\mathrm{Ran}\,(C^n) \subset \mathfrak{D}(T)$, $B^n(\mathfrak{D}(T^n)) \subset \mathfrak{D}(T^{2n})$ and $T^nB^n = B^nT^n$ on $\mathfrak{D}(T^n)$, $TB^n = B^nT$ on $\mathfrak{D}(T)$ by Lemma 4.1.12, it follows that

$$T^nT^nB^nC^{n+1} = T^nB^nT^nC^{n+1} = T^{n-1}B^nTC,\ n \geq 2.$$

By Lemma 4.1.11,

$$\begin{aligned} T^{n-1}B^nTC = T^{n-1}B^n(TC - I) + T^{n-1}B^n &= T^{n-1}B^n - B^nT^{n-1}(I - TC) \\ = TBB - B^n(T_2')^{n-1} &= BTB - B^n(T_2')^{n-1}. \end{aligned}$$

Thus, $C - B = T_2^nC^{n+1} - B^n(T_2')^{n-1}$ and

$$\|C - B\| \leq \|T_2^n\|\|C\|^{n+1} + \|(T_2')^{n-1}\|\|B\|^n,\ \forall\, n \in \mathbb{N}. \tag{4.1.6}$$

Since

$$\lim_{n\to\infty} \|T_2^n\|^{\frac{1}{n}} = \lim_{n\to\infty} \|(T_2')^{n-1}\|^{\frac{1}{n-1}} = 0,$$

it follows from (4.1.6) that $\lim_{n\to\infty} \|C - B\|^{\frac{1}{n}} = 0$. when $C \neq B$. But we have known that $\lim_{n\to\infty} \|C - B\|^{\frac{1}{n}} = 1$ if $C \neq B$. Therefore, $C = B$. □

According to Lemma 4.1.1 and Proposition 4.1.13, we have following definition.

Definition 4.1.14. Let $T \in \mathcal{C}(X)$. Suppose that there is $B \in B(X)$ with $\operatorname{Ran}(B) \subset \mathfrak{D}(T)$ satisfying conditions:

(1) $B = BTB$ and $TB = BT$ on $\mathfrak{D}(T)$.
(2) $\operatorname{Ran}(I - TB) \subset \mathfrak{D}(T)$.
(3) $T\big|_{\operatorname{Ran}(I-TB)}$ is quasi–nilpotent.

Then B is called the generalized Drazin inverse of T and is denoted by T^d.

Clearly, by Theorem 4.1.9, the notation "generalized Drazin inverse" is a generalization of Definition 4.1.6.

Theorem 4.1.15. *Let $T \in \mathcal{C}(X)$. Then the following are equivalent.*

(1) *T^d exists.*
(2) *There is an idempotent operator $Q \in B(X)$ with $\operatorname{Ran}(Q) \subset \mathfrak{D}(T)$ such that*

$$T((I-Q)\mathfrak{D}(T)) \subset (I-Q)\mathfrak{D}(T), \qquad T(\operatorname{Ran}(Q)) \subset \operatorname{Ran}(Q)$$
$$\operatorname{Ker}(T\big|_{(I-Q)\mathfrak{D}(T)}) = \{0\}, \qquad \operatorname{Ran}(T\big|_{(I-Q)\mathfrak{D}(T)}) = (I-Q)X$$

and $T\big|_{\operatorname{Ran}(Q)} \in QN(\operatorname{Ran}(Q))$.
(3) *There is an idempotent operator $Q \in B(X)$ with $\operatorname{Ran}(Q) \subset \mathfrak{D}(T)$ such that $TQ = QT$ on $\mathfrak{D}(T)$ and $\operatorname{Ker}(T + \xi Q) = \{0\}$, $\operatorname{Ran}(T + \xi Q) = X$, $\forall \xi \in \mathbb{C}\backslash\{0\}$.*

Proof. (1)⇒(2) Put $Q = I - TT^d$. Then by Definition 4.1.14,

$$\operatorname{Ran}(Q) \subset \mathfrak{D}(T), \qquad T\big|_{\operatorname{Ran}(Q)} \in QN(B(\operatorname{Ran}(Q))$$
$$T^d = T^dTT^d, \qquad T^dT = TT^d \text{ on } \mathfrak{D}(T).$$

Put $T_1 = T\big|_{(I-Q)\mathfrak{D}(T)}$ and $T_2 = T\big|_{\operatorname{Ran}(Q)}$. Let $x \in \mathfrak{D}(T)$. From $T^dTx = TT^dx$, we get that

$$T(TT^dx) = TT^dTx = (I-Q)Tx \subset \operatorname{Ran}(I-Q),$$

that is, $\operatorname{Ran}(T_1) \subset \operatorname{Ran}(I-Q)$.

Conversely, let $y \in \operatorname{Ran}(I-Q)$. Then $y = TT^dy$ and

$$T^dy = T^dTT^dy = (I-Q)T^dy$$

for $T^dy \in \mathfrak{D}(T)$. Thus, $y \in \operatorname{Ran}(T_1)$.

Now let $x \in \operatorname{Ker} T_1$. Then $Tx = 0$ and $(I - Q)x = x$, $x \in \mathfrak{D}(T)$. So $x = TT^d x = T^d T x = 0$, that is, $\operatorname{Ker} T_1 = \{0\}$.

(2)⇒(3) Since $T(\operatorname{Ran}(Q)) \subset \operatorname{Ran}(Q)$, $T((I-Q)\mathfrak{D}(T)) \subset (I-Q)\mathfrak{D}(T)$ and $\operatorname{Ran}(Q) \subset \mathfrak{D}(T)$, it follows that $TQ = QT$ on $\mathfrak{D}(T)$. Let $\xi \in \mathbb{C}\backslash\{0\}$ and put $C_\xi = T + \xi Q \in \mathcal{C}(X)$. Then

$$C_\xi Q = QTQ + \xi Q, \;\; C_\xi(I-Q) = T(I-Q) \text{ on } \mathfrak{D}(T).$$

Since $\operatorname{Ran}(T\big|_{(I-Q)\mathfrak{D}(T)}) = \operatorname{Ran}(I-Q)$ and $T|\operatorname{Ran}(Q)$ is quasi–nilpotent, it follows that $T\big|_{\operatorname{Ran}(Q)} + \xi I_{\operatorname{Ran}(Q)}$ is invertible in $B(\operatorname{Ran}(Q))$. Thus

$$\begin{aligned}\operatorname{Ran}(C_\xi) &= \operatorname{Ran}(T(I-Q)) + \operatorname{Ran}(QTQ + \xi Q)\\ &= \operatorname{Ran}(I-Q) + \operatorname{Ran}(Q) = X.\end{aligned}$$

Let $x \in \operatorname{Ker} C_\xi$. Then $Tx = -\xi Qx$, $x \in \mathfrak{D}(T)$ and

$$T(I-Q)x = Tx - TQx = Tx - QTx = -\xi Qx - Q(-\xi Qx) = 0.$$

So $(I-Q)x = 0$ and hence $x = (-\xi)^{-1}(T\big|_{\operatorname{Ran}(Q)})x$. This indicates that $x = (-\xi)^n (T\big|_{\operatorname{Ran}(Q)})^n x$ and consequently,

$$\|x\| \le |\xi|^{-n} \|(T\big|_{\operatorname{Ran}(Q)})^n\| \|x\|, \quad \forall n \in \mathbb{N}. \tag{4.1.7}$$

Noting that $\lim_{n\to\infty} \|(T\big|_{\operatorname{Ran}(Q)})^n\|^{\frac{1}{n}} = 0$, we have $x = 0$ from (4.1.7).

(3)⇒(1) By assumption, $C_\xi = T + \xi Q$ has the bounded inverse C_ξ^{-1} and $C_\xi^{-1} Q = Q C_\xi^{-1}$. Put $B_\xi = C_\xi^{-1}(I-Q) = (I-Q)C_\xi^{-1}$. Then $B_\xi \in B(X)$ with $\operatorname{Ran}(B_\xi) \subset \mathfrak{D}(T)$. Note that $QB_\xi = B_\xi Q = 0$. Hence

$$TB_\xi = (C_\xi - \xi Q)B_\xi = I - Q, \quad B_\xi T = I - Q \text{ on } \mathfrak{D}(T),$$

that is, $B_\xi T B_\xi = B_\xi$ and $B_\xi T = T B_\xi$ on $\mathfrak{D}(T)$.

Since $TQ = QTQ$, we have $\operatorname{Ran}(T\big|_{\operatorname{Ran}(Q)}) \subset \operatorname{Ran}(Q)$ and $T\big|_{\operatorname{Ran}(Q)} \in B(\operatorname{Ran}(Q))$. From

$$C_\xi = QTQ + \xi Q + (I-Q)T(I-Q) \text{ on } \mathfrak{D}(T)$$

and $\operatorname{Ran}(C_\xi) = X$, $\operatorname{Ker} C_\xi = \{0\}$, we get that

$$\operatorname{Ran}(T\big|_{\operatorname{Ran}(Q)} + \xi I_{\operatorname{Ran}(Q)}) = \operatorname{Ran}(Q), \;\; \operatorname{Ker}(T\big|_{\operatorname{Ran}(Q)} + \xi I_{\operatorname{Ran}(Q)}) = \{0\}$$

and so that $T\big|_{\operatorname{Ran}(Q)} + \xi I_{\operatorname{Ran}(Q)}$ is invertible in $B(\operatorname{Ran}(Q))$ for any $\xi \in \mathbb{C}\backslash\{0\}$, which means that $\sigma(T\big|_{\operatorname{Ran}(Q)}) = \{0\}$. Therefore,

$$T^d = B_\xi = (T + \xi Q)^{-1}(I-Q), \quad \forall \xi \in \mathbb{C}\backslash\{0\}.$$

□

4.2 The outer generalized inverse with prescribed range and kernel

In this section, we assume that T is a subspace of X and S is closed subspace of Y.

Proposition 4.2.1. *Let $A \in C(X,Y)$ with $T \subset \mathfrak{D}(A)$. Then there is $G \in B(Y,X)$ such that* $\operatorname{Ker} G = S$, $\operatorname{Ran}(G) = T$ *and $GAG = G$ if and only if the following conditions are satisfied:*

(1) $\operatorname{Ker} A \cap T = \{0\}$.
(2) $Y = AT \dotplus S$ *and*
(3) T *is closed in* $\mathfrak{D}(A)$.

Proof. "Only if" part. Put $Q = AG$. Then $Q \in B(Y)$ is idempotent and

$$\operatorname{Ran}(Q) = AT, \ \operatorname{Ker} Q = \operatorname{Ran}(I - Q) = \operatorname{Ker} G = S.$$

Thus $Y = AT \dotplus S$. Let $x \in \operatorname{Ker} A \cap T$. Then $Ax = 0$ and $x = Gy$ for some $y \in Y$. So $x = GAGy = GAx = 0$.

To prove T is closed in $\mathfrak{D}(A)$, let $\{y_n\} \subset T$ and $y_0 \in \mathfrak{D}(A)$ such that $y_n \xrightarrow{\|\cdot\|} y_0$. Choose $\{x_n\} \subset Y$ such that $Gx_n = y_n$. Let $f \in \mathfrak{D}((GA)^*)$. Then

$$(GA)^* f(Gx_n) \to (GA)^* f(y_0) = f(GAy_0).$$

Since $GAG = G$, it follows that

$$(GA)^* f(Gx_n) = f(GAGx_n) = f(Gx_n) \to f(y_0).$$

Thus, $y_0 = GAy_0 \in \operatorname{Ran}(G) = T$ by Proposition 1.2.4 (2).

"If" part. Put $A_0 = A\big|_T : T \to \operatorname{Ran}(A)$. Let $\{x_n, A_0 x_n\} \subset G(A_0)$ and $x_n \xrightarrow{\|\cdot\|} x_0$, $y_n \xrightarrow{\|\cdot\|} y_0$ for some $\{x_0, y_0\} \in X \times Y$. Since $G(A)$ is closed and T is closed in $\mathfrak{D}(A)$, we have $x_0 \in \mathfrak{D}(A)$, $Ax_0 = y_0$ and $x_0 \in T$, that is, $\{x_0, y_0\} \in G(A_0)$. So A_0 is a closed operator with $\mathfrak{D}(A_0) = T$ and $\operatorname{Ran}(A_0) = AT$. Note that $\operatorname{Ran}(A_0)$ is closed and $\operatorname{Ker} A_0 = \{0\}$, it follows from Lemma 1.2.2 that $A_0^{-1} : AT \to T$ is a bounded linear operator. Put $Gy = \begin{cases} A_0^{-1} & y \in AT \\ 0 & y \in S \end{cases}$. It is easy to check that $G \in B(Y,X)$ and $\operatorname{Ker} G = S$, $\operatorname{Ran}(G) = T$ and $GAG = G$ for $Y = AT \dotplus S$. □

Proposition 4.2.2. *Let $A \in C(X,Y)$ with $T \subset \mathfrak{D}(A)$ closed in $\mathfrak{D}(A)$. Then there is at most one $G \in B(Y,X)$ satisfying conditions:*

$$\operatorname{Ran}(G) = T, \ \operatorname{Ker} G = S \text{ and } G = GAG.$$

Proof. Suppose that there is another $G' \in B(Y, X)$ such that $\mathrm{Ran}\,(G') = T$, $\mathrm{Ker}\, G' = S$ and $G'AG' = G'$. Put $Q = AG$ and $Q' = AG'$. Then Q, $Q' \in B(Y)$ are idempotent operators satisfying:

$$\mathrm{Ran}\,(Q) = \mathrm{Ran}\,(Q') = AT, \quad \mathrm{Ran}\,(I_Y - Q) = \mathrm{Ran}\,(I_Y - Q') = S.$$

Thus, $QQ' = Q'$, $Q'Q = Q$, $(I_Y - Q)(I_Y - Q') = I_Y - Q'$ and so that $Q = Q'$. Consequently, $G' = G'AG' = G'AG$.

Put $P = GA$ and $P' = G'A$. Then P, P' are idempotent on $\mathfrak{D}(A)$ and

$$\mathrm{Ran}\,(P) = \mathrm{Ran}\,(G) = \mathrm{Ran}\,(P') = \mathrm{Ran}\,(G') = T.$$

So $GA = P = P'P = G'AGA$ on $\mathfrak{D}(A)$ and consequently,

$$G = GAG = (G'AGA)G = G'AG = G'.$$

□

According to Proposition 4.2.1 and Proposition 4.2.2, we have

Definition 4.2.3. Let $A \in \mathcal{C}(X, Y)$ with $T \subset \mathfrak{D}(A)$. If there exists $G \in B(Y, X)$ with $\mathrm{Ran}\,(G) = T$ and $\mathrm{Ker}\, G = S$ such that $GAG = G$, G is called to be the outer generalized inverse of A with prescribed the range T and the kernel S. We denoted it by $A^{(2)}_{T,S}$.

With different choices of T and S, we can express the generalized Drazin inverse and the Moore–Penrose inverse as the outer generalized inverses with prescribed the range and the kernel as follows.

Corollary 4.2.4. *Let $A \in \mathcal{C}(X)$ such that A^d exists. Put $S = \mathrm{Ker}\, A^d$ and $T = \mathrm{Ran}\,(A^d)$. Then $A^d = A^{(2)}_{T,S}$.*

Proof. We have

$$\mathrm{Ran}\,(A^d) \subset \mathfrak{D}(A), \; A^dAA^d = A^d, \; A^dA = AA^d \text{ on } \mathfrak{D}(A)$$

by Definition 4.1.14. Put $Q = I - AA^d$. Then $\mathrm{Ran}\,(Q) = \mathrm{Ker}\, A^d = S$, $\mathrm{Ran}\,(I - Q) = AT$. So $X = AT \dotplus S$. Let $x \in \mathrm{Ker}\, A \cap T$. Then $Ax = 0$ and $x = A^dy$ for some $y \in X$. Thus, $AA^dy = 0$ and $x = A^dy = A^dAA^dy = 0$, that is, $\mathrm{Ker}\, A \cap T = \{0\}$. To prove $\mathrm{Ran}\,(A^d)$ is closed in $\mathfrak{D}(A)$, let $\{x_n\} \subset \mathrm{Ran}\,(A^d)$ and $x_0 \in \mathfrak{D}(A)$ such that $x_n \xrightarrow{\|\cdot\|} x_0$. Choose $\{y_n\} \subset X$ such that $x_n = A^dy_n$, $n \in \mathbb{N}$. Then

$$AA^dx_n = AA^dA^dy_n = A^dAA^dy_n = A^dy_n = x_n, \; \forall\, n \in \mathbb{N}$$

and so that $AA^dx_0 = x_0 \in \mathfrak{D}(A)$ for $AA^d \in B(X)$. This implies that $x_0 = AA^dx_0 = A^dAx_0 \in \mathrm{Ran}\,(A^d)$. Therefore, $A^{(2)}_{T,S} = A^d$ by Proposition 4.2.1 and Proposition 4.2.2.

□

Corollary 4.2.5. *Let $A \in \mathcal{C}(H,K)$ with $\mathrm{Ran}\,(A)$ closed. Put $S = \mathrm{Ker}\, A^\dagger$ and $T = (\mathrm{Ker}\, A)^\perp \cap \mathfrak{D}(A)$. Then $A^{(2)}_{T,S} = A^\dagger$.*

Proof. Put $Q = I_K - AA^\dagger$. Then from $AA^\dagger A = A$ on $\mathfrak{D}(A)$ and $A^\dagger AA^\dagger = A^\dagger$, we get that

$$\mathrm{Ran}\,(Q) = \mathrm{Ker}\, A^\dagger,\ \mathrm{Ran}\,(I_K - Q) = \mathrm{Ran}\,(AA^\dagger) = A(\mathrm{Ran}\,(A^\dagger)).$$

Noting that

$$\begin{aligned}\mathrm{Ker}\, A^\dagger &= \mathrm{Ran}\,(AA^\dagger)^\perp = \mathrm{Ran}\,(A)^\perp = \mathrm{Ker}\,(A^*) = S\\ \mathrm{Ran}\,(A^\dagger) &= (\mathrm{Ker}\, A)^\perp \cap \mathfrak{D}(A) = T,\end{aligned}$$

we have $AT \oplus S = K$. Clearly, T is closed in $\mathfrak{D}(A)$ and $\mathrm{Ker}\, A \cap T = \{0\}$. So $A^{(2)}_{T,S} = A^\dagger$ by Proposition 4.2.1 and Proposition 4.2.2. □

We now investigate the various expressions of the outer generalized inverse with prescribed range and kernel.

Proposition 4.2.6. *Let $A \in \mathcal{C}(H,K)$ and S, T be closed subspaces in K and H respectively. Suppose that $T \subset \mathfrak{D}(A)$ and $A^{(2)}_{T,S}$ exists. Then $P_{S^\perp}AP_T$ has closed range and $A^{(2)}_{T,S} = (P_{S^\perp}AP_T)^\dagger$.*

Proof. The existence of $A^{(2)}_{T,S}$ implies that $\mathrm{Ker}\, A \cap T = \{0\}$, AT is closed and $K = AT \dotplus S$. Let $P = P_{S,AT}\colon K \to S$ be the idempotent operator. Since $\mathrm{Ran}\,(P) = S$ and $\mathrm{Ran}\,(I_K - P) = AT$, it follows that $PP_S = P_S$, $P_S P = P$ and $(I_Y - P)AT = AT$. Noting that

$$\begin{aligned}(I_K - P)(I_K - P_S) &= I_K + PP_S - P_S - P = I_K - P\\ (I_K - P_S)(I_K - P) &= I_K - P - P_S + P_S P = I_K - P_S,\end{aligned}$$

we have

$$\mathrm{Ran}\,(I_Y - P_S) = (I_K - P_S)(\mathrm{Ran}\,(I_K - P)) = (I_K - P_S)AT$$

and hence $\mathrm{Ran}\,(P_{S^\perp}AP_T) = \mathrm{Ran}\,(P_{S^\perp}) = S^\perp$. Let $x \in T$ and $P_{S^\perp}Ax = 0$. Then $(I_K - P)Ax = Ax$, $Ax = P_S Ax$ and hence $0 = PAx = PP_S Ax = P_S Ax = Ax$. Since $\mathrm{Ker}\, A \cap T = \{0\}$, we have $x = 0$ and consequently, $\mathrm{Ker}\,(P_{S^\perp}AP_T) = T^\perp$ for AP_T is bounded. Therefore, $(P_{S^\perp}AP_T)^\dagger$ exists and

$$\mathrm{Ran}\,((P_{S^\perp}AP_T)^\dagger) = (\mathrm{Ker}\,(P_{S^\perp}AP_T))^\perp = T \tag{4.2.1}$$

$$\mathrm{Ker}\,(P_{S^\perp}AP_T)^\dagger = (\mathrm{Ran}\,(P_{S^\perp}AP_T))^\perp = S. \tag{4.2.2}$$

by Proposition 3.1.2 and Definition 3.1.3. Since

$$(P_{S^\perp}AP_T)^\dagger P_{S^\perp} = (P_{S^\perp}AP_T)^\dagger = P_T(P_{S^\perp}AP_T)^\dagger,$$

by (4.2.1) and (4.2.2), it follows that

$$\begin{aligned}(P_{S^\perp}AP_T)^\dagger &= (P_{S^\perp}AP_T)^\dagger(P_{S^\perp}AP_T)(P_{S^\perp}AP_T)^\dagger \\ &= (P_{S^\perp}AP_T)^\dagger A(P_{S^\perp}AP_T)^\dagger\end{aligned}$$

and so that $A^{(2)}_{T,S} = (P_{S^\perp}AP_T)^\dagger$. □

Proposition 4.2.7. *Let $A \in \mathcal{C}(X,Y)$ with $T \subset \mathfrak{D}(A)$ closed in $\mathfrak{D}(A)$. Let $G \in B(Y,X)$ such that $\operatorname{Ran}(G) = T$ and $\operatorname{Ker} G = S$. Suppose that $A^{(2)}_{T,S}$ exists. Then $AG \in B(Y)$ has the group inverse $(AG)^\#$ and moreover, $A^{(2)}_{T,S} = G(AG)^\#$.*

Proof. We have known $AG \in B(Y)$ and $\operatorname{Ran}(AG) = AT$. Since

$$AT \dotplus S = Y,\ \operatorname{Ker} A \cap T = \{0\},\ \operatorname{Ker} G = S,$$

it follows that $D = AG\big|_{AT}$ has the range AT and the kernel $\{0\}$. Thus $D^{-1} \in B(AT)$.

Let $P\colon Y \to AT$ be the idempotent operator with respect to the decomposition $Y = AT \dotplus S$. Set $C = D^{-1}P \in B(Y)$. It is easy to verify that

$$CAG = P = AGC, \quad AGCAG = AG, \quad CAGC = C.$$

This means that $C = (AG)^\#$. Put $B = G(AG)^\# \in B(Y,X)$. Then

$$BAB = G(AG)^\#(AG)(AG)^\# = G(AG)^\# = B.$$

Note that $Y = AT \dotplus S$, $\operatorname{Ker} G = S$ and $A\big|_T = \operatorname{Ran}(C) = \operatorname{Ran}((AG)^\#)$. We have

$$T = GY = G(AT) = G(\operatorname{Ran}((AG)^\#)) = \operatorname{Ran}(B).$$

From $B = GD^{-1}P$, we obtain that $\operatorname{Ran}(I_Y - P) \subset \operatorname{Ker} B$. Let $y \in \operatorname{Ker} B$. Then $0 = ABy = (AG)D^{-1}Py = Py$, i.e., $y \in \operatorname{Ran}(I_Y - P)$. Thus $\operatorname{Ker} B = \operatorname{Ran}(I_Y - P) = S$.

The above proves that $A^{(2)}_{T,S} = B = G(AG)^\#$. □

Theorem 4.2.8. *Let $A \in \mathcal{C}(X,Y)$ with $T \subset \mathfrak{D}(A)$ closed in $\mathfrak{D}(A)$. Let $G \in B(Y,X)$ satisfy $\operatorname{Ran}(G) = T$ and $\operatorname{Ker} G = S$. If $A^{(2)}_{T,S}$ exists, then the norm limit $\displaystyle\lim_{\substack{\lambda\to 0\\ \lambda\notin\sigma(-AG)}} G(\lambda I_Y + AG)^{-1}$ exists and is equal to $A^{(2)}_{T,S}$.*

Conversely, let $T \subset \mathfrak{D}(A)$ be closed in X and let $G \in B(Y,X)$ such that $\operatorname{Ran}(G) = T$ and $\operatorname{Ker} G = S$. If $\displaystyle\lim_{\substack{\lambda\to 0\\ \lambda\notin\sigma(-AG)}} G(\lambda I_Y + AG)^{-1}$ exists, then $A^{(2)}_{T,S}$ exists and $\displaystyle\lim_{\substack{\lambda\to 0\\ \lambda\notin\sigma(-AG)}} G(\lambda I_Y + AG)^{-1} = A^{(2)}_{T,S}$.

Proof. ($\Rightarrow$) By Proposition 4.2.7, $AG \in B(Y)$ has the group inverse $(AG)^{\#}$. Put $Q = (AG)(AG)^{\#}$. Since

$$\begin{aligned} Q(AG) &= (AG)Q = AG \\ (AG)^{\#}Q &= Q(AG)^{\#} = (AG)^{\#} \\ (AG)(AG)^{\#} &= (AG)^{\#}(AG) = Q, \end{aligned}$$

we have $AG|_{\operatorname{Ran}(Q)}$ is invertible in $B(\operatorname{Ran}(Q))$ with $(AG|_{\operatorname{Ran}(Q)})^{-1} = (AG)^{\#}|_{\operatorname{Ran}(Q)}$. Write $\lambda I_Y + AG$ as

$$\lambda I_Y + AG = \begin{bmatrix} \lambda I_{\operatorname{Ran}(Q)} + (AG)|_{\operatorname{Ran}(Q)} & \\ & \lambda I_{\operatorname{Ran}(I_Y - Q)} \end{bmatrix}. \tag{4.2.3}$$

When $\lambda I_Y + AG$ is invertible in $B(Y)$ and $\lambda \neq 0$, $\lambda I_{\operatorname{Ran}(Q)} + (AG)|_{\operatorname{Ran}(Q)}$ is invertible in $B(\operatorname{Ran}(Q))$ and

$$(\lambda I_Y + AG)^{-1} = \begin{bmatrix} (\lambda I_{\operatorname{Ran}(Q)} + (AG)|_{\operatorname{Ran}(Q)})^{-1} & \\ & \lambda^{-1} I_{\operatorname{Ran}(I_Y - Q)} \end{bmatrix}$$

by (4.2.3). According to the proof of Proposition 4.2.7, $\operatorname{Ran}(I_Y - Q) = S = \operatorname{Ker} G$. So $G(\lambda I_Y + AG)^{-1} = G(\lambda I_{\operatorname{Ran}(Q)} + AG|_{\operatorname{Ran}(Q)})^{-1}Q$ and hence by Corollary 1.4.8,

$$\lim_{\substack{\lambda \to 0 \\ \lambda \notin \sigma(-AG)}} G(\lambda I_Y + AG)^{-1} = G((AG)^{\#}|_{\operatorname{Ran}(Q)})Q = G(AG)^{\#} = A^{(2)}_{T,S}.$$

($\Leftarrow$) Put $W = \lim\limits_{\substack{\lambda \to 0 \\ \lambda \notin \sigma(-AG)}} G(\lambda I_Y + AG)^{-1}$. Since $AG = \lim\limits_{\lambda \to 0}(\lambda I_Y + AG)$, we have

$$WAG = \lim_{\substack{\lambda \to 0 \\ \lambda \notin \sigma(-AG)}} G(\lambda I_Y + AG)^{-1}(\lambda I_Y + AG) = G$$

and hence $\operatorname{Ran}(G) \subset \operatorname{Ran}(W)$. From $W = \lim\limits_{\substack{\lambda \to 0 \\ \lambda \notin \sigma(-AG)}} G(\lambda I_Y + AG)^{-1}$, we get that $\operatorname{Ran}(W) \subset \overline{\operatorname{Ran}(G)}$. So $\operatorname{Ran}(W) = \operatorname{Ran}(G) = T$.

Since $GA \in \mathcal{D}(X, Y)$ with $\mathfrak{D}(GA) = \mathfrak{D}(A)$, it follows from Proposition 1.2.4 (2) that $\mathfrak{D}((GA)^*)$ is total in X^*. Let $f \in \mathfrak{D}((GA)^*)$ and $y \in Y$. Then

$$\begin{aligned} (GA)^* f(Wy) &= \lim_{\substack{\lambda \to 0 \\ \lambda \notin \sigma(-AG)}} (GA)^* f(G(\lambda I_Y + AG)^{-1} y) \\ &= \lim_{\substack{\lambda \to 0 \\ \lambda \notin \sigma(-AG)}} f(GAG(\lambda I_Y + AG)^{-1} y). \end{aligned} \tag{4.2.4}$$

Noting that

$$GAG(\lambda I_Y + AG)^{-1} = G - \lambda G(\lambda I_Y + AG)^{-1},\ Wy \in \mathfrak{D}(A),$$

we have $f(GAWy) = (GA)^* f(Wy) = f(Gy)$ by (4.2.4) and hence $GAW = G$ since $\mathfrak{D}((GA)^*)$ is total.

Similarly, we have $WAW = W$ by using $WAG = G$.

Now we prove that $\operatorname{Ker} W = \operatorname{Ker} G$. Clearly, $\operatorname{Ker} W \subset \operatorname{Ker} G$ by $GAW = G$. Let $y \in \operatorname{Ker} G$ and put $y_\lambda = (\lambda I_Y + AG)^{-1} y$, $\lambda \notin \sigma(-AG)$, $\lambda \neq 0$. Then

$$0 = G(\lambda I_Y + AG) y_\lambda = (\lambda I_X + GA) G y_\lambda. \tag{4.2.5}$$

Let M be in $B(Y)$ such that $M(\lambda I_Y + AG) = (\lambda I_Y + AG)M = I_Y$. Then

$$MAG = I_Y - \lambda M, \quad (I_X - GMA)(\lambda I_X + GA)\xi = \lambda\xi,\ \forall \xi \in \mathfrak{D}(A).$$

Thus $Gy_\lambda = \lambda^{-1}(I_X - GMA)(\lambda I_X + GA)Gy_\lambda = 0$ by (4.2.5) and consequently, $Wy = \lim\limits_{\substack{\lambda \to 0 \\ \lambda \notin \sigma(-AG)}} G(\lambda I_Y + AG)^{-1} y = 0$, i.e., $\operatorname{Ker} G \subset \operatorname{Ker} W$.

Therefore, $A^{(2)}_{T,S} = W = \lim\limits_{\substack{\lambda \to 0 \\ \lambda \notin \sigma(-AG)}} G(\lambda I_Y + AG)^{-1}$. □

Corollary 4.2.9. *Let* $A \in B(H, K)$ *with* $\operatorname{Ran}(A)$ *closed. Then*

$$A^\dagger = \lim_{\lambda \to 0^+} A^*(\lambda I_K + AA^*)^{-1}.$$

Proof. Take $G = A^*$. Then $T = \operatorname{Ran}(G) = (\operatorname{Ker} A)^\perp$ and $S = \operatorname{Ker} G = \operatorname{Ran}(A)^\perp$. So by Corollary 4.2.5 and Theorem 4.2.8,

$$A^\dagger = A^{(2)}_{T,S} = \lim_{\substack{\lambda \to 0 \\ \lambda \notin \sigma(-AA^*)}} A^*(\lambda I_K + AA^*)^{-1}.$$

Note that $AA^* \in B(K)_+$ (cf. Proposition 1.5.16 (1)). Thus,

$$A^\dagger = \lim_{\lambda \to 0^+} A^*(\lambda I_K + AA^*)^{-1} = \lim_{\lambda \to 0^+} (\lambda I_H + A^*A)^{-1} A.$$

□

Corollary 4.2.10. *Let* $A \in Dr(X)$ *with* $\operatorname{ind}(A) = k$. *Then*

$$A^D = \lim_{\substack{\lambda \to 0 \\ \lambda \notin \sigma(-A^{k+1})}} (\lambda I + A^{k+1})^{-1} A^k.$$

Proof. Put $G = A^k$. Then $\operatorname{Ran}(G) = \operatorname{Ran}(A^D)$ and $\operatorname{Ker} G = \operatorname{Ker} A^D$ by Theorem 4.1.5 and Definition 4.1.6. The assertion follows from Corollary 4.2.4 and Theorem 4.2.8. □

Corollary 4.2.11. *Let $A \in \mathcal{C}(X,Y)$ with $T \subset \mathfrak{D}(A)$ closed in $\mathfrak{D}(A)$. Let $G \in B(Y,X)$ such that* $\mathrm{Ran}\,(G) = T$ *and* $\mathrm{Ker}\, G = S$. *Suppose that $A^{(2)}_{T,S}$ exists and* $\mathrm{Re}\,\lambda \geq 0$, $\forall\, \lambda \in \sigma(AG)$. *Then*

$$A^{(2)}_{T,S} = \int_0^{+\infty} G\,\mathrm{e}^{-(AG)t}\mathrm{d}\,t.$$

Proof. By Proposition 4.2.7, AG has the group inverse $(AG)^{\#}$. Put $P = (AG)(AG)^{\#} = (AG)^{\#}(AG)$. Then

$$AG = (AG)P = P(AG), \quad (AG)^{\#} = (AG)^{\#}P = P(AG)^{\#},$$
$$\mathrm{Ker}\, G = \mathrm{Ran}\,(I_Y - P), \qquad A^{(2)}_{T,S} = G(AG)^{\#}.$$

Set $S = (AG)\big|_{\mathrm{Ran}\,(P)}$. Then $AG = \begin{bmatrix} S & \\ & 0 \end{bmatrix}$ and $(AG)^{\#} = \begin{bmatrix} S^{-1} & \\ & 0 \end{bmatrix}$ with respect to $Y = \mathrm{Ran}\,(P) \dotplus \mathrm{Ran}\,(I_Y - P)$, where $S^{-1} = (AG)^{\#}\big|_{\mathrm{Ran}\,(P)}$. Since $\mathrm{Re}\,\lambda \geq 0$, $\forall\, \lambda \in \sigma(AG)$, it follows that $\mathrm{Re}\,\lambda > 0$, $\forall\, \lambda \in \sigma(S)$. Thus $S^{-1} = \displaystyle\int_0^{+\infty} \mathrm{e}^{-tS}\mathrm{d}\,t$ by Proposition 1.4.17 and hence

$$A^{(2)}_{T,S} = G(AG)^{\#} = G\begin{bmatrix} S^{-1} & \\ & I_{\mathrm{Ran}\,(I_Y - P)} \end{bmatrix} = \int_0^{+\infty} G\,\mathrm{e}^{-(AG)t}\mathrm{d}\,t.$$

□

Corollary 4.2.12. *Let $A \in \mathcal{C}(H,K)$ with $T \subset \mathfrak{D}(A)$ closed in H. Let $G \in B(K,H)$ such that* $\mathrm{Ran}\,(G) = T$ *and* $\mathrm{Ker}\, G = S$. *Assume that $A^{(2)}_{T,S}$ exists. Then*

$$A^{(2)}_{T,S} = \int_0^{+\infty} G\,\mathrm{e}^{-(GAG)^*(GAG)t}((GAG)^*G)\,\mathrm{d}\,t.$$

Proof. $\mathrm{Ker}\, A \cap T = \{0\}$ and $\mathrm{Ran}\,(G) = T$ implies that $\mathrm{Ker}\,(AG) = \mathrm{Ker}\, G = S$. Write $AG = \begin{bmatrix} W_1 & 0 \\ W_2 & 0 \end{bmatrix}$ with respect to $K = S^{\perp} \oplus S$, where $W_1 = P_{S^{\perp}}(AG)\big|_{S^{\perp}}$, $W_2 = P_S(AG)\big|_{S^{\perp}}$ and $G = \begin{bmatrix} G_1 & 0 \\ 0 & 0 \end{bmatrix}$ with respect to $K = S^{\perp} \oplus S$ and $H = T \oplus T^{\perp}$, where $G_1 = P_T G|_{S^{\perp}}$. Then W_1 is invertible in $B(S^{\perp})$ since AT is closed. So, $(AG)^{\#} = \begin{bmatrix} W_1^{-1} & 0 \\ W_2 W_1^{-2} & 0 \end{bmatrix}$ and $\mathrm{Ran}\,(G_1) = T$, $\mathrm{Ker}\, G_1 = \{0\}$. Noting that

$$(GAG)^*(GAG) = \begin{bmatrix} (G_1W_1)^*(G_1W_1) & 0 \\ 0 & 0 \end{bmatrix} \tag{4.2.6}$$

and $(G_1W_1)^*(G_1W_1)$ is invertible in $B(S^\perp)$ and is positive, we have

$$((G_1W_1)^*(G_1W_1))^{-1} = \int_0^{+\infty} \mathrm{e}^{-t(G_1W_1)^*(G_1W_1)}\mathrm{d}\,t \tag{4.2.7}$$

by Proposition 1.4.17. From (4.2.6) and (4.2.7), we get that

$$\begin{aligned}\int_0^{+\infty} &\mathrm{e}^{-t(GAG)^*(GAG)}(GAG)^*G\,\mathrm{d}\,t \\ &= \begin{bmatrix}((G_1W_1)^*(G_1W_1))^{-1} & 0 \\ 0 & 0\end{bmatrix}\begin{bmatrix}(G_1W_1)^*G_1 & 0 \\ 0 & 0\end{bmatrix} \\ &= \begin{bmatrix}W_1^{-1} & 0 \\ 0 & 0\end{bmatrix}.\end{aligned}$$

Thus,

$$\begin{aligned}A_{T,S}^{(2)} = G(AG)^{\#} &= \begin{bmatrix}G_1W_1^{-1} & 0 \\ 0 & 0\end{bmatrix} \\ &= \int_0^{+\infty} G\,\mathrm{e}^{-t(GAG)^*(GAG)}(GAG)^*G\,\mathrm{d}\,t.\end{aligned}$$

□

As the end of this section, we will give the reverse law of outer generalized inverses with prescribed ranges and kernels as follows.

Theorem 4.2.13. *Let $A \in B(X,Y)$ and $B \in \mathcal{C}(Z,X)$. Let T_A, S_B be closed subspaces of X and S_A be a closed subspace of Y and T_B be a closed subspace of $\mathfrak{D}(B)$. Suppose that there is $W_A \in B(Y,X)$ and $W_B \in B(X,Z)$ satisfying conditions:*

(1) $\mathrm{Ran}\,(W_A) = T_A$, $\mathrm{Ker}\,W_A = S_A$, $\mathrm{Ran}\,(W_B) = T_B$, $\mathrm{Ker}\,W_B = S_B$.
(2) $A_{T_A,S_A}^{(2)}$ *and* $B_{T_B,S_B}^{(2)}$ *exist.*
(3) $A_{T_A,S_A}^{(2)}ABW_B = BW_B\,A_{T_A,S_A}^{(2)}A$.
(4) $BB_{T_B,S_B}^{(2)}W_AA = W_AA\,BB_{T_B,S_B}^{(2)}$.

Put $T = \mathrm{Ran}\,(W_BW_A)$ and $S = \mathrm{Ker}\,(W_BW_A)$. Then we have

(1) $B_{T_B,S_B}^{(2)}A_{T_A,S_A}^{(2)}$ *and W_BW_A have the same range and kernel.*
(2) $(AB)_{T,S}^{(2)}$ *exists and* $(AB)_{T,S}^{(2)} = B_{T_B,S_B}^{(2)}A_{T_A,S_A}^{(2)}$.

Proof. Put $Q_A = A_{T_A,S_A}^{(2)}A$, $P_A = AA_{T_A,S_A}^{(2)}$, $Q_B = B_{T_B,S_B}^{(2)}B$ and $P_B = BB_{T_B,S_B}^{(2)}$. Then Q_A, P_A and P_B are all bounded and idempotent, Q_B is idempotent on $\mathfrak{D}(B)$ with

$$\mathrm{Ran}\,(Q_A) = \mathrm{Ran}\,(A_{T_A,S_A}^{(2)}) = T_A,\ \mathrm{Ker}\,P_A = \mathrm{Ker}\,A_{T_A,S_A}^{(2)} = S_A$$

$$\mathrm{Ran}\,(Q_B) = \mathrm{Ran}\,(B_{T_B,S_B}^{(2)}) = T_B,\ \mathrm{Ker}\,P_B = \mathrm{Ker}\,B_{T_B,S_B}^{(2)} = S_B.$$

By Proposition 4.2.7, $B^{(2)}_{T_B,S_B} = W_B(BW_B)^{\#}$. Since $Q_ABW_B = BW_BQ_A$ and $BW_B \in Dr(X)$, we have $(BW_B)^{\#}Q_A = Q_A(BW_B)^{\#}$ by Corollary 4.1.10 and so that $B^{(2)}_{T_B,S_B}Q_A = W_BQ_A(BW_B)^{\#}$. This indicates that

$$\mathrm{Ran}\,(B^{(2)}_{T_B,S_B}A^{(2)}_{T_A,S_A}) = B^{(2)}_{T_B,S_B}Q_AX \subset W_BQ_AX = \mathrm{Ran}\,(W_BW_A).$$

By Theorem 4.2.8, $B^{(2)}_{T_B,S_B} = \lim\limits_{\substack{\lambda\to 0\\ \lambda\notin\sigma(-BW_B)}} W_B(\lambda I_X + BW_B)^{-1}$. Thus, W_B $= B^{(2)}_{T_B,S_B}BW_B$ and $W_BQ_A = B^{(2)}_{T_B,S_B}Q_A(BW_B)$, that is,

$$\begin{aligned}\mathrm{Ran}\,(W_B, W_A) &= \mathrm{Ran}\,(W_B, Q_A) \subset \mathrm{Ran}\,(B^{(2)}_{T_B,S_B}Q_Q)\\ &= \mathrm{Ran}\,(B^{(2)}_{T_B,S_B}A^{(2)}_{T_A,S_A}).\end{aligned}$$

So $\mathrm{Ran}\,(B^{(2)}_{T_B,S_B}A^{(2)}_{T_A,S_A}) = \mathrm{Ran}\,(W_BW_A) = T$.

Now we show that $\mathrm{Ker}\,(B^{(2)}_{T_B,S_B}A^{(2)}_{T_A,S_A}) = \mathrm{Ker}\,(W_BW_A) = S$. Let $x \in \mathrm{Ker}\,(W_BW_A)$. Then $W_Ax \in \mathrm{Ker}\,W_B = \mathrm{Ker}\,P_B$, i.e., $P_BW_Ax = 0$. Since

$$\begin{aligned}A^{(2)}_{T_A,S_A} &= \lim_{\substack{\lambda\to 0\\ \lambda\notin\sigma(-AW_A)}} W_A(\lambda I_Y + AW_A)^{-1}\\ &= \lim_{\substack{\lambda\to 0\\ \lambda\notin\sigma(-W_AA)}} (\lambda I_X + W_AA)^{-1}W_A \qquad (4.2.8)\end{aligned}$$

by Theorem 4.2.8 and $P_BW_AAx = W_AAP_BP_BW_Ax = 0$, we have

$$B^{(2)}_{T_B,S_B}A^{(2)}_{T_A,S_A}x = B^{(2)}_{T_B,S_B}P_BA^{(2)}_{T_A,S_A}x = 0.$$

So $\mathrm{Ker}\,(W_BW_A) \subset \mathrm{Ker}\,(B^{(2)}_{T_B,S_B}A^{(2)}_{T_A,S_A})$.

On the other hand, let $x \in \mathrm{Ker}\,(B^{(2)}_{T_B,S_B}A^{(2)}_{T_A,S_A})$. Since $W_A = W_AP_A$, we have $P_BW_Ax = P_BW_AAA^{(2)}_{T_A,S_A}x = W_AABB^{(2)}_{T_B,S_B}A^{(2)}_{T_A,S_A}x = 0$, that is, $W_Ax \in \mathrm{Ker}\,P_B = S_B = \mathrm{Ker}\,W_B$ and hence $x \in \mathrm{Ker}\,(W_BW_A)$. From (4.2.8) and $P_BW_AA = W_AAP_B$, we get that $Q_AP_B = P_BQ_A$. Put $G = B^{(2)}_{T_B,S_B}A^{(2)}_{T_A,S_A} \in B(Y,Z)$ and $\mathrm{Ran}\,(G) = T$, $\mathrm{Ker}\,G = S$. Then

$$\begin{aligned}GABG &= B^{(2)}_{T_B,S_B}Q_AP_BA^{(2)}_{T_A,S_A} = B^{(2)}_{T_B,S_B}P_BQ_AA^{(2)}_{T_A,S_A}\\ &= B^{(2)}_{T_B,S_B}A^{(2)}_{T_A,S_A} = G.\end{aligned}$$

and hence $(AB)^{(2)}_{T,S} = G = B^{(2)}_{T_B,S_B}A^{(2)}_{T_A,S_A}$. □

Remark 4.2.14. Let $A \in B(H,K)$ and $B \in B(L,H)$ with $\mathrm{Ran}\,(A)$ and $\mathrm{Ran}\,(B)$ closed. We take $W_A = A^*$ and $W_B = B^*$ in Theorem 4.2.13. Then

$$A^{(2)}_{\mathrm{Ran}\,(A^*),\mathrm{Ker}\,A^*} = A^{\dagger}, \quad B^{(2)}_{\mathrm{Ran}\,(B^*),\mathrm{Ker}\,B^*} = B^{\dagger}$$

by Corollary 4.2.5. Assume that

$$A^\dagger ABB^* = BB^*A^\dagger A, \quad A^*ABB^\dagger = BB^\dagger A^*A.$$

Then $(AB)^{(2)}_{T,S} = B^\dagger A^\dagger$, where $T = \mathrm{Ran}\,(B^*A^*)$, $S = \mathrm{Ker}\,(B^*A^*)$. So T is closed and hence $\mathrm{Ran}\,(AB)$ is closed too. Therefore, $(AB)^\dagger = (AB)^{(2)}_{T,S} = B^\dagger A^\dagger$ by Corollary 4.2.5, which recovers the "if" part of Corollary 3.2.9.

Corollary 4.2.15. *Let $A \in B(X)$ and $B \in \mathcal{C}(X)$ such that A^d, B^d and $(AB)^d$ exist. Assume that $A^dABB^d = BB^dA^dA$ and $\mathrm{Ran}\,((AB)^d) = \mathrm{Ran}\,(B^dA^d)$, $\mathrm{Ker}\,((AB)^d) = \mathrm{Ker}\,(B^dA^d)$. Then $(AB)^d = B^dA^d$.*

Proof. Take $W_A = A^d$, $W_B = B^d$ in Theorem 4.2.13. Then $(AB)^{(2)}_{T,S} = B^dA^d$ by Corollary 4.2.4 and Theorem 4.2.13, where $T = \mathrm{Ran}\,(B^dA^d)$, $S = \mathrm{Ker}\,(B^dA^d)$. Since $\mathrm{Ran}\,((AB)^d) = T$ and $\mathrm{Ker}\,((AB)^d) = S$, it follows from Corollary 4.2.4 that $(AB)^d = (AB)^{(2)}_{T,S} = B^dA^d$. □

4.3 The stable perturbation of Drazin inverses

Let $T \in \mathcal{C}(X)$ such that $T^\#$ exists, i.e., $T^\# \in B(X)$ satisfies following:

$$\mathrm{Ran}\,(T^\#) \subset \mathfrak{D}(T), \quad TT^\#T = T \text{ on } \mathfrak{D}(T),$$
$$T^\#TT^\# = T^\#, \qquad T^\#T = TT^\# \text{ on } \mathfrak{D}(T).$$

Let $\delta T \in B(X)$ and put $\bar{T} = T + \delta T$. Suppose that $I + T^\#\delta T$ is invertible in $B(X)$ and

$$\Phi_T = I + \delta T(I - TT^\#)\delta T[(I + T^\#\delta T)^{-1}T^\#]^2$$

is also invertible in $B(X)$. Put

$$C_T = (I - TT^\#)\delta T(I + T^\#\delta T)^{-1}T^\#, D_T = (I + T^\#\delta T)^{-1}T^\#\Phi_T^{-1}.$$

Lemma 4.3.1. *Let $T \in \mathcal{C}(X)$ such that $T^\#$ exists. Let $\delta T \in B(X)$ with $\|T^\#\|\|\delta T\| < (1 + \|I - TT^\#\|)^{-1}$. Then Φ_T is invertible in $B(X)$ with*

$$\|\Phi_T^{-1}\| \le \frac{(1 - \|T^\#\|\|\delta T\|)^2}{(1 - \|T^\#\|\|\delta T\|)^2 - \|I - TT^\#\|(\|T^\#\|\|\delta T\|)^2}. \tag{4.3.1}$$

Proof. We have $I + T^\#\delta T$ is invertible in $B(X)$ when $\|T^\#\|\|\delta T\| < (1 + \|I - TT^\#\|)^{-1} < 1$. Moreover,

$$\begin{aligned}\|I - \Phi_T\| &\le \|\delta T\|^2\|I - TT^\#\|\|T^\#\|^2\|(I + T^\#\delta T)^{-1}\|^2 \\ &\le \frac{\|T^\#\|^2\|\delta T\|^2}{(1 - \|T^\#\|\|\delta T\|)^2}\|I - TT^\#\| \\ &< \frac{1}{\|I - TT^\#\|} \le 1.\end{aligned}$$

Thus, Φ_T is invertible in $B(X)$ and

$$\|\Phi_T^{-1}\| \le \sum_{n=0}^{\infty} \|\Phi_T - I\|^n \le \frac{1}{1-\|\Phi_T - I\|}$$
$$\le \frac{(1-\|T^{\#}\|\|\delta T\|)^2}{(1-\|T^{\#}\|\|\delta T\|)^2 - \|I-TT^{\#}\|(\|T^{\#}\|\|\delta T\|)^2}. \qquad \square$$

Lemma 4.3.2. *Let $T \in \mathcal{C}(X)$ such that $T^{\#}$ exists. Let $\delta T \in B(X)$ with $I + T^{\#}\delta T$ and Φ_T invertible in $B(X)$. Put $\bar{T} = T + \delta T \in \mathcal{C}(X)$. If $\bar{T}$ is the stable perturbation of T, i.e., $\mathrm{Ran}\,(\bar{T}) \cap \mathrm{Ker}\, T^{\#} = \{0\}$, then $\bar{T}^{\#}$ exists with $\bar{T}^{\#} = (I + C_T)(D_T + D_T^2\delta T(I - TT^{\#}))(I - C_T)$.*

Proof. Put $P = TT^{\#}$. Then $P \in B(X)$ is idempotent and $\mathrm{Ran}\,(I-P) \subset \mathfrak{D}(T)$ by Definition 4.1.14. With respect to the decomposition

$$\mathfrak{D}(T) = P(\mathfrak{D}(T)) + (I-P)(\mathfrak{D}(T)),$$

T has the form $T = \begin{bmatrix} T_1 & \\ & 0 \end{bmatrix}$, where $\mathfrak{D}(T_1) = P(\mathfrak{D}(T))$, $T_1 = T\big|_{P(\mathfrak{D}(T))}$, $\mathrm{Ran}\,(T_1) = \mathrm{Ran}\,(P)$ and $\mathrm{Ker}\, T_1 = \{0\}$ by Theorem 4.1.15. With respect to the decomposition $X = \mathrm{Ran}\,(P) + \mathrm{Ran}\,(I-P)$, $T^{\#}$ and δT have the forms

$$T^{\#} = \begin{bmatrix} T_1^{-1} & 0 \\ 0 & 0 \end{bmatrix}, \quad \delta T = \begin{bmatrix} \delta_1 & \delta_2 \\ \delta_3 & \delta_4 \end{bmatrix},$$

where $T_1^{-1}\colon \mathrm{Ran}\,(P) \to \mathfrak{D}(T_1)$ is bounded and

$$\delta_1 = P\delta T\big|_{\mathrm{Ran}\,(P)}, \qquad \delta_2 = P\delta T\big|_{\mathrm{Ran}\,(I-P)}$$
$$\delta_3 = (I-P)\delta T\big|_{\mathrm{Ran}\,(P)}, \quad \delta_4 = (I-P)\delta T\big|_{\mathrm{Ran}\,(I-P)}.$$

Then

$$\bar{T} = \begin{bmatrix} T_1 + \delta_1 & \delta_2 \\ \delta_3 & \delta_4 \end{bmatrix} \text{ on } \mathfrak{D}(T), \; (I + T^{\#}\delta T)^{-1}T^{\#} = \begin{bmatrix} (I_1 + T_1^{-1}\delta_1)^{-1}T_1^{-1} & 0 \\ 0 & 0 \end{bmatrix},$$

where $I_1 = I_{\mathrm{Ran}\,(P)}$, $I_2 = I_{\mathrm{Ran}\,(I-P)}$ are identity operators on $\mathrm{Ran}\,(P)$ and $\mathrm{Ran}\,(I-P)$ respectively. So we have

$$\begin{bmatrix} I_1 & 0 \\ -\delta_3(I_1 + T_1^{-1}\delta_1)^{-1}T_1^{-1} & I_2 \end{bmatrix} \begin{bmatrix} T_1 + \delta_1 & \delta_2 \\ \delta_3 & \delta_4 \end{bmatrix} \begin{bmatrix} I_1 & 0 \\ \delta_3(I_3 + T_1^{-1}\delta_1)^{-1}T_1^{-1} & I_2 \end{bmatrix}$$
$$= \begin{bmatrix} (T_1 + \delta_1) + \delta_2\delta_3(I_1 + T_1^{-1}\delta_1)^{-1}T_1^{-1} & \delta_2 \\ \delta_0(\delta_3(I_1 + T_1^{-1}\delta_1)^{-1}T_1^{-1}) & \delta_0 \end{bmatrix} \text{ on } \mathfrak{D}(T),$$

where $\delta_0 = \delta_4 - \delta_3(I_1 + T_1^{-1}\delta_1)^{-1}T_1^{-1}\delta_2$. Noting that

$$(I-P)\delta T(I-P) - (I-P)\delta T(I + T^{\#}\delta T)^{-1}T^{\#}\delta T(I-P) = \begin{bmatrix} 0 & 0 \\ 0 & \delta_0 \end{bmatrix},$$

and $(I-P)\xi = (I-T^{\#}T)\xi$, $\forall\, \xi \in \mathfrak{D}(T)$, it follows from Theorem 2.2.5 that $\delta_0 = 0$. We also note that

$$I \pm C_T = \begin{bmatrix} I_1 & 0 \\ \pm\delta_3(I_1 + T_1^{-1}\delta_1)^{-1}T^{-1} & I_2 \end{bmatrix},$$

$$\Phi_T = \begin{bmatrix} I_1 + \delta_2\delta_3[(I_1 + T_1^{-1}\delta_1)^{-1}T_1^{-1}]^2 & 0 \\ 0 & 0 \end{bmatrix}.$$

So $(I - C_T)\bar{T}(I + C_T) = P\Phi_T(I + T^{\#}\delta T)P + P\delta T(I - P)$ on $\mathfrak{D}(T)$ and moreover, we can check that

$$\begin{aligned}(I - C_T)\bar{T}^{\#}(I + C_T) &= [(I - C_T)\bar{T}(I + C_T)]^{\#} \\ &= (I + T^{\#}\delta T)^{-1}T^{\#}\Phi_T^{-1} + [(I + T^{\#}\delta T)^{-1}T^{\#}\Phi_T^{-1}]^2\delta T(I - P) \\ &= D_T + D_T^2\delta T(I - TT^{\#}),\end{aligned}$$

by using identities $(I - P)(I + T^{\#}\delta T)^{-1}T^{\#} = 0$, $\Phi_T(I - P) = I - P$ and $(I - C_T)^{-1} = (I + C_T)$. □

By means of Lemma 4.3.2, we can establish a perturbation analysis for group inverses under stable perturbation as follows.

Theorem 4.3.3. *Let $T \in \mathcal{C}(X)$ such that $T^{\#}$ exists. Let $\delta T \in B(X)$ and put $\bar{T} = T + \delta T \in \mathcal{C}(X)$. Suppose that $\|T^{\#}\|\|\delta T\| < (1 + \|I - TT^{\#}\|)^{-1}$ and* $\operatorname{Ran}(\bar{T}) \cap \operatorname{Ker} T^{\#} = \{0\}$. *Then $\bar{T}^{\#}$ exists and*

$$\|\bar{T}^{\#}\| \le \frac{\|T^{\#}\|}{[1 - (1 + \|I - TT^{\#}\|)\|T^{\#}\|\|\delta T\|]^2}$$

$$\frac{\|\bar{T}^{\#} - T^{\#}\|}{\|T^{\#}\|} \le \frac{(1 + 2\|I - TT^{\#}\|)\|T^{\#}\|\|\delta T\|}{[1 - (1 + \|I - TT^{\#}\|)\|T^{\#}\|\|\delta T\|]^2}.$$

Proof. By Lemma 4.3.1 and Lemma 4.3.2, $\bar{T}^{\#}$ exists and

$$\bar{T}^{\#} = (I + C_T)D_T + (I + C_T)D_T^2\delta T(I - TT^{\#})(I - C_T) \tag{4.3.2}$$

$$\|\bar{T}^{\#}\| \le (1 + \|C_T\|)\|D_T\| + (1 + \|C_T\|)^2\|D_T\|^2\|\delta T\|\|I - TT^{\#}\| \tag{4.3.3}$$

for $D_T(I \pm C_T) = D_T$. Put $T^{\pi} = I - TT^{\#}$. We have by (4.3.1).

$$\|D_T\| \le \frac{(1 - \|T^{\#}\|\|\delta T\|)\|T^{\pi}\|}{(1 - (1 + \|T^{\pi}\|)\|T^{\#}\|\|\delta T\|)(1 + (\|T^{\pi}\| - 1)\|T^{\#}\|\|\delta T\|)}$$

Since $1 + \|C_T\| \le \dfrac{1 + (\|T^{\pi}\| - 1)\|T^{\#}\|\|\delta T\|}{1 - \|T^{\#}\|\|\delta T\|}$, we deduce that

$$\|\bar{T}^{\#}\| \le \frac{(1 - \|T^{\#}\|\|\delta T\|)\|T^{\#}\|}{(1 - (1 + \|T^{\pi}\|)\|T^{\#}\|\|\delta\|)^2} \le \frac{\|T^{\#}\|}{(1 - (1 + \|T^{\pi}\|)\|T^{\#}\|\|\delta T\|)^2}$$

by (4.3.3). Finally, by (4.3.2),

$$\|\bar{T}^{\#} - T^{\#}\| \le \|D_T - T^{\#}\| + \|C_T D_T\| + (1 + \|C_T\|)^2 \|D_T\|^2 \|\delta T\| \|T^{\pi}\|.$$

Now we have

$$\begin{aligned}
\|D_T - T^{\#}\| &= \|((I + T^{\#}\delta T)^{-1} T^{\#} - T^{\#})\Phi_T^{-1} + T^{\#}(\Phi_T^{-1} - I)\| \\
&\le \frac{\|T^{\#}\|^2 \|\delta T\|}{1 - \|T^{\#}\| \|\delta T\|} \|\Phi_T^{-1}\| + \|T^{\#}\| \|\Phi_T^{-1}\| \|\Phi_T - I\| \\
&\le \frac{\|T^{\#}\|^2 \|\delta T\|}{1 - (1 + \|T^{\pi}\|) \|T^{\#}\| \|\delta T\|} \\
\|C_T D_T\| &\le \frac{\|T^{\pi}\| \|T^{\#}\| \|\delta T\|}{1 - \|T^{\#}\| \|\delta T\|} \frac{(1 - \|T^{\#}\| \|\delta T\|) \|T^{\#}\|}{1 - 2\|T^{\#}\| \|\delta T\| - (\|T^{\pi}\|^2 - 1) \|T^{\#}\| \|\delta T\|^2} \\
&\le \frac{\|T^{\#}\|^2 \|T^{\pi}\| \|\delta T\|}{1 - (1 + \|T^{\pi}\|) \|T^{\#}\| \|\delta T\|}.
\end{aligned}$$

Therefore

$$\|\bar{T}^{\#} - T^{\#}\| \le \frac{(1 + 2\|T^{\pi}\|) \|T^{\#}\|^2 \|\delta T\|}{(1 - (1 + \|T^{\pi}\|) \|T^{\#}\| \|\delta T\|)^2}.$$

□

Using Theorem 4.3.3, we can deal with the perturbation analysis of Drazin invertible operators. To do it, we need some preparations.

Let $T \in Dr(X)$ with $\text{ind}(T) = k$ and put $T^{\pi} = I - TT^D$. Then

$$\text{Ran}\,(T^k) = \text{Ran}\,(T^D) = \text{Ran}\,(I - TT^{\pi}), \ \text{Ker}\,(T^k) = \text{Ker}\,(T^D) = \text{Ran}\,(T^{\pi})$$

by Theorem 4.1.5. Let $\delta T \in B(X)$ and set $\bar{T} = T + \delta T$ and $\delta T^j = \bar{T}^j - T^j$, $j = 1, \cdots, n$. Set $K_D(T) = \|T\| \|T^D\|$ and $\epsilon_T = \dfrac{\|\delta T\|}{\|T\|}$.

From $\delta T^j = (T + \delta T)\delta T^{j-1} + \delta T T^{j-1}, j = 2, \cdots, n,$, we get that

$$\begin{aligned}
&\|\delta T^j\| \le (\|T\| + \|\delta T\|) \|\delta T^{j-1}\| + \|T\|^{j-1} \|\delta T\| \\
&\frac{\|\delta T^j\|}{\|T\|^j} \le (1 + \epsilon_T) \frac{\|\delta T^{j-1}\|}{\|T\|^{j-1}} + \epsilon_T, \ \ j = 2, \cdots, n.
\end{aligned}$$

Consequently, $\|\delta T^n\| \le \|T\|^n ((1 + \epsilon_T)^{n-1} + \cdots + 1)\epsilon_T$. Let $\epsilon_T \le 1$. Then

$$\|\delta T^n\| \le [(1 + \epsilon_A)^n - 1] \|T\|^n < (2^n - 1) \|T\|^n \epsilon_T. \tag{4.3.4}$$

Lemma 4.3.4. *Let $T \in B(X)$ with $0 \in \sigma(T)$. Suppose that there is $n \in N$ such the T^n has the group inverse. Then $T \in Dr(X)$ with $\text{ind}(T) \le n$.*

Proof. By Theorem 4.1.9, $0 \in \sigma(T^n)$ is an isolated point and so that 0 is also an isolated point in $\sigma(T)$ by Proposition 1.4.15. Choose open subsets U_1, U_2 in $\mathbb{C}$ such that $U_1 \cap U_2 = \emptyset$ and $0 \in U_1$, $\sigma(T)\backslash\{0\} \subset U_2$. Let Γ_1 (resp. Γ_2) be any contour surrounding $\{0\}$ (resp. $\sigma(T))\backslash\{0\}$ in U_1 (resp. U_2). Put $h(z) = z$, $\forall\, z \in \mathbb{C}$ and

$$f(z) = \begin{cases} 1 & z \in U_1 \\ 0 & z \in U_2 \end{cases}, \quad g(z) = \begin{cases} 0 & z \in U_1 \\ \frac{1}{z} & z \in U_2 \end{cases}$$

Then f and g are holomorphic on $U_1 \cup U_2$ and

$$P = f(T) = \frac{1}{2\pi i}\oint_{\Gamma_1+\Gamma_2} f(\lambda)(\lambda I - T)^{-1}\mathrm{d}\,\lambda = \frac{1}{2\pi i}\oint_{\Gamma_1} (\lambda I - T)^{-1}\mathrm{d}\,\lambda$$

is an idempotent operator in $B(X)$ $(f^2 = f)$. Moreover, we have

$$TP = PT,\ g(T)P = Pg(T),\ Tg(T) = g(T)T = I - P$$

$(1 - f = hg)$ and $\sigma(TP) = \sigma((hf)(T)) = \{0\}$ by Proposition 1.4.15. Put $X_1 = \mathrm{Ran}\,(I - P)$, $X_2 = \mathrm{Ran}\,(P)$. Then $X = X_1 \dotplus X_2$ and T has the form $T = \begin{bmatrix} T_1 & \\ & T_2 \end{bmatrix}$ under the decomposition $X = X_1 \dotplus X_2$, where $T_1 = T\big|_{X_1}$ and $T_2 = T\big|_{X_2}$ with T_1 invertible in $B(X_1)$ and T_2 quasi–nilpotent in $B(X_2)$.

Assume that $(T^n)^{\#} = \begin{bmatrix} C_1 & C_2 \\ C_3 & C_4 \end{bmatrix}$ with respect to $X = X_1 \dotplus X_2$. Then by the definition of $(T^n)^{\#}$, we have $C_2 = 0$, $C_3 = 0$ and

$$C_1 = (T_1^n)^{-1},\ T_2^nC_4 = C_4T_2^n,\ C_4 = C_4T_2^nC_4,\ T_2^n = T_2^nC_4T_2^n.$$

So $C_4^k = C_4^{2k}(T_2^k)^n$ and $\|C_4^k\| \le \|C_4^k\|^2\|T_2^k\|^n$, $\forall\, k \in \mathbb{N}$. If $C_4 \neq 0$, then

$$1 \le \|C_4^k\|\|T_2^k\|^n \le \|C_2\|^k\|T_2^k\|^n,\ \forall\, k \in \mathbb{N}$$

and hence $\|T_2^k\|^{1/k} \ge \|C_4\|^{-\frac{1}{n}}$, $\forall\, k \in \mathbb{N}$. But this is impossible since $\lim_{k\to\infty} \|T_2^k\|^{1/k} = 0$. Therefore, $T_2^n = 0$ and $T^D = \begin{bmatrix} T_1^{-1} & \\ & 0 \end{bmatrix}$ with $\mathrm{ind}(T) \le n$ by Definition 4.1.6 and Remark 4.1.7. □

Now we give the Perturbation analysis of the Drazin invertible operator as follows.

Proposition 4.3.5. *Let $T \in Dr(X)$ with* $\mathrm{ind}(T) = k$ *and let $\delta T \in B(X)$. Put $\bar{T} = T + \delta T$. Suppose that $K_D^k(T)\epsilon_T < \dfrac{1}{(2^k-1)(1+\|T^{\pi}\|)}$ and*

$\operatorname{Ran}(\bar{T}^k) \cap \operatorname{Ker}(T^D) = \{0\}$. *Then* $\bar{T} \in Dr(X)$ *with* $\operatorname{ind}(\bar{T}) \leq k$ *and*

$$\|\bar{T}^D\| \leq \frac{2^{k-1}K_D^{k-1}(T)\|T^D\|}{[1-(2^k-1)(1+\|T^\pi\|)K_D^k(T)\epsilon_T]^2}$$
$$\frac{\|\bar{T}^D - T^D\|}{\|T^D\|} \leq \frac{2^{k-1}(2^k-1)(1+2\|T^\pi\|)K_D^{2k-1}(T)\epsilon_T}{[1-(2^k-1)(1+\|T^\pi\|)K_D^k(T)\epsilon_T]^2} + (2^{k-1}-1)K_D^{k-1}(T)\epsilon_T.$$

Proof. Noting that $K_D(T) \geq \|TT^D\| \geq 1$, we have $\epsilon_T < 1$. Since T^k has the group inverse $(T^k)^\# = (T^D)^k$ and

$$\|\delta T^k\|\|(T^k)^\#\| \leq (2^k-1)K_D^k(T)\epsilon_T < \frac{1}{1+\|T^\pi\|}$$

and $\operatorname{Ker}(T^k)^\# = \operatorname{Ker} T^D = \operatorname{Ker}(T^k)$ by Theorem 4.1.5, it follows from Theorem 4.3.3 that $(\bar{T}^k)^\#$ exists. Thus $\bar{T} \in Dr(X)$ with $\operatorname{ind}(\bar{T}) \leq k$ by Lemma 4.3.4. Since $\bar{T}^D = (\bar{T}^k)^\#\bar{T}^{k-1}$ and $T^D = (T^k)^\# T^{k-1}$, we have

$$\|\bar{T}^D\| \leq \|(\bar{T}^k)^\#\|\|T\|^{k-1}(1+\epsilon_T)^{k-1} < 2^{k-1}\|T\|^{k-1}\|(\bar{T}^k)^\#\|$$
$$\|\bar{T}^D - T^D\| \leq \|(\bar{T}^k)^\# - (T^k)^\#\|\|\bar{T}\|^{k-1} + \|(T^k)^\#\|\|\delta T^{k-1}\|$$
$$< 2^{k-1}\|(\bar{T}^k)^\# - (T^k)^\#\|\|T\|^{k-1} + (2^{k-1}-1)\|T^D\|K_D^{k-1}(T)\epsilon_T.$$

Applying Theorem 4.3.3 to $(\bar{T}^k)^\#$ and $(T^k)^\#$, we obtain the assertion. □

Let $A, B \in \mathcal{C}(X)$ with $\mathfrak{D}(A) = \mathfrak{D}(B)$. Suppose that $A^\#$ and $B^\#$ exist. We want to express $B^\#$ in terms of A, B and $A^\#$ when B is a stable perturbation of A.

Lemma 4.3.6. *Let* $A, B \in \mathcal{C}(X)$ *with* $\mathfrak{D}(A) = \mathfrak{D}(B)$ *such that* $A^\#$ *and* $B^\#$ *exist. Then the following statements are equivalent.*

(1) $X = \operatorname{Ran}(B) \dotplus \operatorname{Ker}(A^\#) = \operatorname{Ran}(A) \dotplus \operatorname{Ker}(B^\#)$.
(2) $K(A,B) = BB^\# + AA^\# - I$ *is invertible in* $B(X)$.
(3) $\operatorname{Ran}(B) \cap \operatorname{Ker}(A^\#) = \{0\}$ *and* $I + (B-A)A^\#$ *is invertible in* $B(X)$.

Proof. (1)⇒(2) Let $\xi \in \operatorname{Ker}(K(A,B))$. Then

$$BB^\#\xi = (I - AA^\#)\xi \in \operatorname{Ker}(A^\#), \ BB^\#\xi = (I - AA^\#)\xi = 0$$

for $\operatorname{Ran}(B) \cap \operatorname{Ker}(A^\#) = \{0\}$. Thus $B^\#\xi = 0$ and $\xi \in \operatorname{Ran}(A)$. Since $\operatorname{Ran}(A) \cap \operatorname{Ker}(B^\#) = \{0\}$, we have $\xi = 0$, that is, $K(A,B)$ is injective. From $X = \operatorname{Ran}(B) + \operatorname{Ker}(A^\#) = \operatorname{Ran}(A) + \operatorname{Ker}(B^\#)$, we get that

$$\operatorname{Ran}(I - BB^\#) = (I - BB^\#)\operatorname{Ker}(A^\#), \ \operatorname{Ran}(BB^\#) = BB^\#(\operatorname{Ran}(A)).$$

So for any $x \in X$, there are $y_1, y_2 \in X$ such that

$$(I - BB^{\#})x = (I - BB^{\#})(I - AA^{\#})y_1, \ BB^{\#}x = BB^{\#}AA^{\#}y_2$$

for $\mathrm{Ran}\,(I - AA^{\#}) = \mathrm{Ker}\,(A^{\#})$ and $\mathrm{Ran}\,(AA^{\#}) = \mathrm{Ran}\,(A)$. Put $z = -(I - AA^{\#})y_1 + AA^{\#}y_2$. Then

$$\begin{aligned} K(A,B)z &= (BB^{\#} + AA^{\#} - I)(-(I - AA^{\#})y_1 + AA^{\#}y_2) \\ &= (I - BB^{\#})(I - AA^{\#})y_1 + BB^{\#}AA^{\#}y_2 \\ &= (I - BB^{\#})x + BB^{\#}x = x, \end{aligned}$$

that is, $\mathrm{Ran}\,(K(A,B)) = X$. Therefore, $K(A,B)$ is invertible in $B(X)$.

(2)$\Rightarrow$(3) Let $\xi \in \mathrm{Ran}\,(B) \cap \mathrm{Ker}\,(A^{\#})$. Then $BB^{\#}\xi = \xi$ and $AA^{\#}\xi = 0$. So $K(A,B)\xi = 0$ and hence $\xi = 0$, that is, $\mathrm{Ran}\,(B) \cap \mathrm{Ker}\,(A^{\#}) = \{0\}$.

Since $\mathrm{Ran}\,(A^{\#}) \subset \mathfrak{D}(A) = \mathfrak{D}(B)$, $A^{\#} \in B(X)$ and $B, A \in C(X)$, it follows that $I+(B-A)A^{\#} \in B(X)$. To prove $I+(B-A)A^{\#}$ is invertible, we need to show that $\mathrm{Ker}\,(I+(B-A)A^{\#}) = \{0\}$ and $\mathrm{Ran}\,(I+(B-A)A^{\#}) = X$.

Let $\xi \in \mathrm{Ker}\,(I + (B - A)A^{\#})$. Then $(I - AA^{\#})\xi + BA^{\#}\xi = 0$ and so that $BA^{\#}\xi \in \mathrm{Ran}\,(B) \cap \mathrm{Ker}\,(A^{\#}) = \{0\}$. Thus, $AA^{\#}\xi = \xi$ and $BA^{\#}\xi = 0$. Set $\eta = A^{\#}\xi \in \mathfrak{D}(A)$. Then $\eta \in \mathrm{Ker}\,B = \mathrm{Ker}\,(B^{\#})$ and

$$AA^{\#}\eta = A^{\#}A\eta = A^{\#}AA^{\#}\xi = A^{\#}\xi = \eta.$$

So $K(A,B)\eta = 0$ and consequently, $\eta = 0$ and $\xi = A\eta = 0$.

Note that $K(A,B) = I - (I - AA^{\#}) - (I - BB^{\#})$ is invertible and $\mathrm{Ran}\,(I - AA^{\#}) \subset \mathfrak{D}(A)$, $\mathrm{Ran}\,(I - BB^{\#}) \subset \mathfrak{D}(A)$, we have $K(A,B)\mathfrak{D}(A) = \mathfrak{D}(A)$. Thus,

$$BK(A,B)\xi = BAA^{\#}\xi = BA^{\#}A\xi, \ \forall\, \xi \in \mathfrak{D}(A)$$

and, for any $x \in X$, there is $\eta \in \mathfrak{D}(A)$ such that $BB^{\#}x = BK(A,B)\eta = BA^{\#}A\eta$. Put $\xi = A\eta - (I - AA^{\#})x$. Then

$$\begin{aligned} (I + (B - A)A^{\#})\xi &= (I - AA^{\#})\xi + BA^{\#}\xi = -(I - AA^{\#})x + BA^{\#}A\eta \\ &= BB^{\#}x - (I - AA^{\#})x = K(A,B)x, \end{aligned}$$

that is, $\mathrm{Ran}\,(I + (B - A)A^{\#}) = X$.

(3)$\Rightarrow$(1) By Theorem 2.2.13, $\mathrm{Ran}\,(A^{\#}) \cap \mathrm{Ker}\,B = \{0\}$ and

$$X = \mathrm{Ran}\,(B) + \mathrm{Ker}\,(A^{\#}), \ \mathfrak{D}(A) = \mathrm{Ran}\,(A^{\#}) + \mathrm{Ker}\,B.$$

We should prove that

$$X = \mathrm{Ran}\,(A) + \mathrm{Ker}\,(B^{\#}) \text{ and } \mathrm{Ran}\,(A) \cap \mathrm{Ker}\,(B^{\#}) = \{0\}.$$

Let $\xi \in \mathrm{Ran}\,(A) \cap \mathrm{Ker}\,(B^{\#})$. Then

$$AA^{\#}\xi = \xi, \quad \xi \in \mathrm{Ker}\,(B^{\#}) = \mathrm{Ker}\,B \subset \mathfrak{D}(A).$$

So $\xi = A^{\#}A\xi \in \mathrm{Ran}\,(A^{\#}) \cap \mathrm{Ker}\, B = \{0\}$, i.e., $\mathrm{Ran}\,(A) \cap \mathrm{Ker}\,(B^{\#}) = \{0\}$. Since $\mathrm{Ran}\,(A^{\#}) = AA^{\#}\mathfrak{D}(A) \subset \mathrm{Ran}\,(A)$, we have

$$\mathfrak{D}(A) = \mathrm{Ran}\,(A^{\#}) + \mathrm{Ker}\,(B) \subset \mathrm{Ran}\,(A) + \mathrm{Ker}\,(B^{\#}). \tag{4.3.5}$$

Since for any $x \in X$, $(I - AA^{\#})x \in \mathfrak{D}(A)$, it follows from (4.3.5) that $(I - AA^{\#})x = Ax_1 + x_2$ for some $x_1 \in \mathfrak{D}(A)$ and $x_2 \in \mathrm{Ker}\, B^{\#}$. Thus,

$$x = AA^{\#}x + (I - AA^{\#})x = A(A^{\#}x + x_1) + x_2 \in \mathrm{Ran}\,(A) + \mathrm{Ker}\,(B^{\#}),$$

and consequently, $X = \mathrm{Ran}\,(A) + \mathrm{Ker}\,(B^{\#})$. □

Theorem 4.3.7. *Let $A \in \mathcal{C}(X)$ such that $A^{\#}$ exists and let $\delta A \in B(X)$. Suppose that $B = A + \delta A$ has the group inverse $B^{\#}$ and $K(A, B)$ is invertible in $B(X)$. Then*

(1) *$I + A^{\#}\delta A$ is invertible in $B(X)$ and*

$$I + (I + A^{\#}\delta A)^{-1}A^{\#}\delta A(I - AA^{\#})\delta A(I + A^{\#}\delta A)^{-1}A^{\#}$$

is also invertible in $B(X)$.

(2) *B has the form $B = \begin{bmatrix} B_1 & B_1S_2 \\ S_1B_1 & S_1B_1S_2 \end{bmatrix}$ with respect to the decomposition $X = \mathrm{Ran}\,(A) \dotplus \mathrm{Ker}\, A$, where $B_1 \in \mathcal{C}(\mathrm{Ran}\,(A))$ with $\mathfrak{D}(B_1) = AA^{\#}(\mathfrak{D}(A))$, $\mathrm{Ran}\,(B_1) = \mathrm{Ran}\,(A)$, $\mathrm{Ker}\, B_1 = \{0\}$ and*

$$S_1 \in B(\mathrm{Ran}\,(A), \mathrm{Ker}\, A), \quad S_2 \in B(\mathrm{Ker}\, A, \mathrm{Ran}\,(A))$$
$$\mathrm{Ran}\,(S_2) \subset \mathfrak{D}(B_1), \qquad S_2S_1 + I_1 \in GL(B(\mathrm{Ran}\,(T))).$$

(3) $B^{\#} = \begin{bmatrix} S^{-1}B_1^{-1}S^{-1} & S^{-1}B_1^{-1}S^{-1}S_2 \\ S_1S^{-1}B_1^{-1}S^{-1} & S_1S^{-1}B_1^{-1}S^{-1}S_2 \end{bmatrix}$, *where $S = I_1 + S_2S_1$, and*

$$I - BB^{\#} = \begin{bmatrix} I_1 - (I + S_2S_1)^{-1} & -(I_1 + S_2S_1)^{-1}S_2 \\ -S_1(I_1 + S_2S_1)^{-1} & I_2 - S_1(I_1 + S_2S_1)^{-1}S_2 \end{bmatrix}.$$

Proof. Let $A = \begin{bmatrix} A_1 & 0 \\ 0 & 0 \end{bmatrix}$, $\delta A = \begin{bmatrix} \delta_1 & \delta_2 \\ \delta_3 & \delta_4 \end{bmatrix}$ with respect to the decomposition

$$\mathfrak{D}(A) = AA^{\#}(\mathfrak{D}(A)) + \mathrm{Ran}\,(I - AA^{\#}), \quad X = \mathrm{Ran}\,(A) \dotplus \mathrm{Ker}\, A,$$

respectively. Since $K(A, B)$ is invertible in $B(X)$, it follows from Lemma 4.3.6 that $\mathrm{Ran}\,(B) \cap \mathrm{Ker}\,(A^{\#}) = \{0\}$ and $I + \delta AA^{\#}$ (also $I + A^{\#}\delta A$) is invertible in $B(X)$. Thus, from $I + A^{\#}\delta A = \begin{bmatrix} I_1 + A_1^{-1}\delta_1 & A_1^{-1}\delta_2 \\ 0 & I_2 \end{bmatrix}$, we get that $I_1 + A_1^{-1}\delta_1$ is invertible in $B(\mathrm{Ran}\,(T))$, where $A_1^{-1}\colon \mathrm{Ran}\,(AA^{\#}) \to AA^{\#}(\mathfrak{D}(A))$ is bounded. Put

$$\varphi(A_1) = A_1 + \delta_1 + \delta_2\delta_3(I_1 + A_1^{-1}\delta_1)^{-1}A_1^{-1}.$$

Since $(I-C_A)B(I+C_A) = \begin{bmatrix} \varphi(A_1) & \delta_2 \\ 0 & 0 \end{bmatrix}$ by Lemma 4.3.2 and $\begin{bmatrix} \varphi(A_1) & \delta_2 \\ 0 & 0 \end{bmatrix}$ has the group inverse $\begin{bmatrix} \varphi(A_1) & \delta_2 \\ 0 & 0 \end{bmatrix}^{\#}$, we can deduce that $\varphi(A_1)$ has the group inverse $(\varphi(A_1))^{\#}$ and $\varphi(A_1)(\varphi(A_1)^{\#})\delta_2 = \delta_2$. Thus

$$(I - C_A)B^{\#}(I + C_A) = \begin{bmatrix} (\varphi(A_1))^{\#} & ((\varphi(A_1))^{\#})^2\delta_2 \\ 0 & 0 \end{bmatrix} \tag{4.3.6}$$

$$(I - C_A)BB^{\#}(I + C_A) = \begin{bmatrix} \varphi(A_1)(\varphi(A_1))^{\#} & (\varphi(A_1))^{\#}\delta_2 \\ 0 & 0 \end{bmatrix} \tag{4.3.7}$$

and hence

$$\begin{aligned}(I - C_A)K(A, B)(I + C_A) &= (I - C_A)BB^{\#}(I + C_A) \\ &\quad + (I - C_A)AA^{\#}(I + C_A) - I \\ &= \begin{bmatrix} \varphi(A_1)(\varphi(A_1))^{\#} & (\varphi(A_1))^{\#}\delta_2 \\ -\delta_3(I_1 + A_1^{-1}\delta_1)^{-1}A_1^{-1} & -I_2 \end{bmatrix}\end{aligned}$$

by (4.3.6) and (4.3.7). So the invertibility of $K(A, B)$ indicates that

$$\rho(A_1) = \varphi(A_1)(\varphi(A_1))^{\#} - (\varphi(A_1))^{\#}\delta_2\delta_3(I_1 + A_1^{-1}\delta_1)^{-1}A_1^{-1}. \tag{4.3.8}$$

is invertible in $B(\mathrm{Ran}\,(A))$.

Now from (4.3.8), $\rho(A_1)x = (\varphi(A_1))^{\#}(A_1 + \delta_1)x$, $\forall x \in \mathfrak{D}(A_1)$. Since $\rho(A_1)$ is invertible and $\mathrm{Ran}\,(A_1 + \delta_1) = \mathrm{Ran}\,(A)$, $\mathrm{Ker}\,(A_1 + \delta_1) = \{0\}$, we obtain that $\mathrm{Ran}\,((\varphi(A_1))^{\#}) = \mathrm{Ran}\,(A)$ and $\mathrm{Ker}\,(\varphi(A_1)) = \{0\}$. This implies that $\mathrm{Ran}\,(\varphi(A_1)) = \mathrm{Ran}\,(A)$ and $\mathrm{Ker}\,(\varphi(A_1)) = \{0\}$. Moreover, from

$$\varphi(A_1) = [I_1 + \delta_2\delta_3(I_1 + A_1^{-1}\delta_1)^{-1}A_1^{-1}(I_1 + A_1^{-1}\delta_1)^{-1}A_1^{-1}](A_1 + \delta_1).$$

we get that the range (resp. the kernel) of

$$\Phi(A_1) = I_1 + \delta_2\delta_3(I_1 + A_1^{-1}\delta_1)^{-1}A_1^{-1}(I_1 + A_1^{-1}\delta_1)^{-1}A_1^{-1}$$

is $\mathrm{Ran}\,(A)$ (resp. $\{0\}$). So $\Phi(A_1)$ is invertible in $B(\mathrm{Ran}\,(A))$ and hence

$$I_1 + (I_1 + A_1^{-1}\delta_1)^{-1}A_1^{-1}\delta_2\delta_3(I_1 + A_1^{-1}\delta_1)^{-1}A_1^{-1}$$

is invertible in $B(\mathrm{Ran}\,(A))$ by Proposition 1.4.13. Put $B_1 = A_1 + \delta_1, S_1 = \delta_3(I_1 + A_1^{-1}\delta_1)^{-1}A_1^{-1} \in B(\mathrm{Ran}\,(A), \mathrm{Ker}\,A)$. $S_2 = (I_1 + A_1^{-1}\delta_1^{-1})A_1^{-1}\delta_2 \in B(\mathrm{Ker}\,A, \mathrm{Ran}\,(A))$ with $\mathrm{Ran}\,(S_2) \subset \mathfrak{D}(A_1)$ and $S = S_2S_1 + I_1$ is invertible by above argument. Therefore,

$$B = (I + C_A)\begin{bmatrix}\varphi(A_1) & \delta_2\\ 0 & 0\end{bmatrix}(I - C_A) = \begin{bmatrix}B_1 & B_1S_2\\ S_1B_1 & S_1B_1S_2\end{bmatrix}$$

$$\begin{aligned}B^{\#} &= \begin{bmatrix}I_1 & 0\\ S_1 & I_2\end{bmatrix}\begin{bmatrix}B_1(I_1+S_2S_1) & B_1S_2\\ 0 & 0\end{bmatrix}^{\#}\begin{bmatrix}I_1 & 0\\ -S_1 & I_2\end{bmatrix}\\ &= \begin{bmatrix}I_1 & 0\\ S_1 & I_2\end{bmatrix}\begin{bmatrix}S^{-1}B_1^{-1} & S^{-1}B_1^{-1}S^{-1}S_2\\ 0 & 0\end{bmatrix}\begin{bmatrix}I_1 & O\\ -S_1 & I_2\end{bmatrix}\\ &= \begin{bmatrix}S^{-1}B_1^{-1}S^{-1} & S^{-1}B_1^{-1}S^{-1}S_2\\ S_1S^{-1}B_1^{-1}S^{-1} & S_1S^{-1}B_1^{-1}S^{-1}S_2\end{bmatrix}\end{aligned}$$

$$\begin{aligned}I - BB^{\#} &= \begin{bmatrix}I_1 & \\ & I_2\end{bmatrix} - \begin{bmatrix}I_1 & 0\\ S_1 & I_2\end{bmatrix}\begin{bmatrix}B_1S & B_1S_2\\ 0 & 0\end{bmatrix}\begin{bmatrix}B_1S & B_1S_2\\ 0 & 0\end{bmatrix}^{\#}\begin{bmatrix}I_1 & 0\\ -S_1 & I_2\end{bmatrix}\\ &= \begin{bmatrix}I_1 - S^{-1} & -S^{-1}S_2\\ -S_2S^{-1} & I_2 - S_1S^{-1}S_2\end{bmatrix}\end{aligned}$$

□

Let $T \in Dr(X)$ with $k = \operatorname{ind}(T)$. Put $T^{\pi} = I - TT^D$ and $T_C = T(I - T^{\pi})$, $T_N = TT^{\pi}$. Then $T = T_C + T_N$. Moreover, from the definition of T^D, we can deduce that

$$T^s = T_C^s,\ s \geq k,\ T_N^k = 0,\ (T_C)^{\#} = T^D,\ (T_C^l)^{\#} = ((T_C)^{\#})^l,\ \forall\, l \geq 1.$$

Theorem 4.3.8. *Let A, $B \in Dr(X)$ with $s = \operatorname{ind}(A)$ and $t = \operatorname{ind}(B)$. Suppose that $K(A, B) = AA^D + BB^D - I$ is invertible in $B(X)$. Then*

(1) $I+(A^D)^l(B^k - A^l)$ *is invertible in* $B(X)$, $\operatorname{Ran}(B^k) \cap \operatorname{Ker}((A^D)^l) = \{0\}$, $\forall\, k \geq t,\ l \geq 1$.
(2) $W_{k,l} = (I + Y_{k,l}Z_{k,l})(I - Z_{k,l})$ *is invertible in* $B(X)$, *where* $E_{k,l} = B^k - A^l$ *and*

$$\begin{aligned}Y_{k,l} &= (I + (A^D)^l E_{k,l})^{-1}(A^D)^l E_{k,l} A^{\pi}\\ Z_{k,l} &= A^{\pi} E_{k,l}(A^D)^l (I + E_{k,l}(A^D)^l)^{-1}.\end{aligned}$$

(3) $B^D = W_{k,l}^{-1}(I + (A^D)^{l+1}E_{k+1,l+1})^{-1}A^D(I + (A^D)^l E_{k,l})W_{k,l}$, $k \geq t$, $l \geq 1$.
(4) $B^{\pi} = W_{k,l}^{-1}(I + (A^D)^l E_{k,l})^{-1}A^{\pi}(I + (A^D)^l E_{k,l})W_{k,l}$, $k \geq t$, $l \geq 1$.

Proof. We have

$$AA^D = A_C(A_C)^{\#} = A_C^l(A_C^l)^{\#},\ BB^D = B_C(B_C)^{\#} = B_C^k(B_C^k)^{\#},\ l,\ k \in \mathbb{N}$$

and hence $K(A, B) = K(A_C^l, B_C^k) = K(A_C, B_C)$ is invertible in $B(X)$. Thus

$$\operatorname{Ran}(B^k) \cap \operatorname{Ker}((A^D)^l) = \operatorname{Ran}(B_C^k) \cap \operatorname{Ker}((A_C^l)^{\#}) = \{0\}$$

and $I+(A_C^l)^{\#}(B_C^k-A_C^l)$ is invertible in $B(X)$ by Lemma 4.3.6 and

$$I+(I+(A_C^l)^{\#}(B_C^k-A_C^l))^{-1}(A_C^l)^{\#}(B_C^k-A_C^l)(I-A_C^l(A_C^l)^{\#}) \times (B_C^k-A_C^l)(I+(A_C^l)^{\#}(B_C^k-A_C^l))^{-1}(A_C^l)^{\#} \quad (4.3.9)$$

is also invertible in $B(X)$ by Theorem 4.3.7. Note that

$$B^k=B_C^k,\ (A_C^l)^{\#}=(A^D)^l,\ A_C^l(A_C^l)^{\#}=AA^D=A^l(A^D)^l,\ k\geq t,\ l\geq 1.$$

So we obtain that

$$I+(A^D)^l(B^k-A^l)=I+(A_C^l)^{\#}B_C^k-(A_C^l)^{\#}A_C^l$$

is invertible and $I+Y_{k,l}Z_{k,l}$ is invertible by (4.3.9). Since $K(A_C,B_C)$ is invertible, it follows from Theorem 4.3.7 that with respect to the decomposition $X=\operatorname{Ran}(A_C)\dotplus\operatorname{Ker}(A_C)$, B_C has the form

$$B_C=\begin{bmatrix} B_1 & B_1S_2\\ S_1B_1 & S_1B_1S_2\end{bmatrix},$$

where $B_1\in B(\operatorname{Ran}(A_C))$ is invertible, $S_1\in B(\operatorname{Ran}(A_C),\operatorname{Ker}(A_C))$, $S_2\in B(\operatorname{Ker}(A_C),\operatorname{Ran}(A_C))$ with $I_1+S_2S_1$ invertible in $B(\operatorname{Ran}(A_C))$.

Let $A_C=\begin{bmatrix} A_1 & 0\\ 0 & 0\end{bmatrix}$ under the decomposition $X=\operatorname{Ran}(A_C)\dotplus\operatorname{Ker}(A_C)$, where $A_1\in B(\operatorname{Ran}(A_C))$ is invertible. Then

$$\begin{aligned}
&(I+(A^D)^{l+1}E_{k+1,l+1})^{-1}A^D(I+(A^D)^lE_{k,l})\\
&=(I-AA^D+(A_C^{\#})^{l+1}(B_C)^{k+1})^{-1}A_C^{\#}(I+(A_C^{\#})^lB_C^k-AA^D)\\
&=\left(\begin{bmatrix} 0 & \\ & I_2\end{bmatrix}+\begin{bmatrix} A_1^{-(l+1)} & \\ & 0\end{bmatrix}\begin{bmatrix} (B_1S)^kB_1 & (B_1S)^kB_1S_2\\ S_1(B_1S)^kB_1 & S_1(B_1S)^kB_1S_2\end{bmatrix}\right)^{-1}\\
&\quad\times\begin{bmatrix} A_1^{-1} & 0\\ 0 & 0\end{bmatrix}\left(\begin{bmatrix} 0 & \\ & I_2\end{bmatrix}+\begin{bmatrix} A_1^{-l} & \\ & 0\end{bmatrix}\begin{bmatrix} (B_1S)^{k-1}B_1 & (B_1S)^{k_1}B_1S_2\\ S_1(B_1S)^{k-1}B_1 & S_1(B_1(S))^{k-1}B_1S_2\end{bmatrix}\right)\\
&=\begin{bmatrix} B_1^{-1}(B_1S)^{-1}B_1 & B_1^{-1}(B_1S)^{-1}B_1S_2\\ 0 & 0\end{bmatrix}
\end{aligned}$$

$$\begin{aligned}
Y_{k,l}&=\begin{bmatrix} A_1^{-l}(B_1S)^{k-1}B_1 & A_1^{-l}(B_1S)^{k-1}B_1S_2\\ 0 & I_2\end{bmatrix}^{-1}\begin{bmatrix} A_1^{-1} & 0\\ 0 & 0\end{bmatrix}\\
&\quad\times\left(\begin{bmatrix} (B_1S)^{k-1}B_1 & (B_1S)^{k-1}B_1S_2\\ S_1(B_1S)^{k-1}B_1 & S_1(B_1S)^{k-1}B_1S_2\end{bmatrix}-\begin{bmatrix} A_1^l & 0\\ 0 & o\end{bmatrix}\right)\begin{bmatrix} 0 & 0\\ 0 & I_2\end{bmatrix}\\
&=\begin{bmatrix} 0 & S_2\\ 0 & 0\end{bmatrix}
\end{aligned}$$

$$\begin{aligned}
Z_{k,l}&=\begin{bmatrix} 0 & \\ & I_2\end{bmatrix}\begin{bmatrix} (B_1S)^{k-1}B_1A-1^{-l}-I_1 & 0\\ S_1(B_1S)^{k-1}B-1A_1^{-l} & 0\end{bmatrix}\begin{bmatrix} (B_1S)^{k-1}B_1A_1^{-l} & 0\\ S_1(B_1S)^{k-1}B_1A_1^{-l} & I_2\end{bmatrix}^{-1}\\
&=\begin{bmatrix} 0 & 0\\ S_1 & 0\end{bmatrix}.
\end{aligned}$$

Thus, $W_{k,l} = (I + Y_{k,l}Z_{k,l})(I - Z_{k,l}) = \begin{bmatrix} S & 0 \\ 0 & I_2 \end{bmatrix} \begin{bmatrix} I_1 & 0 \\ -S_1 & I_2 \end{bmatrix}$ and so that

$$\begin{aligned} &W_{k,l}^{-1}(I + (A^D)^{l+1}E_{k+1,l+1})^{-1}A^D(I + (A^D)^l E_{k,l})W_{k,l} \\ &= \begin{bmatrix} I_1 & 0 \\ S_1 & I_2 \end{bmatrix} \begin{bmatrix} S^{-1} & 0 \\ 0 & I_2 \end{bmatrix} \begin{bmatrix} B_1^{-1}S^{-1} & B_1^{-1}S^{-1}S_2 \\ 0 & 0 \end{bmatrix} \begin{bmatrix} S & 0 \\ 0 & I_2 \end{bmatrix} \begin{bmatrix} I_1 & 0 \\ -S_1 & I_2 \end{bmatrix} \\ &= \begin{bmatrix} S^{-1}B_1^{-1}S^{-1} & S^{-1}B_1^{-1}S^{-1}S_2 \\ S_1S^{-1}B_1^{-1}S^{-1} & S_1S^{-1}B_1^{-1}S^{-1}S_2 \end{bmatrix} \\ &= (B_C)^{\#} = B^D. \end{aligned}$$

by Theorem 4.3.7 (3). □

Corollary 4.3.9. *Let $A \in Dr(X)$ and $B \in B(X)$. Suppose that there are $k_1, k_2, l_1, l_2 \in \mathbb{N}$ with $k_2 < k_1$ such that the following conditions are satisfied:*

(1) *$I + (A^D)^{l_1}E_{k_1,l_1}$ and $I + Y_{k_1,l_1}Z_{k_1,l_1}$ are invertible in $B(X)$.*
(2) *$B^{k_1}(I + (A^D)^{l_1}E_{k_1,l_1})^{-1}A^{\pi} = 0$.*
(3) *$I + (A^D)^{l_2}E_{k_2,l_2}$ is not invertible in $B(X)$ or $I + (A^D)^{l_2}E_{k_2,l_2}$ is invertible in $B(X)$, but $B^{k_2}(I + (A^D)^{l_2}E_{k_2,l_2})^{-1}A^{\pi} \neq 0$.*

Then $B \in Dr(X)$ with $k_2 < \operatorname{ind}(B) \leq k_1$.

Proof. Let $A = A_C + A_N$. Since $(A_C)^{\#} = A^D$, $(A_C^l)^{\#} = (A^D)^l$, $l \geq 1$, we have $I + (A_C^l)^{\#}(B^k - A_C^l) = I + (A^D)^l E_{k,l}$. Thus, Condition (1) and Condition (2) indicate that $I + (A_C^{l_1})^{\#}(B^{k_1} - A_C^{l_1})$ is invertible and

$$0 = B^{k_1}(I + (A_C^{l_1})^{\#}(B^{k_1} - A_C^{l_1}))^{-1}(I - A_C^{l_1}(A_C^{l_1})^{\#}). \qquad (4.3.10)$$

By Theorem 2.2.3, (4.3.10) implies that $\operatorname{Ran}(B^{k_1}) \cap \operatorname{Ker}(A_C^{l_1})^{\#} = \{0\}$. By Condition (1),

$$\begin{aligned} I + (I + (A_C^{l_1})^{\#}(B^{k_1} - A_C^{l_1}))^{-1}(A_C^{l_1})^{\#}(B^{k_1} - A_C^{l_1})(I - A_C^{l_1}(A_C^{l_1})^{\#}) \\ \times (B^{k_1} - A_C^{l_1})(I + (A_C^{l_1})^{\#}(B^{k_1} - A_C^{l_1}))^{-1}(A_C^{l_1})^{\#} \\ = I + Y_{k_1,l_1}Z_{k_1,l_1} \end{aligned}$$

is invertible in $B(X)$. Thus $(B^{k_1})^{\#}$ exists by Lemma 4.3.2 and hence $B \in Dr(X)$ with $\operatorname{ind}(B) \leq k_1$ by Lemma 4.3.4.

Let $B = B_C + B_N$. If $k_2 \geq \operatorname{ind}(B)$, then $B^{k_2} = B_C^{k_2}, B^{k_1} = B_C^{k_1}$. Thus from the invertibility of $I + (A_C^{l_1})^{\#}(B_C^{k_1} - A_C^{l_1})$ and $\operatorname{Ran}(B_C^{k_1}) \cap \operatorname{Ker}(A_C^{l_1})^{\#} = \{0\}$, we get that by Lemma 4.3.6

$$K(A_C^{l_1}, B_C^{k_1}) = A_C^{l_1}(A_C^{l_1})^{\#} + B_C^{k_1}(B_C^{k_1})^{\#} - I = A_C(A_C)^{\#} + B_C(B_C)^{\#} - I$$

is invertible. Note that $K(A_C^{l_2}, B_C^{l_2}) = K(A_C, B_C)$. It follows from Lemma 4.3.6 that $\operatorname{Ran}(B_C^{k_2}) \cap \operatorname{Ker}((A_C^{l_2})^{\#}) = \{0\}$ and $I + (A_C^{l_2})^{\#}(B_C^{k_2} - A_C^{l_2})$ is invertible in $B(X)$. Since $\operatorname{Ran}(B_C^{k_2}) \cap \operatorname{Ker}((A_C^{l_2})^{\#}) = \{0\}$ and

$$I + (A^D)^{l_2}(B^{k_2} - A^{l_2}) = I + (A_C^{l_2})^{\#}(B_C^{k_2} - A_C^{l_2}) \in GL(B(X)),$$

we have

$$B^{k_2}(I + (A^D)^{l_2}(B^{k_2} - A^{l_2}))^{-1}(I - AA^D) = 0 \tag{4.3.11}$$

by Theorem 2.2.3. But (4.3.11) and that $(I+(A^D)^{l_2}(B^{k_2}-A^{l_2}))$ is invertible in $B(X)$ contradict to Condition (3). Therefore, $k_2 < \operatorname{ind}(B)$. □

Corollary 4.3.10. *Let $\{A_n\} \subset Dr(X)$ and $A \in Dr(X)$ with $s = \operatorname{ind}(A)$ such that $A_n \xrightarrow{\|\cdot\|} A$ as $n \to \infty$. Suppose that $t = \sup_{n\in\mathbb{N}} \operatorname{ind}(A_n) < \infty$. Then the following conditions are equivalent:*

(1) $A_n^D \xrightarrow{\|\cdot\|} A^D$ *as* $n \to \infty$.
(2) $A_n^\pi \xrightarrow{\|\cdot\|} A^\pi$ *as* $n \to \infty$.
(3) $K(A, A_n)$ *is invertible in* $B(X)$ *for* n *large enough.*
(4) $\operatorname{Ran}((A_n)^l) \cap \operatorname{Ker}((A^D)^l) = \{0\}$ *for* n *large enough, where* $l = \max\{s, t\}$.

Proof. (1)⇒(2) From $A_n \xrightarrow{\|\cdot\|} A$ and $A_n^D \xrightarrow{\|\cdot\|} A^D$ as $n \to \infty$ and

$$\begin{aligned}\|A_n^\pi - A^\pi\| &= \|A_n(A_n^D - A^D) + (A_n - A)A^D\| \\ &\le \|A_n\|\|A_n^D - A^D\| + \|A^D\|\|A_n - A\|,\end{aligned}$$

we get that $A_n^\pi \xrightarrow{\|\cdot\|} A^\pi$ as $n \to \infty$.

(2)⇒(3) Choose $n_0 \in \mathbb{N}$ such that

$$\|K(A, A_n) - (2AA^D - I)\| < \frac{1}{\|2AA^D - I\|}, \ \forall\, n > n_0.$$

Since $2AA^D - I$ is invertible in $B(X)$ with $(2AA^D - I)^{-1} = 2AA^D - I$, it follows that

$$\|(2AA^D - I)K(A, A_n) - I\| < \|2AA^D - I\| \frac{1}{\|2AA^D - I\|} = 1.$$

Thus $K(A, A_n)$ is invertible, $\forall\, n > n_0$.

(3)⇒(4) The implication (3)⇒(4) comes from Theorem 4.3.8.

(4)⇒(1) Put $\delta A = A_n - A$, $\epsilon_A = \dfrac{\|\delta A\|}{\|A\|}$ and choose $n_1 \in \mathbb{N}$ such that

$$K_D^l(A)\epsilon_A < \frac{1}{(2^l - 1)(1 + \|A^\pi\|)}, \ \forall\, n > n_1.$$

Then $\|A_n^D - A^D\| \to 0$ $(n \to \infty)$ by Proposition 4.3.5. □

4.4 The perturbation analysis of the outer generalized inverse with prescribed range and kernel

Let T be a closed subspace of X and S be a closed subspace of Y. Let $A \in B(X,Y)$ such that $\operatorname{Ker} A \cap T = \{0\}$ and $AT \dotplus S = Y$. Thus $A^{(2)}_{T,S}$ exists and moreover,

$$\operatorname{Ran}(A^{(2)}_{T,S}) = T, \ \operatorname{Ker} A^{(2)}_{T,S} = S, \ A^{(2)}_{T,S} A A^{(2)}_{T,S} = A^{(2)}_{T,S}$$

by Proposition 4.2.1. In this section, we will give the upper bounds for $\|\bar{A}^{(2)}_{T',S'}\|$ and $\|\bar{A}^{(2)}_{T',S'} - A^{(2)}_{T,S}\|$ when $\delta(T',T)$, $\delta(S',S)$ and $\|\bar{A} - A\|$ are given, where T', S' are closed subspaces and $\bar{A} = A + \delta A \in B(X,Y)$.

Throughout the section, we keep the symbols T, S, A and $A^{(2)}_{T,S}$.

Lemma 4.4.1. *Let P, $Q \in B(X)$ be idempotent. Then*

$$\begin{aligned}\|P - Q\| \le \|I - P\|\|Q\|\delta(\operatorname{Ran}(Q), \operatorname{Ran}(P)) \\ + \|P\|\|I - Q\|\delta(\operatorname{Ker} Q, \operatorname{Ker} P).\end{aligned}$$

Proof. We have

$$\|P - Q\| = \|(I - P)Q - P(I - Q)\| \le \|(I - P)Q\| + \|P(I - Q)\|.$$

Let $y \in \operatorname{Ran}(P)$, $x \in X$. Then

$$\|(I - P)Qx\| = \|(I - P)(Qx - y)\| \le \|I - P\|\|Qx - y\|.$$

Thus,

$$\begin{aligned}\|(I - P)Qx\| &\le \|I - P\|\operatorname{dist}(Qx, \operatorname{Ran}(P)) \\ &\le \|I - P\|\|Qx\|\delta(\operatorname{Ran}(Q), \operatorname{Ran}(P))\end{aligned}$$

and consequently,

$$\|(I - P)Q\| \le \|I - P\|\|Q\|\delta(\operatorname{Ran}(Q), \operatorname{Ran}(P)). \tag{4.4.1}$$

Replacing $I - P$ by P and Q by $I - Q$ in (4.4.1), we get that

$$\begin{aligned}\|P(I - Q)\| &\le \|P\|\|I - Q\|\delta(\operatorname{Ran}(I - Q), \operatorname{Ran}(I - P)) \\ &= \|P\|\|I - Q\|\delta(\operatorname{Ker} Q, \operatorname{Ker} P).\end{aligned}$$

So the assertion follows. □

Put $P = A^{(2)}_{T,S}A$ and $Q = AA^{(2)}_{T,S}$. Then P, Q are idempotent with $\operatorname{Ran}(P) = T$, $\operatorname{Ran}(Q) = AT$, $\operatorname{Ker} Q = S$. Set $A_T = A\big|_T \in B(T, AT)$. Then $\operatorname{Ran}(A_T) = AT$ and $\operatorname{Ker} A_T = \{0\}$.

Lemma 4.4.2. *Keep the symbols as above. Then*

$$\|(A_T)^{-1}\|^{-1} \le \gamma(AP) \le \|P\|\|(A_T)^{-1}\|^{-1}$$

$$\|(A_T)^{-1}\| \le \|A^{(2)}_{T,S}\| \le \|Q\|\|(A_T)^{-1}\|$$

Proof. By the definition of $A^{(2)}_{T,S}$ (see the proof of Proposition 4.2.1, $A^{(2)}_{T,S} = (A_T)^{-1}Q$. Thus $\|A^{(2)}_{T,S}\| \le \|Q\|\|(A_T)^{-1}\|$. Let $x \in AT$. Then $(A_T)^{-1}x = A^{(2)}_{T,S}x$ and hence $\|(A_T)^{-1}\| \le \|A^{(2)}_{T,S}\|$.

From

$$\|x\| = \|(A_T)^{-1}A_Tx\| \le \|(A_T)^{-1}\|\|A_Tx\|, \ \forall\, x \in T,$$

we have

$$\begin{aligned}\|APx\| &\ge \|(A_T)^{-1}\|^{-1}\|Px\| = \|(A_T)^{-1}\|^{-1}\|x-(I-P)x\| \\ &\ge \|(A_T)^{-1}\|^{-1}\operatorname{dist}(x, \operatorname{Ker} P), \quad \forall\, x \in X.\end{aligned}$$

Since $\operatorname{Ker} A \cap T = \{0\}$, it follows that $\operatorname{Ker}(AP) = \operatorname{Ker} P$. Thus,

$$\|(AP)x\| \ge \|(A_T)^{-1}\|^{-1}\operatorname{dist}(x, \operatorname{Ker}(AP)), \ \forall\, x \in X.$$

and so that $\gamma(AP) \ge \|(A_T)^{-1}\|^{-1}$.

Let $x \in X$. Then for any $y \in \operatorname{Ker} P$, $\|Px\| = \|P(x-y)\| \le \|P\|\|x-y\|$. Consequently, $\|Px\| \le \|P\|\operatorname{dist}(x, \operatorname{Ker} P)$, $\forall\, x \in X$. Furthermore,

$$\begin{aligned}\|(AP)x\| &\ge \gamma(AP)\operatorname{dist}(x, \operatorname{Ker}(AP)) = \gamma(AP)\operatorname{dist}(x, \operatorname{Ker} P) \\ &\ge \|P\|^{-1}\gamma(AP)\|Px\|, \quad \forall x \in X.\end{aligned}$$

This implies that $\|(A_T)^{-1}y\| \le \|P\|(\gamma(AP))^{-1}\|y\|$, $\forall\, y \in AT$. Therefore, $\gamma(AP) \le \|P\|\|(A_T)^{-1}\|^{-1}$. □

Lemma 4.4.3. *Set $\kappa_{T,S}(A) = \|A^{(2)}_{T,S}\|\|A\|$. Let T' be a closed subspace of X such that $\delta(T',T) < \dfrac{1}{1+\kappa_{T,S}(A)}$. Then AT' is closed in Y and $\operatorname{Ker} A \cap T' = \{0\}$ and moreover,*

$$\delta(AT, AT') \le \kappa_{T,S}(A)\delta(T,T'), \ \delta(AT', AT) \le \frac{\kappa_{T,S}(A)\delta(T',T)}{1-(1+\kappa_{T,S}(A))\delta(T',T)}.$$

Proof. For any $t \in T$, $t' \in T'$,

$$\operatorname{dist}(At, AT') \le \|At - At'\| \le \|A\|\|t-t'\|, \tag{4.4.2}$$

$$\operatorname{dist}(At', AT) \le \|At' - At\| \le \|A\|\|t'-t\|. \tag{4.4.3}$$

Thus, from (4.4.2) and (4.4.3), we get that

$$\operatorname{dist}(At, AT') \le \|A\|\operatorname{dist}(t, T') \le \|A\|\|t\|\delta(T,T'), \tag{4.4.4}$$

$$\operatorname{dist}(At', AT) \le \|A\|\operatorname{dist}(t', T) \le \|A\|\|t'\|\delta(T',T). \tag{4.4.5}$$

Since $\|t\| \le \|A^{(2)}_{T,S}At\| \le \|A^{(2)}_{T,S}\|\|At\|$, it follows from (4.4.4) that $\delta(AT, AT') \le \kappa_{T,S}(A)\delta(T,T')$.

Now for any $\epsilon > 0$, choose $t_\epsilon \in T$ such that

$$\|t' - t_\epsilon\| < \operatorname{dist}(t', T) + \epsilon \le \|t'\|\delta(T', T) + \epsilon.$$

Then from $\|At' - At_\epsilon\| \le \|A\|\|t' - t_\epsilon\|$, we have

$$\begin{aligned}
\|At'\| &\ge \|At_\epsilon\| - \|A\|\|t' - t_\epsilon\| \\
&> \|(A_T)^{-1}\|^{-1}\|t_\epsilon\| - \|A\|\|t'\|\delta(T', T) - \|A\|\epsilon \\
&\ge \|(A_T)^{-1}\|^{-1}(\|t'\| - \|t' - t_\epsilon\|) - \|A\|\|t'\|\delta(T', T) - \|A\|\epsilon \\
&\ge \|(A_T)^{-1}\|^{-1}\|t'\| - (\|(A_T)^{-1}\|^{-1} + \|A\|)\|t'\|\delta(T', T) \\
&\quad - (\|(A_T)^{-1}\|^{-1} + \|A\|)\epsilon.
\end{aligned}$$

Let $\epsilon \to 0^+$, we have, for any $t' \in T'$,

$$\|At'\| \ge \left[\|(A_T)^{-1}\|^{-1} - (\|(A_T)^{-1}\|^{-1} + \|A\|)\delta(T', T)\right]\|t'\|. \tag{4.4.6}$$

Thus, AT' is closed and $\operatorname{Ker} A \cap T' = \{0\}$, when

$$\delta(T', T) < \frac{1}{1 + \kappa_{T,S}(A)} \le \frac{1}{1 + \|(A_T)^{-1}\|\|A\|}$$

by Lemma 4.4.2. In this case,

$$\|t'\| \le \frac{\|(A_T)^{-1}\|\|At'\|}{1 - (1 + \|A_T^{-1}\|\|A\|)\delta(T', T)} \le \frac{\|A\|^{-1}\kappa_{T,S}(A)\|At'\|}{1 - (1 + \kappa_{T,S}(A))\delta(T', T)}. \tag{4.4.7}$$

So $\delta(AT', AT) \le \dfrac{\kappa_{T,S}(A)\delta(T', T)}{1 - (1 + \kappa_{T,S}(A))\delta(T', T)}$ by (4.4.5) and (4.4.7). □

Lemma 4.4.4. *Let P be idempotent in $B(X)$. Suppose that there is a closed subspace V in X such that $\hat{\delta}(V, \operatorname{Ran}(P)) < (1 + \|P\|)^{-1}$. Then there is an idempotent operator P_V in $B(X)$ satisfying following conditions:*

(1) $\|P_V\| \le (2 + \|P\|)\|P\|, \|I - P_V\| \le (1 + \|P\|)^2$.
(2) $\operatorname{Ran}(P_V) = V$, $\operatorname{Ker}(P_V) = \operatorname{Ker} P$.
(3) $\|P - P_V\| \le (1 + \|P\|)^2\|P\|\delta(\operatorname{Ran}(P), V)$.

Proof. We first show that $V \cap \operatorname{Ker} P = \{0\}$. If $V \cap \operatorname{Ker} P \ne \{0\}$, there is $x_0 \in V \cap \operatorname{Ker} P$ with $\|x_0\| = 1$. Thus $\operatorname{dist}(x_0, \operatorname{Ran}(P)) < (1 + \|P\|)^{-1}$. Choose $y_0 \in \operatorname{Ran}(P)$ such that

$$\|x_0 - y_0\| < \operatorname{dist}(x_0, \operatorname{Ran}(P)) + \frac{1}{1 + \|P\|} - \operatorname{dist}(x_0, \operatorname{Ran}(P)) = \frac{1}{1 + \|P\|}.$$

Then

$$1 = \|x_0\| = \|(I - P)(x_0 - y_0)\| \le \|I - P\|\|x_0 - y_0\| < \frac{1 + \|P\|}{1 + \|P\|},$$

but this is impossible. So $V \cap \operatorname{Ker} P = \{0\}$.

Now we prove that $X = V + \operatorname{Ker} P$. Let $x \in X$ with $\|x\| = 1$. Set $x_0 = x$ and $y_0 = (I - P)x_0$. If $Px_0 = 0$, then $x_0 = 0 + x_0$. Suppose that $Px_0 \neq 0$. Choose $z_0' \in V$ such that $\left\| \frac{Px_0}{\|Px_0\|} - z_0' \right\| < \frac{1}{1 + \|P\|}$. Set $z_0 = \|Px_0\| z_0' \in V$. Then

$$\|Px_0 - z_0\| < \frac{\|Px_0\|}{1 + \|P\|} \leq \frac{\|P\|}{1 + \|P\|}.$$

Put $x_1 = x_0 - (z_0 + y_0) = Px_0 - z_0$. Then

$$\|x_1\| \leq \frac{\|P\|}{1 + \|P\|}, \ \|y_0\| \leq 1 + \|P\|, \ \|z_0\| \leq \|P\| + \frac{\|P\|}{1 + \|P\|}.$$

By induction, we can construct $\{z_n\} \subset V$, $\{y_n\} \subset \operatorname{Ker} P$ and

$$x_n = x_{n-1} - (z_{n-1} + y_{n-1}), \ y_n = (I - P)x_n, \ \forall\, n \in \mathbb{N},$$

such that

$$\|x_n\| \leq \Big(\frac{\|P\|}{1 + \|P\|} \Big)^n, \ \|y_n\| \leq (1 + \|P\|)\|x_n\| \tag{4.4.8}$$

$$\|z_n\| \leq \Big(\|P\| + \frac{\|P\|}{1 + \|P\|} \Big) \|x_n\|, \quad \forall\, n \in \mathbb{N}. \tag{4.4.9}$$

Since $\sum\limits_{n=0}^{\infty} \|y_n\| < \infty$, $\sum\limits_{n=0}^{\infty} \|z_n\| < \infty$ and X is completed, $\operatorname{Ker} P$, V are closed, it follows that $y = \sum\limits_{n=0}^{\infty} y_n \in \operatorname{Ker} P$, $z = \sum\limits_{n=0}^{\infty} z_n \in V$. Thus,

$$\begin{aligned} z + y &= \sum_{n=1}^{\infty} z_{n-1} + \sum_{n=1}^{\infty} y_{n-1} = -\sum_{n=1}^{\infty} (x_n - x_{n-1}) \\ &= x_0 - \lim_{n \to \infty} x_n = x. \end{aligned} \tag{4.4.10}$$

Therefore $X = V + \operatorname{Ker} P$.

Let P_V be the idempotent operator given by $P_V x = z$ with respect to the decomposition $x = z + y$ given in (4.4.10). Then $(I - P_V)x = y$ and hence $\operatorname{Ran}(P_V) = V$, $\operatorname{Ker} P_V = \operatorname{Ker} P$. By (4.4.8), (4.4.9) and (4.4.10),

$$\|P_V x\| = \|z\| \leq \Big(\|P\| + \frac{\|P\|}{1 + \|P\|} \Big) \sum_{n=0}^{\infty} \|x_n\| \leq (2 + \|P\|)\|P\|,$$

$$\|(I - P_V)x\| = \|y\| \leq (1 + \|P\|) \sum_{n=0}^{\infty} \|x_n\| \leq (1 + \|P\|)^2,$$

$\forall\, x \in X$ with $\|x\| = 1$. Consequently,

$$\begin{aligned} \|P_V - P\| &\leq \|P\| \|I - P_V\| \delta(\operatorname{Ran}(P), V) \\ &\leq (1 + \|P\|)^2 \|P\| \delta(\operatorname{Ran}(P), V). \end{aligned}$$

□

Theorem 4.4.5. *Let $A \in B(X,Y)$ and $\bar{A} = A + \delta A \in B(X,Y)$. Let S be a closed subspace in Y and T be a closed subspace in X. Assume that $A^{(2)}_{T,S}$ exists. Put $P = A^{(2)}_{T,S}A$, $Q = AA^{(2)}_{T,S}$. Let S' (resp. T') be a closed subspace in Y (resp. X) such that*

$$\hat{\delta}(S',S) < \frac{1}{1+(1+\|Q\|)^2}, \ \hat{\delta}(T',T) < \frac{1}{1+(2+\|Q\|)\kappa_{T,S}(A)}.$$

Then $A^{(2)}_{T',S'}$ exists and $\|A^{(2)}_{T',S'}\| \le \dfrac{(1+\|Q\|)^2\|A^{(2)}_{T,S}\|}{1-(1+\kappa_{T,S}(A))\delta(T'T)}$,

$$\begin{aligned}\frac{\|A^{(2)}_{T',S'}-A^{(2)}_{T,S}\|}{\|A^{(2)}_{T,S}\|} &\le \frac{(1+\kappa_{T,S}(A))(1+\|P\|)^2\|P\|(1+(1+\|Q\|)^2)^2}{1-(1+\kappa_{T,S}(A))\hat{\delta}(T,T')} \\ &\quad \times \hat{\delta}(T',T) + \kappa_{T,S}(A)(1+\|Q\|)^2\|Q\|\hat{\delta}(T',T) \\ &\quad + (1+(1+\|Q\|)^2)^2(1+\|Q\|)^2\hat{\delta}(S',S).\end{aligned}$$

Suppose that $\|\delta A\|\|A^{(2)}_{T,S}\| < \dfrac{1-(1+\kappa_{T,S}(A))\hat{\delta}(T',T)}{(1+(1+\|Q\|)^2)^2}$. *Then $\bar{A}^{(2)}_{T',S'}$ exists and*

$$\|\bar{A}^{(2)}_{T',S'}\| \le \frac{(1+(1+\|Q\|)^2)^2\|A^{(2)}_{T,S}\|}{1-(1+\kappa_{T,S}(A))\hat{\delta}(T',T)-(1+(1+\|Q\|)^2)^2\kappa_{T,S}\epsilon_A},$$

$$\begin{aligned}\frac{\|\bar{A}^{(2)}_{T',S'}-A^{(2)}_{T,S}\|}{\|A^{(2)}_{T,S}\|} &\le \frac{(1+(1+\|Q\|)^2)^2}{1-(1+\kappa_{T,S}(A))\hat{\delta}(T',T)}\Big[(1+\kappa_{T,S}(A))(1+\|P\|)^2\|P\| \\ &\times \hat{\delta}(T,T') + \frac{(1+(1+\|Q\|)^2)^2\kappa_{T,S}(A)\epsilon_A}{1-(1+\kappa_{T,S}(A))\hat{\delta}(T',T)-(1+(1+\|Q\|)^2)^2\kappa_{T,S}(A)\epsilon_A}\Big] \\ &+ (1+\|Q\|)^2\big[\kappa_{T,S}(A)\|Q\|\hat{\delta}(T',T)+(1+(1+\|Q\|)^2)^2\hat{\delta}(S',S)\big].\end{aligned}$$

Proof. We have $\hat{\delta}(T',T) < \dfrac{1}{1+\kappa_{T,S}(A)} \le \dfrac{1}{1+\|P\|}$. So by Lemma 4.4.3, AT' is closed in Y and $\operatorname{Ker} A \cap T' = \{0\}$ and moreover,

$$\delta(AT',AT) \le \frac{\kappa_{T,S}(A)\delta(T',T)}{1-(1+\kappa_{T,S}(A))\delta(T',T)} < \frac{1}{1+\|Q\|},$$

$$\delta(AT,AT') \le \kappa_{T,S}(A)\delta(T,T') < \frac{1}{1+\|Q\|}.$$

It follows from Lemma 4.4.4 that there is an idempotent operator $Q' \in B(Y)$ such that $\operatorname{Ran}(Q') = AT'$, $\operatorname{Ker} Q' = \operatorname{Ker} Q = S$,

$$\|Q'\| \le (2+\|Q\|)\|Q\|, \ \|I_Y - Q'\| \le (1+\|Q\|)^2$$
$$\|Q'-Q\| \le (1+\|Q\|)^2\|Q\|\delta(AT,AT').$$

Consequently, $\hat{\delta}(S,S') < \dfrac{1}{1+\|I_Y - Q\|}$.

Since $S = \mathrm{Ran}\,(I-Q')$, by applying Lemma 4.4.4 to S and S', we can find an idempotent operator $Q_0'' \in B(Y)$ with $\mathrm{Ran}\,(Q_0'') = S'$, $\mathrm{Ker}\,Q_0'' = \mathrm{Ker}\,(I-Q') = \mathrm{Ran}\,(Q') = AT'$ such that

$$\begin{aligned}&\|Q_0''\| \le (1+\|I_Y - Q'\|)^2\|I_Y - Q'\|\delta(S,S')\\ &\|Q_0''\| \le (2+\|I_Y - Q'\|)\|I_Y - Q'\|,\ \ \|I_Y - Q_0''\| \le (1+\|I_Y - Q'\|)^2.\end{aligned}$$

Put $Q'' = I_Y - Q_0''$. Then $\mathrm{Ran}\,(Q'') = AT'$, $\mathrm{Ker}\,Q'' = S'$, i.e., $AT' \dot{+} S' = Y$. Therefore, $A^{(2)}_{T',S'}$ exists. Furthermore,

$$\begin{aligned}\|Q''-Q\| &\le \|Q_0'' - (I_Y - Q')\| + \|Q'-Q\|\\ &\le (1+\|Q\|)^2\|Q\|\delta(AT,AT') + (1+(1+\|Q\|)^2)^2(1+\|Q\|)^2\delta(S,S')\\ &\le (1+\|Q\|)^2[\|Q\|\kappa_{T,S}(A)\delta(T,T') + (1+(1+\|Q\|)^2)^2\delta(S,S')].\end{aligned}$$

Now from (4.4.6), we get that

$$\|A^{(2)}_{T',S'}x\| \le \frac{\|(A_T)^{-1}\|\|AA^{(2)}_{T',S'}x\|}{1-(1+\|(A_T)^{-1}\|\|A\|)\delta(T',T)},\ \ \forall x \in Y. \tag{4.4.11}$$

Noting that $AA^{(2)}_{T',S'}$ is an idempotent operator with

$$\mathrm{Ran}\,(AA^{(2)}_{T',S'}) = AT' = \mathrm{Ran}\,(Q''),\ \ \mathrm{Ker}\,(AA^{(2)}_{T',S'}) = S' = \mathrm{Ker}\,(Q''),$$

we have $AA^{(2)}_{T',S'} = Q''$. Thus, by Lemma 4.4.2 and (4.4.11),

$$\|A^{(2)}_{T',S'}\| \le \frac{\|Q''\|\|A^{(2)}_{T',S'}\|}{1-(1+\kappa_{T,S}(A))\delta(T',T)} \le \frac{(1+(1+\|Q\|)^2)^2\|A^{(2)}_{T,S}\|}{1-(1+\kappa_{T,S}(A))\hat{\delta}(T',T)}.$$

By Lemma 4.4.4, we can choose an idempotent operator $P' \in B(X)$ with $\mathrm{Ran}\,(P') = T'$, $\mathrm{Ker}\,P' = \mathrm{Ker}\,P$ such that

$$\begin{aligned}&\|P'\| \le (2+\|P\|)^2,\ \ \ \|I_X - P'\| \le (1+\|P\|)^2\\ &\|P'-P\| \le (1+\|P\|)^2\|P\|\hat{\delta}(T',T).\end{aligned}$$

Let $y \in Y$ and put $x' = A^{(2)}_{T',S'}y$, $x = A^{(2)}_{T,S}y$. Then $x \in T$ and $x' \in T'$. Note that $A^{(2)}_{T',S'} = (A_{T'})^{-1}Q''$, $A^{(2)}_{T,S} = (A_T)^{-1}Q$. Thus, $Ax = Qy$, $Ax' = Q''y$ and hence

$$A(x-Px') + A(I-P)x' = (Q-Q'')y. \tag{4.4.12}$$

(4.4.12) implies that

$$\begin{aligned}\|(A_T)^{-1}\|^{-1}\|x-Px'\| &\le \|A_T(x-Px')\|\\ &\le \|A\|\|P'-P\|\|x'\| + \|Q-Q''\|\|y\|.\end{aligned}$$

Thus,

$$\begin{aligned}\|x - x'\| &\le \|x - Px'\| + \|(P' - P)x'\| \le \|x - Px'\| + \|P' - P\|\|x'\| \\ &\le (\kappa_{T,S}(A) + 1)\|P' - P\|\|A^{(2)}_{T',S'}\|\|y\| + \|A^{(2)}_{T,S}\|\|Q - Q''\|\|y\|.\end{aligned}$$

and consequently,

$$\begin{aligned}\|A^{(2)}_{T',S'} - A^{(2)}_{T,S}\| \le{}& (1 + \kappa_{T,S}(A))(1 + \|P\|)^2\|P\|\|A^{(2)}_{T',S'}\|\hat{\delta}(T', T) \\ &+ \|A^{(2)}_{T,S}\|(1 + \|Q\|)^2\big[\|Q\|\kappa_{T,S}(A)\hat{\delta}(T', T) \\ &+ (1 + (1 + \|Q\|)^2)^2\hat{\delta}(S', S)\big].\end{aligned}$$

Finally, we have

$$\begin{aligned}\frac{\|A^{(2)}_{T',S'} - A^{(2)}_{T,S}\|}{\|A^{(2)}_{T,S}\|} \le{}& \frac{(1 + \kappa_{T,S}(A))(1+\|P\|)^2\|P\|(1+(1+\|Q\|)^2)^2}{1 - (1 + \kappa_{T,S}(A))\hat{\delta}(T', T)}\hat{\delta}(T', T) \\ &+(1 + \|Q\|)^2\big[\kappa_{T,S}(A)\|Q\|\hat{\delta}(T', T) + (1 + (1 + \|Q\|)^2)^2\hat{\delta}(S', S)\big].\end{aligned}$$

If $\|\delta A\|\|A^{(2)}_{T,S}\| < \dfrac{1 - (1 + \kappa_{T,S}(A))\hat{\delta}(T', T)}{(1 + (1 + \|Q\|)^2)^2}$, then $\|\delta A\|\|A^{(2)}_{T',S'}\| < 1$ and $I_X + A^{(2)}_{T',S'}\delta A \in GL(B(X))$. Set $G = (I_X + A^{(2)}_{T',S'}\delta A)^{-1}A^{(2)}_{T',S'}$. Then $\mathrm{Ran}\,(G) = T'$, $\mathrm{Ker}\, G = S'$ and

$$\begin{aligned}G\bar{A}G &= (I_X + A^{(2)}_{T',S'}\delta A)^{-1}A^{(2)}_{T',S'}(A + \delta A)A^{(2)}_{T',S'}(I_X + \delta A A^{(2)}_{T',S'})^{-1} \\ &= (I_X + A^{(2)}_{T',S'}\delta A)^{-1}(I_X + \delta A A^{(2)}_{T',S'})(I_X + \delta A A^{(2)}_{T',S'})^{-1} \\ &= G.\end{aligned}$$

So $\bar{A}^{(2)}_{T',S'} = G = (I_X + A^{(2)}_{T',S'}\delta A)^{-1}A^{(2)}_{T',S'}$ and consequently,

$$\begin{aligned}\|\bar{A}^{(2)}_{T',S'}\| &\le \frac{\|A^{(2)}_{T',S'}\|}{1 - \|A^{(2)}_{T',S'}\|\|\delta A\|} \\ &\le \frac{(1 + (1 + \|Q\|)^2)^2\|A^{(2)}_{T,S}\|}{1 - (1 + \kappa_{T,S}(A))\hat{\delta}(T', T) - (1 + (1 + \|Q\|)^2)^2\|A^{(2)}_{T,S}\|\|\delta A\|}\end{aligned}$$

and

$$\begin{aligned}\|\bar{A}^{(2)}_{T',S'} - A^{(2)}_{T,S}\| &\le \|\bar{A}^{(2)}_{T',S'} - A^{(2)}_{T',S'}\| + \|A^{(2)}_{T',S'} - A_{T,S}\| \\ &\le \frac{\|A^{(2)}_{T',S'}\|^2\|\delta A\|}{1 - \|A^{(2)}_{T',S'}\|\|\delta A\|} + \|A^{(2)}_{T',S'} - A^{(2)}_{T,S}\|.\end{aligned}$$

It follows that

$$\frac{\|\bar{A}^{(2)}_{T',S'}-A^{(2)}_{T,S}\|}{\|A^{(2)}_{T,S}\|} \le \frac{(1+(1+\|Q\|)^2)^2}{1-(1+\kappa_{T,S}(A))\hat{\delta}(T',T)}\Big[(1+\kappa_{T,S}(A))(1+\|P\|)^2\|P\|$$

$$\times\ \hat{\delta}(T,T') + \frac{(1+(1+\|Q\|)^2)^2\kappa_{T,S}(A)\epsilon_A}{1-(1+\kappa_{T,S}(A))\hat{\delta}(T',T)-(1+(1+\|Q\|)^2)^2\kappa_{T,S}(A)\epsilon_A}\Big]$$

$$+ (1+\|Q\|)^2\big[\kappa_{T,S}(A)\|Q\|\hat{\delta}(T',T) + (1+(1+\|Q\|)^2)^2\hat{\delta}(S',S)\big]. \qquad \square$$

Remark 4.4.6. Suppose that $A^{(2)}_{T,S}$ and $A^{(2)}_{T',S'}$ exist. Then applying Lemma 4.4.2 to (4.4.11), we have

$$\|A^{(2)}_{T',S'}\| \le \frac{\|A^{(2)}_{T,S}\|\|AA^{(2)}_{T',S'}\|}{1-(1+\kappa_{T,S}(A))\hat{\delta}(T',T)}$$

when $\hat{\delta}(T',T) < (1+\kappa_{T,S}(A))^{-1}$. Let P' be in the proof of Theorem 4.4.5 and replace Q'' by $AA^{(2)}_{T',S'}$ in (4.4.12). Then

$$\|A^{(2)}_{T',S'} - A^{(2)}_{T,S}\| \le (1+\kappa_{T,S}(A))(1+\|P\|)^2\|P\|\|A^{(2)}_{T',S'}\|\hat{\delta}(T',T)$$
$$+ \|AA^{(2)}_{T',S'} - AA^{(2)}_{T,S}\|\|A^{(2)}_{T,S}\|,$$

where $P = A^{(2)}_{T,S}A$. Put $Q = AA^{(2)}_{T,S}$, $Q' = AA^{(2)}_{T',S'}$. By Lemma 4.4.1 and Lemma 4.4.3,

$$\|Q'-Q\| \le \|I_X - Q'\|\|Q\|\delta(AT,AT') + \|Q'\|\|I_X-Q\|\delta(S,S')$$
$$\le (\|Q'-Q\| + \|I_X-Q\|)\|Q\|\kappa_{T,S}(A)\delta(T,T')$$
$$+ (\|Q'-Q\| + \|Q\|)\|I_X-Q\|\delta(S,S').$$

Consequently,

$$\|Q'-Q\| \le \frac{\|I_X-Q\|\|Q\|(\kappa_{T,S}(A)\hat{\delta}(T,T') + \hat{\delta}(S',S))}{1-\|Q\|\kappa_{T,S}(A)\hat{\delta}(T',T) - \|I_X-Q\|\hat{\delta}(S',S)}. \tag{4.4.13}$$

when $\|Q\|\kappa_{T,S}(A)\hat{\delta}(T',T) + \|I_X-Q\|\hat{\delta}(S',S) < 1$. Using (4.4.13), we can estimate $\|A^{(2)}_{T',S'}\|$ and $\|A^{(2)}_{T',S'} - A^{(2)}_{T,S}\|$.

If we consider the perturbation analysis for $A^{(2)}_{T,S}$ on Hilbert spaces, we have more simple expressions of upper bounds than them in Theorem 4.4.5 as follows.

Theorem 4.4.7. *Let $A \in B(H,K)$ and $\bar{A} = A + \delta A \in B(H,K)$. Let T (resp. S) be a closed subspace in H (resp. K) such that $A^{(2)}_{T,S}$ exists. Let T' (resp. S') be a closed subspace in H (resp. K) such that*

$$\hat{\delta}(T',T) < \frac{1}{(1+\kappa_{T,S}(A))^2}, \quad \hat{\delta}(S',S) < \frac{1}{1+(1+\kappa_{T,S}(A))^2}.$$

Then $A^{(2)}_{T',S'}$ *exists and*

$$\|A^{(2)}_{T',S'}\| \le \frac{\|A^{(2)}_{T,S}\|}{1-\kappa_{T,S}(A)(\hat{\delta}(S',S)+\hat{\delta}(T',T))}$$

$$\frac{\|A^{(2)}_{T',S'}-A^{(2)}_{T,S}\|}{\|A^{(2)}_{T,S}\|} \le \frac{1+\sqrt{5}}{2}\frac{\|A\|\|A^{(2)}_{T,S}\|(\hat{\delta}(S',S)+\hat{\delta}(T',T))}{1-\kappa_{T,S}(A)(\hat{\delta}(S',S)+\hat{\delta}(T',T))}.$$

Moreover, if $\|\delta A\|\|A^{(2)}_{T,S}\| < 1-\kappa_{T,S}(A)(\hat{\delta}(S',S)+\hat{\delta}(T',T))$*, then* $\bar{A}^{(2)}_{T',S'} = (I_H + A^{(2)}_{T',S'},\delta A)^{-1}A^{(2)}_{T',S'}$ *and*

$$\|A^{(2)}_{T',S'}\| \le \frac{\|A^{(2)}_{T,S}\|}{1-\kappa_{T,S}(A)(\hat{\delta}(S',S)+\hat{\delta}(T',T))-\kappa_{T,S}(A)\epsilon_A}$$

$$\begin{aligned}\frac{\|\bar{A}^{(2)}_{T',S'}-A^{(2)}_{T,S}\|}{\|A^{(2)}_{T,S}\|} &\le \frac{\kappa_{T,S}(A)\epsilon_A}{1-\kappa_{T,S}(A)(\hat{\delta}(S',S)+\hat{\delta}(T',T))-\kappa_{T,S}(A)\epsilon_A}\\ &\quad\times\frac{1}{1-\kappa_{T,S}(A)(\hat{\delta}(S',S)+\hat{\delta}(T',T))}\\ &\quad+\frac{1+\sqrt{5}}{2}\frac{\kappa_{T,S}(A)(\hat{\delta}(S',S)+\hat{\delta}(T',T))}{1-\kappa_{T,S}(A)(\hat{\delta}(S',S)+\hat{\delta}(T',T))}.\end{aligned}$$

Proof. According to the proof of Theorem 4.4.5, $A^{(2)}_{T',S'}$ exists. Put $B = P_{S^\perp}AP_T$, $\bar{B} = P_{S'^\perp}AP_{T'}$, where $P_{T'}$, P_T and $P_{S'^\perp}$, $P_{S^\perp}$ are orthogonal projections from H onto T', T and from K onto $S'^\perp$, $S^\perp$, respectively. We have

$$\delta B = \bar{B} - B = (P_{S'^\perp} - P_{S^\perp})AP_{T'} + P_{S^\perp}A(P_{T'} - P_T).$$

Since $\|P_{S'^\perp} - P_{S^\perp}\| = \hat{\delta}(S',S)$ and $\|P_{T'} - P_T\| = \hat{\delta}(T',T)$ by Lemma 1.3.7, it follows that $\|\delta B\| \le \|A\|(\hat{\delta}(S',S)+\hat{\delta}(T',T))$ and consequently,

$$\|A^{(2)}_{T,S}\|\|\delta B\| \le \kappa_{T,S}(A)(\hat{\delta}(S',S)+\hat{\delta}(T',T)) < 1.$$

Since $\mathrm{Ran}\,(B)$ and $\mathrm{Ran}\,(\bar{B})$ are closed and $A^{(2)}_{T,S} = B^\dagger$, $A^{(2)}_{T',S'} = \bar{B}^\dagger$ by Proposition 4.2.6 and $\mathrm{Ran}\,(\bar{B})\cap\mathrm{Ker}\,(B^\dagger) \subset S'^\perp \cap S = \{0\}$ (for $\|P_{S'} - P_S\| = \hat{\delta}(S',S) < 1$), it follows from Theorem 3.3.5 that

$$\begin{aligned}\|A^{(2)}_{T',S'}\| &= \|\bar{B}^\dagger\| \le \|(I_H + B^\dagger\delta B)^{-1}\|\|B^\dagger\|\\ &\le \frac{\|A^{(2)}_{T,S}\|}{1-\kappa_{T,S}(A)(\hat{\delta}(S',S)+\hat{\delta}(T',T))}\end{aligned}$$

$$\frac{\|A^{(2)}_{T',S'} - A^{(2)}_{T,S}\|}{\|A_{T,S}\|} \le \frac{1+\sqrt{5}}{2}\|(I_H + B^\dagger \delta B)^{-1}\|\|B^\dagger\|\|\delta B\|$$
$$\le \frac{1+\sqrt{5}}{2}\frac{\|A^{(2)}_{T,S}\|(\hat{\delta}(S',S)+\hat{\delta}(T',T))}{1-\kappa_{T,S}(A)(\hat{\delta}(S',S)+\hat{\delta}(T',T))}.$$

When $\|A^{(2)}_{T,S}\|\|\delta A\| < 1 - \kappa_{T,S}(A)(\hat{\delta}(S',S)+\hat{\delta}(T',T))$,

$$I_H + A^{(2)}_{T',S'}\delta A \in GL(B(H)) \text{ and } A^{(2)}_{T',S'} = (I_H + A^{(2)}_{T',S'}\delta A)^{-1}A^{(2)}_{T',S'}.$$

Thus,

$$\|\bar{A}^{(2)}_{T',S'}\| \le \frac{\|A^{(2)}_{T',S'}\|}{1-\|A^{(2)}_{T',S'}\|\|\delta A\|}$$
$$\le \frac{\|A^{(2)}_{T,S}\|}{1-\kappa_{T,S}(A)(\hat{\delta}(S',S)+\hat{\delta}(T',T))-\kappa_{T,S}(A)\epsilon_A}$$

$$\frac{\|\bar{A}^{(2)}_{T',S'} - A^{(2)}_{T,S}\|}{\|A^{(2)}_{T,S}\|} \le \frac{\|\bar{A}^{(2)}_{T',S'} - A^{(2)}_{T',S'}\|}{\|A^{(2)}_{T,S}\|} + \frac{\|A^{(2)}_{T',S'} - A^{(2)}_{T,S}\|}{\|A^{(2)}_{T,S}\|}$$
$$\le \frac{\kappa_{T,S}(A)\epsilon_A}{1-\kappa_{T,S}(A)(\hat{\delta}(S',S)+\hat{\delta}(T',T))-\kappa_{T,S}(A)\epsilon_A}$$
$$\times \frac{1}{1-\kappa_{T,S}(A)(\hat{\delta}(S',S)+\hat{\delta}(T',T))}$$
$$+\frac{1+\sqrt{5}}{2}\frac{\|A^{(2)}_{T,S}\|(\hat{\delta}(S',S)+\hat{\delta}(T',T))}{1-\kappa_{T,S}(A)(\hat{\delta}(S',S)+\hat{\delta}(T',T))}.$$

□

4.5 Drazin inverses and generalized Drazin inverses in Banach algebras

Let $\mathcal{A}$ be a Banach algebra and let $N(\mathcal{A})$ (resp. $QN(\mathcal{A})$) denote the set of all nilpotent (resp. quasinilpotent) elements in $\mathcal{A}$. Throughout the section, we always assume that $\mathcal{A}$ is a Banach algebra and $\tilde{\mathcal{A}} = \{x+\lambda \,|\, x \in \mathcal{A}, \lambda \in \mathbb{C}\}$.

Definition 4.5.1. Let $a \in \mathcal{A}\backslash\{0\}$. If there is $b \in \mathcal{A}$ such that

$$ab = ba,\ b = bab,\ a - a^2 b \in QN(\mathcal{A}), \tag{4.5.1}$$

then a is called to be generalized Drazin invertible and b is called a generalized Drazin inverse of a, which is denoted by a^d.

When $a - a^2b \in N(\mathcal{A})$, a is called to be Drazin invertible and b is a Drazin inverse of a, denoted by a^D. The order of $a - a^2b$ is called the index of a. We denote it by $\text{ind}(a)$.

When $\text{ind}(a) = 1$, the b in (4.5.1) is called the group inverse of a and is denoted by $a^\#$. Let $Dr(\mathcal{A})$ (resp. $Drg(\mathcal{A})$) denote the set of all Drazin (resp. generalized Drazin) invertible elements in $\mathcal{A}$.

Remark 4.5.2. (1) Comparing Definition 4.5.1 with Definition 4.1.6 and Definition 4.1.14, we can find that the definition of the Drazin inverse or the generalized Drazin inverse is same when we consider the element in $\mathcal{A} = B(X)$.

(2) Using the same method as it in the proof of Proposition 4.1.13, we get that the generalized Drazin inverse or the Drazin inverse of an element in $\mathcal{A}\backslash\{0\}$ is unique.

According to Definition 4.5.1, we have basic properties of the Drazin inverse or the generalized Drazin inverse as follows.

Proposition 4.5.3. *Let $a \in Dr(\mathcal{A})$ with $\text{ind}(a) = k$. Then*

(1) $(a^D)^D = a$ *iff* $\text{ind}(a) = 1$.
(2) $(a^l)^\#$ *exists for any* $l \geq k$ *and* $a^D = (a^l)^\# a^{l-1}$, $(a^D)^l = (a^l)^\#$.
(3) $a^{l+1}a^D = a^l, \forall\, l \geq k$.
(4) $(xax^{-1})^D = xa^Dx^{-1}$, $\forall\, x \in GL(\tilde{\mathcal{A}})$.

Proposition 4.5.4. *Let $a \in Drg(A)$. Then*

(1) $(a^d)^l = (a^l)^d$, $\forall\, l \geq 1$.
(2) $(a^d)^d = a$ *iff* $\text{ind}(a) = 1$.
(3) $(xax^{-1})^d = xa^dx^{-1}$, $\forall\, x \in GL(\tilde{\mathcal{A}})$.

Now we characterize the existence of the generalized Drazin inverse in a unital Banach algebra as follows.

Theorem 4.5.5. *Let $\mathcal{A}$ be a unital Banach algebra and let $a \in \mathcal{A}\backslash\{0\}$. Then the following statements are equivalent.*

(1) $a \in Drg(\mathcal{A})$.
(2) *There is an idempotent element p in $\mathcal{A}$ with $pa = ap$ such that $pap \in GL(p\mathcal{A}p)$ and $(1-p)a(1-p) \in QN((1-p)\mathcal{A}(1-p))$.*
(3) $0 \notin \sigma(a)$ *or* $0 \in \sigma(a)$ *is an isolated point.*

Proof. (1)$\Rightarrow$(2) Put $p = aa^d$. Since

$$aa^d = a^d a,\ a^d = a^d a a^d, \text{ and } a(1 - aa^d) \in QN(\mathcal{A}),$$

we have $pa^d = a^d p = a^d$ and $papa^d = a^d pap = p$. So $a^d \in p\mathcal{A}p$ and $pap \in GL(p\mathcal{A}p)$ with $(pap)^{-1} = a^d$. From $a(1 - aa^d) \in QN(\mathcal{A})$, we obtain that

$$\|((1-p)a(1-p))^n\|^{1/n} = \|(a(1-p))^n\|^{1/n} \to 0 \quad (n \to \infty),$$

that is, $(1-p)a(1-p) \in QN((1-p)\mathcal{A}(1-p))$.

(2)$\Rightarrow$(3) If $p = 1$, then $a \in GL(\mathcal{A})$. Suppose that $p \neq 1$. Set $a_C = pap$, $a_N = (1-p)a(1-p)$. Then $a = a_C + a_N$ and $a_C a_N = a_N a_C = 0$. Put $\mathcal{A}_1 = p\mathcal{A}p$, and $\mathcal{A}_2 = (1-p)\mathcal{A}(1-p)$. Then $\sigma(a) = \sigma_{\mathcal{A}_1}(a_C) \cup \sigma_{\mathcal{A}_2}(a_N)$. Since $a_N \in QN(\mathcal{A}_2)$ and a_C is invertible in $\mathcal{A}_1$, it follows that $\sigma(a) = \sigma_{\mathcal{A}_1}(a_C) \cup \{0\}$ and $0 \notin \sigma_{\mathcal{A}_1}(a_C)$. Thus, $0 \in \sigma(a)$ is an isolated point.

(3)$\Rightarrow$(1) Choose open subset U_1, U_2 in $\mathbb{C}$ such that $U_1 \cap U_2 = \emptyset$ and $0 \in U_1$, $-1 \notin U_1$, $\sigma(a)\backslash\{0\} \subset U_2$. Let Γ_1 (resp. Γ_2) be any contour surrounding $\{0\}$ (resp. $\sigma(a)\backslash\{0\}$) in U_1 (resp. U_2). Put $h(z) = z$, $\forall\, z \in \mathbb{C}$ and

$$f(z) = \begin{cases} 1 & z \in U_1 \\ 0 & z \in U_2 \end{cases}, \quad g(z) = \begin{cases} 0 & z \in U_1 \\ \dfrac{1}{z} & z \in U_2 \end{cases}.$$

Then f, g are holomorphic on $U_1 \cup U_2$ and

$$p = f(a) = \frac{1}{2\pi i}\oint_{\Gamma_1+\Gamma_2} f(\lambda)(\lambda - a)^{-1}\mathrm{d}\,\lambda = \frac{1}{2\pi i}\oint_{\Gamma_1} (\lambda - a)^{-1}\mathrm{d}\,\lambda \quad (4.5.2)$$

is an idempotent element in $\mathcal{A}$. Put

$$b = g(a) = \frac{1}{2\pi i}\oint_{\Gamma_1+\Gamma_2} g(\lambda)(\lambda - a)^{-1}\mathrm{d}\,\lambda = \frac{1}{2\pi i}\oint_{\Gamma_2} \lambda^{-1}(\lambda - a)^{-1}\mathrm{d}\,\lambda. \quad (4.5.3)$$

Then $ab = ba$ and

$$\begin{aligned} bab = (g^2 h)(a) &= \frac{1}{2\pi i}\oint_{\Gamma_1+\Gamma_2} g^2(\lambda)h(\lambda)(\lambda - a)^{-1}\mathrm{d}\,\lambda \\ &= \frac{1}{2\pi i}\oint_{\Gamma_2} \lambda^{-1}(\lambda - a)^{-1}\mathrm{d}\,\lambda \\ &= b. \end{aligned}$$

Finally, by Proposition 1.4.15 (3),

$$\begin{aligned} \sigma(a(1-ab)) &= \sigma((h(1-hg))(a)) = (h(1-hg))(\sigma(a)) \\ &= \{h(\lambda)(1 - h(\lambda)g(\lambda)) | \lambda \in \sigma(a)\} \\ &= \{0\}. \end{aligned}$$

Therefore, $a \in Drg(\mathcal{A})$ with $a^d = b$. □

Let $a \in Drg(\mathcal{A})$ and put $a^\pi = 1 - aa^d$. (4.5.2) and (4.5.3) imply that $a^\pi = p$. So we call a^π is the spectral projection of a at $\{0\}$.

Corollary 4.5.6. *Let $\mathcal{A}$ be a unital Banach algebra and let $a \in Drg(\mathcal{A})$.*

(1) *If $c \in \mathcal{A}$ and $ca = ac$, then $ca^d = a^d c$.*
(2) *a^d is in the closed subalgebra of $\mathcal{A}$ generated by 1 and a.*
(3) *$a + a^\pi \in GL(\mathcal{A})$ and $a^d = (a + a^\pi)^{-1}(1 - a^\pi)$*

Proof. (1) and (2) come from the proof of Theorem 4.5.5.

Now we show (3). Keep the symbols in the proof of Theorem 4.5.5. We have $a + a^\pi = (h + f)(a)$. So

$$\sigma(a + a^\pi) = (h + f)(\sigma(a)) = (\sigma(a)\backslash\{0\}) \cup \{1\}$$

by Proposition 1.4.13 (3) and hence $a + a^\pi \in GL(\mathcal{A})$.

Noting that $h(z) + f(z) \neq 0$ and $(h(z) + f(z))^{-1}(1 - f(z)) = g(z)$, $\forall\, z \in U_1 \cup U_2$, we have

$$a^d = b = g(a) = (h + f)^{-1}(a)(1 - f(a)) = (a + a^\pi)^{-1}(1 - a^\pi).$$

□

Corollary 4.5.7. *Let $a \in \mathcal{A}$ such that $(a^n)^\#$ exists for some $n \in N$. Then $a \in Dr(\mathcal{A})$ with $\operatorname{ind}(a) \leq n$.*

Proof. By Corollary 4.5.6 (2), $(a^n)^\#$ is in the closed subalgebra of $\mathcal{A}$ generated by 1 and a^n. Thus $(a^n)^\# a = a(a^n)^\#$. Put $b = (a^n)^\# a^{n-1}$. Then it is easy to check that $ab = ba$ and

$$bab = (a^n)^\# a^{n-1} a (a^n)^\# a^{n-1} = b,$$
$$[a(1 - ab)]^n = a^n(1 - a^n(a^n)^\#) = 0.$$

So $a \in Dr(\mathcal{A})$ with $\operatorname{ind}(a) \leq n$. □

Similar to Theorem 4.5.5, we have

Proposition 4.5.8. *Let $\mathcal{A}$ be a unital Banach algebra and let $a \in \mathcal{A}$. The following statements are equivalent:*

(1) *$a \in Dr(\mathcal{A})$.*
(2) *There is an idempotent element p in $\mathcal{A}$ with $ap = pa$ such that $pap \in GL(p\mathcal{A}p)$ and $(1 - p)a(1 - p) \in N((1 - p)\mathcal{A}(1 - p))$.*
(3) *$0 \notin \sigma(a)$ or $0 \in \sigma(a)$ is an isolated point and $\lambda = 0$ is a pole of $\lambda \mapsto (\lambda - a)^{-1}$, $\lambda \in \mathbb{C}\backslash\sigma(a)$.*

Proof. (1)⇒(2) Put $p = aa^D$. Then from

$$aa^D = a^D a,\ a^D = a^D aa^D,\ a(1 - aa^D) \in N(\mathcal{A}),$$

we get that $pa^D = a^D p = a^D$, $(pap)^{-1} = a^D$ in $p\mathcal{A}p$ and

$$((1-p)a(1-p))^k = (a(1-p))^k = 0,\ \ k = \mathrm{ind}(a).$$

(2)⇒(3) Put $a_C = pap$ and $a_N = (1-p)a(1-p)$. Then $a = a_C + a_N$, $a_C a_N = a_N a_C = 0$ and $a_N^s = 0$ for some $s \in \mathbb{N}$. Since a_C is invertible in $p\mathcal{A}p$, we have

$$(\lambda p - a_C)^{-1} = a_C^{-1}(\lambda a_C^{-1} - p)^{-1} = -\sum_{n=0}^{\infty} a_C^{-(n+1)}\lambda^n,\ |\lambda| < \|a_C^{-1}\|^{-1}$$

$$(\lambda(1-p) - a_C)^{-1} = \lambda^{-1}(1 - p - \lambda^{-1}a_N)^{-1} = \sum_{n=0}^{s-1} \frac{a_N^n}{\lambda^{n+1}},\ \ |\lambda| > 0$$

by Lemma 1.4.7 (3) and consequently,

$$\begin{aligned}(\lambda - a)^{-1} &= (\lambda p - a_C)^{-1} + (\lambda(1-p) - a_N)^{-1}\\ &= -\sum_{n=0}^{\infty} a_C^{-(n+1)}\lambda^n + \sum_{n=0}^{s-1}\frac{a_N^n}{\lambda^{n+1}},\ \ 0 < |\lambda| < \|a_C^{-1}\|^{-1}.\end{aligned}$$

This means that $\lambda = 0$ is a pole of $\lambda \mapsto (\lambda - a)^{-1}$, $\lambda \in \mathbb{C}\backslash\sigma(a)$.

(3)⇒(1) By Theorem 4.5.5, we obtain that $a \in Drg(\mathcal{A})$ and $aa^d aaa^d$ is invertible in $GL(aa^d\mathcal{A}aa^d)$ with the inverse a^d and $a^\pi aa^\pi \in QN(a^\pi\mathcal{A}a^\pi)$. Let $0 < |\lambda| < \|a^d\|^{-1}$. Then

$$(\lambda - a)^{-1} = -\sum_{n=0}^{\infty}(a^d)^{n+1}\lambda^n + \sum_{n=0}^{\infty}\frac{(a^\pi aa^\pi)^n}{\lambda^{n+1}},\ \ 0 < |\lambda| < \|a^d\|^{-1}. \quad (4.5.4)$$

Since $\lambda = 0$ is a pole of $(\lambda - a)^{-1}$, it follows that there is $k_0 \in \mathbb{N}$ such that $(a^\pi aa^\pi)^k = 0$, $\forall\, k \geq k_0$, that is, $aa^\pi \in N(\mathcal{A})$. Therefore, $a \in Dr(\mathcal{A})$. □

For the group inverse, we have more deep algebraic characterization as follows.

Theorem 4.5.9. *Let $\mathcal{A}$ be a unital algebra and $a \in Gi(\mathcal{A})$. Then the following conditions are equivalent:*

(1) *$a^{\#}$ exists.*
(2) *$aa^+ + a^+a - 1$ is invertible in A for some a^+.*
(3) *$a^2a^+ + 1 - aa^+ \in GL(\mathcal{A})$ for some a^+.*
(4) *$a^2a^+ + 1 - aa^+ \in GL(\mathcal{A})$ for any a^+.*

Proof. (1)⇒(2) Take $a^+ = a^\#$, then we get the assertion.

(2)⇒(3) Set $V = a^+a + aa^+ - 1 \in GL(A)$. Then

$$aV = a^2a^+, \ aa^+V = aa^+a^+a = Va^+a, \ a^+aV = Vaa^+.$$

Thus, $w = a^2a^+ + 1 - aa^+ = aV + 1 - aa^+$ and

$$\begin{aligned}(V^{-1}a^+ + 1 - aa^+)w &= V^{-1}a^+aV + (1 - aa^+)aV + V^{-1}a^+(1 - aa^+) \\ &\quad + 1 - aa^+ \\ &= aa^+ + 1 - aa^+ = 1, \\ w(V^{-1}a^+ + 1 - aa^+) &= aa^+ + aV(1 - aa^+) + (1 - aa^+)V^{-1}a^+ + 1 - aa^+ \\ &= 1 + a(1 - a^+a)V + V^{-1}(1 - a^+a)a^+ = 1,\end{aligned}$$

i.e., $a^2a^+ + 1 - aa^+ \in GL(\mathcal{A})$ with $(a^2a^+ + 1 - aa^+)^{-1} = V^{-1}a^+ + 1 - aa^+$.

(3)⇒(1) Set $p = aa^+$ and put $a_1 = pap$, $a_2 = pa(1-p)$. Then $a = a_1 + a_2$ (for $(1-p)a = 0$) and

$$w = a^2a^+ + 1 - aa^+ = (a_1 + a_2)p + 1 - p = a_1 + 1 - p \in GL(\mathcal{A}).$$

So $a_1 \in GL(p\mathcal{A}p)$ and consequently, $a^\# = a_1^{-1} + a_1^{-2}a_2$.

Since the implication (4)⇒(3) is clearly, we now prove that (1)⇒(4).

For any generalized inverse a^+ of a, put $w = a^2a^+ + 1 - aa^+$ and

$$\begin{aligned} b_1 &= (1 - a^\pi)a^+(1 - a^\pi), & b_2 &= (1 - a^\pi)a^+a^\pi \\ b_3 &= a^\pi a^+(1 - a^\pi), & b_4 &= a^\pi a^+ a^\pi. \end{aligned}$$

Then $b = b_1 + b_2 + b_3 + b_4$ and

$$aba = (ab_1 + ab_2)a = ab_1a = a. \tag{4.5.5}$$

Since

$$aa^\# = a^\#a = 1 - a^\pi, \ (1 - a^\pi)a(1 - a^\pi) = a, \ (1 - a^\pi)a^\#(1 - a^\pi) = a^\#,$$

it follows that $a \in GL((1 - a^\pi)\mathcal{A}(1 - a^\pi))$ with $a^{-1} = a^\#$. Thus, from (4.5.5), we have $b_1 = a^{-1} = a^\#$ in $(1 - a^\pi)\mathcal{A}(1 - a^\pi)$. Consequently,

$$\begin{aligned} w &= a(aa^+) + 1 - aa^+ = a(ab_1 + ab_2) + 1 - (ab_1 + ab_2) \\ &= a(1 - a^\pi) + a^2b_2 + 1 - (1 - a^\pi) - ab_2 \\ &= a + a^\pi + (a^2b_2 - ab_2) \end{aligned}$$

is invertible in $\mathcal{A}$ with $w^{-1} = a^\# + a^\pi - (a^2b_2 - ab_2)$. □

Using Theorem 4.5.8, we can characterize the existence of the group inverse or the Drazin inverse in the matrix algebra over a untial commutative Banach algebra easily.

Corollary 4.5.10 ([Huang (1993)]). *Let $\mathcal{A}$ be a unital commutative Banach and let M_A be its spectral space. Suppose that $\mathcal{A}$ is semi-simple and M_A is connected. Let $A \in M_n(\mathcal{A})$ $(n \geq 2)$.*

(1) $A^{\#}$ *exists iff* $\operatorname{rank}(\hat{A}(\varphi)) = \operatorname{rank}(\hat{A}(\varphi))^2 = m \leq n$, $\forall\, \varphi \in M_A$;
(2) $A \in Dr(M_n(\mathcal{A}))$ *iff there is* $k \in \mathbb{N}$ *such that* $\operatorname{rank}(\hat{A}(\varphi))^k = \operatorname{rank}(\hat{A}(\varphi))^{k+1} =$ *constant,* $\forall\, \varphi \in M_A$,

where $\hat{A}(\varphi) = (\hat{a_{ij}}(\varphi))_{n\times n}$, $\forall\, \varphi \in M_A$ *for* $A = (a_{ij})_{n\times n} \in M_n(\mathcal{A})$.

Proof. For each $\varphi \in M_{\mathcal{A}}$, let $\hat{A}(\varphi) \in M_n(\mathbb{C})$ have the Jordan form

$$\hat{A}(\varphi) = X_\varphi J_\varphi X_\varphi^{-1} = X_\varphi \begin{bmatrix} J_1^{(\varphi)} & \\ & J_0^{(\varphi)} \end{bmatrix} X_\varphi^{-1}, \tag{4.5.6}$$

where $J_0^{(\varphi)}$ and $J_1^{(\varphi)}$ are parts of J_φ corresponding to zero and nonzero eigenvalues (cf. P164, Theorem 8 of [Ben-Israel and Greville (2003)]). From (4.5.6), we deduce that $\hat{A}(\varphi)$ has the group inverse iff $J_0^{(\varphi)} = 0$ and iff $\operatorname{rank}(\hat{A}(\varphi)) = \operatorname{rank}(\hat{A}(\varphi))^2, \forall \varphi \in M_{\mathcal{A}}$.

Similarly, $\operatorname{ind}(\hat{A}(\varphi)) = k$, $\forall\, \varphi \in M_{\mathcal{A}}$ iff $\operatorname{rank}(\hat{A}(\varphi))^k = \operatorname{rank}(\hat{A}(\varphi))^{k+1}$, $\forall\, \varphi \in M_{\mathcal{A}}$.

Now we prove (1). If $A^{\#}$ exists, then $A \in Gi(M_n(A))$ and $(\hat{A}(\varphi))^{\#}$ exists, for every $\varphi \in M_{\mathcal{A}}$. Thus $\operatorname{rank}(\hat{A}(\varphi)) =$ constant, $\forall\, \varphi \in M_A$ by Theorem 2.5.8 and $\operatorname{rank}(\hat{A}(\varphi)) = \operatorname{rank}(\hat{A}(\varphi))^2 =$ constant, $\forall\, \varphi \in M_A$ by above argument.

Conversely, if $\operatorname{rank}(\hat{A}(\varphi)) = \operatorname{rank}(\hat{A}(\varphi))^2 =$ constant, $\forall\, \varphi \in M_{\mathcal{A}}$, then $A \in Gi(M_n(\mathcal{A}))$ by Theorem 2.5.8 and $(\hat{A}(\varphi))^{\#}$ exists for each $\varphi \in M_{\mathcal{A}}$. Let B be a generalized inverse of A in $M_n(\mathcal{A})$. Then

$$\hat{A}(\varphi)\hat{B}(\varphi)\hat{A}(\varphi) = \hat{B}(\varphi),\ \ \hat{B}(\varphi)\hat{A}(\varphi)\hat{B}(\varphi) = \hat{B}(\varphi).$$

It follows from Theorem 4.5.9 that

$$(\hat{A}(\varphi))^2\hat{B}(\varphi) + I_n - \hat{A}(\varphi)\hat{B}(\varphi) \in GL(M_n(\mathbb{C})),\ \forall\, \varphi \in M_A$$

and so that $A^2B + I_n - AB \in GL(M_n(\mathcal{A}))$ by Proposition 1.4.22. Therefore $A^{\#}$ exists by Theorem 4.5.9.

(2) Noting that from $\operatorname{rank}(\hat{A}(\varphi))^k = \operatorname{rank}(\hat{A}(\varphi))^{k+1}$, $\forall\, \varphi \in M_{\mathcal{A}}$, we get that $\operatorname{rank}(\hat{A}(\varphi))^k = \operatorname{rank}(\hat{A}(\varphi))^{2k}$, $\forall\, \varphi \in M_{\mathcal{A}}$, So by applying (1) to A^k, we can obtain the assertion. □

4.6 Stable perturbation of Drazin inverses

In this section, we assume that $\mathcal{A}$ is always a unital Banach. For $a \in \mathcal{A}\backslash\{0\}$, let L_a and R_a be linear operators in $B(\mathcal{A})$, given by $L_a(x) = ax$, $R_a(x) = xa$, $\forall\, x \in \mathcal{A}$ respectively. Let $a \in Dr(\mathcal{A})$ with $\mathrm{ind}(a) = 1$. Then

$$\mathrm{Ker}\,(L_a) = (1 - a^{\#}a)\mathcal{A} = \mathrm{Ker}\,(L_{a^{\#}}),\ \mathrm{Ker}\,(R_a) = \mathcal{A}(1 - a^{\#}a) = \mathrm{Ker}\,(R_{a^{\#}}).$$

Lemma 4.6.1. *Let $a \in \mathcal{A}\backslash\{0\}$. Then*

(1) $\sigma(L_a) \cup \sigma(R_a) = \sigma(a)$.
(2) $a \in Dr(\mathcal{A})$ *with* $\mathrm{ind}(a) = 1 \iff$ *both* $(L_a)^{\#}$ *and* $(R_a)^{\#}$ *exist.*

Proof. (1) Obviously, $\sigma(L_a) \subset \sigma(a)$ and $\sigma(R_a) \subset \sigma(a)$.

Now let $\lambda \in \sigma(a)$. If $\lambda \notin \sigma(L_a)$ and $\lambda \notin \sigma(R_a)$, then there are $S_1, S_2 \in B(\mathcal{A})$ such that $L_{(a-\lambda)}S_1 = I_{\mathcal{A}} = R_{(a-\lambda)}S_2$. Thus

$$(a - \lambda)S_1(1) = S_2(1)(a - \lambda) = 1$$

and hence $a - \lambda \in GL(\mathcal{A})$. But it contradicts to the assumption that $\lambda \in \sigma(a)$. Therefore, $\sigma(a) \subset \sigma(L_a) \cup \sigma(R_a)$.

(2) It is clear that $(L_a)^{\#}$ and $(R_a)^{\#}$ exist when $a^{\#}$ exists. Conversely, if $(L_a)^{\#}$ and $(R_a)^{\#}$ exist, then $0 \notin \sigma(a)$ or $0 \in \sigma(a)$ is an isolated point by (1) and Theorem 4.1.9 (4). So $a \in GL(\mathcal{A})$ or $a \in Drg(\mathcal{A})$ by Theorem 4.5.5 and consequently, $(L_a)^{\#} = L_{a^d}$. This implies that $a^d = a^{\#}$. □

Now using L_a, R_a and Lemma 4.6.1 and Theorem 2.4.7, we can generalized Lemma 4.3.2, Theorem 4.3.3, Lemma 4.3.6, Theorem 4.3.7, Theorem 4.3.8 and Corollary 4.3.9 to the case of the unital Banach algebra as follows.

Lemma 4.6.2. *Let $a \in Dr(\mathcal{A})$ with $\mathrm{ind}(a) = 1$ and let $\delta a \in \mathcal{A}$ with $1 + a^{\#}\delta a \in GL(\mathcal{A})$. Put $\bar{a} = a + \delta a \in \mathcal{A}$ and*

$$\Phi(a) = 1 + \delta a(1 - aa^{\#})\delta a[(1 + a^{\#}\delta a)^{-1}a^{\#}]^2.$$

If $\bar{a}\mathcal{A} \cap (1 - aa^{\#})\mathcal{A} = \{0\}$ and $\Phi(a) \in GL(\mathcal{A})$, then $\bar{a}^{\#}$ exists and

$$\bar{a}^{\#} = (1 + C(a))(D(a) + D^2(a)\delta a(1 - aa^{\#}))(1 - C(a)),$$

here $C(a) = (1 - aa^{\#})\delta a(1 + a^{\#}\delta a)^{-1}a^{\#}$, $D(a) = (1 + a^{\#}\delta a)^{-1}a^{\#}\Phi^{-1}(a)$.

Using Lemma 4.6.2, we can deduce the upper bounds of $\|\bar{a}^{\#}\|$ and $\dfrac{\|\bar{a}^{\#} - a^{\#}\|}{\|a^{\#}\|}$ respectively when $\bar{a}$ is a stable perturbation of a as follows.

Theorem 4.6.3. *Let $a \in Dr(\mathcal{A})$ with $\mathrm{ind}(a) = 1$ and let $a \in \mathcal{A}$ such that $\|a^{\#}\|\|\delta a\| < (1 + \|a^{\pi}\|)^{-1}$ ($a^{\pi} = 1 - aa^{\#}$). Suppose that $\bar{a}\bar{\mathcal{A}} \cap (1 - aa^{\#})\mathcal{A} = \{0\}$. Then $\bar{a}^{\#}$ exists and*

$$\|\bar{a}^{\#}\| \leq \frac{\|a^{\#}\|}{[1 - (1 + \|a^{\pi}\|)\|a^{\#}\|\|\delta a\|]^2},$$
$$\frac{\|\bar{a}^{\#} - a^{\#}\|}{\|a^{\#}\|} \leq \frac{(1 + 2\|a^{\pi}\|)\|a^{\#}\|\|\delta a\|}{[1 - (1 + \|a^{\pi}\|)\|a^{\#}\|\|\delta a\|]^2}.$$

Let a and $\bar{a} = a + \delta a \in \mathcal{A}$ and put $\delta a^j = (a + \delta a)^j - a^j$, $j \in \mathbb{N}$. Set $\epsilon_a = \dfrac{\|\delta a\|}{\|a\|}$. Then $\|\delta a^j\| < (2^j - 1)\|a\|^j \epsilon_a$, $j \in \mathbb{N}$ when $\epsilon_a < 1$ (cf. (4.3.4)).

Corollary 4.6.4. *Let $a \in Dr(\mathcal{A})$ with $\mathrm{ind}(a) = n$ and $\delta a \in \mathcal{A}$. Put $\bar{a} = a + \delta a$, $\kappa_D(a) = \|a\|\|a^D\|$. Suppose that $\kappa_D^n(a)\epsilon_a < \dfrac{1}{(2^n - 1)(1 + \|a^{\pi}\|)}$ and $\bar{a}^n\mathcal{A} \cap a^{\pi}\mathcal{A} = \{0\}$. Then $\bar{a} \in Dr(\mathcal{A})$ with $\mathrm{ind}(\bar{a}) \leq n$ and*

$$\|\bar{a}^D\| \leq \frac{2^{n-1}K_D^{n-1}(a)\|a^D\|}{[1 - (2^n - 1)(1 + \|a^{\pi}\|)\kappa_D^n(a)\epsilon_a]^2}$$
$$\frac{\|\bar{a}^D - a^D\|}{\|a^D\|} \leq \frac{2^{n-1}(2^n - 1)(1 + 2\|a^{\pi}\|)\kappa_D^{2n-1}(a)\epsilon_a}{[1 - (2^n - 1)(1 + \|a^{\pi}\|)\kappa_D^n(a)\epsilon_a]^2} + (2^{n-1} - 1)\kappa_D^{n-1}(a)\epsilon_a$$

Proof. We have $a^D = (a^n)^{\#}a^{n-1}$ and $aa^D = a^n(a^n)^{\#}$, $(a^n)^{\#} = (a^D)^n$. Noting that $\kappa_D(a) \geq \|aa^D\| \geq 1$, $\epsilon_a < 1$,

$$\|\delta a^n\|\|(a^n)^{\#}\| < (2^n - 1)\|a\|^n\|(a^D)^n\|\epsilon_a < \frac{1}{1 + \|a^{\pi}\|},$$

and $\bar{a}^n\mathcal{A} \cap (1 - a^n(a^n)^{\#})\mathcal{A} = \bar{a}^n\mathcal{A} \cap a^{\pi}\mathcal{A} = \{0\}$, it follows from Theorem 4.6.3 and Corollary 4.5.7 that $\bar{a} \in Dr(\mathcal{A})$ with $\mathrm{ind}(\bar{a}) \leq n$ and

$$\begin{aligned}\|\bar{a}^D\| &= \|(\bar{a}^n)^{\#}(\bar{a})^{n-1}\| \leq \|(\bar{a}^n)^{\#}\|\|a\|^{n-1}(1 + \epsilon_a)^{n-1} \\ &< 2^{n-1}\|a\|^{n-1}\|(\bar{a}^n)^{\#}\|, \\ \|\bar{a}^D - a^D\| &= \|(\bar{a}^n)^{\#}(\bar{a})^{n-1} - (a^n)^{\#}a^{n-1}\| \leq \|(\bar{a}^n)^{\#} - (a^n)^{\#}\|\|\bar{a}\|^{n-1} \\ &\quad + \|(a^n)^{\#}\|\|\delta a^{n-1}\| \\ &< 2^{n-1}\|a\|^{n-1}\|(\bar{a}^n)^{\#} - (a^n)^{\#}\| + (2^{n-1} - 1)\|a^D\|\kappa_D^{n-1}(a)\epsilon_a.\end{aligned}$$

Finally, applying Theorem 4.6.3 to $(a^n)^{\#}$ and $(\bar{a}^n)^{\#}$, we obtain the assertion. □

Lemma 4.6.5. *Let a, $b \in \mathcal{A}$ such that $a^{\#}$ and $b^{\#}$ exist. Then the following statements are equivalent:*

(1) $\mathcal{A} = b\mathcal{A} \dotplus (1 - aa^{\#})\mathcal{A} = a\mathcal{A} \dotplus (1 - bb^{\#})\mathcal{A}$, $\mathcal{A} = \mathcal{A}b \dotplus \mathcal{A}(1 - aa^{\#}) = \mathcal{A}a \dotplus \mathcal{A}(1 - bb^{\#})$.
(2) $K(a, b) = aa^{\#} + bb^{\#} - 1 \in GL(\mathcal{A})$.
(3) $b\mathcal{A} \cap (1 - aa^{\#})\mathcal{A} = \{0\}$ *and* $1 + a^{\#}(b - a) \in GL(\mathcal{A})$.

Let $a \in Dr(\mathcal{A})$ with $\mathrm{ind}(a) = 1$. Then for any $b \in \mathcal{A}$, we have

$$b = aa^{\#}baa^{\#} + aa^{\#}b(1 - aa^{\#}) + (1 - aa^{\#})baa^{\#} + (1 - aa^{\#})b(1 - aa^{\#}).$$

Put

$$\begin{aligned} b_1 &= aa^{\#}baa^{\#} \in aa^{\#}\mathcal{A}aa^{\#}, \\ b_2 &= aa^{\#}b(1 - aa^{\#}) \in aa^{\#}\mathcal{A}(1 - aa^{\#}), \\ b_3 &= (1 - aa^{\#})baa^{\#} \in (1 - aa^{\#})\mathcal{A}aa^{\#}, \\ b_4 &= (1 - aa^{\#})b(1 - aa^{\#}) \in (1 - aa^{\#})\mathcal{A}(1 - aa^{\#}). \end{aligned}$$

Then b can be expressed as the form $b = \begin{bmatrix} b_1 & b_2 \\ b_3 & b_4 \end{bmatrix}$.

Lemma 4.6.6. *Let* $a, b \in Dr(\mathcal{A})$ *with* $\mathrm{ind}(a) = \mathrm{ind}(b) = 1$. *Suppose that* $K(a, b)$ *is invertible in* $\mathcal{A}$. *Then*

(1) $1 + a^{\#}(b - a) \in GL(\mathcal{A})$ *and*

$$1 + (1 + a^{\#}(b - a))^{-1}a^{\#}(b - a)(1 - aa^{\#})(b - a)(1 + a^{\#}(b - a))^{-1}a^{\#}$$

is also invertible in $\mathcal{A}$.
(2) b *has the form* $b = \begin{bmatrix} b_1 & b_1 s_2 \\ s_1 b_1 & s_1 b_1 s_2 \end{bmatrix}$, *where* $b_1 \in GL(aa^{\#}\mathcal{A}aa^{\#})$, $s_2 \in aa^{\#}\mathcal{A}(1 - aa^{\#})$, $s_1 \in (1 - aa^{\#})\mathcal{A}aa^{\#}$ *with* $aa^{\#} + s_2 s_1 \in GL(aa^{\#}\mathcal{A}aa^{\#})$.
(3) $b^{\#}$ *has the form*

$$b^{\#} = \begin{bmatrix} s^{-1}b_1^{-1}s^{-1} & s^{-1}b_1^{-1}s^{-1}s_2 \\ s_1 s^{-1}b_1^{-1}s^{-1} & s_1 s^{-1}b_1^{-1}s^{-1}s_2 \end{bmatrix},$$

$b^{\pi} = 1 - bb^{\#}$ *has the form*

$$b^{\pi} = \begin{bmatrix} aa^{\#} - s^{-1} & -s^{-1}s_2 \\ -s_1 s^{-1} & a^{\pi} - s_1 s^{-1} s_2 \end{bmatrix},$$

where $s = aa^{\#} + s_2 s_1$.

Theorem 4.6.7. *Let* $a, b \in Dr(X)$ *with* $s = \mathrm{ind}(a)$ *and* $t = \mathrm{ind}(b)$. *Suppose that* $aa^D + bb^D - 1$ *is invertible in* $\mathcal{A}$. *Then*

(1) $1 + (a^D)^l(b^k - A^l)$ *is invertible in* $\mathcal{A}$, $b^k\mathcal{A} \cap a^{\pi}\mathcal{A} = \{0\}$, $\forall\, k \geq t$, $l \geq 1$.

(2) $w_{k,l} = (1 + y_{k,l}z_{k,l})(1 - z_{k,l})$ *is invertible in* $\mathcal{A}$*, where* $E_{k,l} = b^k - a^l$ *and*

$$y_{k,l} = (1 + (a^D)^l E_{k,l})^{-1}(a^D)^l E_{k,l}a^\pi$$
$$z_{k,l} = a^\pi E_{k,l}(a^D)^l(1 + E_{k,l}(a^D)^l)^{-1}.$$

(3) $b^D = w_{k,l}^{-1}(1 + (a^D)^{l+1}E_{k+1,l+1})^{-1}a^D(1 + (a^D)^l E_{k,l})w_{k,l}$, $k \geq t$, $l \geq 1$.

(4) $b^\pi = w_{k,l}^{-1}(1 + (a^D)^l E_{k,l})^{-1}a^\pi(1 + (a^D)^l E_{k,l})w_{k,l}$, $k \geq t$, $l \geq 1$.

Corollary 4.6.8. *Let* $a \in \mathcal{A}$ *and* $b \in \mathcal{A}$*. Suppose that there are* k_1*,* k_2*,* l_1*,* $l_2 \in \mathbb{N}$ *with* $k_2 < k_1$ *such that the follow conditions are satisfied.*

(1) $1 + (a^D)^{l_1}E_{k_1,l_1}$ *and* $1 + y_{k_1,l_1}z_{k_1,l_1}$ *are invertible in* $\mathcal{A}$.

(2) $b^{k_1}(1 + (a^D)^{l_1}E_{k_1,l_1})^{-1}a^\pi = 0$.

(3) $1 + (a^D)^{l_2}E_{k_2,l_2}$ *is not invertible in* $\mathcal{A}$ *or* $1 + (a^D)^{l_2}E_{k_2,l_2}$ *is invertible in* $\mathcal{A}$*, but* $b^{k_2}(1 + (a^D)^{l_2}E_{k_2,l_2})^{-1}a^\pi \neq 0$.

Then $b \in Dr(\mathcal{A})$ *with* $k_2 < \mathrm{ind}(b) \leq k_1$.

For the perturbation of generalized Drazin inverses, we have following result:

Proposition 4.6.9. *Let* $a \in Drg(\mathcal{A})$ *and* $\delta a \in \mathcal{A}$ *such that* $\bar{a} = a + \delta a \in Drg(\mathcal{A})$*. Suppose that* $\|(a + a^\pi)^{-1}\|(\|\delta a\| + \|\bar{a}^\pi - a^\pi\|) < 1$*. Then*

$$\|\bar{a}^d\| \leq \frac{\|a^d\| + \|(a + a^\pi)^{-1}\|\|\bar{a}^\pi - a^\pi\|}{1 - \|(a + a^\pi)^{-1}\|(\|\delta a\| + \|\bar{a}^\pi - a^\pi\|)},$$
$$\|\bar{a}^d - a^d\| \leq \frac{\|(a + a^\pi)^{-1}\|(\|\delta a\|\|a^d\| + (1 + \|a^d\|)\|\bar{a}^\pi - a^\pi\|)}{1 - \|(a + a^\pi)^{-1}\|(\|\delta a\| + \|\bar{a}^\pi - a^\pi\|)}.$$

Proof. Since $a + a^\pi$ and $\bar{a} + \bar{a}^\pi$ are invertible and

$$a^d = (a + a^\pi)^{-1}(1 - a^\pi),\ \bar{a}^d = (\bar{a} + \bar{a}^\pi)^{-1}(1 - \bar{a}^\pi)$$

by Corollary 4.5.6 (3), we have

$$\begin{aligned}\bar{a}^d &= \big[1+(a+a^\pi)^{-1}(\delta a+\bar{a}^\pi-a^\pi)\big]^{-1}(a+a^\pi)^{-1}(1-\bar{a}^\pi)\\ &= \big[1+(a+a^\pi)^{-1}(\delta a+\bar{a}^\pi-a^\pi)\big]^{-1}\big[a^d+(a+a^\pi)^{-1}(a^\pi-\bar{a}^\pi)\big]. \qquad (4.6.1)\end{aligned}$$

So $\|\bar{a}^d\| \leq \dfrac{\|a^d\| + \|(a + a^\pi)^{-1}\|\|\bar{a}^\pi - a^\pi\|}{1 - \|(a + a^\pi)^{-1}\|(\|\delta a\| + \|\bar{a}^\pi - a^\pi\|)}$ and

$$\begin{aligned}\|\bar{a}^d - a^d\| &= \Big\|\Big(\big[1 + (a + a^\pi)^{-1}(\delta a + \bar{a}^\pi - a^\pi)\big]^{-1} - 1\Big)a^d\\ &\quad + \big[1 + (a + a^\pi)^{-1}(\delta a + \bar{a}^\pi - a^\pi)\big]^{-1}(a + a^\pi)^{-1}(a^\pi - \bar{a}^\pi)\Big\|\\ &\leq \frac{\|(a + a^\pi)^{-1}\|(\|\delta a\|\|a^d\| + (1 + \|a^d\|)\|\bar{a}^\pi - a^\pi\|)}{1 - \|(a + a^\pi)^{-1}\|(\|\delta a\| + \|\bar{a}^\pi - a^\pi\|)}.\end{aligned}$$

□

Remark 4.6.10. (1) Assume that $a \in Drg(\mathcal{A})$ and $\delta a \in \mathcal{A}$ satisfy following condition:

$$a^\pi \delta a = \delta a a^\pi = 0 \text{ and } \|a^d \delta a\| < 1. \tag{*}$$

Put $a^\pi = 1 - aa^d$ and

$$a_C = aaa^d \in GL(aa^d \mathcal{A} aa^d), \ a_N = a(1 - aa^d) \in QN(a^\pi \mathcal{A} a^\pi).$$

Then $\bar{a} = a + \delta a = (a_C + \delta a) + a_N$. When $\|a^d \delta a\| < 1$, $(a_C + \delta a)^{-1} = (aa^d + a^d \delta a)^{-1} a^d$ and $\bar{a}^d = (1 + a^d \delta a)^{-1} a^d$, $\bar{a}\bar{a}^d = aa^d$. In this case, Proposition 4.6.9 can be improved as

$$\|\bar{a}^d\| \le \frac{\|a^d\|}{1 - \|a^d \delta a\|}, \ \|\bar{a}^d - a^d\| \le \frac{\|a^d\|^2 \|\delta a\|}{1 - \|a^d \delta a\|}. \tag{4.6.2}$$

(4.6.2) appears in [Rakočević and Wei (2001)].

(2) Consider following condition which is slightly weaker than (*).

$$\bar{a} \in Drg(\mathcal{A}) \text{ and } \bar{a}^\pi = a^\pi. \tag{**}$$

From (**) we have $\bar{a} a^\pi = a^\pi \bar{a}$, i.e., $\delta a a^\pi = a^\pi \delta a$ and $a + \delta a + a^\pi \in GL(\mathcal{A})$. Note that

$$\begin{aligned} 1 + (a + a^\pi)^{-1} \delta a &= 1 - a^\pi + (a + a^\pi)^{-1} (1 - a^\pi) \delta a (1 - a^\pi) \\ &\quad + a^\pi + (a + a^\pi)^{-1} a^\pi \delta a a^\pi \\ &= 1 - a^\pi + a^d \delta a + a^\pi + (a + a^\pi)^{-1} a^\pi \delta a a^\pi. \end{aligned}$$

So $a + \delta a + a^\pi \in GL(\mathcal{A})$ implies that $1 - a^\pi + a^d \delta a (1 - a^\pi)$ is invertible in $(1 - a^\pi)\mathcal{A}(1 - a^\pi)$. Hence, $1 + \delta a a^d$ and $1 + a^d \delta a$ are invertible in $\mathcal{A}$. Put $b = (1 + a^d \delta a)^{-1} a^d$. Then

$$\begin{aligned} b(a + \delta a) &= (1 + a^d \delta a)^{-1} (1 + a^d \delta a + a^d a - 1) = a^d a = aa^d \\ (a + \delta a) b &= (aa^d - 1 + 1 + \delta a a^d)(1 + \delta a a^d)^{-1} = aa^d, \end{aligned}$$

since $a^\pi (1 + \delta a a^d) = (1 + a^d \delta a) a^\pi = a^\pi$. Moreover,

$$\begin{aligned} b(a + \delta a) b &= aa^d (1 + a^d \delta a)^{-1} a^d = aa^d a^d (1 + \delta a^d)^{-1} = b \\ (a + \delta a)(1 - (a + \delta a) b) &= \bar{a}(1 - \bar{a}\bar{a}^d) \in QN(\mathcal{A}). \end{aligned}$$

Thus, $\bar{a}^d = b = (1 + a^d \delta a)^{-1} a^d$ and consequently,

$$\|\bar{a}^d\| \le \frac{\|a^d\|}{1 - \|a^d \delta a\|}, \ \|\bar{a}^d - a^d\| \le \frac{\|a^d\|^2 \|\delta a\|}{1 - \|a^d \delta a\|} \tag{4.6.3}$$

(cf. [Castro-González *et al.* (2002b)]).

Finally, we consider the continuity of Drazin inverses and generalized Drazin inverses in a unital Banach algebra.

Proposition 4.6.11. *Let $\{a_n\} \subset Drg(\mathcal{A})$ and $a \in Drg(\mathcal{A})$. Suppose that $\lim_{n\to\infty} \|a_n - a\| = 0$. Then the following conditions are equivalent:*

(1) $\lim_{n\to\infty} \|a_n^d - a^d\| = 0$.
(2) $\sup_{n\in\mathbb{N}} \|a_n^d\| < +\infty$.
(3) $\lim_{n\to\infty} \|a_n^\pi - a^\pi\| = 0$

Proof. Since the implication (1)⇒(2) is obvious and the implication (3)⇒(1) comes directly from Proposition 4.6.9, we have to prove (2)⇒(3).

Set $K = \sup_{n\geq 0} \|a_n^d\| < +\infty$, where $a_0 = a$. Put $a_C^{(n)} = a_n(1 - a_n^\pi)$ and $a_N^{(n)} = a_n a_n^\pi$. Then $a_C^{(n)}$ is invertible in $(1 - a_n^\pi)\mathcal{A}(1 - a_n^\pi)$ with the inverse a_n^d and $a_N^{(n)}$ is quasi–nilpotent in $a_n^\pi \mathcal{A} a_n^\pi$. Then for any $0 < |\lambda| < 1/K$, $\lambda a_n^\pi - a_N^{(n)}$ is invertible in $a_n^\pi \mathcal{A} a_n^\pi$, and

$$\lambda(1 - a_n^\pi) - a_C^{(n)} = -a_C^{(n)}(1 - a_n^\pi - \lambda a_n^d)$$

is invertible in $(1 - a_n^\pi)\mathcal{A}(1 - a_n^\pi)$ with the inverse $-(1 - a_n^\pi - \lambda a_n^d)^{-1} a_n^d$ since $\|\lambda a_n^d\| < 1$, $n \geq 0$. Thus, for every $n \geq 0$,

$$(\lambda - a_n)^{-1} = -\sum_{k=0}^{\infty} (a_n^d)^{k+1} \lambda^k + \sum_{k=0}^{\infty} (a_n a_n^\pi)^k \lambda^{-k-1}, \quad 0 < |\lambda| < 1/K.$$

Choose $r \in (0, K^{-1})$ and put $M = \max_{|\lambda|=r} \|(\lambda - a)^{-1}\|$. Pick $n_0 \in \mathbb{N}$ such that $\|a_n - a\| < \dfrac{1}{2M}$, $\forall\, n \geq n_0$. Then for any $n \geq n_0$ and $|\lambda| = r$,

$$\|(\lambda - a_n) - (\lambda - a)\| = \|a_n - a\| < \frac{1}{2\|(\lambda - a)^{-1}\|}.$$

Thus

$$\|(\lambda - a_n)^{-1} - (\lambda - a)^{-1}\| \leq 2M^2 \|a_n - a\|, \; \forall\, n \geq n_0, \; |\lambda| = r$$

by Corollary 1.4.8 and hence

$$\begin{aligned} \|a_n^\pi - a^\pi\| &= \left\| \frac{1}{2\pi i} \oint_{|\lambda|=r} [(\lambda - a_n)^{-1} - (\lambda - a)^{-1}] \mathrm{d}\,\lambda \right\| \\ &\leq \frac{1}{2\pi} 2M^2 \|a_n - a\| (2\pi r) \to 0 \;\; (\text{as } n \to \infty). \end{aligned}$$

□

Combining Proposition 4.6.11 with Corollary 4.6.4 and Theorem 4.6.7, we have

Corollary 4.6.12. *Let* $\{a_n\} \subset Dr(\mathcal{A})$ *and* $a \in Dr(\mathcal{A})$ *with* $s = \mathrm{ind}(A)$ *such that* $\lim\limits_{n\to\infty} \|a_n - a\| = 0$. *Suppose that* $t = \sup\limits_{n\in\mathbb{N}} \mathrm{ind}(a_n) < \infty$. *Then the following conditions are equivalent:*

(1) $a_n^D \xrightarrow{\|\cdot\|} a^D$ *as* $n \to \infty$.
(2) $\sup\limits_{n\in\mathbb{N}} \|a_n^D\| < +\infty$.
(3) $a_n^\pi \xrightarrow{\|\cdot\|} a^\pi$ *as* $n \to \infty$.
(4) $K(a, a_n)$ *is invertible in* $\mathcal{A}$ *for* n *large enough.*
(5) $a_n^l \mathcal{A} \cap (1 - aa^D)\mathcal{A} = \{0\}$ *for* n *large enough, where* $l = \max\{s, t\}$.

4.7 Representations for the Drazin inverses of sums, products and block matrices

Up to now, there are many results about the representations of $(a+b)^D$ (or $(a+b)^d$) and $\begin{bmatrix} a & b \\ c & d \end{bmatrix}^D$ (or $\begin{bmatrix} a & b \\ c & d \end{bmatrix}^d$) under various conditions, where a, b, c, d are in a Banach algebra or are some bounded operators on a Banach space. In this section, we present some basic and important results about the representations of $(ab)^t$, $(a+b)^t$ and $\begin{bmatrix} a & b \\ 0 & c \end{bmatrix}^t$, $(t = D,\ d)$ for certain a, b, c in a unital Banach algebra $\mathcal{A}$. In the following, we always assume that $\mathcal{A}$ is a unital Banach algebra.

Lemma 4.7.1. *Let* $a,\ b \in \mathcal{A}\backslash\{0\}$ *with* $ab = ba$.

(1) *If* $a,\ b \in Dr(\mathcal{A})$, *then* $ab \in Dr(\mathcal{A})$ *and* $(ab)^D = b^D a^D$ *with* $\mathrm{ind}(ab) \leq \max\{\mathrm{ind}(a), \mathrm{ind}(b)\}$.
(2) *If* $a,\ b \in Drg(\mathcal{A})$, *then* $ab \in Drg(\mathcal{A})$ *and* $(ab)^d = b^d a^d$.

Proof. (1) From $ab = ba$, we get that by Corollary 4.5.6,
$$a^D b = ba^D,\ ab^D = b^D a \text{ and } a^D b^D = b^D a^D.$$

Thus,
$$\begin{aligned} b^D a^D abb^D a^D &= b^D bb^D a^D aa^D = b^D a^D \\ ab(1 - abb^D a^D) &= ab(1 - b^D b) + abb^D b(1 - a^D a) \\ &= ab(1 - b^D b) + b^2 b^D a(1 - a^D a). \end{aligned}$$

Let $k_1 = \text{ind}(a)$, $k_2 = \text{ind}(b)$ and set $k = \max\{k_1, k_2\}$. Note that

$$ab(1 - b^D b)b^2 b^D a(1 - a^D a) = 0, \; b^{k_2}(1 - b^D b) = a^{k_1}(1 - a^D a) = 0.$$

So we have

$$(ab(1 - abb^D a^D))^k = a^k[b(1 - bb^D)]^k + b^2 b^D [a(1 - aa^D)]^k = 0.$$

Therefore, $(ab)^D = b^D a^D$ with $\text{ind}(ab) \leq \max\{\text{ind}(a), \text{ind}(b)\}$.

(2) When $a, b \in Drg(\mathcal{A})$, we have $ab^d = b^d a$, $a^d b = ba^d$, $a^d b^d = b^d a^d$. Thus, $b^d a^d (ab) = (ab) b^d a^d$, $b^d a^d (ab) b^d a^d = b^d a^d$. Since $a(1 - a^d a)$, $b(1 - b^d b) \in QN(\mathcal{A})$ and

$$ab[1 - (ab)b^d a^d] = ab(1 - b^d b) + b^2 b^d a(1 - a^d a),$$

it follows from Corollary 1.4.12 (2) that $ab[1 - (ab)b^d a^d] \in QN(\mathcal{A})$. Thus, $ab \in Drg(A)$ and $(ab)^d = b^d a^d$. □

Lemma 4.7.2. *Let $a, b \in \mathcal{A}\backslash\{0\}$ with $ab = ba = 0$.*

(1) *If $a, b \in Dr(\mathcal{A})$, then $(a + b)^D = a^D + b^D$.*
(2) *If $a, b \in Drg(\mathcal{A})$, then $(a + b)^d = a^d + b^d$.*

Proof. (1) $ab = ba = 0$ implies that $a^D b = ba^D = 0$ and $ab^D = b^D a = 0$. Thus

$$(a^D + b^D)(a + b) = (a + b)(a^D + b^D) = aa^D + bb^D.$$
$$(a^D + b^D)(a + b)(a^D + b^D) = (a^D a + b^D b)(a^D + b^D) = a^D + b^D.$$

Let $k = \max\{\text{ind}(a), \text{ind}(b)\}$. Since

$$\begin{aligned}(a + b)(1 - (a + b)(a^D + b^D)) &= (a + b)(1 - aa^D - bb^D) \\ &= a(1 - aa^D) + b(1 - bb^D),\end{aligned}$$

it follows that $(a + b)^k[1 - (a + b)(a^D + b^D)] = 0$. Thus $a + b \in Dr(\mathcal{A})$ and $(a + b)^D = a^D + b^D$.

(2) From $ab = ba = 0$. we get that $a^d b = ba^d = 0$ and $ab^d = b^d a = 0$. Thus.

$$(a^d + b^d)(a + b) = (a + b)(a^d + b^d) = a^d a + b^d b,$$
$$(a^d + b^d)(a + b)(a^d + b^d) = (a^d + b^d)(aa^d + bb^d) = a^d + b^d,$$
$$(a + b)[1 - (a + b)(a^d + b^d)] = a(1 - aa^d) + b(1 - bb^d) \in QN(\mathcal{A}),$$

that is, $a + b \in Dr(A)$ and $(a + b)^d = a^d + b^d$. □

Proposition 4.7.3. *Let $a, b \in \mathcal{A}\backslash\{0\}$ with $ab = ba$.*

(1) *If* $a,\ b \in Dr(\mathcal{A})$ *with* $1 + a^D b \in Dr(\mathcal{A})$, *then* $a + b \in Dr(\mathcal{A})$ *and*

$$(a+b)^D = (1+a^D b)^D a^D + a^\pi \sum_{i=0}^{k-1} (-1)^i (b^D)^{i+1} a^i$$
$$= (1+a^D b)^D a^D + a^\pi b(1+aa^\pi b)^{-1},$$

where $k = \mathrm{ind}(a)$.

(2) *If* $a,\ b \in Drg(\mathcal{A})$ *with* $1 + a^d b \in Drg(\mathcal{A})$, *then* $a + b \in Drg(\mathcal{A})$ *and*

$$(a+b)^d = (1+a^d b)^d a^d + a^\pi b^d (1+aa^\pi b^d)^{-1}$$
$$= (1+a^d b)^d a^d + a^\pi \sum_{i=0}^{\infty} (-1)^i (b^d)^{i+1} (aa^\pi)^i.$$

Proof. Put $a_1 = a(1-a^\pi)$, $a_2 = aa^\pi$. Then $a_1 \in GL((1-a^\pi)\mathcal{A}(1-a^\pi))$ with $a_1^{-1} = a^D$ and $a_2 \in QN(a^\pi \mathcal{A} a^\pi)$. From $ab = ba$, we have

$$ab^D = b^D a, \ a^D b = ba^D \text{ and } a^D b^D = b^D a^D.$$

Thus, $a^\pi b = ba^\pi$ and $a^\pi b^D = b^D a^\pi$. Put $b_1 = b(1-a^\pi)$ and $b_2 = ba^\pi$. Then $b_1 \in Dr((1-a^\pi)\mathcal{A}(1-a^\pi))$, $b_2 \in Dr(a^\pi \mathcal{A} a^\pi)$ by Lemma 4.7.1 (1). Since

$$1 + a^D b = 1 - a^\pi + a_1^{-1} b_1 + a^\pi \in Dr(\mathcal{A})$$

implies that $1 - a^\pi + a_1^{-1} b_1 \in Dr((1-a^\pi)\mathcal{A}(1-a^\pi))$, we have

$$a_1 + b_1 = a_1(1 - a^\pi + a_1^{-1} b_1) \in Dr((1-a^\pi)\mathcal{A}(1-a^\pi))$$
$$(a_1+b_1)^D = (1 - a^\pi + a_1^{-1} b_1)^D a_1^{-1} = (1-a^\pi)(1+a^D b)^D a^D$$
$$= (1+a^D b)^D a^D$$

by Lemma 4.7.1 (1).

Now we check that $(a_2+b_2)^D = b_2^D(a^\pi + a_2 b_2^D)^{-1}$. Since $a_2 \in N(a^\pi \mathcal{A} a^\pi)$ and $b_2^D a_2 = a_2 b_2^D$, it follows that $a_2 b_2^D \in N(a^\pi \mathcal{A} a^\pi)$ and hence $a^\pi + a_2 b_2^D \in GL(a^\pi \mathcal{A} a^\pi)$. Clearly, $a_2 + b_2$ commutes with $(a^\pi + a_2 b_2^D)^{-1} b_2^D$ and

$$(a^\pi + a_2 b_2^D)^{-1} b_2^D (a_2+b_2)(a^\pi + a_2 b_2^D)^{-1} b_2^D = (a^\pi + a_2 b_2^D)^{-1} b_2^D.$$

Put $m = \mathrm{ind}(a) + \mathrm{ind}(b_2)$. Then

$$(a_2+b_2)^m (a^\pi - b_2 b_2^D) = [a_2(a^\pi - b_2 b_2^D) + b_2(a^\pi - b_2 b_2^D)]^m$$
$$= \sum_{i=0}^{m} \binom{i}{k} a_2^i [b_2(a^\pi - b_2 b_2^D)]^{m-i} = 0. \qquad (4.7.1)$$

Since $a_2 b_2 = b_2 a_2$ and $a_2 b_2^D = b_2^D a_2$, we have by (4.7.1),

$$(a_2+b_2)^{m+1} b_2^D (a^\pi + a_2 b_2^D)^{-1} = (a_2+b_2)^m (a_2 b_2^D + b_2 b_2^D)(a_2^\pi + a_2 b_2^D)^{-1}$$
$$= (a_2+b_2)^m - (a_2+b_2)^m (a^\pi - b_2 b_2^D)(a^\pi + a_2 b_2^D)^{-1}$$
$$= (a_2+b_2)^m$$

and hence $(a_2+b_2)^D = b_2^D(a^\pi + a_2b_2^D)^{-1} = a^\pi b^D(1+aa^\pi b^D)^{-1}$.

Noting that $(a_1+b_1)(a_2+b_2) = (a_2+b_2)(a_1+b_1) = 0$, we get that by Lemma 4.7.2 (1),

$$\begin{aligned}(a+b)^D &= (a_1+b_1)^D + (a_2+b_2)^D = (1+a^Db)^Da^D + a^\pi b^D(1+aa^\pi b^D)^{-1}\\ &= (1+a^Db)^Da^D + a^\pi\sum_{i=0}^{k-1}(-1)^i(b^D)^{i+1}(aa^\pi)^i.\end{aligned}$$

(2) Put $a_1 = a(1-a^\pi)$, $a_2 = aa^\pi$, $b_1 = b(1-a^\pi)$ and $b_2 = ba^\pi$. Then

$$\begin{aligned}&a_1 \in GL((1-a^\pi)\mathcal{A}(1-a^\pi)), \quad a_2 \in QN(a^\pi\mathcal{A}a^\pi),\\ &b_1 \in Drg((1-a^\pi)\mathcal{A}(1-a^\pi)), \quad b_2 \in Drg(a^\pi\mathcal{A}a^\pi)\end{aligned}$$

by Lemma 4.7.1 (2).

Similar to the proof of (1), we have $a_2b_2^d \in QN(a^\pi\mathcal{A}a^\pi)$, a_2+b_2 commutes with $(a^\pi + a_2b_2^d)^{-1}b_2^d$ and $(a_1+b_1)^d = (1+a^db)a^d$,

$$[(a^\pi + a_2b_2^d)^{-1}b_2^d]^2(a_2+b_2) = (a^\pi + a_2b_2^d)^{-1}b_2^d.$$

Note that

$$\begin{aligned}&(a_2+b_2)[a^\pi - (a_2+b_2)(a^\pi + a_2b_2^d)^{-1}b_2^d]\\ &\quad = a_2(a^\pi - b_2b_2^d)(a^\pi + a_2b_2^d)^{-1} + b_2(a^\pi - b_2b_2^d)(a^\pi + a_2b_2^d)^{-1}\end{aligned}$$

and $a_2(a^\pi - b_2b_2^d)(a^\pi + a_2b_2^d)^{-1}$, $b_2(a^\pi - b_2b_2^d)(a^\pi + a_2b_2^d)^{-1}$ are quasi-nilpotent. So $(a_2+b_2)\big[a^\pi - (a_2+b_2)(a^\pi + a_2b_2^d)^{-1}b_2^d\big] \in QN(a^\pi\mathcal{A}a^\pi)$ by Corollary 1.4.12 (2) and consequently,

$$\begin{aligned}(a+b)^d &= (a_1+b_1)^d + (a_2+b_2)^d = (1+a^db)^da^d + a^\pi b^d(1+aa^\pi b^d)^{-1}\\ &= (1+a^db)^da^d + a^\pi\sum_{i=0}^{\infty}(-1)^i(b^d)^{i+1}(aa^\pi)^i.\end{aligned}$$

□

Let $a, b \in Dr(\mathcal{A})$ and $c \in \mathcal{A}$. Put $\hat{M} = \begin{bmatrix} a & c\\ 0 & b\end{bmatrix}$. We now compute $\hat{M}^D$ by means of Corollary 4.6.8. Set $M = \begin{bmatrix} a & 0\\ 0 & b\end{bmatrix}$ Then for any $k, l \in N$,

$$\hat{M}^k - M^l = \begin{bmatrix} a^k - a^l & x\\ 0 & b^k - b^l\end{bmatrix}, \ (M^D)^l = \begin{bmatrix} (a^D)^l & 0\\ 0 & (b^D)^l\end{bmatrix}$$

$$I_2 + (M^D)^l(\hat{M}^k - M^l) = \begin{bmatrix} 1 - a^Da + (a^D)^la^k & (a^D)^lx_k\\ 0 & 1 - b^Db + (b^D)^lb^k\end{bmatrix},$$

where $x_k = \sum\limits_{i=0}^{k-1} a^i c b^{k-1-i}$, $I_2 = \begin{bmatrix} 1 & \\ & 1 \end{bmatrix}$.

Note that $1 - a^D a + (a^D)^l a^k$, $1 - b^D b + (b^D)^l b^k$ are invertible in $\mathcal{A}$ with

$$(1 - a^D a + (a^D)^l a^k)^{-1} = 1 - a^D a + (a^D)^k a^l,$$
$$(1 - b^D b + (b^D)^l b^k)^{-1} = 1 - b^D b + (b^D)^k b^l.$$

So $I_2 + (M^D)^l(\hat{M}^k - M^l)$ is invertible in $M_2(\mathcal{A})$ and

$$[I_2 + (M^D)^l(\hat{M}^k - M^l)]^{-1}(M^D)^l = \begin{bmatrix} (a^D)^k & -(a^D)^k x_k (b^D)^k \\ 0 & (b^D)^k \end{bmatrix}.$$

Moreover, for k, $l \in \mathbb{N}$,

$$\begin{aligned} y_{k,l} &= [I_2 + (M^D)^l(\hat{M}^k - M^l)]^{-1}(M^D)^l(\hat{M}^k - M^l)(I_2 - MM^D) \\ &= \begin{bmatrix} 0 & (a^D)^k x_k (1 - bb^D) \\ 0 & 0 \end{bmatrix}, \\ z_{k,l} &= (I_2 - MM^D)(\hat{M}^k - M^l)(M^D)^l[I_2 + (\hat{M}^k - M^l)(M^D)^l]^{-1} \\ &= \begin{bmatrix} 0 & (1 - a^p a) x_k (b^D)^k \\ 0 & 0 \end{bmatrix}, \\ w_{k,l} &= (I_2 + y_{k,l} z_{k,l})(I_2 - z_{k,l}) = \begin{bmatrix} 1 & -(1 - a^D a) x_k (b^D)^k \\ 0 & 1 \end{bmatrix} \end{aligned}$$

and

$$\hat{M}^k[I_2 + (M^D)^l(\hat{M}^k - M^l)]^{-1}M^\pi = \begin{bmatrix} a^k(1 - aa^D) & (1 - aa^D)x_k(1 - bb^D) \\ 0 & b^k(1 - bb^D) \end{bmatrix}.$$

Let $n = \mathrm{ind}(a)$, $m = \mathrm{ind}(b)$ and put $k = n + m$. Then

$$a^k(1 - aa^D) = b^k(1 - bb^D) = 0, \; (1 - aa^D)x_k(1 - bb^D) = 0.$$

Thus $\hat{M}^D$ exists with $\mathrm{ind}(\hat{M}) \le \mathrm{ind}(n) + \mathrm{ind}(b)$ and

$$\begin{aligned} \hat{M}^D &= w_{k,l}^{-1}[I_2 + (a^D)^{l+1}(\hat{M}^{k+1} - M^{l+1})]^{-1}M^D \\ &\quad \times [I_2 + (a^D)^l(\hat{M}^k - M^l)]w_{k,l} \\ &= w_{k,l}^{-1} \begin{bmatrix} a^D & (a^D)^{k+1}(x_k - x_{k+1}b^D) \\ 0 & b^D \end{bmatrix} w_{k,l} \\ &= \begin{bmatrix} a^D & (a^D)^{k+1}(x_k - x_{k+1}b^D) + (1 - aa^D)x_k(b^D)^{k+1} \\ 0 & b^D \end{bmatrix}. \end{aligned}$$

Furthermore,

$$(1-aa^D)x_k(b^D)^{k+1} = (1-aa^D)\sum_{i=0}^{n-1} a^i cb^{n+m-1-i}(b^D)^{m+n+1}$$
$$= (1-aa^D)\sum_{i=0}^{n-1} a^i c(b^D)^{i+2}$$
$$(a^D)^{k+1}(x_k - x_{k+1}b^D) = (a^D)^{k+1}(\sum_{i=0}^{k-1} a^i cb^{k-1-i})(1-bb^D) - (a^D)^{k+1}a^k cb^D$$
$$= (\sum_{i=0}^{m-1}(a^D)^{i+2}cb^l)(b-bb^D) - a^D cb^D.$$

In short, we have

Theorem 4.7.4. *Let* $a, b \in Dr(\mathcal{A})$ *with* $\operatorname{ind}(a) = n$, $\operatorname{ind}(b) = m$. *Then* $\begin{bmatrix} a & c \\ 0 & b \end{bmatrix} \in Dr(M_2(\mathcal{A}))$ *and* $\begin{bmatrix} a & c \\ 0 & b \end{bmatrix}^D = \begin{bmatrix} a^D & y \\ 0 & b^D \end{bmatrix}$, *where*

$$y = \Big(\sum_{i=0}^{m-1}(a^D)^{i+2}cb^i\Big)(1-bb^D) + (1-aa^D)\Big(\sum_{i=0}^{n-1} a^i c(b^D)^{i+2}\Big) - a^D cb^D.$$

Theorem 4.7.5. *Let* $a, b \in Drg(\mathcal{A})$ *and* $c \in \mathcal{A}$. *Put* $M = \begin{bmatrix} a & c \\ 0 & b \end{bmatrix}$. *Then* $M \in Drg(M_2(A))$ *and* $M^d = \begin{bmatrix} a^d & z \\ 0 & b^d \end{bmatrix}$, *where*

$$z = \sum_{n=0}^{\infty}(a^d)^{n+2}c(bb^\pi)^n + \sum_{n=0}^{\infty}(aa^\pi)^n c(b^d)^{n+2} - a^d cb^d.$$

Proof. If $\lambda \notin \sigma(M)$, then there is $G = \begin{bmatrix} g_1 & g_2 \\ g_3 & g_4 \end{bmatrix} \in M_2(\mathcal{A})$ such that

$$\begin{bmatrix} \lambda - a & -c \\ 0 & \lambda - b \end{bmatrix}\begin{bmatrix} g_1 & g_2 \\ g_3 & g_4 \end{bmatrix} = \begin{bmatrix} g_1 & g_2 \\ g_3 & g_4 \end{bmatrix}\begin{bmatrix} \lambda - a & -c \\ 0 & \lambda - b \end{bmatrix} = \begin{bmatrix} 1 & \\ & 1 \end{bmatrix}.$$

Then $(\lambda - b)g_4 = g_4(\lambda - b) = 1$ and hence $g_3 = 0$,

$$g_2 = (\lambda - a)^{-1}c(\lambda - b)^{-1},\ (\lambda - a)g_1 = g_1(\lambda - a) = 1,$$

that is, $\lambda \notin \sigma(a) \cup \sigma(b)$.

Conversely, if $\lambda \notin \sigma(a) \cup \sigma(b)$, then it is easy to check to

$$\begin{bmatrix} \lambda - a & -c \\ 0 & \lambda - b \end{bmatrix}^{-1} = \begin{bmatrix} (\lambda - a)^{-1} & (\lambda - a)^{-1}c(\lambda - b)^{-1} \\ 0 & (\lambda - b)^{-1} \end{bmatrix}.$$

Thus, $\sigma(M) = \sigma(a) \cup \sigma(b)$.

Since 0 is an isolated point of $\sigma(a)$ and $\sigma(b)$, 0 is an isolated point of $\sigma(M)$. Thus, $M \in Drg(M_2(\mathcal{A}))$. Pick $\delta > 0$ such that for any λ with $0 < |\lambda| < \delta$,

$$(\lambda I_2 - M)^{-1} = -\sum_{n=0}^{\infty}(M^d)^{n+1}\lambda^n + \sum_{n=0}^{\infty}(MM^\pi)^n\lambda^{-(n+1)},$$

$$(\lambda - a)^{-1} = -\sum_{n=0}^{\infty}(a^d)^{n+1}\lambda^n + \sum_{n=0}^{\infty}(aa^\pi)^n\lambda^{-(n+1)},$$

$$(\lambda - b)^{-1} = -\sum_{n=0}^{\infty}(b^d)^{n+1}\lambda^n + \sum_{n=0}^{\infty}(bb^\pi)^n\lambda^{-(n+1)},$$

by (4.5.4). Comparing the constant terms of Laurent expansions on the both side of

$$(\lambda I_2 - M)^{-1} = \begin{bmatrix} (\lambda - a)^{-1} & (\lambda - a)^{-1}c(\lambda - b)^{-1} \\ 0 & (\lambda - b)^{-1} \end{bmatrix},$$

we obtain that $M^d = \begin{bmatrix} a^d & z \\ 0 & b^d \end{bmatrix}$, where

$$z = \sum_{n=0}^{\infty}(a^d)^{n+2}c(bb^\pi)^n + \sum_{n=0}^{\infty}(aa^\pi)^n c(b^d)^{n+2} - a^d cb^d.$$

□

Lemma 4.7.6. *Let $a, b \in \mathcal{A}\backslash\{0\}$.*

(1) *If $ab \in Dr(\mathcal{A})$, then $ba \in Dr(\mathcal{A})$ and*

$$(ba)^D = b((ab)^D)^2 a, \quad |\mathrm{ind}(ab) - \mathrm{ind}(ba)| \le 1.$$

(2) *If $ab \in Drg(\mathcal{A})$, then $ba \in Drg(\mathcal{A})$ and $(ba)^d = b[(ab)^d]^2 a$.*

Proof. (1) Set $k = \mathrm{ind}(ab)$ and $c = (ab)^D$. Then

$$(ab)^k c(ab) = (ab)^k, \quad c(ab) = (ab)c, \quad c(ab)c = c.$$

Put $d = bc^2a$. Then $d(ba) = bc^2aba = bca$, $(ba)d = babc^2a = bca$. So $a(ba) = (ba)d$, $d(ab)d = bcabc^2a = bc^2a = d$ and

$$\begin{aligned}(ba)^{k+1}d(ba) &= b(ab)^k ad(ba) = b(ab)^k abca \\ &= b(ab)^k c(ab)a = b(ab)^k a = (ba)^{k+1}.\end{aligned}$$

Therefore, $ba \in Dr(\mathcal{A})$, $(ba)^D = d$ with $\mathrm{ind}(ba) \le k + 1$.

Let $n = \mathrm{ind}(ba)$. Similar arguments as above show that $k = \mathrm{ind}(ab) \le n + 1$. Thus, $k - 1 \le \mathrm{ind}(ba) \le 1 + k$.

(2) Put $x = b[(ab)^d]^2 a$. Then $x(ba) = (ba)x$, $x(ba)x = x$ since

$$(ab)^d(ab) = (ab)(ab)^d \text{ and } (ab)^d(ab)(ab)^d = (ab)^d.$$

Now for any $n \in \mathbb{N}$,

$$\begin{aligned}(ba)^{n+1}(1 - xba) &= b(ab)^n a - b(ab)^n ab(ab)^d a \\ &= b(ab)^n[1 - (ab)(ab)^d]a \\ \|(ba)^{n+1}(1 - xba)\| &\leq \|a\|\|b\|\|(ab)^n[1 - (ab)(ab)^d]\|. \end{aligned} \tag{4.7.2}$$

Since $ab[1-(ab)(ab)^d] \in QN(\mathcal{A})$, it follows from (4.7.2) that $(ba)(1-xba) \in QN(\mathcal{A})$. Therefore, $ba \in Drg(\mathcal{A})$ and $(ba)^d = b[(ab)^d]^2 a$. □

Proposition 4.7.7. *Let $a, b \in \mathcal{A}\backslash\{0\}$ with $ab = 0$.*

(1) *If $a, b \in Dr(\mathcal{A})$ with $m = \mathrm{ind}(a)$, $n = \mathrm{ind}(b)$, then $a + b \in Dr(\mathcal{A})$ and*

$$(a + b)^D = \Big(\sum_{i=0}^{m-1}(b^D)^{i+1}a^i\Big)(1 - aa^D) + (1 - bb^D)\Big(\sum_{i=0}^{n-1} b^i(a^D)^{i+1}\Big).$$

(2) *If $a, b \in Drg(\mathcal{A})$, then $a + b \in Drg(\mathcal{A})$ and*

$$(a + b)^d = \sum_{n=0}^{\infty}(bb^\pi)^n(a^d)^{n+1} + \sum_{n=0}^{\infty}(b^d)^{n+1}(aa^\pi)^n.$$

Proof. Put $x = \begin{bmatrix} b & 1 \\ 0 & 0 \end{bmatrix}$ and $y = \begin{bmatrix} 1 & 0 \\ a & 0 \end{bmatrix}$. Then

$$xy = \begin{bmatrix} a + b & 0 \\ 0 & 0 \end{bmatrix} \text{ and } yx = \begin{bmatrix} b & 1 \\ 0 & a \end{bmatrix}.$$

Note that $yx \in Dr(M_2(\mathcal{A}))$ when $a, b \in Dr(\mathcal{A})$ and $yx \in Drg(M_2(\mathcal{A}))$ when $a, b \in Drg(\mathcal{A})$ by Theorem 4.7.4 and Theorem 4.7.5 respectively. So $a + b \in Dr(\mathcal{A})$ (resp. $Drg(\mathcal{A})$) when $a, b \in Dr(\mathcal{A})$ (resp. $Drg(\mathcal{A})$) by Lemma 4.7.6. The assertions follow from Lemma 4.7.6, Theorem 4.7.4 and Theorem 4.7.5. □

4.8 Notes

Section 4.1 is based on [Taylor (1958)], [Caradus (1978)], [Nashed (1976b)] and [Koliha and Tran (2001)].

Section 4.2. The existence of $A_{T,S}^{(2)}$ for the bounded operator A on Banach spaces is mentioned in [Djordjević and Stanimirović (2001)] and [Djordjević *et al.* (2004)]. The definition of $A_{T,S}^{(2)}$ for the densely–defined closed operator A is new. The main results in this section such as Proposition 4.2.7, Theorem 4.2.8 and Theorem 4.2.13 generalize Theorem 1.3 of [Djordjević *et al.* (2004)], Theorem 1 of [Du and Xue (2009)] and Theorem 3.3 of [Djordjević (2001)] respectively.

Section 4.3. Lemma 4.3.2, Theorem 4.3.3 and Proposition 4.3.5 come from [Xue and Chen (2007)]. Lemma 4.3.6, Theorem 4.3.7 and Theorem 4.3.8 generalize corresponding results in [Castro-González and Vélez-Cerrada (2008)] and [Xu *et al.* (2010a)].

Section 4.4 is based on [Du and Xue (2011)]. Lemma 4.4.4 comes from Theorem 11 (P100) of [Muller (2007)]. It is not easy to present the perturbation analysis of the outer generalized inverse with with prescribed range and kernel. Because this type of generalized inverses involves three variables: an operator–argument and two subspace–arguments. There are a few papers concerning this problem. Some interesting results about this can be found in [Djordjević and Stanimirović (2001)] and [Djordjević and Wei (2004)]. We completely solve this problem in [Du and Xue (2011)].

Section 4.5 is based on [Koliha (1996)]. Theorem 4.5.9 comes from [Patrício and Costa (2009)] with a different proof.

Section 4.6. Theorem 4.6.3 is Theorem 4.2 of [Xue (2007c)]. Proposition 4.6.9 is a rough result in the estimation of the perturbation of generalized Drazin inverse. Because it contains the term $\|\bar{a}^{\pi} - a^{\pi}\|$ which is hard to be estimated. [Castro-González *et al.* (2002a)] contains some better estimations. Proposition 4.6.11 comes from [Rakočević (1999)]. There are more results about the continuity of the Drazin inverse in [Rakočević (1999)].

Section 4.7 is based on [Campbell and Meyer (1979)], [Deng and Wei (2011)], [Ding and Wei (2002)] and [Hartwig *et al.* (2001)]. For more results about the representations of Drazin inverses of sums and matrices, please see [Castro-González *et al.* (2009)], [Castro-González and Koliha (2004)], [Castro-González and Martínez-Serrano (2010)].

Chapter 5

Miscellaneous Applications

In this chapter, we present some applications of the stable perturbation of generalized inverses, Moore–Penrose inverses and Drazin inverses. These applications include

(1) perturbation analysis of the generalized Bott–Duffin inverse.
(2) perturbation of the frame sequence in Hilbert space.
(3) the distribution of null spaces of a vector–valued holomorphic mapping.
(4) estimations of the condition numbers of a Moore–Penrose inverse and a Drazin inverse.
(5) the local linearization of a C^1–mapping.

5.1 The Generalized Bott–Duffin inverse of an operator on a Hilbert Space

The Bott–Duffin inverse was first introduced by Bott and Duffin in [Bott and Duffin (1953)]. Ben–Israel and Greville have developed many properties and applications of the Bott–Duffin inverse in [Ben-Israel and Greville (2003)]. Later, Chen, Y. defined the generalized Bott–Duffin inverse of a square matrix and gave some properties and applications in [Chen (1990)]. Some time latter, Chen, G., Liu, G., Xue, Y. and Zhang, X. established the perturbation theory of generalized Bott–Duffin inverses of L–zero matrices and presented the expression of the generalized Bott–Duffin inverses in their papers [Chen *et al.* (2002)], [Chen *et al.* (2003)], [Xue and Chen (2002)] and [Zhang *et al.* (2005)].

In this section, we first define the Bott–Duffin inverse and the generalized Bott–Duffin inverse of an operator on a Hilbert space, then establish the perturbation theory of the generalized Bott–Duffin inverse and finally present a perturbation analysis to the problem

$$\min \|x\| \quad \text{subject to} \quad \begin{cases} Ax + B^*y = b \\ \quad\ \ Bx \ = d \end{cases} \tag{5.1.1}$$

under certain small perturbation of A and B for some operators A and B.

Let H be a Hilbert space and $A \in B(H)$. Let L be a closed subspace of H. Put $A_L = P_L A|_L$, $A'_L = P_{L^\perp} A|_L$ and $T_L = AP_L + P_{L^\perp}$. Then T_L has the form $T_L = \begin{bmatrix} A_L & 0 \\ A'_L & I_2 \end{bmatrix}$, where I_2 (resp. I_1) denotes the identity operator on $L^\perp$ (resp. L). Since $T_L \begin{bmatrix} I_1 & 0 \\ -A'_L & I_2 \end{bmatrix} = \begin{bmatrix} A_L & \\ & I_2 \end{bmatrix}$ and $C_L = \begin{bmatrix} I_1 & 0 \\ -A'_L & I_2 \end{bmatrix}$ is invertible in $B(H)$, it follows that $\operatorname{Ran}(T_L) = \operatorname{Ran}(A_L) \oplus L^\perp$ and hence $\operatorname{Ran}(T_L)$ is closed iff $\operatorname{Ran}(A_L)$ is closed.

We will keep above symbols throughout the section.

Definition 5.1.1. Let $A \in B(H)$ and let L be a closed subspace of H.

(1) If T_L is invertible in $B(H)$, then $A^{(-1)}_{(L)} = P_L(T_L)^{-1}$ is called the Bott–Duffin inverse of A with respect to L.
(2) In general, if $\operatorname{Ran}(T_L)$ is closed in H, then $A^{(\dagger)}_{(L)} = P_L(T_L)^\dagger$ is called the generalized Bott–Duffin inverse of A with respect to L.

We first characterize the invertibility of T_L as follows.

Proposition 5.1.2. *Let $A \in B(H)$ and L be a closed subspace of H.*

(1) *T_L is invertible in $B(H)$ iff $\operatorname{Ker} A \cap L = \{0\}$ and $H = AL \dotplus L^\perp$.*
(2) *When T_L is invertible in $B(H)$, $A^{(2)}_{L,L^\perp}$ exists and $A^{(2)}_{L,L^\perp} = A^{(\dagger)}_{(L)}$.*

Proof. (1) Clearly, $\operatorname{Ran}(T_L) \subset AL + L^\perp$. On the other hand, for any $x_1 \in L$ and $x_2 \in L^\perp$, put $x = x_1 + x_2$. Then $T_L x = Ax_1 + x_2$ and so that $AL + L^\perp \subset \operatorname{Ran}(T_L)$. Thus $\operatorname{Ran}(T_L) = AL + L^\perp$.

Assume that T_L is invertible. Then $AL + L^\perp = \operatorname{Ran}(T_L) = H$ and $\operatorname{Ker} T_L = \{0\}$. Note that $\operatorname{Ker} T_L = \{0\}$ is equivalent to that $\operatorname{Ker} A_L = \{0\}$. It is easy to check that $\operatorname{Ker} A_L = \{0\}$ if and only if

$$\operatorname{Ker} A \cap L = \{0\} \text{ and } AL \cap L^\perp = \{0\}. \tag{5.1.2}$$

Thus $AL \dotplus L^\perp = H$.

Conversely, if $\operatorname{Ker} A \cap L = \{0\}$ and $H = AL \dotplus L^\perp$. Then $\operatorname{Ker} T_L = \{0\}$ by (5.1.2) and $P_L H = P_L AL$. So $\operatorname{Ran}(T_L) = H$ and consequently, T_L is invertible in $B(H)$.

(2) From $T_L C_L = \begin{bmatrix} A_L & \\ & I_2 \end{bmatrix}$, we get that A_L is invertible when T_L is invertible and moreover,

$$(T_L)^{-1} = \begin{bmatrix} A_L^{-1} & 0 \\ -A'_L A_L^{-1} & I_2 \end{bmatrix}, \; A_{(L)}^{(-1)} = P_L (T_L)^{-1} = \begin{bmatrix} A_L^{-1} & \\ & 0 \end{bmatrix}.$$

Thus, $\operatorname{Ran}(A_{(L)}^{(-1)}) = L$, $\operatorname{Ker}(A_{(L)}^{(-1)}) = L^{\perp}$ and $A_{(L)}^{(-1)} A A_{(L)}^{-1)} = A_{(L)}^{(-1)}$. This shows that $A_{L,L^{\perp}}^{(2)}$ exists and $A_{L,L^{\perp}}^{(2)} = A_{(L)}^{(-1)}$. □

In general, we have

Proposition 5.1.3. *Let $A \in B(H)$ and L be a closed subspace of H such that* $\operatorname{Ran}(T_L)$ *is closed. Then*

(1) $A_{(L)}^{(\dagger)} = \begin{bmatrix} A_L^{\dagger} - (I_1 - A_L^{\dagger} A_L)(A'_L)^* C^{-1} A'_L A_L^{\dagger} & (I_1 - A_L^{\dagger} A_L)(A'_L)^* C^{-1} \\ 0 & 0 \end{bmatrix}$,
where $C = I_2 + A'_L (I_1 - A_L^{\dagger} A_L)(A'_L)^*$ is invertible in $B(L^{\perp})$.
(2) $\|A_{(L)}^{(\dagger)}\| \le \|T_L^{\dagger}\| \le (2\|A\|^2 + 1)^{1/2} \|A_L^{\dagger}\| \le (2\|A\|^2 + 1)^{1/2} \|A_{(L)}^{(\dagger)}\|$.

Proof. (1) Put $X = \begin{bmatrix} A_L^{\dagger} & 0 \\ -A'_L A_L^{\dagger} & I_2 \end{bmatrix}$. Then

$$T_L X = \begin{bmatrix} A_L A_L^{\dagger} & \\ & I_2 \end{bmatrix}, \quad X T_L = \begin{bmatrix} A_L^{\dagger} A_L & 0 \\ A'_L (I_1 - A_L^{\dagger} A_L) & I_2 \end{bmatrix}$$

and $T_L X T_L = T_L$, $X T_L X = X$. Thus,

$$T_L^{\dagger} = (I - O(I - X T_L)) X O(T_L X) \tag{5.1.3}$$
$$= (X T_L + T_L^* X^* - I)^{-1} X (T_L X + X^* T_L^* - I)^{-1} \tag{5.1.4}$$

by Proposition 3.1.10. Simple computation shows that

$$(T_L X + X^* T_L^* - I)^{-1} = \begin{bmatrix} 2A_L^{\dagger} A_L - I_1 & \\ & I_2 \end{bmatrix}$$

$$(X T_L + T_L^* X^* - I)^{-1} = \begin{bmatrix} 2A_L^{\dagger} A_L - I_1 & (I_1 - A_L^{\dagger} A_L)(A'_L)^* \\ A'_L (I_1 - A_L^{\dagger} A_L) & I_2 \end{bmatrix}^{-1}$$

$$= \begin{bmatrix} 2A_L^{\dagger} A_L - I_1 + (I_1 - A_L^{\dagger} A_L)(A'_L)^* C^{-1} A'_L A_L^{\dagger} & (I_1 - A_L^{\dagger} A_L)(A'_L)^* C^{-1} \\ -C^{-1} A'_L (I_1 - A_L^{\dagger} A_L) & C^{-1} \end{bmatrix}.$$

Finally, using (5.1.4), we can obtain the assertion.

(2) By (5.1.3), $\|A^{(\dagger)}_{(L)}\| = \|P_L T_L^\dagger\| \leq \|X\|$. For any $x \in H$, put $x_1 = P_L x$ and $x_2 = P_{L^\perp} x$. Then

$$\begin{aligned}\|Xx\|^2 &= \|A_L^\dagger x_1\|^2 + \| - A'_L A_L^\dagger x_1 + x_2\|^2 \\ &\leq \|A_L^\dagger x_1\|^2 + 2\|A\|^2\|A_L^\dagger\|^2\|x_1\|^2 + 2\|x_2\|^2 \\ &= (2\|A\|^2 + 1)\|A_L^\dagger\|^2\|x_1\|^2 + 2\|x_2\|^2.\end{aligned}$$

Since $\|A_L^\dagger\|\|A\| \geq \|A_L^\dagger\|\|A_L\| \geq \|A_L^\dagger A_L\| = 1$, we have from above

$$\|Xx\|^2 \leq (2\|A\|^2 + 1)\|A_L^\dagger\|^2(\|x_1\|^2 + \|x_2\|^2) = (2\|A\|^2 + 1)\|A_L^\dagger\|^2\|x\|^2$$

and consequently, $\|X\| \leq (2\|A\|^2 + 1)^{1/2}\|A_L^\dagger\|$.

Now let $x \in L$. Then by (1),

$$\begin{aligned}\|A^{(\dagger)}_{(L)} x\|^2 &= \|A_L^\dagger x - (I_1 - A_L^\dagger A_L)(A'_L)^* C^{-1} A'_L A_L^\dagger x\|^2 \\ &= \|A_L^\dagger x\|^2 + \|(I_1 - A_L^\dagger A_L)(A'_L)^* C^{-1} A'_L A_L^\dagger x\|^2 \\ &\geq \|A_L^\dagger x\|^2.\end{aligned}$$

Thus $\|A_L^\dagger\| \leq \|A^{(\dagger)}_{(L)}\|$. □

According Proposition 5.1.3, we can deduce some conditions that $A^{(\dagger)}_{(L)}$ has the simplest expression $A^{(\dagger)}_{(L)} = (P_L A P_L)^\dagger$ as follows.

Corollary 5.1.4. *Let $A \in B(H)$ and L be a closed subspace of H such that* $\mathrm{Ran}\,(T_L)$ *is closed. Then the following statements are equivalent:*

(1) $A^{(\dagger)}_{(L)} = (P_L A P_L)^\dagger$.
(2) $\mathrm{Ker}\, A_L \subset \mathrm{Ker}\,(A'_L)$.
(3) $AL \cap L^\perp = \{0\}$.
(4) $\mathrm{Ker}\, A \cap L = \mathrm{Ker}\,(A_L)$, *i.e.*, $\mathrm{Ker}\,(P_L A P_L) = \mathrm{Ker}\, A P_L$.

Proof. (1)⇔(2) Since $P_L A P_L = \begin{bmatrix} A_L & 0 \\ 0 & 0 \end{bmatrix}$, we have $(P_L A P_L)^\dagger = \begin{bmatrix} A_L^\dagger & 0 \\ 0 & 0 \end{bmatrix}$. Thus, by Proposition 5.1.3 (1), $A^{(\dagger)}_{(L)} = (P_L A P_L)^\dagger$ iff $A'_L(I_1 - A_L^\dagger A_L) = 0$, that is, $\mathrm{Ker}\, A_L \subset \mathrm{Ker}\,(A'_L)$.

(2)⇒(3) Let $x \in AL \cap L^\perp$. Then $x \in L^\perp$ and $x = Ay$ for some $y \in L$. Thus $y \in \mathrm{Ker}\, A_L \subset \mathrm{Ker}\,(A'_L)$, that is, $(I - P_L)Ay = 0$ and consequently, $x = Ay = P_L Ay = 0$.

(3)⇒(4) Obviously, $\mathrm{Ker}\, A \cap L \subset \mathrm{Ker}\,(A_L)$. Let $x \in L$ and $P_L Ax = 0$. Then $Ax \in L^\perp$. It follows from $AL \cap L^\perp = \{0\}$ that $Ax = 0$, i.e., $x \in \mathrm{Ker}\, A \cap L$.

(4)⇒(2). Let $x \in \mathrm{Ker}\, A_L$. Then $x \in \mathrm{Ker}\, A \cap L$ and hence $x \in \mathrm{Ker}\,(A'_L)$. So $\mathrm{Ker}\, A_L \subset \mathrm{Ker}\,(A'_L)$. □

Corollary 5.1.4 indicates following useful notation:

Definition 5.1.5. Let $A \in B(H)$ and L be a closed subspace of H. A is called L–zero, if $AL \cap L^{\perp} = \{0\}$.

Remark 5.1.6. A special kind of L–zero operators is the L–positive semi–defined operator. A self–adjoint operator $A \in B(H)$ is called to be L–positive semi–defined (or L–p.s.d) if $(Ax, x) \geq 0$, $\forall\, x \in L$ and $(Ax, x) = 0$, $x \in L$ implies that $Ax = 0$.

Let $\delta A \in B(H)$ and put $\bar{A} = A + \delta A$. Let $\bar{L}$ be a closed subspace of H. Then $\hat{\delta}(L, \bar{L}) = \|P_L - P_{\bar{L}}\|$ by Proposition 1.3.6. Put $\bar{T}_{\bar{L}} = \bar{A}P_{\bar{L}} + P_{\bar{L}^{\perp}}$. Then

$$\|\bar{T}_{\bar{L}} - T_L\| = \|\delta A P_{\bar{L}} + (A - I)(P_{\bar{L}} - P_L)\| \leq \|\delta A\| + \|A - I\|\hat{\delta}(L, \bar{L}).$$

Proposition 5.1.7. *Let A, $\bar{A} = A + \delta A \in B(H)$ and L, $\bar{L}$ be closed subspaces of H. Suppose that* $\mathrm{Ran}\,(T_L)$ *is closed and $\bar{T}_{\bar{L}}$ is the stable perturbation of T_L. Further assume that $\|T_L^{\dagger}\|(\|\delta A\| + \|A - I\|\hat{\delta}(L, \bar{L})) < 1$. Then*

$$\|\bar{A}^{(\dagger)}_{(\bar{L})}\| \leq \|\bar{T}_{\bar{L}}^{\dagger}\| \leq \frac{(\|A\|^2 + 1)^{1/2}\|A^{(\dagger)}_{(L)}\|}{1 - \|T_L^{\dagger}\|(\|\delta A\| + \|A - I\|\hat{\delta}(L, \bar{L}))}$$

$$\frac{\|\bar{A}^{(\dagger)}_{(\bar{L})} - A^{(\dagger)}_{(L)}\|}{\|A^{(\dagger)}_{(L)}\|} \leq \frac{1 + \sqrt{5}}{2}\,\frac{(\|A\|^2 + 1)\|A^{(\dagger)}_{(L)}\|(\|\delta A\| + \|A - I\|\hat{\delta}(L, \bar{L})}{1 - \|T_L^{\dagger}\|(\|\delta A\| + \|A - I\|\hat{\delta}(L, \bar{L}))} + (\|A\|^2 + 1)^{1/2}\hat{\delta}(L, \bar{L}).$$

Proof. Put $\Delta = \bar{T}_{\bar{L}} - T_L$. Then $\|T_L^{\dagger}\|(\|\delta A\| + \|A - I\|\hat{\delta}(L, \bar{L})) < 1$ implies that $I + T_L^{\dagger}\Delta$ is invertible. So by Theorem 3.3.5,

$$\|\bar{T}_{\bar{L}}^{\dagger}\| \leq \frac{\|T_L^{\dagger}\|}{1 - \|T_L^{\dagger}\|\|\Delta\|}, \quad \frac{\|\bar{T}_{\bar{L}}^{\dagger} - T_L^{\dagger}\|}{\|T_L^{\dagger}\|} \leq \frac{1 + \sqrt{5}}{2}\,\frac{\|T_L^{\dagger}\|\|\Delta\|}{1 - \|T_L^{\dagger}\|\|\Delta\|}. \tag{5.1.5}$$

Note that

$$\|\bar{A}^{(\dagger)}_{(\bar{L})} - A^{(\dagger)}_{(L)}\| \leq \|P_{\bar{L}} - P_L\|\|T_L^{\dagger}\| + \|\bar{T}_{\bar{L}}^{\dagger} - T_L^{\dagger}\|. \tag{5.1.6}$$

So combining (5.1.5) and (5.1.6) with Proposition 5.1.3, we can obtain the assertions. □

Now Let $A \in B(H)$ be L–zero. Then $\mathrm{Ker}\,(P_LAP_L) = \mathrm{Ker}\,(AP_L)$ by Corollary 5.1.4. Thus, if $\mathrm{Ran}\,(T_L)$ is closed (or equivalently, $\mathrm{Ran}\,(P_LAP_L)$ is closed in L), then $\gamma(P_LAP_L) > 0$ and

$$\begin{aligned}\|AP_Lx\| \geq \|P_LAP_Lx\| &\geq \gamma(P_LAP_L)\mathrm{dist}\,(x, \mathrm{Ker}\,(P_LAP_L)) \\ &= \gamma(P_LAP_L)\mathrm{dist}\,(x, \mathrm{Ker}\,(AP_L)), \qquad \forall\, x \in H.\end{aligned}$$

This means that $\gamma(AP_L) \geq \gamma(P_L AP_L) > 0$ and $\|(AP_L)^\dagger\| \leq \|(P_L AP_L)^\dagger\|$ by Proposition 3.1.9.

Let $\bar{A} = A + \delta A \in B(H)$ and $\bar{L}$ be a closed subspace of H. Then

$$\|\bar{A}P_{\bar{L}} - AP_L\| \leq \|\delta A\| + \|A\|\hat{\delta}(L, \bar{L})$$
$$\|P_{\bar{L}}\bar{A}P_{\bar{L}} - P_L AP_L\| \leq \|\delta A\| + 2\|A\|\hat{\delta}(L, \bar{L}).$$

Set $\epsilon_A = \|\delta A\|\|A\|^{-1}$ and $\kappa_{gBD}(A) = \|A^{(\dagger)}_{(L)}\|\|A\|$ when $\mathrm{Ran}\,(T_L)$ is closed.

Theorem 5.1.8. *Let A, $\bar{A} = A + \delta A \in B(H)$ and let L, $\bar{L}$ be closed subspaces of H. Suppose that A is L-zero,* $\mathrm{Ran}\,(T_L)$ *is closed and $\bar{A}P_{\bar{L}}$ is the stable perturbation of AP_L. Assume that $\kappa_{gBD}(A)(\epsilon_A + 2\hat{\delta}(L, \bar{L})) < \frac{1}{2}$. Then $\bar{A}$ is $\bar{L}$-zero and*

$$\|\bar{A}^{(\dagger)}_{(\bar{L})}\| \leq \frac{\|A^{(\dagger)}_{(L)}\|}{1 - \kappa_{gBD}(A)(\epsilon_A + 2\hat{\delta}(L, \bar{L}))}$$
$$\frac{\|\bar{A}^{(\dagger)}_{(\bar{L})} - A^{(\dagger)}_{(L)}\|}{\|A^{(\dagger)}_{(L)}\|} \leq \frac{1+\sqrt{5}}{2}\,\frac{\kappa_{gBD}(A)(\epsilon_A + 2\hat{\delta}(L, \bar{L}))}{1 - \kappa_{gBD}(A)(\epsilon_A + 2\hat{\delta}(L, \bar{L}))}.$$

Proof. Since A is L–zero, it follows from Corollary 5.1.4 that $\mathrm{Ker}\, AP_L = P_L AP_L$. Thus, by Proposition 1.3.2 (4), we have

$$\begin{aligned}\delta(&\mathrm{Ker}\,(P_{\bar{L}}\bar{A}P_{\bar{L}}), \mathrm{Ker}\,(\bar{A}P_{\bar{L}})) \\ &\leq \delta(\mathrm{Ker}\,(P_{\bar{L}}\bar{A}P_{\bar{L}}), \mathrm{Ker}\,(P_L AP_L)) + 2\delta(\mathrm{Ker}\,(P_L AP_L), \mathrm{Ker}\,(\bar{A}P_{\bar{L}})) \\ &= \delta(\mathrm{Ker}\,(P_{\bar{L}}\bar{A}P_{\bar{L}}), \mathrm{Ker}\,(P_L AP_L)) + 2\delta(\mathrm{Ker}\,(AP_L), \mathrm{Ker}\,(\bar{A}P_{\bar{L}})).\end{aligned}$$

Note that $\|(AP_L)^\dagger\|\|(\bar{A}P_{\bar{L}} - AP_L)\| \leq \kappa_{gBD}(A)(\epsilon_A + 2\hat{\delta}(L, \bar{L})) < 1$. So $I + (AP_L)^\dagger(\bar{A}P_{\bar{L}} - AP_L)$ is invertible. Since $\mathrm{Ker}\,(AP_L) \cap (\mathrm{Ker}\,(\bar{A}P_{\bar{L}}))^\perp = \{0\}$, it follows from Theorem 3.3.3, Theorem 1.3.8 and Corollary 1.3.9 that

$$\begin{aligned}\delta(\mathrm{Ker}\,(AP_L), \mathrm{Ker}\,(\bar{A}P_{\bar{L}})) &\leq \|(AP_L)^\dagger\|\|(\bar{A}P_{\bar{L}} - AP_L)\| \\ &\leq \kappa_{gBD}(A)(\epsilon_A + \hat{\delta}(L, \bar{L})).\end{aligned}$$

Meanwhile, by Lemma 1.3.5,

$$\begin{aligned}\delta(\mathrm{Ker}\,(P_{\bar{L}}\bar{A}P_{\bar{L}}), \mathrm{Ker}\,(P_L AP_L)) &\leq \|(P_L AP_L)^\dagger\|\|P_{\bar{L}}\bar{A}P_{\bar{L}} - P_L AP_L\| \\ &\leq \kappa_{gBD}(A)(\epsilon_A + 2\hat{\delta}(L, \bar{L})).\end{aligned}$$

Therefore, we have

$$\begin{aligned}\delta(\mathrm{Ker}\,(P_{\bar{L}}\bar{A}P_{\bar{L}}), \mathrm{Ker}\,(\bar{A}P_{\bar{L}})) &\leq \kappa_{gBD}(A)(2\epsilon_A + 3\hat{\delta}(L, \bar{L})) \\ &< 1. \end{aligned} \tag{5.1.7}$$

Since $\operatorname{Ker}(\bar{A}P_{\bar{L}}) \subset \operatorname{Ker}(P_{\bar{L}}\bar{A}P_{\bar{L}})$, $\delta(\operatorname{Ker}(P_{\bar{L}}\bar{A}P_{\bar{L}}), \operatorname{Ker}(P_L A P_L)) = 1$ when $\operatorname{Ker}(\bar{A}P_{\bar{L}}) \subsetneqq \operatorname{Ker}(P_{\bar{L}}\bar{A}P_{\bar{L}})$ by Proposition 1.3.3. But this contradicts to (5.1.7). So $\operatorname{Ker}(\bar{A}P_{\bar{L}}) = \operatorname{Ker}(P_{\bar{L}}\bar{A}P_{\bar{L}})$ and hence $\bar{A}$ is $\bar{L}$–zero by Corollary 5.1.4.

Now $P_{\bar{L}}\bar{A}P_{\bar{L}}$ is the stable perturbation of $P_L A P_L$. Applying Theorem 3.3.3 to $P_{\bar{L}}\bar{A}P_{\bar{L}}$ and $P_L A P_L$, we can get the rest assertions. □

Let H, K be Hilbert spaces. Let $A \in B(H)$ and $B \in B(H, K)$ with $\operatorname{Ran}(B)$ closed. Let $b \in H$ and $d \in \operatorname{Ran}(B)$. We now solve the linear systems

$$\begin{cases} Ax + B^* y = b \\ \qquad Bx = d \end{cases} \tag{5.1.8}$$

as follows. The general solution of the second equation in (5.1.8) is $x = B^\dagger d + z$, $\forall\, z \in \operatorname{Ker} B = L$. Thus, we have by the first equation in (5.1.8),

$$Az + B^* y = b - AB^\dagger d. \tag{5.1.9}$$

Note that $\operatorname{Ran}(B^*) = (\operatorname{Ker} B)^\perp = L^\perp$ and set $u = z + B^* y$. Then (5.1.9) can be written as

$$(AP_L + P_{L^\perp})u = b - AB^\dagger d. \tag{5.1.10}$$

Now suppose that $\operatorname{Ran}(T_L)$ is closed, $AL \cap L^\perp = 0$ and

$$b - AB^\dagger d \in \operatorname{Ran}(AP_L + P_{L^\perp}) = AL + L^\perp.$$

Then the general solution of (5.1.10) is

$$u = (AP_L + P_{L^\perp})^\dagger (b - AB^\dagger d) + w_1,$$

where $w_1 \in \operatorname{Ker}(AP_L + P_{L^\perp}) = \operatorname{Ker} A \cap L$ is arbitrary. Consequently, the general solution of (5.1.8) is

$$\begin{cases} x = \ B^\dagger d + A^{(\dagger)}_{(L)}(b - AB^\dagger d) + w_1 \\ y = \ (B^*)^\dagger P_{L^\perp}(AP_L + P_{L^\perp})^\dagger (b - AB^+ d) + w_2, \end{cases}$$

where $w_2 \in \operatorname{Ker} B^*$ is arbitrary. Noting that $\operatorname{Ran}(B^\dagger) = L^\perp$ and

$$\begin{aligned} \operatorname{Ran}(A^{(\dagger)}_{(L)}) &= \operatorname{Ran}((P_L A P_L)^\dagger) = (\operatorname{Ker}(P_L A P_L))^\perp \\ &= (\operatorname{Ker} A \cap L + L^\perp)^\perp \subset (\operatorname{Ker} A \cap L)^\perp \cap L, \end{aligned}$$

by Corollary 5.1.4 when A is L–zero, we get that the solution of Problem (5.1.1) is

$$x = B^\dagger d + A^{(\dagger)}_{(L)}(b - AB^\dagger d) \tag{5.1.11}$$

In the following, we consider the perturbation analysis of the problem (5.1.1) when A is L–zero and self–adjoint.

Lemma 5.1.9. *Let $A \in B(H)$ with $A^* = A$ and L be a closed subspace of H. Suppose that* $\mathrm{Ran}\,(T_L)$ *is closed and A is L-zero. Then $AL + L^\perp =$* $\mathrm{Ran}\,(A) + L^\perp$.

Proof. In fact, $AL + L^\perp = \mathrm{Ran}\,(T_L) = (\mathrm{Ker}\,(P_L A + P_{L^\perp}))^\perp$ and $\mathrm{Ker}\,(P_L A + P_{L^\perp}) = \mathrm{Ker}\,A \cap L$ by Corollary 5.1.4. So we have

$$AL + L^\perp = (\mathrm{Ker}\,A \cap L)^\perp \supset (\mathrm{Ker}\,A)^\perp + L^\perp \supset \mathrm{Ran}\,(A) + L^\perp.$$

Since $AL + L^\perp \subset \mathrm{Ran}\,(A) + L^\perp$, we obtain the assertion. □

Corollary 5.1.10. *Let $A \in B(H)$ with $A^* = A$ and* $\mathrm{Ran}\,(T_L)$ *closed and let $B \in B(H, K)$. Let $b \in H$ and $d \in K$. Suppose that A is L-zero and* $b \in \mathrm{Ran}\,([A\,B^*])$, $d \in \mathrm{Ran}\,(B)$, *where* $L = \mathrm{Ker}\,B$. *Then Problem* (5.1.1) *has the solution* (5.1.11).

Proof. By Lemma 5.1.9,

$$AL + L^\perp = \mathrm{Ran}\,(A) + L^\perp = \mathrm{Ran}\,(A) + \mathrm{Ran}\,(B^*) = \mathrm{Ran}\,([A\,B^*]).$$

Thus $b - AB^\dagger d \in \mathrm{Ran}\,([A\,B^*]) = AL + L^\perp$. □

Now let Problem (5.1.12)

$$\min \|x\| \quad \text{subject to} \quad \begin{cases} \bar{A}x + \bar{B}^* y = \bar{b} \\ \bar{B}x = \bar{d} \end{cases}, \tag{5.1.12}$$

is the perturbation of Problem (5.1.1), where $\bar{A} = A + \delta A$, $\bar{B} = B + \delta B \in B(H)$, $\bar{b} = b + \delta b \in H$ and $\bar{d} = d + \delta d \in K$ with $\bar{b} \in \mathrm{Ran}\,([\bar{A}\,\bar{B}^*])$, $\bar{d} \in \mathrm{Ran}\,(\bar{B})$. Assume that $\bar{A}^* = \bar{A}$, $\mathrm{Ran}\,(\bar{A})$ is closed and $\bar{A}$ is $\bar{L}$–zero, where $\bar{L} = \mathrm{Ker}\,\bar{B}$. Suppose that $\mathrm{Ran}\,(\bar{B})$ is closed. Then the solution of Problem (5.1.12) is

$$\bar{x} = \bar{B}^\dagger \bar{d} + \bar{A}^{(\dagger)}_{(\bar{L})}(\bar{b} - \bar{A}\bar{B}^\dagger \bar{d}).$$

Set $\epsilon_b = \dfrac{\|\delta b\|}{\|b\|}$, $\epsilon_d = \dfrac{\|\delta d\|}{\|d\|}$ and $\epsilon_A = \dfrac{\|\delta A\|}{\|A\|}$, $\epsilon_B = \dfrac{\|\delta B\|}{\|B\|}$. Put $\kappa(B) = \|B^\dagger\|\|B\|$ when $\mathrm{Ran}\,(B)$ is closed.

Theorem 5.1.11. *Let A, $\bar{A} = A + \delta A \in B(H)$ be L-zero and self-adjoint and Let B, $\bar{B} = B + \delta B \in B(H, K)$ with* $L = \mathrm{Ker}\,B$ *and let b, $\bar{b} = b + \delta b \in H$, $d, \bar{d} = d + \delta d \in K$ with* $b \in \mathrm{Ran}\,([A\,B^*])$, $d \in \mathrm{Ran}\,(B)$ *and*

$\bar{b} \in \mathrm{Ran}\,([\bar{A}\,\bar{B}^*])$, $\bar{d} \in \mathrm{Ran}\,(\bar{B})$. *Assume that* $\kappa_{gBD}(A)(\epsilon_A + 2\kappa(B)\epsilon_B) < \dfrac{1}{2}$ *and that both* $\mathrm{Ran}\,(T_L)$ *and* $\mathrm{Ran}\,(B)$ *are all closed. Let* $\bar{A}P_{\bar{L}}$ *be the stable perturbation of* AP_L *and* $\bar{B}$ *be the stable perturbation of* B. *Let* x, $\bar{x}$ *be the solutions of Problem* (5.1.1) *and Problem* (5.1.12) *respectively. Then*

$$\begin{aligned}\frac{\|\bar{x}-x\|}{\|x\|} &\le \left(1 + \frac{\kappa_{gBD}(A)}{1-\kappa_{gBD}(A)(\epsilon_A + 2\kappa(B)\epsilon_B)}\right)\frac{\kappa(B)(\epsilon_d + 2\epsilon_B)}{1-\kappa(B)\epsilon_B} \\ &\quad + \frac{\kappa_{gBD}(A)}{1-\kappa_{gBD}(A)(\epsilon_A + 2\kappa(B)\epsilon_B)}\left(\frac{\|b\|}{\|P_L b\|}(\epsilon_b + \kappa(B)\epsilon_B) + \kappa(B)\epsilon_B\right. \\ &\quad \left. + \frac{\epsilon_A\kappa(B)}{1-\kappa(B)\epsilon_B}(1+\epsilon_d)\right) + \frac{2\kappa_{gBD}(A)(\epsilon_A + 2\kappa(B)\epsilon_A)}{1-\kappa_{gBD}(A)(\epsilon_A + 2\kappa(B)\epsilon_B)}.\end{aligned}$$

Proof. Since $\bar{B}$ is the stable perturbation of B and that $\|B^\dagger\|\|\delta B\| = \kappa(B)\epsilon_B < 1$ implies that $I_H + B^\dagger \delta B$ is invertible in $B(H)$, it follows from Theorem 3.3.3, Theorem 1.3.8 and Corollary 1.3.9 that $\bar{B}^\dagger$ exists and

$$\|\bar{B}^\dagger\| \le \frac{\|B^\dagger\|}{1-\kappa(B)\epsilon_B}, \ \hat{\delta}(L,\bar{L}) = \|\bar{B}^\dagger\bar{B} - B^\dagger B\| \le \|\delta B\| \min\{\|\bar{B}^\dagger\|, \|B^\dagger\|\}. \tag{5.1.13}$$

Thus $\bar{A}$ is $\bar{L}$–zero and $\mathrm{Ran}\,(P_{\bar{L}}\bar{A}P_{\bar{L}})$ is closed by Theorem 5.1.8. So $\bar{A}^{(\dagger)}_{(\bar{L})} = (P_{\bar{L}}\bar{A}P_{\bar{L}})^\dagger$ and consequently,

$$\begin{aligned}\bar{x} - x &= \bar{B}^\dagger\bar{d} - B^\dagger d + \bar{A}^{(\dagger)}_{(\bar{L})}(\bar{b} - \bar{A}\bar{B}^\dagger\bar{d}) - A^{(\dagger)}_{(L)}(b - AB^\dagger d) \\ &= (I - \bar{A}^{(\dagger)}_{(\bar{L})}A)(\bar{B}^\dagger\bar{d} - B^\dagger d) + \bar{A}^{(\dagger)}_{(\bar{L})}\delta b \\ &\quad - \bar{A}^{(\dagger)}_{(\bar{L})}\delta A\bar{B}^\dagger\bar{d} + (\bar{A}^{(\dagger)}_{(\bar{L})} - A^{(\dagger)}_{(L)})(b - AB^\dagger d),\end{aligned}$$

where $\bar{L} = \mathrm{Ker}\,\bar{B}$. Since $d \in \mathrm{Ran}\,(B)$ and

$$\bar{B}^\dagger - B^\dagger = -\bar{B}^\dagger\delta BB^\dagger - (B^\dagger B - \bar{B}^\dagger\bar{B})B^\dagger + \bar{B}^\dagger(I_K - BB^\dagger),$$

it follows from (5.1.13) that

$$\begin{aligned}\|\bar{B}^\dagger\bar{d} - B^\dagger d\| &\le \|\bar{B}^\dagger\delta d\| + \|\bar{B}^\dagger d - B^\dagger d\| \\ &\le \frac{\|B^\dagger\|}{1-\kappa(B)\epsilon_B}(\|\delta d\| + 2\|\delta B\|\|B^\dagger d\|).\end{aligned}$$

Now $b - AB^\dagger d \in \mathrm{Ran}\,([A, B^*]) = AL + L^\perp$ implies that

$$P_L(I - AA^{(\dagger)}_{(L)})(b - AB^\dagger d) = 0.$$

So from the identity

$$\begin{aligned}\bar{A}^{(\dagger)}_{(\bar{L})} - A^{(\dagger)}_{(L)} &= -\bar{A}^{(\dagger)}_{(\bar{L})}(P_{\bar{L}}\bar{A}P_{\bar{L}} - P_LAP_L)A^{(\dagger)}_{(L)} - \Big[(P_LAP_L)^\dagger(P_LAP_L) \\ &\quad - (P_{\bar{L}}\bar{A}P_{\bar{L}})^\dagger(P_{\bar{L}}\bar{A}P_{\bar{L}})\Big]A^{(\dagger)}_{(L)} + \bar{A}^{(\dagger)}_{(\bar{L})}(P_{\bar{L}} - P_L) \\ &\quad + \bar{A}^{(\dagger)}_{(\bar{L})}\Big[P_L - (P_LAP_L)(P_LAP_L)^\dagger\Big],\end{aligned}$$

we get that

$$\begin{aligned}\|(\bar{A}^{(\dagger)}_{(\bar{L})} - A^{(\dagger)}_{(L)})(b - AB^\dagger d)\| &\le \|\bar{A}^{(\dagger)}_{(L)}(P_{\bar{L}} - P_L)(b - AB^\dagger d)\| \\ &\quad + 2\|\bar{A}^{(\dagger)}_{(\bar{L})}\|(\|\delta A\| + 2\|A\|\|P_{\bar{L}} - P_L\|)\|A^{(\dagger)}_{(L)}(b - AB^\dagger d)\| \\ &\le 2\frac{\kappa_{gBD}(A)(\epsilon_A + 2\kappa(B)\epsilon_B)}{1 - \kappa_{gBD}(A)(\epsilon_A + 2\kappa(B)\epsilon_B)}\|A^{(\dagger)}_{(L)}(b - AB^\dagger d)\| \\ &\quad + \frac{\kappa_{gBD}(A)\kappa(B)\epsilon_B}{1 - \kappa_{gBD}(A)(\epsilon_A + 2\kappa(B)\epsilon_B)}\left(\frac{\|b\|}{\|A\|} + \|B^\dagger d\|\right)\end{aligned}$$

by Theorem 5.1.8. Therefore,

$$\begin{aligned}\|\bar{x} - x\| \le \; & \frac{\|B^\dagger\|(\|\delta d\| + 2\|\delta B\|\|B^\dagger d\|)}{1 - \kappa(B)\epsilon_B}\left[1 + \frac{\kappa_{gBD}(A)}{1 - \kappa_{gBD}(A)(\epsilon_A + 2\kappa(B)\epsilon_B)}\right] \\ & + \frac{\|A^{(\dagger)}_{(L)}\|}{1 - \kappa_{gBD}(A)(\epsilon_A + 2\kappa(B)\epsilon_B)}\left[\|\delta b\| + \frac{\|\delta A\|\|B^\dagger\|}{1 - \kappa(B)\epsilon_B}(\|d\| + \|\delta d\|)\right] \\ & + 2\frac{\kappa_{gBD}(A)(\epsilon_A + 2\kappa(B)\epsilon_B)}{1 - \kappa_{gBD}(\epsilon_A + 2\kappa(B)\epsilon_B)}\|A^{(\dagger)}_{(L)}(b - AB^\dagger d)\| \\ & + \frac{\kappa_{gBD}(A)\kappa(B)\epsilon_B}{1 - \kappa_{gBD}(A)(\epsilon_A + 2\kappa(B)\epsilon_B)}\left[\frac{\|b\|}{\|A\|} + \|B^\dagger d\|\right].\end{aligned}$$

Since $\operatorname{Ran}(B^\dagger) = L^\perp$ and $\operatorname{Ran}(A^{(\dagger)}_{(L)}) \subset L$, we have

$$\|B^\dagger d\| \le \|x\|, \quad \|A^{(\dagger)}_{(L)}(b - AB^\dagger d)\| \le \|x\|.$$

From $Bx = d$ and $P_L Ax = P_L b$, we get that $\dfrac{\|d\|}{\|B\|} \le \|x\|$, $\dfrac{\|P_L b\|}{\|A\|} \le \|x\|$. So finally we get the assertion. □

5.2 Some applications of Moore–Penrose inverses in frame theory

This section is based on [Christensen (1999)] and [Christensen (2003)].

Let H be a Hilbert space with the inner product $(\cdot,\cdot)$.

Definition 5.2.1. A sequence $\{f_n\}$ in H is a frame for H if there are constants A, $B > 0$ such that

$$A\|f\|^2 \le \sum_{n=1}^{\infty} |(f, f_n)|^2 \le B\|f\|^2, \ \forall f \in H.$$

A and B are called frame bounds.

Definition 5.2.2. Let $\{f_n\}$ be a sequence in H. $\{f_n\}$ is a frame sequence if $\{f_n\}$ is a frame for $\overline{span}\{f_n\}$.

$\{f_n\}$ is a Bessel sequence if there is a constant $B > 0$ such that

$$\sum_{n=1}^{\infty} |(f, f_n)|^2 \le B\|f\|^2, \quad \forall\, f \in H.$$

Remark 5.2.3. A frame sequence $\{f_n\}$ is a Bessel sequence. In fact, for any $f \in H$, we can find $f_1' \in \overline{span}\{f_n\}$ and $f_2' \in (\overline{span}\{f_n\})^{\perp}$ such that $f = f_1' + f_2'$. So $(f, f_n) = (f_1', f_n)$, $\forall\, n \in \mathbb{N}$ and consequently,

$$\sum_{n=1}^{\infty} |(f, f_n)|^2 = \sum_{n=1}^{\infty} |(f_1', f_n)|^2 \le B\|f_1'\|^2 \le B\|f\|^2.$$

Lemma 5.2.4. *Suppose that $\{f_n\}$ be a Bessel sequence in H. Then the linear operator $T\colon l^2 \to K$ given by $T(\{a_n\}) = \sum_{n=1}^{\infty} a_n f_n$ is bounded with $\|T\| \le \sqrt{B}$, $\overline{\text{Ran}\,(T)} = \overline{span}\{f_n\}$ and $T^* f = \{(f, f_n)\} \in l^2$, $\forall\, f \in H$.*

Proof. For any n, $m \in \mathbb{N}$ with $m < n$ and $f \in H$, $\{a_n\} \in l^2$, we have

$$\left|\left(f, \sum_{k=m}^{n} a_k f_k\right)\right|^2 = \left|\sum_{k=m}^{n} \bar{a}_k (f, f_k)\right|^2 \le \left(\sum_{k=m}^{n} |a_k|^2\right)\left(\sum_{k=m}^{n} |(f, f_k)|^2\right)$$
$$\le B\left(\sum_{k=m}^{n} |a_k|^2\right)\|f\|^2$$

and consequently,

$$\left\|\sum_{k=m}^{n} a_k f_k\right\| \le \sqrt{B}\left(\sum_{k=m}^{n} |a_k|^2\right)^{1/2}. \tag{5.2.1}$$

Since $\sum_{n=1}^{\infty} |a_n|^2 < \infty$, it follows from (5.2.1) that $\sum_{k=m}^{n} a_k f_k$ is a Cauchy sequence in H. Thus $\sum_{k=1}^{n} a_k f_k$ converges to an element $T(\{a_n\})$ in H in norm and so that $\|T(\{a_n\})\| \le \sqrt{B}\|\{a_n\}\|$, i.e., $\|T\| \le \sqrt{B}$.

Obviously, from the definition of T, we obtain that $\overline{\text{Ran}\,(T)} = \overline{span}\{f_n\}$.

For any $\{a_n\} \in l^2$ and $f \in H$, we have

$$(T^* f, \{a_n\}) = (f, T(\{a_n\})) = \sum_{n=1}^{\infty} \bar{a}_n (f, f_n)$$
$$= (\{(f, f_n)\}, \{a_n\})$$

for $\sum_{n=1}^{\infty} |(f, f_n)|^2 \le B\|f\|^2$. This means that $T^* f = \{(f, f_n)\}$, $\forall\, f \in H$. $\square$

Definition 5.2.5. The operator defined in Lemma 5.2.4 is called the pre–frame operator associated with the Bessel sequence $\{f_n\}$. $S = TT^* \in B(H)$ is called the frame operator.

Clearly, the frame operator S has the form $Sf = \sum\limits_{n=1}^{\infty} (f, f_n)f_n, \forall f \in H$.

Proposition 5.2.6. *Let $\{f_n\}$ be a Bessel sequence and T be the pre–frame operator associated with $\{f_n\}$. Then*

(1) *$\{f_n\}$ is a frame for $\overline{span}\{f_n\}$ iff $\operatorname{Ran}(T)$ is closed.*
(2) *$\{f_n\}$ is a frame for H iff $\operatorname{Ran}(T) = H$.*

Proof. (1) Noting that

$$0 \le (Sf, f) = \sum_{n=1}^{\infty} |(f, f_n)|^2, \quad \forall f \in H,$$

we get that $\operatorname{Ker} S = (\overline{span}\{f_n\})^{\perp}$.

If $\operatorname{Ran}(T)$ is closed, then so is $\operatorname{Ran}(T^*)$ by Theorem 1.2.17. Consequently, $\gamma(T) = \gamma(T^*) > 0$ and

$$\|T^*f\| \ge \gamma(T)\|f\|, \ \forall f \in (\operatorname{Ker} T^*)^{\perp} = (\operatorname{Ker} S)^{\perp} = \overline{span}\{f_n\}.$$

So we have, for any $f \in \overline{span}\{f_n\}$,

$$\gamma^2(T)\|f\|^2 \le (TT^*f, f) = \sum_{n=1}^{\infty} |(f, f_n)|^2 \le \|S\|\|f\|^2, \tag{5.2.2}$$

that is, $\{f_n\}$ is a frame for $\overline{span}\{f_n\}$.

Conversely, if $\{f_n\}$ is a frame for $\overline{span}\{f_n\}$, then

$$\|T^*f\|^2 = (Sf, f) = \sum_{n=1}^{\infty} |(f, f_n)|^2 \ge A\|f\|^2, \ \forall f \in \overline{span}\{f_n\}.$$

and hence $\gamma(T) \ge \sqrt{A} > 0$, i.e., $\operatorname{Ran}(T)$ is closed.

(2) We have $\|T^*f\|^2 = (Sf, f) \ge A\|f\|^2, \forall f \in H$ when $\{f_n\}$ is a frame for H. Thus, $\operatorname{Ker} T^* = \{0\}$, $\operatorname{Ran}(T^*)$ is closed and $\operatorname{Ran}(T)$ is closed too. Therefore, $\operatorname{Ran}(T) = (\operatorname{Ker} T^*)^{\perp} = H$.

Conversely, if $\operatorname{Ran}(T) = H$, then $H = \overline{span}\{f_n\}$. So (5.2.2) holds for every $f \in H$, that is, $\{f_n\}$ is a frame for H. □

We now discuss the perturbation of a frame sequence in a Hilbert space.

Let $\{f_n\}$ be a frame for $\overline{span}\{f_n\}$ and let T be the pre–frame operator for $\{f_n\}$. Put $S = TT^*$. Then $\operatorname{Ran}(T)$ and $\operatorname{Ran}(S)$ are all closed by Proposition 5.2.6 and moreover, $T^\dagger = T^*S^\dagger$ by Proposition 3.1.8. So

$$T^\dagger f = \{(S^\dagger f, f_n)\} = \{(f, S^\dagger f_n)\}, \quad \forall f \in H.$$

We now give two kinds of perturbation results about the frame sequence $\{f_n\} \subset H$ as follows.

Theorem 5.2.7. *Let $\{f_n\}$, T and S be as above. Let $\{g_n\} \subset H$ be a Bessel sequence and let G be the pre-frame operator for $\{g_n\}$. Suppose that $\sum_{n=1}^{\infty} \|S^\dagger f_n\| \|g_n - f_n\| = C < 1$ and $\overline{span}\{g_n\} \cap (\overline{span}\{f_n\})^\perp = \{0\}$. Then $\{g_n\}$ is a frame for $\overline{span}\{g_n\}$.*

Proof. We have $(G-T)T^\dagger f = \sum_{n=1}^{\infty} (f, S^\dagger f_n)(g_n - f_n)$, $\forall f \in H$ by Lemma 5.2.4. Thus,

$$\|(G-T)T^\dagger f\| \le \sum_{n=1}^{\infty} \|f\| \|S^\dagger f_n\| \|g_n - f_n\| = C\|f\|, \quad \forall f \in H$$

and hence $\|(G-T)T^\dagger\| \le C < 1$. So $I + (G-T)T^\dagger$ and $I + T^\dagger(G-T)$ are all invertible in $B(H)$. Note that

$$\text{Ran}\,(T) = \overline{span}\{f_n\} \text{ and } \overline{\text{Ran}\,(G)} = \overline{span}\{g_n\}$$

by Lemma 5.2.4. So $\text{Ran}\,(G) \cap \text{Ran}\,(A)^\perp = \{0\}$ by the second assumption and consequently, $\text{Ran}\,(G)$ is closed with $\|G^\dagger\| \le \frac{\|A^\dagger\|}{1-C}$ by Theorem 3.3.5 and furthermore, $\{g_n\}$ is a frame for $\overline{span}\{g_n\}$ by Proposition 5.2.6 and its lower bound is $\gamma^2(G) = \frac{1}{\|G^\dagger\|^2} \ge \frac{(1-C)^2}{\|T^\dagger\|^2}$ by (5.2.2). □

Theorem 5.2.8. *Let $\{f_n\} \subset H$ be a frame sequence with bounds A, B. Let $\{g_n\} \subset H$ and assume that there exists numbers λ_1, $\mu \in \mathbb{R}_+$ and $\lambda_2 \in [0,1)$ such that*

$$\Big\|\sum_{k=1}^{n} c_k(g_k - f_k)\Big\| \le \lambda_1 \Big\|\sum_{k=1}^{n} c_k f_k\Big\| + \lambda_2 \Big\|\sum_{k=1}^{n} c_k g_k\Big\| + \mu\Big(\sum_{k=1}^{n} |c_k|^2\Big)^{1/2} \quad (5.2.3)$$

for all $c_1, \cdots, c_n \in \mathbb{C}$, $\forall n \in \mathbb{N}$. Then $\{g_n\}$ is a Bessel sequence with upper bound $\left(1 + \frac{\lambda_1 + \lambda_2 + \mu B^{-\frac{1}{2}}}{1-\lambda_2}\right)^2 B$.

Let G and T be the pre-frame operators for $\{g_n\}$ and $\{f_n\}$ respectively. If $\gamma = \delta(\text{Ker}\,T, \text{Ker}\,G) < 1$ and $\lambda_1 + \frac{\mu}{A^{1/2}(1-\gamma^2)^{1/2}} < 1$, then $\{g_n\}$ is a frame sequence with lower bound $(1-\gamma^2)\left(1 - \frac{\lambda_1 + \lambda_2 + \frac{\mu}{A^{1/2}(1-\gamma^2)^{1/2}}}{1+\lambda_2}\right)^2 A$.

Proof. From (5.2.3), we get that, for any $c_1, \cdots, c_n \in \mathbb{C}$,

$$\Big\|\sum_{k=1}^{n} c_k g_k\Big\| \le \Big\|\sum_{k=1}^{n} c_k(g_k - f_k)\Big\| + \Big\|\sum_{k=1}^{n} c_k f_k\Big\|$$
$$\le (1+\lambda_1)\Big\|\sum_{k=1}^{n} c_k f_k\Big\| + \lambda_2\Big\|\sum_{k=1}^{n} c_k g_k\Big\| + \mu\Big(\sum_{k=1}^{n} |c_k|^2\Big)^{1/2},$$

so

$$\Big\|\sum_{k=1}^{n} c_k g_k\Big\| \le \frac{1+\lambda_1}{1-\lambda_2}\Big\|\sum_{k=1}^{n} c_k f_k\Big\| + \frac{\mu}{1-\lambda_2}\Big(\sum_{k=1}^{n} |c_k|^2\Big)^{1/2}, \ \forall \{c_k\}_1^n. \quad (5.2.4)$$

Then for any $c_m, \cdots, c_n$ $(m < n)$, we have, by (5.2.1) and (5.2.4),

$$\Big\|\sum_{k=m}^{n} c_k g_k\Big\| \le \Big(\frac{1+\lambda_1}{1-\lambda_2}\sqrt{B} + \frac{\mu}{1-\lambda_2}\Big)\Big(\sum_{k=m}^{n} |c_k|^2\Big)^{1/2}. \quad (5.2.5)$$

This means that $\sum\limits_{k=m}^{n} c_k g_k$ is a Cauchy sequence when $\{c_n\} \in l^2$. Thus, we can define a linear operator $G\colon l^2 \to H$ by $G(\{c_n\}) = \sum\limits_{n=1}^{\infty} c_k g_k$ with $\|G\| \le \dfrac{(1+\lambda_1)\sqrt{B}+\mu}{1-\lambda_2}$ by letting $m = 1$ and $n \to \infty$ in (5.2.5).

For any $g \in H$ and $k \in \mathbb{N}$, set $c_n = \begin{cases} (g, g_n) & n \le k \\ 0 & n > k \end{cases}$. Then

$$|(G(\{c_n\}), g)| = \sum_{i=1}^{k} |(g, g_i)|^2 \le \|G(\{c_n\})\|\|g\| \le \|G\|\Big(\sum_{i=1}^{k} |(g, g_i)|^2\Big)^{1/2}\|g\|$$

and consequently, we get that

$$\sum_{n=1}^{\infty} |(g, g_n)|^2 \le \|G\|^2\|g\|^2 \le \|g\|^2\Big(1 + \frac{\lambda_1 + \lambda_2 + \mu B^{-\frac{1}{2}}}{1-\lambda_2}\Big)^2 B, \ \forall g \in H.$$

Now from (5.2.3), we have

$$\|(G-T)(\{c_n\})\| \le \lambda_1\|T(\{c_n\})\| + \lambda_2\|G(\{c_n\})\| + \mu\|(\{c_n\})\|, \ \forall \{c_n\} \in l^2.$$

Thus, for any $\{c_n\} \in l^2$,

$$\|G(\{c_n\})\| \ge \|T(\{c_n\})\| - \|(G-T)(\{c_n\})$$
$$\ge (1-\lambda_1)\|T(\{c_n\})\| - \lambda_2\|G(\{c_n\})\| - \mu\|\{c_n\}\|$$

and so that

$$\|G(\{c_n\})\| \ge \frac{1-\lambda_1}{1+\lambda_2}\|T(\{c_n\})\| - \frac{\mu}{1+\lambda_2}\|\{c_n\}\|.$$

Therefore,

$$\gamma(G) \geq \frac{1-\lambda_1}{1+\lambda_2}\gamma(T)(1-\gamma^2)^{1/2} - \frac{\mu}{1+\lambda_2} \tag{5.2.6}$$

by Lemma 1.3.6. Since $\gamma(T) \geq \sqrt{A} > 0$ (see the proof of Proposition 5.2.6), it follows from (5.2.6) that $\gamma(G) > 0$ when $\gamma < 1$ and $\lambda_1 + \frac{\mu}{A^{1/2}(1-\gamma^2)^{1/2}} < 1$. In this case, $\{g_n\}$ is a frame sequence with lower bound

$$\gamma^2(G) \geq (1-\gamma^2)\left(1 - \frac{\lambda_1 + \lambda_2 + \frac{\mu}{A^{1/2}(1-\gamma^2)^{1/2}}}{1+\lambda_2}\right)^2 A$$

by Proposition 5.2.6 and (5.2.6). □

5.3 Generalized Cowen–Douglas mappings

Let H be a separable complex Hilbert space and D be a bounded domain in $\mathbb{C}$. According to [Cowen and Douglas (1978)], the subset $B_n(D)$ in $B(H)$ consists of operators T satisfying:

(1) $D \subset \sigma(T)$ and $\dim \operatorname{Ker}(T-z) = n$ $(1 \leq n < \infty)$, $\forall\, z \in D$.
(2) $\operatorname{Ran}(T-z) = H$ for all $z \in D$.
(3) $\bigvee_{z \in D} \operatorname{Ker}(T-z) = H$.

An operator in $B_n(D)$ is called the Cowen–Douglas operator. Such operator has been widely investigated from various aspects, such as the unitary equivalence, similar equivalence of two Cowen–Douglas operators and the irreducibility, strongly irreducibility of a Cowen–Douglas operator etc. Please see [Cowen and Douglas (1978)], [Jiang and Wang (1998)], [Xue (1989)] and [Xue and Wang (1999)] for details.

In this section, we will introduce a notation called the generalized Cowen–Douglas mapping which is a generalization of the Cowen–Douglas operator on Hilbert space. Throughout the section, X is always a separable Banach space.

Definition 5.3.1. Let Ω be a domain in $\mathbb{C}^n$ and $T\colon \Omega \to B(X)$ be a holomorphic mapping. If T satisfies following conditions:

(1) for any $\omega \in \Omega$, $T(\omega) \in Gib(X)$,
(2) there is $k \in \mathbb{N}$ such that $\dim \operatorname{Ker}(T(\omega)) = k$, $\forall\, \omega \in \Omega$,
(3) $T(\omega)\dfrac{\partial T(\omega)}{\partial \omega_j} = \dfrac{\partial T(\omega)}{\partial \omega_j}T(\omega)$, $j = 1, \cdots, n$, $\forall\, \omega = (\omega_1, \cdots, \omega_n) \in \Omega$,

then we call T is a generalized Cowen–Douglas mapping over Ω.

We write $G_k(\Omega, X)$ to denote set of all generalized Cowen–Douglas mappings over Ω.

Lemma 5.3.2. *Let $T \in G_k(\Omega, X)$. Then for any $\omega_0 \in \Omega$, there is an open neighborhood $U(\omega_0)$ of ω_0 in Ω and a holomorphic mapping $\omega \mapsto X(\omega)$ of $U(\omega_0)$ to $B(X)$ with $X(\omega_0) = I$ such that $X(\omega)$ is invertible and $\operatorname{Ker} T(\omega) = X(\omega) \operatorname{Ker} T(\omega_0)$, $\forall\, \omega \in U(\omega_0)$.*

Proof. Choose an open neighborhood $U(\omega_0)$ of ω in Ω such that

$$\|T(\omega) - T(\omega_0)\| < \|T^+(\omega_0)\|^{-1}, \quad \forall\, \omega \in U(\omega_0).$$

Put $Z(\omega) = I + T^+(\omega_0)(T(\omega) - T(\omega_0))$, $\omega \in U(\omega_0)$. Then $Z(\omega)$ is holomorphic and $\|I - Z(\omega)\| < 1$, $\forall\, \omega \in U(\omega_0)$. Thus,

$$X(\omega) = (Z(\omega))^{-1} = \sum_{n=0}^{\infty} (I - Z(\omega))^n, \ X(\omega_0) = I$$

is invertible for each $\omega \in U(\omega_0)$ and is holomorphic on $U(\omega_0)$.

Since $\dim \operatorname{Ker}(T(\omega)) = \dim \operatorname{Ker}(T(\omega_0)) = k$ and $Z(\omega)$ is invertible in $B(X)$, $\forall\, \omega \in U(\omega_0)$, it follows from Corollary 2.2.8 and Corollary 2.2.7 that $\operatorname{Ker} T(\omega) = X(\omega) \operatorname{Ker} T(\omega_0)$, $\forall\, \omega \in U(\omega_0)$. □

The following corollary is a generalization of Corollary 1.13 in [Cowen and Douglas (1978)].

Corollary 5.3.3. *Let $T \in G_k(\Omega, X)$ and let Γ be a subset of Ω containing at least one interior point. Then $\bigvee_{\omega \in \Gamma} \operatorname{Ker} T(\omega) = \bigvee_{\omega \in \Omega} \operatorname{Ker} T(\omega)$.*

Proof. Put $X_T = \bigvee_{\omega \in \Omega} \operatorname{Ker} T(\omega)$ and $X_0 = \bigvee_{\omega \in \Gamma} \operatorname{Ker} T(\omega)$. Obviously, $X_0 \subset X_T$. If $X_0 \neq X_T$, then there is $x_0 \in X_T \backslash X_0$ and $f_0 \in (X_T)^*$ such that $f_0(x_0) = 1$ and $f_0|_{X_0} \equiv 0$. Let f be an extension of f_0 on X with $\|f\| = \|f_0\|_{X_T}$. Set $\Delta = \{\omega \in \Omega |\ f|_{\operatorname{Ker} T(\omega)} \equiv 0\}$. Then $\Gamma \subset \Delta$ and Δ° (the interior of Δ) is nonempty.

Let $\omega_0 \in \Omega$ and $\{\omega_n\} \subset \Delta^\circ$ such that $\lim\limits_{n \to \infty} \omega_n = \omega_0$. By Lemma 5.3.2, there is an open neighborhood $U(\omega_0)$ of ω_0 in Ω and a holomorphic invertible mapping $\omega \mapsto X(\omega)$ of $U(\omega_0)$ to $B(X)$ with $X(\omega_0) = I$ such that $\operatorname{Ker} T(\omega) = X(\omega) \operatorname{Ker} T(\omega_0)$, $\forall\, \omega \in U(\omega_0)$. Thus, for any $\xi \in \operatorname{Ker} T(\omega_0)$, $\xi(\omega) = X(\omega)\xi$ is holomorphic on $U(\omega_0)$. Since $U(\omega_0) \cap \Delta^\circ$ is a nonempty open subset in $U(\omega_0)$ and $f(\xi(\omega)) = 0$, $\forall\, \omega \in U(\omega_0) \cap \Delta^\circ$, it follows from Theorem A.2.2 (1) that $f(\xi(\omega))$ vanishes on $U(\omega_0)$. Consequently, $f|_{\operatorname{Ker} T(\omega_0)} \equiv 0$, i.e., Δ° is closed in Ω. Since Δ° is open in Ω and Ω is connected, we have $\Delta^\circ = \Omega = \Delta$, a contradiction. □

Let $T \in G_k(\Omega, X)$. Since $T(\omega)\dfrac{\partial T(\omega)}{\partial \omega_j} = \dfrac{\partial T(\omega)}{\partial \omega_j}T(\omega)$, it follows that $\dfrac{\partial T(\omega)}{\partial \omega_j}(\operatorname{Ker} T(\omega)) \subset \operatorname{Ker} T(\omega)$, $\forall\, \omega \in \Omega$ and $j = 1, \cdots, n$. So $D_j^T(\omega) \triangleq \dfrac{\partial T(\omega)}{\partial \omega_j}\Big|_{\operatorname{Ker} T(\omega)} \in B(\operatorname{Ker} T(\omega))$, $\forall\, \omega \in \Omega$ and $j = 1, \cdots, n$.

Definition 5.3.4. Let $T \in G_k(\Omega, X)$. We call T is nonsingular if for any $\omega_0 \in \Omega$, there is $i \in \{1, \cdots, n\}$ such that $D_i^T(\omega_0)$ is invertible in $B(\operatorname{Ker} T(\omega_0))$.

Corollary 5.3.5. *Let $T \in G_k(\Omega, X)$ be nonsingular. Then for each $\omega_0 \in \Omega$, there is an open neighborhood $V(\omega_0)$ of ω_0 in Ω and an $i \in \{1, \cdots, n\}$ such that $D_i^T(\omega)$ is invertible in $B(\operatorname{Ker} T(\omega))$, $\forall\, \omega \in V(\omega_0)$.*

Proof. Since T is nonsingular, there is $i \in \{1, \cdots, n\}$ such that $D_i^T(\omega_0)$ is invertible in $B(\operatorname{Ker} T(\omega_0))$. By Lemma 5.3.2, there is an open neighborhood $U(\omega_0)$ of ω_0 in Ω and a holomorphic invertible mapping $\omega \mapsto X(\omega)$ of $U(\omega_0)$ to $B(X)$ such that $X(\omega_0) = I$ and $\operatorname{Ker} T(\omega) = X(\omega) \operatorname{Ker} T(\omega_0)$, $\forall\, \omega \in U(\omega_0)$.

Put $H(\omega) = X^{-1}(\omega) D_i^T(\omega) X(\omega) \in B(\operatorname{Ker} T(\omega_0))$, $\omega \in U(\omega_0)$. Note that $\dim \operatorname{Ker}(T(\omega_0)) < \infty$ and

$$H(\omega)\xi = X^{-1}(\omega)\frac{\partial T(\omega)}{\partial \omega_i}X(\omega)\xi, \quad \forall\, \xi \in \operatorname{Ker}(T(\omega_0)),\ \omega \in U(\omega_0).$$

So $\lim\limits_{\omega \to \omega_0} \|H(\omega) - H(\omega_0)\| = 0$. Since $H(\omega_0)$ is invertible, it follows that there is an open neighborhood $V(\omega_0)$ of ω_0 in $U(\omega_0)$ such that

$$\|H(\omega) - H(\omega_0)\| < \frac{1}{\|H^{-1}(\omega_0)\|}, \quad \forall\, \omega \in V(\omega_0).$$

Then $H(\omega)$ is invertible for any $\omega \in V(\omega_0)$. Therefore, $D_i^T(\omega)$ is invertible in $B(\operatorname{Ker} T(\omega))$, $\forall\, \omega \in V(\omega_0)$. □

Let $\alpha = (\alpha_1, \cdots, \alpha_n) \in \mathbb{Z}_+^n$. Set $|\alpha| = \sum\limits_{i=1}^n \alpha_i$ and $D^\alpha = \dfrac{\partial^{|\alpha|}}{\partial \omega_1^{\alpha_1} \cdots \partial \omega_n^{\alpha_n}}$, for $\omega = (\omega_1, \cdots, \omega_n) \in \Omega$.

Let $\omega_0 = (\omega_1^0, \cdots, \omega_n^0) \in \Omega$ and let $T \in G_k(\Omega, X)$. Choose a basis $\{r_j\}_1^k$ for $\operatorname{Ker} T(\omega_0)$. By Lemma 5.3.2, there is an open neighborhood $U(\omega_0)$ of ω_0 in ω and a holomorphic invertible mapping $X(\omega)$ of $U(\omega_0)$ to $B(X)$ with

$X(\omega_0) = I$ and $\operatorname{Ker} T(\omega) = X(\omega)\operatorname{Ker} T(\omega_0)$, $\forall\, \omega \in U(\omega_0)$. Put $r_j(\omega) = X(\omega)r_j$, $j = 1, \cdots, k$, $\omega \in U(\omega_0)$. Then $\{r_j(\omega)\}_1^k$ is a basis for $\operatorname{Ker} T(\omega)$, $\forall\, \omega \in U(\omega_0)$.

From $T(\omega)r_j(\omega) = 0$, $j = 1, \cdots, n$, $\forall\, \omega \in U(\omega_0)$ we get that

$$\frac{\partial T(\omega)}{\partial \omega_i} r_j(\omega) + T(\omega)\frac{\partial r_j(\omega)}{\partial \omega_i} = 0, \quad i = 1, \cdots, n, \ j = 1, \cdots, k.$$

Thus,

$$T^2(\omega)\frac{\partial r_j(\omega)}{\partial \omega_i} = 0 \quad \text{and} \quad \frac{\partial T(\omega)}{\partial \omega_i} r_j(\omega) = -T(\omega)\frac{\partial r_j(\omega)}{\partial \omega_i}. \tag{5.3.1}$$

We can deduce from (5.3.1) that for any $q \in \mathbb{Z}_+$ and $\alpha = (\alpha_1, \cdots, \alpha_n) \in \mathbb{Z}_+^n$,

$$T^{|\alpha|+1}(\omega)D^\alpha r_j(\omega) = 0, \quad T^q(\omega)\frac{\partial^q r_j(\omega)}{\partial \omega_i^q} = (-1)^q q!\Big(\frac{\partial T(\omega)}{\partial \omega_i}\Big)^q r_j(\omega), \tag{5.3.2}$$

$i = 1, \cdots, n$, $j = 1, \cdots, k$ and $\omega = (\omega_1, \cdots, \omega_n) \in U(\omega_0)$.

Lemma 5.3.6. *Let $T \in G_k(\Omega, X)$ and suppose that T is nonsingular. Given $q \geq 0$. Then for each $\omega_0 \in \Omega$, there is an $i \in \{1, \cdots, n\}$, an open neighborhood $W(\omega_0)$ of ω_0 in Ω and a sequence of holomorphic mappings $\{r_j(\omega)\}_1^k$ on $W(\omega_0)$ such that for each $\omega \in W(\omega_0)$, $\Big\{\frac{\partial^s r_j(\omega)}{\partial \omega_i^s}\Big| 0 \leq s \leq q,\ 1 \leq j \leq k\Big\}$ forms a basis for* $\operatorname{Ker} T^{q+1}(\omega)$.

Proof. By Corollary 5.3.5 and above argument, we can find $i \in \{1, \cdots, n\}$, an open neighborhood $W(\omega_0)$ of ω_0 in Ω and holomorphic mappings $r_1(\omega), \cdots, r_k(\omega)$ on $W(\omega_0)$ such that for each $\omega \in W(\omega_0)$, $D_i^T(\omega)$ is invertible in $B(\operatorname{Ker} T(\omega))$ and $\{r_1(\omega), \cdots, r_k(\omega)\}$ forms a basis for $\operatorname{Ker} T(\omega)$.

Fixed $\omega \in W(\omega_0)$ and assume that for any $0 \leq p < q$,

$$\Big\{\frac{\partial^s r_j(\omega)}{\partial \omega_i^s}\Big| 0 \leq s \leq p,\ 1 \leq j \leq k\Big\}$$

is linearly independent. Let $\{c_{s,j} | 0 \leq s \leq p+1,\ 1 \leq j \leq k\} \subset \mathbb{C}$ such that

$$\sum_{s=0}^{p+1}\sum_{j=1}^{s} c_{s,j}\frac{\partial^s r_j(\omega)}{\partial \omega_i^s} = 0. \tag{5.3.3}$$

Applying (5.3.2) to (5.5.3), we get that

$$\begin{aligned} 0 &= T^{p+1}(\omega)\Big(\sum_{s=0}^{p+1}\sum_{j=1}^{k} c_{s,j}\frac{\partial^s r_j(\omega)}{\partial \omega_i^s}\Big) \\ &= (-1)^{(p+1)}(p+1)!(D_i^T(\omega))^{p+1}\Big(\sum_{j=1}^{k} c_{p+1,j} r_j(\omega)\Big). \end{aligned}$$

Thus, $c_{p+1,j} = 0,\ j = 1, \cdots, k$ and consequently, $c_{s,j} = 0,\ 0 \le s \le p$, $1 \le j \le k$ by (5.3.3) and the induction hypothesis.

Now assume that $\left\{\frac{\partial^s r_j(\omega)}{\partial \omega_i^s}\,\middle|\, 0 \le s \le p-1,\, 1 \le j \le k\right\}$ forms a basis for $\operatorname{Ker} T^p(\omega)$ $(p \ge 1)$. Let $\eta \in \operatorname{Ker} T^{p+1}(\omega)$. Then $T^p(\omega)\eta \in \operatorname{Ker} T(\omega)$. Noting that $\left\{T^p(\omega)\frac{\partial^p r_j(\omega)}{\partial \omega_i^p}\,\middle|\, 1 \le j \le k\right\}$ is contained in $\operatorname{Ker} T(\omega)$ by (5.3.2) and is also linearly independent by (5.3.2) and that $D_i^T(\omega)$ is invertible in $B(\operatorname{Ker} T(\omega))$, we can find $\{g_j\}_1^k \subset \mathbb{C}$ such that

$$T^p(\omega)\eta = \sum_{j=1}^{k} g_j T^p(\omega)\frac{\partial^p r_j(\omega)}{\partial \omega_i^p}.$$

Thus, $\eta - \sum\limits_{j=1}^{k} g_j \frac{\partial^p r_j(\omega)}{\partial \omega_i^q} \in \operatorname{Ker} T^p(\omega)$. Therefore, by the induction hypothesis, η can be expressed as a linear combinations of elements in

$$\left\{\frac{\partial^s r_j(\omega)}{\partial \omega_i^s}\,\middle|\, 0 \le s \le p,\, 1 \le j \le k\right\}.$$

□

From Lemma 5.3.6, we can obtain

Corollary 5.3.7. *Let T be in Lemma 5.3.6. Then* $\dim \operatorname{Ker} T^q(\omega) = qk$, $q \ge 1$ *and* $\forall\, \omega \in \Omega$.

Now we present the main result of this section as follows:

Theorem 5.3.8. *Let $T \in G_k(\Omega, X)$ and let $\omega_0 \in \Omega$. Put*

$$X_T = \bigvee_{\omega \in \Omega} \operatorname{Ker} T(\omega) \text{ and } K_T = \bigvee_{s=1}^{\infty} \operatorname{Ker} T^s(\omega_0).$$

Then $X_T \subset K_T$ and $T(\omega_0)X_T \subset K_T$. In addition, if T is nonsingular, then $X_T = K_T$ and $T(\omega_0)X_T = K_T$.

Proof. (1) By Lemma 5.3.2, we have an open neighborhood $U(\omega_0)$ of ω_0 in Ω and holomorphic mappings $r_1(\omega), \cdots, r_k(\omega)$ of $U(\omega_0)$ to X such that $\{r_j(\omega)\}_1^k$ forms a basis for $\operatorname{Ker} T(\omega)$, $\forall\, \omega \in U(\omega_0)$. According to Theorem A.2.2 (2), there exists an open neighborhood $U_0(\omega_0)$ of ω_0 in $U(\omega_0)$ such that

$$r_j(\omega) = \sum_{\alpha} \frac{D^\alpha r_j(\omega_0)}{\alpha!}(\omega_1 - \omega_1^0)^{\alpha_1} \cdots (\omega_n - \omega_n^0)^{\alpha_n}, \tag{5.3.4}$$

$\forall\,\omega = (\omega_1, \cdots, \omega_n) \in U_0(\omega_0)$, $1 \le j \le k$, where $\alpha\,! = \alpha_1\,! \cdots \alpha_n\,!$. Then we have $r_j(\omega) \in K_T$ by (5.3.2) and (5.3.4), $\forall\, 1 \le j \le k$ and $\omega \in U_0(\omega_0)$. Thus $X_T = \bigvee\limits_{\omega \in U_0(\omega_0)} \operatorname{Ker} T(\omega) \subset K_T$ by Corollary 5.3.3.

Since $T(\omega_0)\big[D^\alpha r_j(\omega_0)\big] \subset \operatorname{Ker} T^{|\alpha|}(\omega_0)$ by (5.3.2), $\forall\, j = 1, \cdots, k$ and $\alpha \in \mathbb{Z}_+^n$ with $|\alpha| > 0$, it follows from (5.3.4) that $T(\omega_0) r_j(\omega) \subset K_T$ and hence $T(\omega_0) X_T \subset K_T$, $\forall\, \omega \in U_0(\omega_0)$.

Now suppose that T is nonsingular. Let i, $W(\omega_0)$ and $\{r_j(\omega)\}_1^k$ be as in Lemma 5.3.6. Pick $R_1, \cdots, R_n > 0$ such that

$$\sigma = \{(\omega_1, \cdots, \omega_n) \in \Omega \,|\, |\omega_s - \omega_s^0| = R_j,\ 1 \le s \le n\} \subset W(\omega_0).$$

Then

$$\frac{\partial^m r_j}{\partial \omega_i^m}\Big|_{\omega=\omega_0} = \frac{m!}{(2\pi i)^n} \oint_\sigma \frac{r_j(\omega)\, d\omega_1 \cdots d\omega_n}{(\omega_1 - \omega_1^0) \cdots (\omega_{i_0} - \omega_i^0)^{m+1} \cdots (\omega_n - \omega_n^0)} \tag{5.3.5}$$

by Theorem A.2.2 (2), $m = 0, 1, \cdots$, $j = 1, \cdots, k$. Since

$$\Big\{ \frac{\partial^m r_j}{\partial \omega_i^m}\Big|_{\omega=\omega_0} \;\Big|\; 0 \le m \le q,\ 1 \le j \le k \Big\}$$

forms a basis for $T^{q+1}(\omega_0)$ by Lemma 5.3.6, it follows from (5.3.5) that $K_T \subset X_T$ and so that $X_T = K_T$.

Note that the set $\Big\{ T(\omega_0)\Big(\frac{\partial^s r_j}{\partial \omega_i^s}\Big|_{\omega=\omega_0} \Big) \,\Big|\, 0 \le s \le q,\ 1 \le j \le k \Big\}$ consisting of qk vectors is linearly independent by Lemma 5.3.6 and contained in $\operatorname{Ker} T^q(\omega_0)$. Since $\dim \operatorname{Ker} T^q(\omega_0) = qk$ by Corollary 5.3.7, it follows that it forms a basis for $\operatorname{Ker} T^q(\omega_0)$. So, $K_T \subset \overline{T(\omega_0) X_T} \subset K_T$ by (5.3.5). Consider the operator $T_X = T(\omega_0)\big|_{X_T} \in B(X_T, K_T)$. Since $\operatorname{Ker} T(\omega_0) \subset X_T$, we have $\operatorname{Ker} T_X = \operatorname{Ker} T(\omega_0)$. Using the fact that $\operatorname{Ran}(T(\omega_0))$ is closed in X, we obtain that, for any $z \in X_T$,

$$\|T_X(z)\| = \|T(\omega_0) z\| \ge \gamma(T(\omega_0)) \operatorname{dist}(z, \operatorname{Ker} T_X).$$

Therefore, $\operatorname{Ran}(T_X)$ is closed in K_T and consequently, $T(\omega_0) X_T = K_T$. □

Since $T \in Gib(H)$ iff $\operatorname{Ran}(T)$ is closed in H, we have

Corollary 5.3.9. *Let $T\colon \Omega \to B(H)$ be a holomorphic mapping satisfying following:*

(1) $T(\omega) \dfrac{\partial T(\omega)}{\partial \omega_i} = \dfrac{\partial T(\omega)}{\partial \omega_i} T(\omega)$, $\forall\, \omega \in \Omega$, $i = 1, \cdots, n$.
(2) *T is nonsingular on Ω.*
(3) $\operatorname{Ran}(T(\omega))$ *is closed and* $\dim \operatorname{Ker} T(\omega) = n\, (1 \le n \le \infty)$, $\forall\, \omega \in \Omega$.

Then $\bigvee\limits_{\omega \in \Omega} \operatorname{Ker} T(\omega) = \bigvee\limits_{k=1}^{\infty} \operatorname{Ker} T^k(\omega_0)$, $\forall\, \omega \in \Omega$.

5.4 Condition numbers related to Moore–Penrose inverses and Drazin inverses

Let X be a Banach space and let $A \in GL(B(X))$. Suppose that A has a small perturbation $\delta A \in B(X)$ such that $A+\delta A \in GL(B(X))$. To measure the sensitivity of A^{-1} under the perturbation δA, one defines the condition number $\mathrm{Cond}(A)$ as

$$\mathrm{Cond}(A) = \lim_{\epsilon \to 0^+} \sup_{\|\delta A\| \le \epsilon \|A\|} \frac{\|(A+\delta A)^{-1} - A^{-1}\|}{\epsilon \|A^{-1}\|}.$$

Let $b \in X \backslash \{0\}$ and $\delta b \in X$. Let A and $A + \delta A$ be as above. Then the solutions of the operator equations

$$Ax = b, \qquad (A+\delta A)x = b + \delta b$$

are $x = A^{-1}b$ and $x = (A+\delta A)^{-1}(b+\delta b)$ respectively. To measure the sensitivity of the solution of $Ax = b$, one defines the condition number $\mathrm{Cond}(A, b)$ as

$$\mathrm{Cond}(A,b) = \lim_{\epsilon \to 0^+} \sup_{\substack{\|\delta A\| \le \epsilon \|A\| \\ \|\delta b\| \le \epsilon \|b\|}} \frac{\|(A+\delta A)^{-1}(b+\delta b) - A^{-1}b\|}{\epsilon \|A^{-1}b\|}.$$

When $X = \mathbb{C}^n$ and the matrix norm is induced by the vector norm on $\mathbb{C}^n$, it is well-known that

$$\mathrm{Cond}(A) = \|A\|\|A^{-1}\|, \quad \mathrm{Cond}(A,b) = \|A\|\|A^{-1}\| + \frac{\|A^{-1}\|\|b\|}{\|A^{-1}b\|}$$

(cf. [Bartels (1991)], [Demmel (1987)] and [Higham and Higham (1992)]).

For general Banach space X, we also have

Proposition 5.4.1. *Let $A \in GL(B(X))$ and let $b \in X \backslash \{0\}$. Then*

$$\mathrm{Cond}(A) = \|A\|\|A^{-1}\|, \quad \mathrm{Cond}(A,b) = \|A\|\|A^{-1}\| + \frac{\|A^{-1}\|\|b\|}{\|A^{-1}b\|}.$$

Proof. Put $\kappa_A = \|A^{-1}\|\|A\|$. For any $\epsilon \in (0, \kappa_A^{-1})$, let $\delta A \in B(X)$ and $\delta b \in X$ with $\|\delta A\| \le \epsilon \|A\|$ and $\|\delta b\| \le \epsilon \|b\|$. Then $\|A^{-1}\|\|\delta A\| < 1$. Thus, $A + \delta A \in GL(B(X))$ with $(A+\delta A)^{-1} = \sum\limits_{n=0}^{\infty} (-1)^n (A^{-1}\delta A)^n A^{-1}$. Simple computation shows that

$$(A+\delta A)^{-1}-A^{-1}=\sum_{n=1}^{\infty}(-1)^n(A^{-1}\delta A)^nA^{-1}, \tag{5.4.1}$$

$$\begin{aligned}(A+\delta A)^{-1}(b+\delta b)-A^{-1}b &= \sum_{n=1}^{\infty}(-1)^n(A^{-1}\delta A)^nA^{-1}b \\ &\quad+\sum_{n=0}^{\infty}(-1)^n(A^{-1}\delta A)^nA^{-1}\delta b. \end{aligned}\tag{5.4.2}$$

Thus, by (5.4.1) and (5.4.2) respectively, we get that

$$\frac{\|(A+\delta A)^{-1}-A^{-1}\|}{\|A^{-1}\|}\le\frac{\epsilon\|A^{-1}\|\|A\|}{1-\epsilon\|A^{-1}\|\|A\|}$$

$$\frac{\|(A+\delta A)^{-1}(b+\delta b)-A^{-1}b\|}{\|A^{-1}b\|}\le\frac{\epsilon\|A^{-1}\|}{1-\epsilon\|A^{-1}\|\|A\|}\left(\|A\|+\frac{\|\|b\|}{\|A^{-1}b\|}\right)$$

and consequently,

$$\text{Cond}(A)\le\|A\|\|A^{-1}\|,\quad \text{Cond}(A,b)\le\|A\|\|A^{-1}\|+\frac{\|A^{-1}\|\|b\|}{\|A^{-1}b\|}.$$

On the other hand, from (5.4.1), we have

$$\|(A+\delta A)^{-1}-A^{-1}\|\ge\|A^{-1}\delta AA^{-1}\|-\sum_{n=2}^{\infty}\|A^{-1}\|^{n+1}\|\delta A\|^n. \tag{5.4.3}$$

Since $\|A^{-1}\|=\sup_{x\in S(X)}\|A^{-1}x\|=\|(A^{-1})^*\|=\sup_{x^*\in S(X^*)}\|(A^{-1})^*x^*\|$, we can find $x_\epsilon\in S(X)$ and $x_\epsilon^*\in S(X^*)$ such that

$$\|A^{-1}x_\epsilon\|\ge\|A^{-1}\|-\epsilon,\quad \|(A^{-1})^*x_\epsilon^*\|\ge\|A^{-1}\|-\epsilon,$$

$(\epsilon<\|A^{-1}\|)$. Put $\delta A(x)=\epsilon\|A\|x_\epsilon^*(x)x_\epsilon$, $\forall\, x\in X$. Then $\|\delta A\|=\epsilon\|A\|$ and

$$A^{-1}\delta AA^{-1}(x)=\epsilon\|A\|(A^{-1}x_\epsilon^*(A^{-1}x)x_\epsilon)=\epsilon\|A\|A^{-1}(A^{-1})^*x_\epsilon^*(x)x_\epsilon.$$

So $\|A^{-1}\delta AA^{-1}\|\ge(\|A^{-1}\|-\epsilon)^2\epsilon\|A\|$ and hence

$$\begin{aligned}\text{Cond}(A)&\ge\lim_{\epsilon\to0^+}\left(\frac{\|A\|(\|A^{-1}\|-\epsilon)^2}{\|A^{-1}\|}-\sum_{n=2}^{\infty}\|A^{-1}\|^n\|A\|^n\epsilon^{n-1}\right)\\ &=\|A^{-1}\|\|A\|\end{aligned}$$

by (5.4.3).

Now by (5.4.2),

$$\begin{aligned}\|(A+\delta A)^{-1}(b+\delta b)-A^{-1}b\|&\ge\|A^{-1}\delta AA^{-1}b+A^{-1}\delta b\|\\ &\quad-\sum_{n=2}^{\infty}\|A^{-1}\|^n\|\delta A\|^n\|A^{-1}b\|-\sum_{n=1}^{\infty}\|A^{-1}\|^{n+1}\|\delta A\|^n\|\delta b\|.\end{aligned}\tag{5.4.4}$$

Set $\delta A(x) = \epsilon\|A\| f_0(x) x_\epsilon$, $\forall x \in X$ and $\delta b = \epsilon\|b\| x_\epsilon$ in (5.4.4), where $f_0 \in S(X^*)$ with $f_0(A^{-1}b) = \|A^{-1}b\|$. So we get that

$$\begin{aligned}\mathrm{Cond}(A,b) &\geq \lim_{\epsilon\to 0^+}\Big(\|A\| + \frac{\|b\|}{\|A^{-1}b\|}\Big)(\|A^{-1}\| - \epsilon)\\ &= \|A^{-1}\|\|A\| + \frac{\|A^{-1}\|\|b\|}{\|A^{-1}b\|}.\end{aligned}$$

□

If we replace the invertibility of A in Proposition 5.4.1 by Moore–Penrose or Drazin invertibility of A, we have following definitions:

Definition 5.4.2. Let H, K be Hilbert spaces and let $A \in B(H,K)$ with $\mathrm{Ran}\,(A)$ closed. The Moore–Penrose condition number of A is given by

$$\mathrm{Cond}_{mp}(A) = \lim_{\epsilon\to 0^+}\sup\left\{\frac{\|(A+\delta A)^\dagger - A^\dagger\|}{\epsilon\|A^\dagger\|}\,\middle|\, \begin{array}{c}\mathrm{Ran}\,(A+\delta A)\cap \mathrm{Ran}\,(A)^\perp = \{0\}\\ \|\delta A\| \leq \epsilon\|A\|\end{array}\right\}.$$

Let b, $\delta b \in \mathrm{Ran}\,(A)$. The condition number related to the equation $Ax = b$ is defined as

$$\begin{aligned}&\mathrm{Cond}_{mp}(A,b)\\ &\quad= \lim_{\epsilon\to 0^+}\sup\left\{\frac{\|(A+\delta A)^\dagger(b+\delta b) - A^\dagger b\|}{\epsilon\|A^\dagger b\|}\,\middle|\, \begin{array}{c}\mathrm{Ran}\,(A+\delta A)\cap \mathrm{Ran}\,(A)^\perp = \{0\}\\ \|\delta A\| \leq \epsilon\|A\|,\ \|\delta b\| \leq \epsilon\|b\|\end{array}\right\}.\end{aligned}$$

Definition 5.4.3. Let X be Banach spaces and let $A \in Dr(X)$ with $\mathrm{ind}(A) = k$. The Drazin condition number of A is given by

$$\mathrm{Cond}_D(A) = \lim_{\epsilon\to 0^+}\sup\left\{\frac{\|(A+\delta A)^D - A^D\|}{\epsilon\|A^D\|}\,\middle|\, \begin{array}{c}\mathrm{Ran}\,((A+\delta A)^k)\cap \mathrm{Ker}\,(A^D) = \{0\}\\ \|\delta A\| \leq \epsilon\|A\|\end{array}\right\}.$$

The following theorems will give the bounds of $\mathrm{Cond}_{mp}(A)$, $\mathrm{Cond}_{mp}(A,b)$ and $\mathrm{Cond}_D(A)$.

Theorem 5.4.4. *Let H, K be Hilbert spaces and $A \in B(H,K)$ with $\mathrm{Ran}\,(A)$ closed. Let $b \in \mathrm{Ran}\,(A)\backslash\{0\}$. Set $\kappa_{mp}(A) = \|A^\dagger\|\|A\|$. Then*

$$\kappa_{mp}(A) \leq \mathrm{Cond}_{mp}(A) \leq \frac{1+\sqrt{5}}{2}\kappa_{mp}(A)$$
$$\kappa_{mp}(A) + \frac{\|A^\dagger\|\|b\|}{\|A^\dagger b\|} \leq \mathrm{Cond}_{mp}(A,b) \leq 2\kappa_{mp}(A) + \frac{\|A^\dagger\|\|b\|}{\|A^\dagger b\|}.$$

Proof. For any $0 < \epsilon < (\|A^\dagger\|\|A\|)^{-1}$ and $\|\delta A\| \leq \epsilon\|A\|$, we have $\|A^\dagger\|\|\delta A\| < 1$. Since $\mathrm{Ran}\,(A+\delta A)\cap\mathrm{Ran}\,(A)^\perp = \{0\}$, it follows from

Theorem 3.3.5 that

$$\frac{\|(A+\delta A)^\dagger - A^\dagger\|}{\|A^\dagger\|} \le \frac{1+\sqrt{5}}{2}\frac{\|A^\dagger\|\|\delta A\|}{1-\|A^\dagger\|\|\delta A\|} \le \frac{1+\sqrt{5}}{2}\frac{\epsilon\|A^\dagger\|\|A\|}{1-\epsilon\|A^\dagger\|\|A\|}. \tag{5.4.5}$$

(5.4.5) implies that $\mathrm{Cond}_{mp}(A) \le \frac{1+\sqrt{5}}{2}\|A^\dagger\|\|A\|$.

Choose $x \in S(H)$ such that $\|A^\dagger x\| > \|A^\dagger\| - \epsilon$ ($\epsilon < \|A^\dagger\|$). Set $x_1 = \frac{AA^\dagger x}{\|A^\dagger Ax\|}$. Then $x_1 \in \mathrm{Ran}\,(A)$ with $\|x_1\| = 1$ and $\|A^\dagger x_1\| = \|Ax\|\|AA^\dagger x\|^{-1} > \|A^\dagger\| - \epsilon$.

Similarly, we can find $y_1 \in \mathrm{Ran}\,(A^*) = (\mathrm{Ker}\,A)^\perp$ with $\|y_1\| = 1$ such that $\|(A^*)^\dagger y_1\| > \|(A^*)^\dagger\| - \epsilon$.

Put $\delta A(x) = \epsilon\|A\|(x, y_1)x_1$, $\forall x \in H$. Then $\mathrm{Ran}\,(\delta A) \subset \mathrm{Ran}\,(A)$, $\mathrm{Ker}\,A \subset \mathrm{Ker}\,\delta A$ and moreover,

$$\|\delta A\| = \epsilon\|A\|, \quad \|A^\dagger\delta AA^\dagger\| > \epsilon\|A\|(\|A^\dagger\| - \epsilon)^2.$$

Thus, by Proposition 3.4.1, $(A+\delta A)^\dagger = (I + A^\dagger\delta A)^{-1}A^\dagger$ and consequently,

$$\|(A+\delta A)^\dagger - A^\dagger\| \ge \|A^\dagger\delta AA^\dagger\| - \sum_{n=2}^{\infty}\|A^\dagger\|^n\|\delta A\|^n \ge \epsilon\|A\|(\|A^\dagger\| - \epsilon)^2 - O(\epsilon^2). \tag{5.4.6}$$

(5.4.6) gives $\mathrm{Cond}_{mp}(A) \ge \|A^\dagger\|\|A\|$.

Since $b \in \mathrm{Ran}\,(A)$, it follows from Theorem 3.3.7 that, for $\|\delta b\| \le \epsilon\|b\|$ and $\|\delta A\| \le \epsilon\|A\|$ with $\mathrm{Ran}\,(A+\delta A) \cap \mathrm{Ran}\,(A)^\perp = \{0\}$,

$$\begin{aligned}\frac{\|(A+\delta A)^\dagger(b+\delta b) - A^\dagger b\|}{\|A^\dagger b\|} &\le \frac{\|A^\dagger\|}{1-\|A^\dagger\|\|\delta A\|}\left[\frac{\|\delta b\|}{\|A^\dagger b\|} + \|\delta A\|\right] + \|A^\dagger\|\|\delta A\| \\ &\le \frac{\epsilon\|A^\dagger\|}{1-\epsilon\kappa_{mp}(A)}\left[\frac{\|b\|}{\|A^\dagger b\|} + \|A\|\right] + \epsilon\kappa_{mp}(A),\end{aligned}$$

which implies that $\mathrm{Cond}_{mp}(A, b) \le 2\kappa_{mp}(A) + \frac{\|A^\dagger\|\|b\|}{\|A^\dagger b\|}$.

Let x_1 be as above. Put $\delta b = \epsilon\|b\|x_1 \in \mathrm{Ran}\,(A)$ and define

$$\delta A(x) = \frac{\epsilon\|A\|}{\|A^\dagger b\|}(x, A^\dagger b)x_1, \quad \forall x \in H.$$

Then $\|\delta A\| = \epsilon\|A\|$ and $\operatorname{Ran}(\delta A) \subset \operatorname{Ran}(A)$, $\operatorname{Ker} A \subset \operatorname{Ker}\delta A$ ($\operatorname{Ran}(A^\dagger) = (\operatorname{Ker} A)^\perp$). Thus, $(A+\delta A)^\dagger = (I_H + A^\dagger \delta A)^{-1}A^\dagger$ and hence

$$\begin{aligned}
&\|(A+\delta A)^\dagger(b+\delta b) - A^\dagger b\| \\
&\quad = \|[(I_H + A^\dagger\delta A)^{-1} - I_H]A^\dagger b + (I_H + A^\dagger\delta A)^{-1}A^\dagger\delta b\| \\
&\quad \geq \|A^\dagger\delta A A^\dagger b + A^\dagger\delta b\| - \sum_{n=2}^{\infty}\|A^\dagger\|^n\|\delta A\|^n\|A^\dagger b\| \\
&\quad\quad - \sum_{n=1}^{\infty}\|A^\dagger\|^n\|\delta A\|^n\|A^\dagger\delta b\| \\
&\quad \geq \epsilon\|A^\dagger x_1\|(\|A\|\|A^\dagger b\| + \|b\|) - O(\epsilon^2).
\end{aligned}$$

Therefore, $\operatorname{Cond}_{mp}(A,b) \geq \kappa_{mp}(A) + \dfrac{\|A^\dagger\|\|b\|}{\|A^\dagger b\|}$. □

Theorem 5.4.5. *Let X be a Banach space and let $A \in Dr(X)$ with $\operatorname{ind}(A) = n \geq 1$. Put $\kappa_D(A) = \|A^D\|\|A\|$. Then*

$$\frac{\kappa_D(A)}{\|AA^D\|^2} \leq \operatorname{Cond}_D(A) \leq \big[n - 1 + n(1+2\|A^\pi\|)\kappa_D^n(A)\big]\kappa_D^{n-1}(A).$$

Proof. For any positive number $\epsilon < \dfrac{1}{(2^n-1)(1+\|A^\pi\|)\kappa_D^n(A)}$ and any $\delta A \in B(X)$ with $\|\delta A\| \leq \epsilon\|A\|$, we have $\kappa_D(A)\epsilon_A \leq \dfrac{1}{(2^n-1)(1+\|A^\pi\|)}$, where $A^\pi = I - AA^D$ and $\epsilon_A = \|A\|^{-1}\|\delta A\|$.

Assume that $\operatorname{Ran}((A+\delta A)^n) \cap \operatorname{Ker}(A^D) = \{0\}$. Then $A+\delta A \in Dr(X)$ with $\operatorname{ind}(A+\delta A) \leq n$ and

$$\|((A+\delta A)^n)^\# - (A^n)^\#\| \leq \frac{(1+2\|A^\pi\|)\|(A^D)^n\|\|\delta A^n\|}{\big[1-(1+\|A^\pi\|)\|(A^D)^n\|\|\delta A^n\|\big]^2}$$

by Theorem 4.3.3 and Lemma 4.3.4, where

$$\delta A^k = (A+\delta A)^k - A^k \text{ and } \|\delta A^k\| \leq \big[(1+\epsilon_A)^k - 1\big]\|A\|^k, \ \forall k \in \mathbb{N}$$

(see (4.3.4)). Since $(A+\delta A)^D = ((A+\delta A)^n)^\#$ and $A^D = (A^n)^\#$, we have

$$\begin{aligned}
\|(A+\delta A)^D - A^D\| &\leq \|((A+\delta A)^n)^\# - (A^n)^\#\|\|A+\delta A\|^{n-1} \\
&\quad + \|(A^n)^\#\|\|\delta A^{n-1}\| \\
&\leq \frac{(1+2\|A^\pi\|)\|A^D\|^{2n}\big[(1+\epsilon_A)^n - 1)\big]\|A\|^{2n-1}}{\big[1-(1+\|A^\pi\|)\|(A^D)^n\|\big[(1+\epsilon_A)^n-1)\big]\|A\|^n\big]^2} \\
&\quad \times (1+\epsilon_A)^{n-1} + \|(A^D)^n\|\big[(1+\epsilon_A)^{n-1} - 1\big]\|A\|^{n-1}.
\end{aligned}$$

We can deduce from above that

$$\begin{aligned}\mathrm{Cond}_D(A) &\le n(1+2\|A^\pi\|)\kappa_D^{2n-1}(A)+(n-1)\kappa_D^{n-1}(A)\\ &= \left[n-1+n(1+2\|A^\pi\|)\kappa_D^n(A)\right]\kappa_D^{n-1}(A)\end{aligned}$$

by using $\lim\limits_{\epsilon\to 0^+}\dfrac{(1+\epsilon)^k-1}{\epsilon}=k,\ k\in\mathbb{N}$.

For $\epsilon<\kappa_D^{-1}(A)$, choose $x_\epsilon\in S(X)$ and $f_\epsilon\in S(X^*)$ such that

$$\|A^Dx_\epsilon\|>\|A^D\|-\epsilon,\ \ \|(A^D)^*f_\epsilon\|>\|(A^D)^*\|-\epsilon=\|A^D\|-\epsilon,$$

Put $\delta A(x)=\dfrac{\epsilon\|A\|}{\|AA^D\|^2}f_\epsilon(AA^Dx)AA^Dx_\epsilon,\ \forall x\in X$. Then $\|\delta A\|\le\epsilon\|A\|$, $AA^D\delta AAA^D=\delta A$ and

$$\|A^D\delta AA^D\|>\frac{\epsilon\|A\|}{\|AA^D\|^2}(\|A^D\|-\epsilon)^2$$

by using $AA^D=A^DA$ and $A^D=A^DAA^D$. In this case, by Remark 4.6.10 (1), $A+\delta A\in Dr(X)$ with $(A+\delta A)^D=(I+A^D\delta A)^{-1}A^D$. Thus, we have

$$\begin{aligned}\|(A+\delta A)^D-A^D\| &\ge \|A^D\delta AA^D\|-O(\epsilon^2)\\ &> \frac{\epsilon\|A\|}{\|AA^D\|^2}(\|A^D\|-\epsilon)^2-O(\epsilon^2)\end{aligned}$$

and consequently, $\mathrm{Cond}_D(A)\ge\|AA^D\|^{-2}\kappa_D(A)$. □

5.5 The local conjugacy of a C^1–mapping

A series of Ma's papers, such as [Ma (1999)], [Ma (2000)], [Ma (2001)] and [Ma (2008)], gives many important applications of the stable perturbation theory of generalized inverses in nonlinear functional analysis and infinite–dimensional manifolds. This section is based on [Berger (1977)] and [Ma (1999)]. Throughout the section, X, Y are always Banach spaces.

Definition 5.5.1. Let U be an open subset of X. Let $f\colon U\to Y$ be a given mapping. Let $u_0\in U$. We say f is differentiable at u_0 provided there is a bounded linear operator $Df(u_0)\colon X\to Y$ such that for any $\epsilon>0$, there is $\delta>0$ such that for every $u\in(O(u_0,\delta)\backslash\{u_0\})\cap U$

$$\|f(u)-f(u_0)-Df(u_0)(u-u_0)\|<\epsilon\|u-u_0\|,$$

that is, $\lim\limits_{u\to u_0}\dfrac{f(u)-f(u_0)-Df(u_0)(u-u_0)}{\|u-u_0\|}=0.$

If f is differentiable at each $u_0\in U$, the mapping $Df\colon U\to B(X,Y)$, $u\mapsto Df(u)$ is called the derivative of f. Moreover, if Df is a continuous mapping, we say f is of class C^1 .

Definition 5.5.2. Let U and V be open subsets of X and Y respectively. If the mapping $f\colon U \to V$ is bijective and both f and f^{-1} are all C^1–class, we say f is a diffeomorphism.

Definition 5.5.3. Let U be an open subset of X and $f\colon U \to Y$ is of class C^1. Let $x_0 \in U$. We say f can be locally linearized at x_0 or f is locally conjugate to $f'(x_0)$ near x_0, if there exist two neighborhoods U_1 at x_0 and V_1 at 0 with two diffeomorphisms $u\colon U_1 \to u(U_1) \subset X$ and $v\colon V_1 \to v(V_1) \subset Y$ such that

(1) $u(x_0) = 0$, $v(0) = f(x_0)$ and $u'(x_0) = I_X$, $v'(0) = I_Y$.
(2) $f(x) = v(f'(x_0)u(x))$, $\forall\, x \in U_1$.

Theorem 5.5.4. *Let $x_0 \in X$ and U be an open neighborhood of x_0 in X. Let $f\colon U \to Y$ of class C^1. Suppose that $f'(x_0) \in Gib(X,Y)$. Then f is locally conjugate to $f'(x_0)$ near x_0 iff*

$$V(f,x_0) = \{x \in U \mid \mathrm{Ran}\,(f'(x)) \cap \mathrm{Ker}\,(f'(x_0))^+ = \{0\}\}$$

is a neighborhood of x_0.

Proof. Suppose that f is locally conjugate to $f'(x_0)$ near x_0. Then there are two diffeomnphisms $u\colon U_0 \to u(U_0)$ and $v\colon V_0 \to v(V_0)$ with

$$u(x_0) = 0,\ u'(x_0) = I_X, \text{and } v(0) = f(x_0),\ v'(0) = I_Y$$

such that $f(x) = v(f'(x_0)u(x))$, $\forall\, x \in U_0$. Then

$$f'(x) = v'(f'(x_0)u(x))f'(x_0)u'(x), \quad \forall\, x \in U_0.$$

Set $g(x) = (u')^{-1}(x)(f'(x_0))^+(v')^{-1}(f'(x_0)u(x))$, $\forall\, x \in U_0$. Then

$$\begin{aligned}
f'(x)g(x)f'(x) &= v'(f'(x_0)u(x))f'(x_0)u'(x)(u')^{-1}(x)(f'(x_0))^+ \\
&\quad \times (v')^{-1}(f'(x)u(x))v'(f'(x_0)u(x))f'(x_0)u'(x) \\
&= v'(f'(x_0)u(x))f'(x_0)(f'(x_0))^+f'(x_0)u'(x) \\
&= v'(f'(x_0)u(x))f'(x_0)u'(x) \\
&= f'(x) \\
g(x)f'(x)g(x) &= (u')^{-1}(x)(f'(x_0))^+(v')^{-1}(f'(x_0)u(x))v'(f'(x_0)u(x)) \\
&\quad \times f'(x_0)u'(x)(u')^{-1}(x)(f'(x_0))^+(v')^{-1}(f'(x_0)u(x)) \\
&= (u')^{-1}(x)(f'(x_0))^+f'(x_0)(f'(x_0))^+(v')^{-1}(f'(x_0)u(x)) \\
&= g(x).
\end{aligned}$$

Thus $(f'(x))^+ = g(x)$, $\forall\, x \in U_0$. Note that $\lim\limits_{x\to x_0} g(x) = (f'(x_0))^+$ and

$$\begin{aligned}\delta(\mathrm{Ran}\,(f'(x)), \mathrm{Ran}\,(f'(x_0))) &\le \gamma^{-1}(f'(x))\|f'(x) - f'(x_0)\| \\ &\le \|(f'(x))^+\|\|f'(x) - f'(x_0)\|\end{aligned}$$

by Lemma 1.3.5 (2) and Theorem 2.1.4. Choose an open subset U_1 in U_0 such that $x_0 \in U_1$ and

$$\|(f'(x))^+\|\|f'(x) - f'(x_0)\| < \frac{1}{\|I_Y - f'(x_0)(f'(x_0))^+\|}, \quad \forall\, x \in U_0.$$

Then $\mathrm{Ran}\,(f'(x)) \cap \mathrm{Ker}\,(f'(x_0))^+ = \{0\}$, $\forall\, x \in U_1$ by Proposition 2.2.10. So $x_0 \in U_1 \subset V(f, x_0)$.

Conversely, pick an open neighborhood U_0 of x_0 in $V(f, x_0)$. Set

$$u(x) = (f(x_0))^+(f(x)-f(x_0))+(I_X-(f'(x_0))^+f'(x_0))(x-x_0), \ \forall\, x \in U_0.$$

Then $u(x_0) = 0$ and

$$u'(x) = (f'(x_0))^+f'(x) + I_X - (f'(x_0))^+f'(x_0), \ \forall\, x \in U_0. \tag{5.5.1}$$

Thus u is of C^1 and $u'(x_0) = I_X$ by (5.5.1). Applying Inverse Function Theorem (Theorem A.3.1) to $u'(x)$, we can find $O(0, r) \subset X$ and $U_0' = u^{-1}(O(0, r)) \subset U_0$ such that $u\colon U_0' \to O(0, r)$ is a diffeomorphism and $u'(x) \in GL(B(X))$, $\forall\, x \in U_0'$.

Since $I_X + (f'(x_0))^+(f'(x) - f(x_0)) = u'(x)$ is invertible in $B(X)$ by (5.5.1) and $\mathrm{Ran}\,(f'(x)) \cap \mathrm{Ker}\,(f'(x_0)) = \{0\}$, $\forall\, x \in U_0'$, it follows from Corollary 2.2.7, that

$$\mathrm{Ker}\, f'(x) = (u'(x))^{-1}\mathrm{Ker}\, f'(x_0), \quad \forall\, x \in U_0'. \tag{5.5.2}$$

Note that $(f'(x_0))^+(f(x) - f(x_0))$ is continuous at x_0. So, we can choose an open neighborhood U_0'' of x_0 in U_0' such that

$$(f'(x_0))^+(f(x) - f(x_0)) \in O(0, r), \quad \forall\, x \in U_0''.$$

Fix $x \in U_0''$ and set

$$y_1 = (f'(x_0))^+(f(x) - f(x_0)), \ y_2 = (I_X - (f'(x_0))^+f'(x_0))(x - x_0).$$

Noting that, for each $t \in [0, 1]$,

$$y_1 + (1 - t)y_2 = t(f'(x_0))^+(f(x) - f(x_0)) + (1 - t)u(x) \in O(0, r),$$

we can define a differentiable mapping $\varphi\colon [0, 1] \to Y$ by

$$\varphi(t) = f \circ u^{-1}(y_1 + (1 - t)y_2), \quad \forall\, t \in [0, 1].$$

Then

$$\varphi'(t) = (f' \circ u^{-1})(y_1 + (1 - t)y_2)(u^{-1})'(y_1 + (1 - t)y_2)(-y_2), \ \forall\, t \in [0, 1].$$

Set $x_1 = u^{-1}(y_1 + (1-t)y_2) \in U_0'$. Since $u^{-1}(u(x)) = x$, $\forall\, x \in U_0'$, we have $(u^{-1})'(u(x))u'(x) = I_X$ and hence

$$(u^{-1})'(y_1 + (1-t)y - 2) = (u'(x_1))^{-1}.$$

Note that $y_2 \in \operatorname{Ker} f'(x_0)$. So $(u'(x_1))^{-1}y_2 \in \operatorname{Ker} f'(x_1)$ by (5.5.2). Thus

$$\varphi'(t) = f'(x_1)(u'(x_1))^{-1}y_2 = 0, \ \forall\, t \in [0,1]$$

and consequently, $\varphi(0) = \varphi(1)$, that is,

$$f \circ u^{-1}(f^{-1}(x_0)(f(x) - f(x_0))) = f \circ u^{-1}(u(x)) = f(x), \ \forall\, x \in U_0''. \quad (5.5.3)$$

Now choose $O_Y(0,\rho) = \{y \in Y \mid \|y\| < \rho\}$ such that

$$(f'(x_0))^+ x \in u(U_0'') \subset O(0,r), \quad \forall\, x \in O_Y(0,\rho)$$

and define

$$v(x) = (f \circ u^{-1})((f'(x_0))^+ x) + (I_Y - f'(x_0)(f'(x_0))^+)x, \ \forall\, x \in O_Y(0,\rho).$$

Then $v(0) = 0$, $f(u^{-1}(0)) = f(x_0)$ and

$$\begin{aligned} v'(x) = f' \circ (u^{-1}((f'(x_0))^+ x)u' \circ u^{-1}((f'(x_0))^+ x)(f'(x_0))^+ \\ + I_Y - f'(x_0)(f'(x_0))^+, \ \forall\, x \in O_Y(0,\rho). \end{aligned}$$

Consequently, $v'(0) = f'(x_0)(f'(x_0))^+ + I_Y - f'(x_0)(f'(x_0))^+ = I_Y$ and hence, by Theorem A.3.1, there is $O_Y(0,\rho') \subset O_Y(0,\rho)$ such that $v\colon O_Y(0,\rho') \to v(O_Y(0,\rho'))$ is a diffeomorphism.

Pick $O(x_0,\delta_0) \subset U_0''$ such that $f'(x_0)u(x) \in O_Y(0,\rho')$, $\forall\, x \in O(x_0,\delta_0)$ since $u(x_0) = 0$, $f'(x_0) \in B(X,Y)$ and $u(x)$ is continuous at x_0. Then for any $x \in O(x_0,\delta_0)$,

$$\begin{aligned} (v \circ f'(x_0) \circ u)(x) &= v \circ f'(x_0)((f'(x_0))^+(f(x) - f(x_0)) \\ &\quad + (I_X - (f'(x_0))^+ f(x_0))(x - x_0)) \\ &= v(f'(x_0)(f'(x_0))^+(f(x) - f(x_0))) \\ &= f \circ u^{-1}((f'(x_0))^+ f'(x_0)(f(x_0))^+(f(x) - f(x_0))) \\ &\quad + (I_Y - f'(x_0)(f'(x_0))^+)f'(x_0)(f'(x_0))^+(f(x) - f(x_0)) \\ &= f \circ u^{-1}((f'(x_0))^+(f(x) - f(x_0))) \\ &= f(x). \end{aligned}$$

by (5.5.3). Take $U_1 = O(x_0,\delta_0)$ and $V_1 = O_Y(0,\rho)$. Then $u\colon U_1 \to u(U_1)$ and $v\colon V_1 \to v(V_1)$ are diffeomorphic. $\square$

Remark 5.5.5. Let M be a topological space and $h\colon M \to B(X,Y)$ be a mapping. According to [Ma (1999)], if $h(x_0) \in Gib(X,Y)$ and $V(h,x_0)$ is a neighborhood of x_0, we say $h(x)$ is locally fine at x_0. In other words, for each $x \in V(h,x_0)$, $h(x)$ is a stable perturbation of $h(x_0)$.

Thus, Theorem 5.5.4 says that f is locally conjugate to $f'(x_0)$ near x_0 iff $f(x)$ is locally fine at x_0.

Combining Theorem 5.5.4 with Corollary 2.2.8, we have

Corollary 5.5.6. *Let U be an open subset of X and $f\colon U \to Y$ be a C^1-mapping. If one of the following conditions holds, then for each $x \in U$, f is locally conjugate to $f'(x)$ near x.*

(1) $\dim \operatorname{Ker} f'(x)$ *is constant and* $f'(x) \in Gib(X,Y)$, $\forall\, x \in U$.
(2) $\dim \operatorname{Ran}(f'(x))$ *is constant and* $f'(x) \in Gib(X,Y)$, $\forall\, x \in U$.
(3) $\dim \operatorname{Ker}(f'(x))^*$ *is constant and* $f'(x) \in Gib(X,Y)$, $\forall x \in U$.
(4) $\operatorname{Ran}(f'(x))$ *is closed, and both* $\dim \operatorname{Ker} f'(x)$ *and* $\dim \operatorname{Coker} f'(x)$ *are all constant,* $\forall\, x \in U$.

Chapter 6

Some Related Topics

This chapter consists of the perturbation of reduced minimum modulus, the density of Moore–Penrose invertible elements in some C^*–algebras, properties of reduced minimum modulus of an element in a C^*–algebra and the lifting of Moore–Penrose invertible elements from quotient C^*–algebras as well as Fredholm elements and their indices.

6.1 Perturbation of reduced minimum modulus

This section is based on [Xue (2008a)]. In this section, we first give a new expression of the reduced minimum modulus of a closed operator and then present the perturbation analysis of the reduced minimum modulus of a closed operator under the perturbation of a bounded operator. Throughout the section, X and Y are always Banach spaces.

Lemma 6.1.1. *Let $T \in \mathcal{C}(X,Y)$. Then there is a sequence of operators $\{A_n\} \subset B(X,Y)$ such that*

(1) $\operatorname{Ker} T \subsetneqq \operatorname{Ker}(T+A_n)$, $\operatorname{Ran}(T+A_n) \subset \operatorname{Ran}(T)$, $\forall\, n \geq 1$.
(2) $\lim\limits_{n\to\infty} \|A_n\| = \gamma(T)$.

Moreover, if $\operatorname{Ran}(T)$ *is closed, then* $\overline{\operatorname{Ran}(T+A_n)} \neq \operatorname{Ran}(T)$, $\forall\, n \geq 1$.

Proof. According to the definition of $\gamma(T)$, we can find $\{x_n\} \subset \mathfrak{D}(T)$ such that

$$\lim_{n\to\infty} \|Tx_n\| = \gamma(T) \text{ and } \operatorname{dist}(x_n, \operatorname{Ker} T) = 1, \ \forall\, n \geq 1.$$

Thus there is a sequence $\{f_n\} \subset X^*$ with $\|f_n\| = 1$ and

$$f_n(x_n) = \operatorname{dist}(x_n, \operatorname{Ker} T) = 1, \ f_n(x) = 0, \ \forall\, x \in \operatorname{Ker} T, \ \forall\, n \geq 1.$$

Put $A_nx = -(Tx_n)f_n(x)$, $\forall\, x \in X$. Then $A_n \in B(X,Y)$, $\lim\limits_{n\to\infty}\|A_n\| = \gamma(T)$ and

$$\operatorname{Ker}T \subset \operatorname{Ker}(T+A_n),\ \operatorname{Ran}(T+A_n) \subset \operatorname{Ran}(T),\ \forall\, n \geq 1.$$

Note that $x_n \in \operatorname{Ker}(T+A_n)$ and $x_n \notin \operatorname{Ker}T$. So $\operatorname{Ker}T \subsetneqq \operatorname{Ker}(T+A_n)$, $\forall\, n \geq 1$. This proves (1) and (2).

Now suppose that $\operatorname{Ran}(T)$ is closed. Let $\{x_n\}$, $\{f_n\}$ and $\{A_n\}$ be as above. Define linear functions g_n on $\operatorname{Ran}(T)$ by $g_n(Tx) = f_n(x)$, $x \in \mathfrak{D}(T)$, $n \geq 1$. g_n is well–defined since $f_n(x) = 0$, $\forall\, x \in \operatorname{Ker}T$. Moreover, for any $x \in \mathfrak{D}(T)$, any $z \in \operatorname{Ker}T$ and $\forall\, n \geq 1$, we have

$$|g_n(Tx)| = |f_n(x-z)| \leq \|f_n\|\|x-z\|,$$

thus,

$$|g_n(Tx)| \leq \operatorname{dist}(x, \operatorname{Ker}T) \leq \frac{1}{\gamma(T)}\|Tx\|$$

($\|Tx\| \geq \gamma(T)\operatorname{dist}(x,\operatorname{Ker}T)$), that is, g_n is bounded on $\operatorname{Ran}(T)$ and hence by Proposition 1.1.4, there is $\{\hat{g_n}\} \subset Y^*$ such that

$$\hat{g_n}(y) = g_n(y),\ \forall\, y \in \operatorname{Ran}(T) \text{ and } \|\hat{g_n}\| \leq \frac{1}{\gamma(T)}$$

for any $n \in \mathbb{N}$. Since $\hat{g_n}((T+A_n)x) = 0$, $\forall\, x \in \mathfrak{D}(T) = \mathfrak{D}(T+A_n)$ and $\hat{g_n}(Tx_n) = 1$, we conclude that $\overline{\operatorname{Ran}(T+A_n)} \subsetneqq \operatorname{Ran}(T)$, $\forall\, n \in \mathbb{N}$. □

Let $T \in \mathcal{C}(X,Y)$ and set

$$M_1(T) = \{A \in B(X,Y) \,|\, \overline{\operatorname{Ran}(T+A)} \subset \overline{\operatorname{Ran}(T)},\ \operatorname{Ker}T \subsetneqq \operatorname{Ker}(T+A)\},$$

$$M_2(T) = \{A \in B(X,Y) \,|\, \overline{\operatorname{Ran}(T+A)} \subsetneqq \overline{\operatorname{Ran}(T)},\ \operatorname{Ker}T \subset \operatorname{Ker}(T+A)\}.$$

Theorem 6.1.2. *Let $T \in \mathcal{C}(X,Y)$ and $M_1(T)$, $M_2(T)$ be as above. Then*

$$\begin{aligned}\gamma(T) &= \inf\{\,\|A\| \,|\, A \in M_1(T)\,\}\\ &= \inf\left\{\frac{\|A\|}{\delta(\operatorname{Ker}(T+A), \operatorname{Ker}T)}\,\middle|\, \operatorname{Ker}(T+A) \not\subseteq \operatorname{Ker}T,\ A \in B(X,Y)\right\}\\ &= \inf\{\,\|A\| \,|\, \operatorname{Ker}T \subsetneqq \operatorname{Ker}(T+A),\ A \in B(X,Y)\,\}.\end{aligned}$$

In addition, if $\operatorname{Ran}(T)$ *is closed, then*

$$\begin{aligned}\gamma(T) &= \inf\{\,\|A\| \,|\, A \in M_2(T)\,\}\\ &= \inf\left\{\frac{\|A\|}{\delta(\operatorname{Ran}(T), \operatorname{Ran}(T+A))}\,\middle|\, \operatorname{Ran}(T) \not\subseteq \overline{\operatorname{Ran}(T+A)},\right.\\ &\qquad \left. A \in B(X,Y)\right\}\\ &= \inf\{\,\|A\| \,|\, \overline{\operatorname{Ran}(T+A)} \subsetneqq \operatorname{Ran}(T),\ A \in B(X,Y)\,\}.\end{aligned}$$

Proof. Set

$$S_1(T) = \{A \in B(X,Y) \,|\, \mathrm{Ker}\,(T+A) \not\subseteq \mathrm{Ker}\,T\},$$
$$S_2(T) = \{A \in B(X,Y) \,|\, \overline{\mathrm{Ran}\,(T)} \not\subseteq \overline{\mathrm{Ran}\,(T+A)}\},$$
$$S_3(T) = \{A \in B(X,Y) \,|\, \mathrm{Ker}\,T \subsetneqq \mathrm{Ker}\,(T+A)\},$$
$$S_4(T) = \{A \in B(X,Y) \,|\, \overline{\mathrm{Ran}\,(T+A)} \subsetneqq \overline{\mathrm{Ran}\,(T)}\}.$$

Clearly, $M_1(T) \subset S_3(T) \subset S_1(T)$, $M_2(T) \subset S_4(T) \subset S_2(T)$. By Proposition 1.3.3 (1), $\delta(\mathrm{Ker}\,(T+A), \mathrm{Ker}\,T) = 1$ when $A \in M_1(T)$ or $A \in S_3(T)$; by Proposition 1.3.3 (1) and Proposition 1.3.2 (2), $\delta(\mathrm{Ran}\,(T), \mathrm{Ran}\,(T+A)) = 1$ when $A \in M_2(T)$ or $A \in S_4(T)$. Thus we have by Lemma 1.3.5,

$$\begin{aligned}\gamma(T) &\le \inf\left\{\frac{\|A\|}{\delta(\mathrm{Ran}\,(T), \mathrm{Ran}\,(T+A))} \,\middle|\, A \in S_2(T)\right\}\\ &\le \inf\{\|A\| \,|\, A \in S_4(T)\}\\ &\le \inf\{\|A\| \,|\, A \in M_2(T)\} \end{aligned} \tag{6.1.1}$$

and

$$\begin{aligned}\gamma(T) &\le \inf\left\{\frac{\|A\|}{\delta(\mathrm{Ker}\,(T+A), \mathrm{Ker}\,T)} \,\middle|\, A \in S_1(T)\right\}\\ &\le \inf\{\|A\| \,|\, A \in S_3(T)\}\\ &\le \inf\{\|A\| \,|\, A \in M_1(T)\}. \end{aligned} \tag{6.1.2}$$

On the other hand, by Lemma 6.1.1, there is a sequence of operators $\{A_n\} \subset B(X,Y)$ such that $\lim\limits_{n\to\infty} \|A_n\| = \gamma(T)$ and $\mathrm{Ran}\,(T+A_n) \subset \mathrm{Ran}\,(T)$, $\mathrm{Ker}\,T \subsetneqq \mathrm{Ker}\,(T+A_n)$, $n \ge 1$ and moreover, $\overline{\mathrm{Ran}\,(T+A_n)} \subsetneqq \mathrm{Ran}\,(T)$ if $\mathrm{Ran}\,(T)$ is closed. Since $\{A_n\}_1^\infty \subset M_1(T)$ and $\{A_n\}_1^\infty \subset M_2(T)$ if $\mathrm{Ran}\,(T)$ is closed, it follows that

$$\inf\{\|A\| \,|\, A \in M_1(T)\} \le \gamma(T), \tag{6.1.3}$$
$$\inf\{\|A\| \,|\, A \in M_2(T)\} \le \gamma(T) \text{ (when } \mathrm{Ran}\,(T) \text{ is closed).} \tag{6.1.4}$$

Therefore, combining (6.1.1) with (6.1.4) and (6.1.2) with (6.1.3), we get the results. □

Corollary 6.1.3. *Let $T \in \mathcal{C}(X,Y)$*

(1) *If $\dim \mathrm{Ker}\,T < \infty$, then*

$$\gamma(T) = \inf\{\|A\| \,|\, \dim \mathrm{Ker}\,(T+A) > \dim \mathrm{Ker}\,T,\ A \in B(X,Y)\};$$

(2) *If $\dim \mathrm{Ran}\,(T) < \infty$, then*

$$\gamma(T) = \inf\{\|A\| \,|\, \dim \mathrm{Ran}\,(T) > \dim \mathrm{Ran}\,(T+A),\ A \in B(X,Y)\}.$$

(3) *If* $\mathrm{Ran}\,(T)$ *is closed and* $\mathrm{CodimRan}\,(T) < \infty$*, then*

$$\gamma(T) = \inf\{\|A\| \mid \mathrm{Codim}\overline{\mathrm{Ran}\,(T+A)} > \mathrm{CodimRan}\,(T),\ A \in B(X,Y)\}.$$

Proof. (1) Let $A \in B(X,Y)$ with $\dim\mathrm{Ker}\,(T+A) > \dim\mathrm{Ker}\,T$. If $\dim\mathrm{Ker}\,(T+A) < \infty$, then $\delta(\mathrm{Ker}\,(T+A), \mathrm{Ker}\,T) = 1$ by Proposition 1.3.3 (2). If $\dim\mathrm{Ker}\,(T+A) = \infty$, we can choose a subspace V in $\mathrm{Ker}\,(T+A)$ such that $\dim\mathrm{Ker}\,T < \dim V < \infty$. Then

$$1 \geq \delta(\mathrm{Ker}\,(T+A), \mathrm{Ker}\,T) \geq \delta(V, \mathrm{Ker}\,T) = 1.$$

Noting that

$$M_1(T) \subset \{A \in B(X,Y) \mid \dim\mathrm{Ker}\,(T+A) > \dim\mathrm{Ker}\,T\} \subset S_1(T),$$

we obtain that by Theorem 6.1.2,

$$\gamma(T) \leq \inf\{\|A\| \mid \dim\mathrm{Ker}\,(T+A) > \dim\mathrm{Ker}\,T,\ A \in B(X,Y)\} \leq \gamma(T),$$

i.e., $\gamma(T) = \inf\{\|A\| \mid \dim\mathrm{Ker}\,(T+A) > \dim\mathrm{Ker}\,T,\ A \in B(X,Y)\}$.

(2) The proof is similar to the proof of (1).

(3) Let $A \in B(X,Y)$ with $\mathrm{CodimRan}\,(T) < \mathrm{Codim}\overline{\mathrm{Ran}\,(T+A)}$. Since $\mathrm{Codim}\overline{\mathrm{Ran}\,(T+A)} = \dim\mathrm{Ker}\,(T+A)^*$ and $\mathrm{CodimRan}\,(T) = \dim\mathrm{Ker}\,T^*$ by Remark 1.2.14 and Theorem 1.2.15, it follows from Corollary 6.1.3 (1) that

$$\gamma(T) = \gamma(T^*) \leq \|A^*\| = \|A\|. \tag{6.1.5}$$

Now for any $\epsilon > 0$, we can choose $B \in B(X,Y)$ with $\overline{\mathrm{Ran}\,(T+B)} \subsetneqq \mathrm{Ran}\,(T)$ such that $\gamma(T) > \|B\| - \epsilon$ by Theorem 6.1.2. From

$$\mathrm{Ker}\,T^* = \mathrm{Ran}\,(T)^{\perp} = \{f \in Y^* \mid f(y) = 0, \forall\, y \in \mathrm{Ran}\,(T)\},$$
$$\mathrm{Ker}\,(T+B)^* = \mathrm{Ran}\,(T+B)^{\perp}, \quad \overline{\mathrm{Ran}\,(T+B)} \subsetneqq \mathrm{Ran}\,(T),$$

we deduce that $\mathrm{Ker}\,T^* \subsetneqq \mathrm{Ker}\,(T+B)^*$. Thus, $\mathrm{CodimRan}\,(T) < \mathrm{Codim}\overline{\mathrm{Ran}\,(T+B)}$. This means that

$$\inf\{\|A\| \mid \mathrm{Codim}\overline{\mathrm{Ran}\,(T+A)} > \mathrm{CodimRan}\,(T)\} \leq \|B\| < \gamma(T)+\epsilon. \tag{6.1.6}$$

Combining (6.1.5) with (6.1.6), we can obtain the assertion. □

Corollary 6.1.4. *Let* $T \in GL(B(X))$*. Then*

$$\mathrm{dist}\,(T, B(X)\backslash GL(B(X))) = \|T^{-1}\|^{-1}.$$

Proof. If there exists $A \in B(X)$ such that $\|T - A\| < \|T^{-1}\|^{-1}$. Then $\|I - T^{-1}A\| < 1$ so that $A \in GL(B(X))$. This indicates that

$$\operatorname{dist}(T, B(X)\backslash GL(B(X))) \geq \|T^{-1}\|^{-1}.$$

Now for every $\epsilon > 0$, we can find $S \in B(X)$ such that $\operatorname{Ker} S \neq \{0\}$ and

$$\|T^{-1}\|^{-1} = \gamma(T) > \|T - S\| - \epsilon$$

by Theorem 6.1.2. Since $S \in B(X)\backslash GL(B(X))$, we have

$$\|T^{-1}\|^{-1} \leq \operatorname{dist}(T, B(X)\backslash GL(B(X))) < \|T^{-1}\|^{-1} + \epsilon.$$

The assertion follows. □

Let $T \in \mathcal{C}(X, Y)$ and $A \in B(X, Y)$. In the following, we will consider the relationship between $\gamma(T + A)$ and $\gamma(T)$ and then discuss the continuity of the functional $T \mapsto \gamma(T)$ on $\mathcal{C}(X, Y)$.

Proposition 6.1.5. *$T \in \mathcal{C}(X, Y)$ and $A \in B(X, Y)$. Then*

$$\gamma(T + A) \geq \gamma(T)\frac{1 - \delta(\operatorname{Ker} T, \operatorname{Ker}(T + A))}{1 + \delta(\operatorname{Ker} T, \operatorname{Ker}(T + A))} - \|A\|; \tag{6.1.7}$$

In addition, if $\operatorname{Ran}(T + A)$ *is closed, then*

$$\gamma(T + A) \geq \gamma(T)\frac{1 - \delta(\operatorname{Ran}(T + A), \operatorname{Ran}(T))}{1 + \delta(\operatorname{Ran}(T + A), \operatorname{Ran}(T))} - \|A\|. \tag{6.1.8}$$

Proof. By Theorem 6.1.2, there is $\{B_n\} \subset S_3(T + A)$ (or $\{B_n\} \subset S_4(T + A)$ when $\operatorname{Ran}(T + A)$ is closed) such that $\lim\limits_{n\to\infty} \|B_n\| = \gamma(T + A)$. So $\delta(\operatorname{Ker}(T + A + B_n), \operatorname{Ker}(T + A)) = 1$ (or $\delta(\operatorname{Ran}(T + A), \operatorname{Ran}(T + A + B_n)) = 1$ when $\operatorname{Ran}(T + A)$ is closed), $n = 1, 2, \cdots$. It follows from Lemma 1.3.5 (1) and Proposition 1.3.2 (4) that

$$\begin{aligned}\|A\| + \|B_n\| &\geq \|B_n + A\| \geq \gamma(T)\delta(\operatorname{Ker}(T + A + B_n), \operatorname{Ker} T)\\ &\geq \gamma(T)\frac{\delta(\operatorname{Ker}(T + A + B_n), \operatorname{Ker}(T + A) - \delta(\operatorname{Ker} T, \operatorname{Ker}(T + A))}{1 + \delta(\operatorname{Ker} T, \operatorname{Ker}(T + A))},\end{aligned}$$

$n \geq 1$. Let $n \to \infty$, we obtain the (6.1.7).

When $\operatorname{Ran}(T + A)$ is closed, we have also by Lemma 1.3.5 (2) and Proposition 1.3.2 (4),

$$\begin{aligned}\|A\| + \|B_n\| &\geq \|B_n + A\| \geq \gamma(T)\delta(\operatorname{Ran}(T), \operatorname{Ran}(T + A + B_n))\\ &\geq \gamma(T)\frac{\delta(\operatorname{Ran}(T + A), \operatorname{Ran}(T + A + B_n)) - \delta(\operatorname{Ran}(T + A), \operatorname{Ran}(T))}{1 + \delta(\operatorname{Ran}(T + A), \operatorname{Ran}(T))},\end{aligned}$$

$n \geq 1$. Now let $n \to \infty$, we get the inequality (6.1.8). □

Proposition 6.1.6. *Let $T \in \mathcal{C}(X,Y)$ and $A \in B(X,Y)$. If one of following conditions is satisfied, then $|\gamma(T+A)-\gamma(T)| \le \|A\|$.*

(1) $\dim \mathrm{Ker}\,(T+A) = \dim \mathrm{Ker}\, T < \infty$.
(2) $\dim \mathrm{Ran}\,(T+A) = \dim \mathrm{Ran}\,(T) < \infty$.
(3) $\mathrm{Ran}\,(T)$, $\mathrm{Ran}\,(T+A)$ *are closed and* $\dim \mathrm{Coker}\,(T+A) = \dim \mathrm{Coker}\,(T) < \infty$.

Proof. For any $\epsilon > 0$, there is $C \in B(X,Y)$ such that

$$\dim \mathrm{Ker}\,(T+A) = \dim \mathrm{Ker}\, T < \dim \mathrm{Ker}\,(T+A+C),\ \gamma(T+A) > \|C\| - \epsilon.$$

by Corollary 6.1.3. Thus, by using Corollary 6.1.3 again, we have

$$\gamma(T) \le \|A+C\| \le \|A\| + \|C\| < \gamma(T+A) + \|A\| + \epsilon.$$

Then $\gamma(T) - \gamma(T+A) \le \|A\|$ as $\epsilon \to 0$. Similarly, we have $\gamma(T+A) - \gamma(T) \le \|A\|$. So $|\gamma(T+A)-\gamma(T)| \le \|A\|$.

Similarly, by using Corollary 6.1.3, we can obtain the results when T and $T+A$ satisfy (2) or (3). □

The following corollary present two estimates of the perturbation of $\gamma(\cdot)$ in general case.

Corollary 6.1.7. *Let $T \in \mathcal{C}(X,Y)$ and $A \in B(X,Y)$. Then*

$$|\gamma(T+A)-\gamma(T)| \le \max\{\gamma(T+A),\, \gamma(T)\} \frac{2\,\hat{\delta}(\mathrm{Ker}\, T, \mathrm{Ker}\,(T+A))}{1+\hat{\delta}(\mathrm{Ker}\, T, \mathrm{Ker}\,(T+A))} + \|A\|. \quad (6.1.9)$$

If $\hat{\delta}(\mathrm{Ker}\, T, \mathrm{Ker}\,(T+A)) < 1$, then

$$|\gamma(T+A)-\gamma(T)| \le \frac{4\|A\|}{1-\hat{\delta}(\mathrm{Ker}\, T, \mathrm{Ker}\,(T+A))}. \quad (6.1.10)$$

If $\mathrm{Ran}\,(T)$ and $\mathrm{Ran}\,(T+A)$ are all closed, then

$$|\gamma(T+A)-\gamma(T)| \le \max\{\gamma(T+A),\, \gamma(T)\} \frac{2\,\hat{\delta}(\mathrm{Ran}\,(T), \mathrm{Ran}\,(T+A))}{1+\hat{\delta}(\mathrm{Ran}\,(T), \mathrm{Ran}\,(T+A))} + \|A\|. \quad (6.1.11)$$

If $\mathrm{Ran}\,(T)$ and $\mathrm{Ran}\,(T+A)$ are all closed and $\hat{\delta}(\mathrm{Ran}\,(T), \mathrm{Ran}\,(T+A)) < 1$, then

$$|\gamma(T+A)-\gamma(T)| \le \frac{4\|A\|}{1-\hat{\delta}(\mathrm{Ran}\,(T), \mathrm{Ran}\,(T+A))} \quad (6.1.12)$$

Proof. By (6.1.7),

$$\begin{aligned}\gamma(T)-\gamma(T+A) &\le \gamma(T)\frac{2\delta(\operatorname{Ker}T,\operatorname{Ker}(T+A))}{1+\delta(\operatorname{Ker}T,\operatorname{Ker}(T+A))}+\|A\| \\ &\le \gamma(T)\frac{2\,\hat{\delta}(\operatorname{Ker}T,\operatorname{Ker}(T+A))}{1+\hat{\delta}(\operatorname{Ker}T,\operatorname{Ker}(T+A))}+\|A\|.\end{aligned}$$

Exchange the position of T and $T+A$ in above inequality, we get

$$\gamma(T+A)-\gamma(T) \le \gamma(T+A)\frac{2\,\hat{\delta}(\operatorname{Ker}T,\operatorname{Ker}(T+A))}{1+\hat{\delta}(\operatorname{Ker}T,\operatorname{Ker}(T+A))}+\|A\|.$$

Thus, we have (6.1.9).

By (6.1.7), we have

$$\gamma(T) \ge \gamma(T+A)\frac{1-\delta(\operatorname{Ker}(T+A),\operatorname{Ker}T)}{1+\delta(\operatorname{Ker}(T+A),\operatorname{Ker}T)}-\|A\|.$$

Thus, by Lemma 1.3.5,

$$\begin{aligned}\gamma(T+A)-\gamma(T) &\le \frac{(\gamma(T)+\|A\|)(1+\delta(\operatorname{Ker}(T+A),\operatorname{Ker}T)}{1-\delta(\operatorname{Ker}(T+A),\operatorname{Ker}T)}-\gamma(T) \\ &\le \frac{4\|A\|}{1-\delta(\operatorname{Ker}(T+A),\operatorname{Ker}T)} \le \frac{4\|A\|}{1-\hat{\delta}(\operatorname{Ker}T,\operatorname{Ker}(T+A))}.\end{aligned}$$

Similarly, we also have

$$\gamma(T)-\gamma(T+A) \le \frac{4\|A\|}{1-\hat{\delta}(\operatorname{Ker}T,\operatorname{Ker}(T+A))}.$$

So we get (6.1.10).

The rest of proofs is similar. □

Remark 6.1.8. Markus in [Markus (1959)] showed that if $S, T \in B(X)$ with $\operatorname{Ran}(S)$ and $\operatorname{Ran}(T)$ closed, then

$$\begin{aligned}|\gamma(S)-\gamma(T)| &\le \frac{3\|S-T\|}{1-2\hat{\delta}(\operatorname{Ker}S,\operatorname{Ker}T)}, & \hat{\delta}(\operatorname{Ker}S,\operatorname{Ker}T) &< \frac{1}{2} \\ |\gamma(S)-\gamma(T)| &\le \frac{3\|S-T\|}{1-2\hat{\delta}(\operatorname{Ran}(S),\operatorname{Ran}(T))}, & \hat{\delta}(\operatorname{Ran}(S),\operatorname{Ran}(T)) &< \frac{1}{2}.\end{aligned}$$

(cf. Lemma 3.4 in [Koliha and Rakočević (1998)]). These inequalities may be the earliest estimate pert to the reduced minimum moduli of operators. Much time late after this result, there is an alternate form of the estimate $|\gamma(S)-\gamma(T)|$ given by

$$|\gamma(S)-\gamma(T)| \le \max\{\gamma(S),\gamma(T)\}\hat{\delta}(\operatorname{Ker}S,\operatorname{Ker}T)+\|S-T\|$$

when X, Y are Hilbert spaces and S, $T \in B(X,Y)$ (cf. [Apostol (1985)] or [Chen *et al.* (1996)]). When X, Y are Banach spaces and S, $T \in B(X,Y)$, the above is rewritten as

$$|\gamma(S) - \gamma(T)| \leq 2\max\{\gamma(S), \gamma(T)\}\hat{\delta}(\operatorname{Ker} S, \operatorname{Ker} T) + \|S - T\| \quad (6.1.13)$$

(cf. [Zhu *et al.* (2003)]). Clearly, (6.1.9) implies (6.1.13) when $S = T + A$.

As end of the section, we discuss the behavior of $\lim\limits_{n\to\infty} \gamma(T + A_n)$ for $T \in \mathcal{C}(X,Y)$ and $\{A_n\} \subset B(X,Y)$ with $\lim\limits_{n\to\infty} \|A_n\| = 0$.

Proposition 6.1.9. *Let $T \in \mathcal{C}(X,Y)$ and $\{A_n\} \subset B(X,Y)$. Suppose that $\lim\limits_{n\to\infty} \|A_n\| = 0$ and $\alpha = \inf\limits_{n\geq 1} \gamma(T + A_n) > 0$. Then $\gamma(T) \geq \alpha$.*

Proof. Since $\gamma(T^*) = \gamma(T)$ and $\gamma(T^* + A_n^*) = \gamma(T + A_n)$, $\forall\, n \in \mathbb{N}$, by Corollary 1.2.18, we have

$$\|(T^* + A_n^*)g\| \geq \alpha \operatorname{dist}(g, \operatorname{Ker}(T^* + A_n^*)), \ \forall\, g \in \mathfrak{D}(T^*), \ n \geq 1.$$

Given $g \in \mathfrak{D}(T^*)$. We can find $\{g_n\} \in \operatorname{Ker}(T^* + A_n^*)$ such that

$$\operatorname{dist}(g, \operatorname{Ker}(T^* + A_n^*)) \leq \|g - g_n\| < \operatorname{dist}(g, \operatorname{Ker}(T^* + A_n^*)) + \frac{1}{n} \leq \|g\| + 1,$$

$n \in \mathbb{N}$. Thus, $\|g_n\| \leq 2\|g\| + 1$, $\forall\, n \in \mathbb{N}$ and hence there is a subnet $\{g_\beta\}$ of $\{g_n\}$ and a $g_0 \in Y^*$ such that g_β converges weakly* to g_0 by Theorem 1.1.16. Thus, for any $x \in \mathfrak{D}(T)$, $g_\beta(Tx) \to g_0(Tx)$.

Noting that $\|A_\beta^* g_\beta\| \to 0$ and

$$g_\beta(Tx) = T^* g_\beta(x) = -A_\beta^* g_\beta(x), \quad \forall\, x \in \mathfrak{D}(T),$$

we have $g_0(Tx) = 0$, $\forall\, x \in \mathfrak{D}(T)$ and consequently, $g_0 \in \operatorname{Ran}(T)^\perp = \operatorname{Ker} T^*$ by Theorem 1.2.15. Now for any $y \in Y$,

$$\begin{aligned}|(g - g_n)(y)| \leq \|g - g_n\|\|y\| &< \|y\|\Big(\frac{1}{n} + \operatorname{dist}(g, \operatorname{Ker}(T^* + A_n^*))\Big) \\ &\leq \|y\|\Big(\frac{1}{n} + \alpha^{-1}\|(T + A_n)^* g\|\Big).\end{aligned}$$

So we have

$$|(g - g_0)(y)| \leq \|y\|\alpha^{-1}\|T^* g\|, \quad \forall\, g \in \mathfrak{D}(T^*)$$

and hence

$$\|T^* g\| \geq \alpha \|g - g_0\| \geq \alpha \operatorname{dist}(g, \operatorname{Ker} T^*),$$

which implies that $\gamma(T) = \gamma(T^*) \geq \alpha$. □

From Proposition 6.1.9, we get the following result which is Lemma 1.9 of [Apostol (1976)].

Corollary 6.1.10. *For given $\alpha > 0$, the set $\{T \in B(X,Y) | \gamma(T) \geq \alpha\}$ is norm–closed in $B(X,Y)$.*

Corollary 6.1.11. *$T \in \mathcal{C}(X,Y)$ and $\{A_n\} \subset B(X,Y)$ with $\lim\limits_{n\to\infty} \|A_n\| = 0$.*

(1) *If $\gamma(T) > 0$ and $\operatorname{Ker} T = \{0\}$ or $\operatorname{Ran}(T) = Y$, then $\lim\limits_{n\to\infty} \gamma(T + A_n) = \gamma(T)$.*
(2) *If $\gamma(T) = 0$, then $\lim\limits_{n\to\infty} \gamma(T + A_n) = 0$.*
(3) *If $\operatorname{Ran}(T)$ is closed and $\operatorname{Ker} T \neq \{0\}$, $\operatorname{Ran}(T) \neq Y$, then there is $\{B_n\} \subset B(X,Y)$ with $\lim\limits_{n\to\infty} \|B_n\| = 0$ such that $\lim\limits_{n\to\infty} \gamma(T + B_n) \neq 0$.*

Proof. (1) Assume that $\operatorname{Ker} T = \{0\}$. Then

$$\|(T + A_n)x\| \geq \|Tx\| - \|A_n x\| \geq \gamma(T)\|x\| - \|A_n\|\|x\|, \quad \forall\, x \in \mathfrak{D}(T).$$

So we can obtain that $\operatorname{Ker}(T + A_n) = \{0\}$ when n is large enough such that $\|A_n\| < \gamma(T)$. Consequently, $|\gamma(T + A_n) - \gamma(T)| \leq \|A_n\|$ by Proposition 6.1.6.

If $\operatorname{Ran}(T) = Y$, then $\operatorname{Ker} T^* = \{0\}$ and $\gamma(T^*) = \gamma(T) > 0$. By using above argument to $(T + A_n)^*$ and T^*, we also have

$$|\gamma(T + A_n) - \gamma(T)| = |\gamma((T + A_n)^*) - \gamma(T^*)| \leq \|A_n^*\| = \|A_n\|.$$

(2) If $\lim\limits_{n\to\infty} \gamma(T + A_n) \neq 0$, then there are an $\epsilon_0 > 0$ and a subsequence $\{\gamma(T + A_{n_k})\}$ of $\{\gamma(T + A_n)\}$ such that $\gamma(T + A_{n_k}) \geq \epsilon_0$, $\forall\, k \geq 1$. Thus, $\gamma(T) \geq \epsilon_0$ by Proposition 6.1.9, which contradicts the assumption $\gamma(T) = 0$.

(3) Pick $x_0 \in \operatorname{Ker} T$ with $\|x_0\| = 1$ and $y_0 \in Y \backslash \operatorname{Ran}(T)$ with $\|y_0\| = 1$. Let $x_0^* \in X^*$ such that $\|x_0^*\| = x_0^*(x_0) = 1$ and put

$$B_n(x) = n^{-1} x_0^*(x)\, y_0, \quad \forall\, x \in X,\ n \geq 1.$$

Then

$$\operatorname{Ker}(T + B_n) = \operatorname{Ker} T \cap \operatorname{Ker} B_n \text{ and } \operatorname{Ker}(T + B_n) \subsetneqq \operatorname{Ker} T, \ \forall\, n \geq 1.$$

So $\gamma(T + B_n) \leq \| - B_n\| = n^{-1}$ by Theorem 6.1.2 and hence

$$\lim_{n\to\infty} \gamma(T + B_n) = 0 \neq \gamma(T).$$

□

Combining Lemma 1.3.5, Theorem 6.1.2 with Corollary 6.1.10, we have

Corollary 6.1.12. *Let $\{T_n\} \subset B(X,Y)$ with $\gamma(T_n) > 0$, $\forall\, n \in \mathbb{N}$ and $T \in B(X,Y)$ with $\lim\limits_{n\to\infty} \|T_n - T\| = 0$.*

(1) *If* $\inf\limits_{n\geq 1}\gamma(T_n)>0$, *then* $\gamma(T)>0$ *and* $\lim\limits_{n\to\infty}\gamma(T_n)=\gamma(T)$.

(2) *If* $\gamma(T)>0$, *then* $\lim\limits_{n\to\infty}\gamma(T_n)=\gamma(T)$ *iff* $\inf\limits_{n\geq 1}\gamma(T_n)>0$.

Remark 6.1.13. (1) Harte and Mbekhta proved that if $T\in B(X,Y)$ with $\mathrm{Ran}\,(T)$ closed satisfies condition: $\mathrm{Ker}\,T=\{0\}$ or $\mathrm{Ran}\,(T)=Y$, then $\gamma(\cdot)$ is continuous at T; if T satisfies condition: $\mathrm{Ker}\,T\neq\{0\}$ and $\mathrm{Ran}\,(T)\neq Y$, then $\gamma(\cdot)$ is discontinuous at T (Theorem 9 of [Harte and Mbekhta (1993)]). By Corollary 6.1.11 (2), if $T\in B(X,Y)$ such that $\mathrm{Ran}\,(T)$ is not closed, then $\gamma(\cdot)$ is continuous at T. All these present the continuity of $\gamma(\cdot)$ on $B(X,Y)$.

(2) Let X, Y be Hilbert spaces and let $\{T_n\}\subset B(X,Y)$ and $T\in B(X,Y)$ with $\mathrm{Ran}\,(T)$ and $\mathrm{Ran}\,(T_n)$ are all closed, $n\geq 1$. Assume that $\lim\limits_{n\to\infty}\|T_n-T\|=0$. Then by Corollary 3.3.6, Corollary 6.1.12 (2) can be rewritten as

$$\lim_{n\to\infty}\gamma(T_n)=\gamma(T)\quad\text{iff}\quad\inf_{n\geq 1}\gamma(T_n)>0\quad\text{and iff}$$

$$\mathrm{Ran}\,(T_n)\cap\mathrm{Ran}\,(T)^{\perp}=\{0\}\quad\text{iff}\quad\mathrm{Ker}\,T\cap(\mathrm{Ker}\,T_n)^{\perp}=\{0\}$$

for n large enough since $\gamma(T_n)=\|T_n^{\dagger}\|^{-1}$, $\forall\, n\geq 1$ and $\gamma(T)=\|T^{\dagger}\|^{-1}$.

6.2 Perturbation analysis for the consistent operator equation on Banach spaces

In §3 of Chapter 2, we give the perturbation analysis for the minimal norm solution of the operator equation $Ax=b$ when $A\in Gib(X,Y)$, where X, Y are Banach spaces. In this section, we study the perturbation analysis for the minimal norm solution of the operator equation $Ax=b$ when $A\in B(X,Y)$ (or $\mathcal{C}(X,Y)$) and $\mathrm{Ran}\,(A)$ is closed in Y. Because the closeness of $\mathrm{Ran}\,(T)$ in Y does not indicate that $\mathrm{Ker}\,T$ (or $\mathrm{Ran}\,(T)$) in general admits a complemented closed subspace in X (or Y), that is, A may not have a general inverse, the investigation about it is significant.

This section is mainly based on [Wang *et al.* (2009)]. In this section, we assume that X and Y are Banach space. Let $A\in B(X,Y)$ with $\mathrm{Ran}\,(T)$ closed. The number $\kappa(A)=\|A\|\gamma(A)^{-1}$ introduced in [Chen *et al.* (2004)] is called the generalized condition number of A. We also assume that $b\in\mathrm{Ran}\,(T)\backslash\{0\}$, $\bar{A}=A+\delta A\in B(X,Y)$ and $\bar{b}=b+\delta b\in Y\backslash\{0\}$. Put $\epsilon_A=\dfrac{\|\delta A\|}{\|A\|}$, $\epsilon_b=\dfrac{\|\delta b\|}{\|b\|}$ and set

$$S(A,b)=\{x\in X\,|\,Ax=b\},\quad m(A,b)=\inf\{\|x\|\,|\,x\in S(A,b)\}.$$

Proposition 6.2.1. *Let $A \in B(X,Y)$ with* Ran(A) *closed. Then for any $z \in X$, we have*

$$\frac{1}{\kappa(A)}\frac{\|Az-b\|}{\|b\|} \le \frac{\operatorname{dist}(z,S(A,b))}{m(A,b)} \le \kappa(A)\frac{\|Az-b\|}{\|b\|} \tag{6.2.1}$$

Proof. Let $x \in S(T,b)$. Then for any $y \in \operatorname{Ker} A$,

$$\|Az-b\| = \|A(z-x-y)\| \le \|A\|\|z-x-y\| \tag{6.2.2}$$

$$\|Az-b\| = \|A(z-x)\| \ge \gamma(A)\operatorname{dist}(z-x,\operatorname{Ker} A). \tag{6.2.3}$$

(6.2.2) implies that $\|Az-b\| \le \|A\|\operatorname{dist}(z-x,\operatorname{Ker} A)$. Note that $S(A,b) = x + \operatorname{Ker} A$. So $\operatorname{dist}(z-x,\operatorname{Ker} A) = \operatorname{dist}(z,S(A,b))$. Thus,

$$\frac{\|Az-b\|}{\|A\|} \le \operatorname{dist}(z,S(A,b)) \le \frac{\|Az-b\|}{\gamma(A)}. \tag{6.2.4}$$

Now from $A(x-y)=b$, we get that

$$\|b\| \le \|A\|\|x-y\| \text{ and } \|b\| \ge \gamma(A)\operatorname{dist}(x,\operatorname{Ker} A).$$

Since $m(A,b) = \operatorname{dist}(x,\operatorname{Ker} A)$, it follows that

$$\frac{\|b\|}{\|A\|} \le m(A,b) \le \frac{\|b\|}{\gamma(A)}. \tag{6.2.5}$$

Combining (6.2.5) with (6.2.4), we get (6.2.1). □

Remark 6.2.2. (1) The $Az-b$ in (6.2.1) is called the residual associated with z.

(2) If X is a reflexive Banach space, then there is $x_0 \in S(A,b)$ such that $\operatorname{dist}(z,S(T,b)) = \|z-x_0\|$ by Proposition 1.1.21. Thus (6.2.1) can be written as

$$\frac{1}{\kappa(A)}\frac{\|Az-b\|}{\|b\|} \le \frac{\|z-x_0\|}{\operatorname{dist}(x_0,\operatorname{Ker} A)} \le \kappa(A)\frac{\|Az-b\|}{\|b\|}. \tag{6.2.6}$$

(6.2.6) generalizes Theorem 2.1 of [Ding and Wei (2002)].

Corollary 6.2.3. *Suppose that $\bar b \in$* Ran$(\bar A)$. *Let $\bar x \in S(\bar A,\bar b)$. Then*

$$\frac{1}{\kappa(A)}\frac{\|\delta b-\delta A\bar x\|}{\|b\|} \le \frac{\operatorname{dist}(\bar x,S(A,b))}{m(A,b)} \le \kappa(A)\frac{\|\delta b-\delta A\bar x\|}{\|b\|}.$$

When $\delta A = 0$, Corollary 6.2.3 is Theorem 2.7 of [Chen *et al.* (2004)].

Proposition 6.2.4. *Let $A, \bar A = A+\delta A \in B(X,Y)$ with* Ran(A) *closed. Suppose that X is reflexive and $\kappa(A)\epsilon_A < 1$, $\bar b \in$* Ran$(\bar A)$. *Then for any $\bar x \in S(\bar A,\bar b)$ there is $x \in S(A,b)$ such that*

$$\frac{1}{\kappa(A)(1+\epsilon_A)}\frac{\|\delta b-\delta Ax\|}{\|b\|} \le \frac{\|\bar x-x\|}{\operatorname{dist}(x,\operatorname{Ker} A)} \le \frac{\kappa(A)}{1-\kappa(A)\epsilon_A}\frac{\|\delta b-\delta Ax\|}{\|b\|}.$$

Proof. Pick $x \in S(A,b)$ such that $\|\bar{x}-x\| = \operatorname{dist}(\bar{x}, S(A,b))$. Since $\bar{A}(\bar{x}-x) = \delta b - \delta Ax$, it follows that

$$\begin{aligned}\|\delta b - \delta Ax\| &\geq \|A(\bar{x}-x)\| - \|\delta A\|\|\bar{x}-x\| \\ &\geq \gamma(A)\operatorname{dist}(\bar{x}-x, \operatorname{Ker} A) - \|\delta A\|\|\bar{x}-x\|; \\ \|\delta b - \delta Ax\| &\leq \|A+\delta A\|\|\bar{x}-x\| \leq (\|A\| + \|\delta A\|)\|\bar{x}-x\|.\end{aligned}$$

Noting that $\operatorname{dist}(\bar{x}-x, \operatorname{Ker} A) = \|\bar{x}-x\|$, we have

$$\frac{\|\delta b - \delta Ax\|}{\|A\| + \|\delta A\|} \leq \|\bar{x}-x\| \leq \frac{1}{\gamma(A)(1-\kappa(A)\epsilon_A)}\|\delta b - \delta Ax\|. \tag{6.2.7}$$

Therefore, combining (6.2.5) with (6.2.7), we get the assertion. □

From (6.2.7), we have

$$\frac{\|\bar{x}-x\|}{\|x\|} \leq \frac{1}{\gamma(A)(1-\kappa(A)\epsilon_A)}\Big(\frac{\|\delta b\|}{\|x\|} + \|\delta A\|\Big).$$

Thus, from $\|b\| = \|Ax\| \leq \|A\|\|x\|$, we deduce following corollary

Corollary 6.2.5. *Let* $A, \bar{A} = A + \delta A \in B(X,Y)$ *with* $\operatorname{Ran}(A)$ *closed. Suppose that* X *is reflexive and* $\kappa(A)\epsilon_A < 1$, $\bar{b} \in \operatorname{Ran}(\bar{A})$. *Then for any* $\bar{x} \in S(\bar{A}, \bar{b})$, *there is* $x \in S(A,b)$ *such that*

$$\frac{\|\bar{x}-x\|}{\|x\|} \leq \frac{\kappa(A)}{1-\kappa(A)\epsilon_A}(\epsilon_b + \epsilon_A).$$

Remark 6.2.6. Proposition 6.2.4 generalizes Theorem 2.1 of [Ding (2001)] and Corollary 6.2.5 generalizes Proposition 2.3.9.

Let $A \in C(X,Y)$ and $\delta A \in B(X,Y)$ with $\operatorname{Ran}(T)$ closed. Put $\bar{A} = A + \delta A$. Let $b \in \operatorname{Ran}(A)$ and $\bar{b} = b + \delta b \in Y$ ($\bar{b}$ may not be in $\operatorname{Ran}(\bar{A})$). We consider the perturbation problem (6.2.8) of the consistent operator equation $Ax = b$

$$\|\bar{A}x - \bar{b}\| = \min_{z \in X}\|\bar{A}z - \bar{b}\| = \operatorname{dist}(\bar{b}, \operatorname{Ran}(\bar{A})). \tag{6.2.8}$$

By Proposition 2.3.7, (6.2.8) has solutions and $\hat{S}(\bar{A}, \bar{b})$ (set of its solutions) is closed and convex when X and Y are reflexive and $\operatorname{Ran}(\bar{A})$ is closed.

Remark 6.2.7. Let Y be reflexive. $\hat{S}(\bar{A}, \bar{b})$ can not be expressed as the form

$$\hat{S}(\bar{A}, \bar{b}) = \bar{x} + \operatorname{Ker}\bar{A}, \quad \text{for some } \bar{x} \in \hat{S}(\bar{A}, \bar{b}) \tag{6.2.9}$$

because the vector $\bar{y} \in \operatorname{Ran}(\bar{A})$ with $\|\bar{b} - \bar{y}\| = \operatorname{dist}(\bar{b}, \operatorname{Ran}(\bar{A}))$ is not unique. In order to get the form (6.2.9), we may assume that Y is strictly convex. In this case, there is unique $b_0 \in \operatorname{Ran}(\bar{A})$ such that $\operatorname{dist}(\bar{b}, \operatorname{Ran}(\bar{A})) = \|\bar{b} - b_0\|$. Thus, $\hat{S}(\bar{A}, \bar{b}) = \bar{x}_0 + \operatorname{Ker}\bar{A}$ for some $\bar{x}_0$ with $\bar{A}\bar{x}_0 = b_0$.

Furthermore, if X is strictly convex, then the minimal norm solution to (6.2.8) is unique (i.e., the element $\bar{x}_m \in S(\bar{A}, \bar{b})$ which satisfies $\|\bar{x}_m\| = \min\{\|\bar{x}\| \mid \bar{x} \in S(\bar{A}, \bar{b})\}$ is unique).

Lemma 6.2.8. *Let $A \in \mathcal{C}(X, Y)$ with* $\operatorname{Ran}(A)$ *closed and let $\delta A \in B(X, Y)$ Put $\bar{A} = A + \delta A$. Suppose that X, Y is reflexive and* $\operatorname{Ran}(\bar{A})$ *is closed. Then for any $\bar{x} \in \hat{S}(\bar{A}, \bar{b})$,*

$$\operatorname{dist}(\bar{x}, S(A, b)) \le \gamma^{-1}(A)(2\|\delta b\| + \|\delta A\| m(A, b)) + \gamma^{-1}(A)\|\delta A\|\bar{x}\|.$$

Proof. Take $x_0 \in S(A, b)$. $S(A, b) = \{x_0 + y \mid y \in \operatorname{Ker} A\}$. Thus

$$\begin{aligned}\operatorname{dist}(\bar{x}, S(A, b)) &= \operatorname{dist}(\bar{x} - x_0, \operatorname{Ker} A) \le \gamma^{-1}(A)\|A(\bar{x} - x_0)\| \\ &= \gamma^{-1}(A)\|\bar{A}\bar{x} - \bar{b} + \delta b - \delta A\bar{x}\|.\end{aligned}$$

Since for any $y \in \operatorname{Ker} A$,

$$\|\bar{A}\bar{x} - \bar{b}\| \le \|\bar{A}(x_0 - y) - \bar{b}\| = \|\delta A(x_0 - y) - \delta b\|,$$

we obtain that

$$\operatorname{dist}(\bar{x}, S(A, b)) \le \frac{2\|\delta b\|}{\gamma(A)} + \gamma^{-1}(A)\|\delta A\|(\|x_0 - y\| + \|\bar{x}\|)$$

and so that

$$\operatorname{dist}(\bar{x}, S(T, b)) \le \gamma^{-1}(A)(2\|\delta b\| + \|\delta A\| m(A, b)) + \gamma^{-1}(A)\|\delta A\|\|\bar{x}\|. \qquad \square$$

Lemma 6.2.9. *Let $A \in \mathcal{C}(X, Y)$ with* $\operatorname{Ran}(A)$ *closed and let $\delta A \in B(X, Y)$ with $\gamma^{-1}(A)\|\delta A\| < 1$. Put $\bar{A} = A + \delta A$. Suppose that X, Y are reflexive and $\delta = \delta(\operatorname{Ker} A, \operatorname{Ker}\bar{A}) < \dfrac{1 - \gamma^{-1}(A)\|\delta A\|}{1 + \gamma^{-1}(A)\|\delta A\|}$. Then*

$$\|\bar{x}_m\| \le \frac{[1 + \gamma^{-1}(A)(2\|\delta b\|\|x_m\|^{-1} + \|\delta A\|)](1 + \delta)}{1 - \gamma^{-1}(A)\|\delta A\| - (1 + \gamma^{-1}(A)\|\delta A\|)\delta}\|x_m\|. \tag{6.2.10}$$

where $x_m \in S(A, b)$, $\bar{x}_m \in \hat{S}(\bar{A}, \bar{b})$ such that $\|x_m\| = \min\{\|x\| \mid x \in S(A, b)\}$ *and* $\|\bar{x}_m\| = \min\{\|\bar{x}\| \mid \bar{x} \in \hat{S}(\bar{A}, \bar{b})\}$.

Proof. It follows from Proposition 6.1.5 that $\gamma(\bar{A}) > 0$ when $\delta < \dfrac{1 - \gamma^{-1}(A)\|\delta A\|}{1 + \gamma^{-1}(A)\|\delta A\|}$. So $\bar{x}_m$ exists. Since $S(A, b)$ is closed and convex, we can choose $x_1 \in S(A,b)$ such that $\|\bar{x}_m - x_1\| = \operatorname{dist}(\bar{x}_m, S(A,b))$ by Proposition 1.1.21. Then for any $z \in \operatorname{Ker} \bar{A}$,

$$\|\bar{x}_m\| \le \|\bar{x}_m - z\| \le \|\bar{x}_m - x_1\| + \|x_1 - x_m - z\| + \|x_m\|.$$

Thus, by Lemma 6.2.8,

$$\begin{aligned}\|\bar{x}_m\| &\le \|\bar{x}_m - x_\epsilon\| + \operatorname{dist}(x_1 - x_m, \operatorname{Ker} \bar{A}) + \|x_m\| \\ &\le \|\bar{x}_m - x_1\| + \|x_1 - x_m\|\delta + \|x_m\| \\ &\le \|\bar{x}_m - x_1\|(1+\delta) + \|\bar{x}_m\|\delta + \|x_m\|(1+\delta) \\ &\le [\gamma^{-1}(A)(2\|\delta b\|\|x_m\|^{-1} + \|\delta A\|)\|x_m\| + \gamma^{-1}(A)\|\delta A\|\|\bar{x}_m\|](1+\delta) \\ &\quad + \|\bar{x}_m\|\delta + \|x_m\|(1+\delta).\end{aligned}$$

The above shows that

$$\begin{aligned}&[1 - \delta - \gamma^{-1}(A)\|\delta A\|(1+\delta)]\|\bar{x}_m\| \\ &\qquad \le [1 + \gamma^{-1}(A)(2\|\delta b\|\|x_m\|^{-1} + \|\delta A\|)](1+\delta)\|x_m\|.\end{aligned}$$

This gives (6.2.10). □

Now we present the main result of the section as follows:

Theorem 6.2.10. *Let $A \in C(X,Y)$ with $\operatorname{Ran}(A)$ closed and let $\delta A \in B(X,Y)$ with $\gamma^{-1}(A)\|\delta A\| < 1$. Put $\bar{A} = A + \delta A$. Suppose that X, Y are reflexive and $\delta = \delta(\operatorname{Ker} A, \operatorname{Ker} \bar{A}) < \dfrac{1 - \gamma^{-1}(A)\|\delta A\|}{1 + \gamma^{-1}(A)\|\delta A\|}$. Then*

$$\frac{\operatorname{dist}(\bar{x}_m, S(A,b))}{\|x_m\|} \le \frac{2\gamma^{-1}(A)(\|\delta b\|\|x_m\|^{-1}(1-\delta) + \|\delta A\|)}{1 - \gamma^{-1}(A)\|\delta A\| - (1 + \gamma^{-1}(A)\|\delta A\|)\delta}, \tag{6.2.11}$$

where $\bar{x}_m$ and x_m are defined in Lemma 6.2.9.

Proof. By Lemma 6.2.8 and Lemma 6.2.9, we have

$$\begin{aligned}\frac{\operatorname{dist}(\bar{x}_m, S(A,b))}{\|x_m\|} &\le \gamma^{-1}(A)(2\|\delta b\|\|x_m\|^{-1} + \|\delta T\|) \\ &\quad + \frac{[1 + \gamma^{-1}(A)(2\|\delta b\|\|x_m\|^{-1} + \|\delta A\|)]\gamma^{-1}(A)\|\delta A\|(1+\delta)}{1 - \gamma^{-1}(A)\|\delta A\|\|\delta A\| - (1 + \gamma^{-1}(A)\|\delta A\|)\delta} \\ &\le \frac{2\gamma^{-1}(A)(\|\delta b\|\|x_m\|^{-1}(1-\delta) + \|\delta A\|)}{1 - \gamma^{-1}(A)\|\delta A\| - (1 + \gamma^{-1}(A)\|\delta A\|)\delta}.\end{aligned}$$

□

Remark 6.2.11. In Theorem 6.2.10, if $A \in B(X,Y)$, then $\|A\|\|x_m\| \ge \|b\|$ and $\kappa(A)\epsilon_A = \gamma^{-1}(A)\|\delta A\|$. In this case, (6.2.11) can be rewritten as

$$\frac{\operatorname{dist}(\bar{x}_m, S(A,b))}{\|x_m\|} \le \frac{2\kappa(A)(\epsilon_b(1-\delta) + \epsilon_A)}{1 - \kappa(A)\epsilon_A - (1 + \kappa(A)\epsilon_A)\delta}.$$

Corollary 6.2.12. *With the same assumptions as in Theorem* 6.2.10.

(1) *If* $\operatorname{Ker} A \subset \operatorname{Ker} \bar{A}$ *and* $\gamma^{-1}(A)\|\delta A\| < 1$, *then*

$$\frac{\operatorname{dist}(\bar{x}_m, S(A,b))}{\|x_m\|} \le \frac{2\gamma^{-1}(A)(\|\delta b\|\|x_m\|^{-1} + \|\delta A\|)}{1 - \gamma^{-1}(A)\|\delta A\|}.$$

(2) *If* $\dim \operatorname{Ker} \bar{A} = \dim \operatorname{Ker} A < \infty$ *and* $\gamma^{-1}(A)\|\delta A\| < 1/3$, *then*

$$\frac{\operatorname{dist}(\bar{x}_m, S(A,b))}{\|x_m\|} \le \frac{2\gamma^{-1}(A)(\|\delta b\|\|x_m\|^{-1} + 2\|\delta A\|)}{1 - 3\gamma^{-1}(A)\|\delta A\|}. \tag{6.2.12}$$

Especially, when X, Y *are all finite dimensional and* $\operatorname{rank} \bar{A} = \operatorname{rank} A$, $\kappa(A)\epsilon_A < 1/3$, *then*

$$\frac{\operatorname{dist}(\bar{x}_m, S(A,b))}{\|x_m\|} \le \frac{2\kappa(A)(\epsilon_b + 2\epsilon_A)}{1 - 3\kappa(A)\epsilon_A}. \tag{6.2.13}$$

Proof. (1) If $\operatorname{Ker} A \subset \operatorname{Ker} \bar{A}$, $\delta(\operatorname{Ker} A, \operatorname{Ker} \bar{A}) = 0$. The assertion follows from Theorem 6.2.10.

(2) By Proposition 6.1.6,

$$\gamma(\bar{A}) \ge \gamma(A) - \|\delta A\| = \gamma(A)(1 - \gamma^{-1}(A)\|\delta A\|) > \frac{2}{3}\gamma(A) > 0.$$

So by Lemma 1.3.5,

$$\begin{aligned}\delta(\operatorname{Ker} A, \operatorname{Ker} \bar{A}) &\le \gamma^{-1}(\bar{A})\|\delta A\| < \frac{3}{2}\gamma^{-1}(A)\|\delta A\| \\ &< \frac{1}{2} < \frac{1 - \gamma^{-1}(A)\|\delta A\|}{1 + \gamma^{-1}(A)\|\delta A\|}\end{aligned}$$

when $\gamma^{-1}(A)\|\delta A\| < 1/3$. Therefore, we have

$$\begin{aligned}\frac{\operatorname{dist}(\bar{x}_m, S(A,b))}{\|x_m\|} &\le \frac{2\gamma^{-1}(A)(\|\delta b\|\|x_m\|^{-1}(1-\delta) + \|\delta A\|}{1 - \gamma^{-1}(A)\|\delta A\| - (1 + \gamma^{-1}(A)\|\delta A\|)\delta} \\ &< \frac{2\gamma^{-1}(A)(\|\delta b\|\|x_m\|^{-1} + 2\|\delta A\|)}{1 - 3\gamma^{-1}(A)\|\delta A\|}.\end{aligned}$$

If X, Y are all finite dimensional, then $\operatorname{rank} \bar{A} = \operatorname{rank} A$ implies that $\dim \operatorname{Ker} \bar{A} = \dim \operatorname{Ker} A$. Note that $\|A\|\|x_m\| \ge \|b\|$. So (6.2.13) follows from (6.2.12). □

We now illustrate our results by following example:

Example 6.2.13. Let $X = \mathbb{R}^2$ and let A, $\bar{A}$, b and $\bar{b}$ be

$$A = \begin{bmatrix} 1 & 1 \\ 0 & 0 \end{bmatrix}, \quad \bar{A} = \begin{bmatrix} 1+\epsilon & 1+\epsilon \\ \epsilon & \epsilon \end{bmatrix}, \quad b = \begin{bmatrix} 1 \\ 0 \end{bmatrix}, \quad \bar{b} = \begin{bmatrix} 1+\epsilon \\ 0 \end{bmatrix},$$

where $\epsilon \in (0,1)$. We take the p–norm on X $(p \in \mathbb{N})$.

If $p = 1$, then the minimal norm solutions of

$$Ax = b, \qquad \|\bar{b} - \bar{A}\bar{x}\| = \mathrm{dist}\,(\bar{b}, \mathrm{Ran}\,(\bar{A}))$$

are

$$x_m = \begin{bmatrix} 1 \\ 0 \end{bmatrix} + s \begin{bmatrix} 1 \\ -1 \end{bmatrix}, \quad \bar{x}_m = \begin{bmatrix} 1 \\ 0 \end{bmatrix} + t \begin{bmatrix} 1 \\ -1 \end{bmatrix}, \ -1 \le s, t \le 0.$$

Thus, $\mathrm{dist}\,(\bar{x}_m, S(A,b)) = 0$ and $\dfrac{\|\bar{x}_m - x_m\|}{\|x_m\|} = |s + t|$ is uncertain.

If p is even, then the minimal norm solutions of

$$Ax = b, \qquad \|\bar{b} - \bar{A}\bar{x}\| = \mathrm{dist}\,(\bar{b}, \mathrm{Ran}\,(\bar{A}))$$

are

$$x_m = \begin{bmatrix} \frac{1}{2} \\ \frac{1}{2} \end{bmatrix}, \qquad \bar{x}_m = \begin{bmatrix} \frac{\epsilon(p)}{2} \\ \frac{\epsilon(p)}{2} \end{bmatrix},$$

where $\epsilon(p) = \dfrac{(1+\epsilon)^{\frac{p}{p-1}}}{(1+\epsilon)^{\frac{p}{p-1}} + \epsilon^{\frac{p}{p-1}}}$. We also have $\gamma(A) = \|A\| = 2^{\frac{p-1}{p}}$. Let $\delta A = \bar{A} - A$ and $\delta b = \bar{b} - b$. Then $\epsilon_b = \epsilon$, $\|\delta A\| = 2\epsilon$, $\epsilon_A = 2^{\frac{1}{p}}\epsilon$. Suppose that $\epsilon < 2^{-\frac{1}{p}}$. Since $\mathrm{Ker}\,\bar{A} = \mathrm{Ker}\,A$, we have by Corollary 6.2.3 (1),

$$\frac{\mathrm{dist}\,(\bar{x}_m, S(A,b))}{\|x_m\|} \le \frac{2(1 + 2^{\frac{1}{p}})\epsilon}{1 - 2^{\frac{1}{p}}\epsilon}, \qquad \frac{\|\bar{x}_m - x_m\|}{\|x_m\|} \le 1 - \epsilon(p)$$

when ϵ is small enough.

Remark 6.2.14. The perturbation analysis of minimal norm solutions of the non–consistent perturbation of a consistent operator equation in reflexive Banach spaces is the special case of the perturbation analysis of Problem$(*)$

$$\min \|x\| \quad \text{subject to} \quad \|Ax - b\| = \min\{\|Az - b\| \mid z \in \mathfrak{D}(A)\} \qquad (*)$$

where X, Y are reflexive Banach spaces and $A \in \mathfrak{D}(X,Y)$ with $\mathrm{Ran}\,(A)$ closed. Considering Example 6.2.13, we can yield a problem:

Problem: Let X, Y be strictly convex and reflexive Banach spaces and let $A \in \mathcal{C}(X,Y)$ with $\mathrm{Ran}\,(A)$ closed. Let $\delta A \in B(X,Y)$ with $\gamma^{-1}(A)\|\delta A\| < 1$. Put $\bar{A} = A + \delta A$. Let x_m be the solution of Problem $(*)$ and $\bar{x}_m$ be the solution of the problem

$$\min \|x\| \quad \text{subject to} \quad \|\bar{A}x - \bar{b}\| = \min\{\|\bar{A}z - \bar{b}\| \mid z \in \mathfrak{D}(A)\}.$$

where b, $\bar{b} = b + \delta b \in Y$. Can we give an estimation of the upper bound of $\dfrac{\|\bar{x}_m - x_m\|}{\|x_m\|}$ by means of $\hat{\delta}(\mathrm{Ker}\,A, \mathrm{Ker}\,\bar{A})$?

6.3 Fredholm elements relative to a homomorphism and their indices

This section is based on [Xue (2008b)]. Throughout the section, we always assume that $\mathcal{A}$, $\mathcal{B}$ are unital Banach algebras and $T\colon \mathcal{A}\to\mathcal{B}$ is a continuous unital homomorphism. In [Harte (1982)], Harte introduced the T–Fredholm elements and T–Wely elements in $\mathcal{A}$ as follows.

Put $\mathrm{Fred}_T(\mathcal{A}) = T^{-1}(GL(\mathcal{B}))$ and $\mathrm{Fred}_T^0(\mathcal{A}) = GL(\mathcal{A}) + \mathrm{Ker}\,T$. The elements in $\mathrm{Fred}_T(\mathcal{A})$ are called to be T–Fredholm and in $\mathrm{Fred}_T^0(\mathcal{A})$ are called to be T–Wely.

Definition 6.3.1 ([Rieffel (1983)]). *Let $\mathcal{E}$ be a unital Banach algebra. We say that $\mathcal{E}$ is of topological stable rank one, denoted as $\mathrm{tsr}(\mathcal{E}) = 1$, if $GL(\mathcal{E})$ is dense in $\mathcal{E}$.*

If $\mathcal{E}$ is non–unital and $\mathrm{tsr}(\tilde{\mathcal{E}}) = 1$, we also set $\mathrm{tsr}(\mathcal{E}) = 1$.

Proposition 6.3.2. *Let $\mathcal{A}$, $\mathcal{B}$ and T be as above.*

(1) $\mathrm{Fred}_T^0(\mathcal{A})$ *is open in* $\mathrm{Fred}_T(\mathcal{A})$.
(2) *If* $\mathrm{Fred}_T^0(\mathcal{A})$ *is closed in* $\mathrm{Fred}_T(\mathcal{A})$, $\mathrm{Fred}_T(\mathcal{A})\cap\overline{GL(\mathcal{A})}\subset\mathrm{Fred}_T^0(\mathcal{A})$.
(3) $\mathrm{Fred}_T(\mathcal{A})\cap\overline{GL(\mathcal{A})}\supset\mathrm{Fred}_T^0(\mathcal{A})$ *iff* $\mathrm{tsr}(\mathrm{Ker}\,T) = 1$.

Proof. (1) Since $GL(\mathcal{A})$ is open in $\mathcal{A}$, $GL(\mathcal{A}) + k$ is open in $\mathcal{A}$ for each $k\in\mathrm{Ker}\,T$. Thus $\mathrm{Fred}_T^0(\mathcal{A}) = \{GL(\mathcal{A}) + k\,|\,k\in\mathrm{Ker}\,T\}$ is open in $\mathcal{A}$ and hence is open in $\mathrm{Fred}_T(\mathcal{A})$.

(2) is obivous since $GL(\mathcal{A})\subset\mathrm{Fred}_T^0(\mathcal{A})$.

(3) Put $\mathcal{J} = \mathrm{Ker}\,T$. If $\mathrm{tsr}(\mathcal{J}) = 1$, then for any $a\in GL(\mathcal{A})$ and $k\in\mathcal{J}$,

$$a + k = a(1 + a^{-1}k)\in a\,(\overline{GL(\tilde{\mathcal{J}})})\subset\overline{GL(\mathcal{A})},$$

i.e., $\mathrm{Fred}_T^0(\mathcal{A})\subset\mathrm{Fred}_T(\mathcal{A})\cap\overline{GL(\mathcal{A})}$.

Conversely, for any $k\in\mathcal{J}$ and any $\epsilon\in(0,1)$, there exists $x_\epsilon\in GL(\mathcal{A})$ such that

$$\|1 + k - x_\epsilon\| < \frac{\epsilon}{4(1 + \|1 + k\|)} < \frac{1}{2}.$$

Put $a_\epsilon = x_\epsilon - k$. Then $a_\epsilon\in GL(\mathcal{A})$ and $\|a_\epsilon^{-1}\| < 2$. Set $z_\epsilon = a_\epsilon^{-1}x_\epsilon$. Then $z_\epsilon\in GL(\mathcal{A})$, $T(z_\epsilon) = T(z_\epsilon^{-1}) = 1$, i.e., $z_\epsilon\in GL(\tilde{\mathcal{J}})$ and furthermore,

$$\|1 + k - z_\epsilon\|\le\|1 + k - x_\epsilon\| + \|a_\epsilon^{-1}\|\|1 - a_\epsilon\|\|x_\epsilon\| < \epsilon.$$

Now let $x = \lambda 1 + z\in\tilde{\mathcal{J}}$. If $\lambda = 0$, we put $x_\epsilon = \epsilon 1 + z = \epsilon(1 + \epsilon^{-1}z)$. Then $\|x - x_\epsilon\| < \epsilon$ and $x_\epsilon\in GL(\tilde{\mathcal{J}})$. So $x\in\overline{GL(\tilde{\mathcal{J}})}$. If $\lambda\ne 0$, then $x = \lambda(1 + \lambda^{-1}z)\in\overline{GL(\tilde{\mathcal{J}})}$. Therefore, $\mathrm{tsr}(\mathrm{Ker}\,T) = 1$. $\square$

Definition 6.3.3. Let $\mathcal{E}$ be a unital normed algebra. We say that $\mathcal{E}$ is closed under inverses, if for every $x \in \mathcal{E}$ with $\|1-x\| < 1$, we have $x^{-1} \in \mathcal{E}$.

Lemma 6.3.4. *Let $\mathcal{A}$, $\mathcal{B}$ and T be as above.*

(1) *If* $\mathrm{Fred}_T(\mathcal{A}) \subset Gi(\mathcal{A})$, *then $T(\mathcal{A})$ is closed under inverses.*
(2) *Suppose that $T(\mathcal{A})$ is closed under inverses. Let $b \in \mathcal{B}$ with $\|1-b\| < 1/2$ and $b^{-1} \in T(\mathcal{A})$. Then $b \in T(\mathcal{A})$.*

Proof. (1) Let $b \in T(\mathcal{A})$ such that $\|1-b\| < 1$. Then $b \in GL(\mathcal{B})$. Choose $a \in \mathcal{A}$ such that $b = T(a)$. Since $a \in \mathrm{Fred}_T(\mathcal{A}) \subset Gi(\mathcal{A})$, there is $a_0 \in \mathcal{A}$ such that $aa_0a = a$ and consequently, $b^{-1} = T(a_0)$.

(2) $\|1-b\| < 1/2$ indicates that $b \in GL(\mathcal{B})$ and $\|b^{-1}\| < 2$. So

$$\|1 - b^{-1}\| \le \|b^{-1}\|\|1-b\| < 1.$$

Since $T(\mathcal{A})$ is closed under inverses and $b^{-1} \in T(\mathcal{A})$, it follows that $b = (b^{-1})^{-1} \in T(\mathcal{A})$. □

Proposition 6.3.5. *If $T(\mathcal{A})$ is closed under inverses, then* $\mathrm{Fred}_T^0(\mathcal{A})$ *is closed in* $\mathrm{Fred}_T(\mathcal{A})$.

Proof. Let $x \in \mathrm{Fred}_T(\mathcal{A})$ and $\{x_n\} \subset \mathrm{Fred}_T^0(\mathcal{A})$ with $\lim\limits_{n\to\infty} \|x_n - x\| = 0$. Choose n_0 such that

$$\|T(x_{n_0}) - T(x)\| < \frac{1}{2\|(T(x))^{-1}\|}.$$

Then $\|T(x_{n_0})(T(x))^{-1} - 1\| < 1/2$. Put $b = T(x_{n_0})(T(x))^{-1} \in GL(\mathcal{B})$. Then $b^{-1} \in T(\mathcal{A})$ and $\|1-b^{-1}\| < 1$. So by Lemma 6.3.4 (2), there is $d \in \mathcal{A}$ such that $b = T(d)$. Then $T(dxx_{n_0}^{-1}) = T(xx_{n_0}^{-1}d) = 1$. Set $c = x_{n_0}^{-1}d \in \mathcal{A}$. Then $T(cx) = T(xc) = 1$ and hence $k_1 = xc - 1$ and $k_2 = cx - 1$ are all in $\mathrm{Ker}\,T$. Pick n_1 such that $\|x_{n_1} - x\| < 1/\|c\|$. Then

$$\|1 + k_1 - x_{n_1}c\| = \|(x - x_{n_1})c\| \le \|x - x_{n_1}\|\|c\| < 1$$

so that $g = k_1 - x_{n_1}c \in GL(\mathcal{A})$. Therefore,

$$\begin{aligned} x &= g^{-1}(k_1 - x_{n_1}c)x = g^{-1}k_1x - g^{-1}x_{n_1}(1+k_2) \\ &= g^{-1}k_1x - g^{-1}x_{n_1}k_2 - g^{-1}x_{n_1} \in \mathrm{Fred}_T^0(\mathcal{A}). \end{aligned}$$

□

By Proposition 6.3.2 and Proposition 6.3.5, we have

Corollary 6.3.6. *Let $T\colon \mathcal{A} \to \mathcal{B}$ be a unital continuous homomorphism. Suppose that $T(A)$ is closed under inverses. Then* $\mathrm{Fred}_T(\mathcal{A}) \cap \overline{GL(\mathcal{A})} = \mathrm{Fred}_T^0(\mathcal{A})$ *iff* $\mathrm{tsr}(\mathrm{Ker}\,T) = 1$.

Corollary 6.3.7 ([Harte (1987)]). *Let $T\colon \mathcal{A}\to\mathcal{B}$ be a unital continuous homomorphism. Suppose that $T(A)\subset Gi(\mathcal{A})$ and*

$$1+\operatorname{Ker}T\subset D_d(\mathcal{A})=\{a\in\mathcal{A}|\, a\in aGL(\mathcal{A})a\}.$$

Then $\mathrm{Fred}_T(\mathcal{A})\cap\overline{GL(\mathcal{A})}=\mathrm{Fred}_T^0(\mathcal{A})$.

Proof. We need only to check that $\mathrm{tsr}(\operatorname{Ker}T)=1$ when $1+\operatorname{Ker}T\subset D_d(\mathcal{A})$. Let $x\in\operatorname{Ker}T$. Then there is $y\in GL(\mathcal{A})$ such that

$$(1+x)y(1+x)=1+x. \tag{6.3.1}$$

So $1-y\in\widetilde{\operatorname{Ker}T}$ by (6.3.1). Put $p=y(1+x)$. Then p is idempotent in $\widetilde{\operatorname{Ker}T}$ by (6.3.1).

For any $\epsilon>0$, set $z_\epsilon=y^{-1}(p+\epsilon(1-p))$. Then z_ϵ is invertible in $\widetilde{\operatorname{Ker}T}$ and $\|1+x-z_\epsilon\|<\|y^{-1}(1-p)\|\epsilon$. This indicates that $\mathrm{tsr}(\operatorname{Ker}T)=1$. □

Corollary 6.3.8. *Suppose that $T(A)$ is closed in $\mathcal{B}$. If* $\mathrm{tsr}(\mathcal{A})=1$, *then* $\mathrm{Fred}_T(\mathcal{A})=\mathrm{Fred}_T^0(\mathcal{A})$.

Proof. By Corollary 6.3.6, we should prove that $\mathrm{tsr}(\operatorname{Ker}T)=1$ when $\mathrm{tsr}(\mathcal{A})=1$.

Since $T(\mathcal{A})$ is closed in $\mathcal{B}$, we have $\gamma(T)>0$ and

$$\|Tx\|\geq\gamma(T)\mathrm{dist}\,(x,\operatorname{Ker}T),\quad \forall\, x\in\mathcal{A}. \tag{6.3.2}$$

Let $x\in\operatorname{Ker}T$ and put $M=(1+\|T\|)(1+\gamma^{-1}(T))(1+\|1+x\|)$. Then for any $\epsilon\in(0,1)$, there is $x_\epsilon\in GL(\mathcal{A})$ such that $\|1+x-x_\epsilon\|<\dfrac{\epsilon}{3M}$. Then $\|T(1-x_\epsilon)\|<\dfrac{\|T\|\epsilon}{3M}$ and hence by (6.3.2),

$$\mathrm{dist}\,(1-x_\epsilon,\operatorname{Ker}T)<\gamma^{-1}(T)\|T(1-x_\epsilon)\|<\frac{\gamma^{-1}(T)\|T\|\epsilon}{3M}<\frac{1}{2}$$

Choose $k\in\operatorname{Ker}T$ such that $\|1-x_\epsilon-k\|<\dfrac{\gamma^{-1}(T)\|T\|\epsilon}{3M}$. Then $x_\epsilon+k\in GL(\mathcal{A})$ and $\|(x_\epsilon+k)^{-1}\|<2$. Put $c_\epsilon=(x_\epsilon+k)^{-1}x_\epsilon\in GL(\mathcal{A})$. Since $T(c_\epsilon)=1$, we have $c_\epsilon\in\widetilde{\operatorname{Ker}T}$. Moreover,

$$\begin{aligned}\|1+x-c_\epsilon\|&=\|1+x-x_\epsilon-((x_\epsilon+k)^{-1}-1)x_\epsilon\|\\&\leq\|1+x-x_\epsilon\|+\|(x_\epsilon+k)^{-1}\|\|x_\epsilon+k-1\|\|x_\epsilon\|\\&<\frac{\epsilon}{3}+2\|x_\epsilon+k-1\|(1+\|1+x\|)<\epsilon.\end{aligned}$$

Therefore, $\mathrm{tsr}(\operatorname{Ker}T)=1$. □

Consider following two examples:

Example 6.3.9. Let X be a separable Banach space and let $\mathcal{K}(X)$ denote the Banach algebra of all compact operators on X. Let T be the canonical homomorphism of $B(X)$ onto $B(X)/\mathcal{K}(X)$. Then $\operatorname{Ker} T = \mathcal{K}(X)$. Using the fact that every nonzero point in the spectrum of a compact operator is isolated, we can deduce that $\operatorname{tsr}(\mathcal{K}(X)) = 1$. So by Corollary 6.3.6, $\operatorname{Fred}_T^0(B(X)) = \operatorname{Fred}_T(B(X)) \cap \overline{GL(B(X))}$.

Example 6.3.10. Set $\overline{\mathbf{D}} = \{z \in \mathbb{C} \mid |z| \leq 1\}$. Let $\mathcal{A} = C(\overline{\mathbf{D}})$ and $\mathcal{B} = C(\mathbf{S}^1)$. Let T be the homomorphism from $\mathcal{A}$ onto $\mathcal{B}$ given by $T(f)(z) = f(z)$, $\forall\, z \in \mathbf{S}^1$, $f \in C(\overline{\mathbf{D}})$. Since $\operatorname{Ker} T \cong C_0(\mathbb{R}^2)$ and $\widetilde{\operatorname{Ker} T} \cong C(\mathbf{S}^2)$, it follows from Proposition 1.7 of [Rieffel (1983)] that $\operatorname{tsr}(\operatorname{Ker} T) \neq 1$. By Proposition 6.3.2 and Proposition 6.3.5, $\operatorname{Fred}_T^0(\mathcal{A})$ is both open and closed in $\operatorname{Fred}_T(\mathcal{A})$ and $\operatorname{Fred}_T(\mathcal{A}) \cap \overline{GL(\mathcal{A})} \subsetneqq \operatorname{Fred}_T^0(\mathcal{A})$.

In the rest of the section, we assume that $\operatorname{Ran}(T) = \mathcal{B}$ and $\mathcal{A}$ has an ideal $\mathcal{I}$ associating a linear functional ϕ on $\mathcal{I}$ such that $\overline{\mathcal{I}} = \operatorname{Ker} T = \mathcal{J}$ and

$$\phi(ax) = \phi(xa), \quad \forall\, a \in \mathcal{I},\ x \in \mathcal{A}.$$

The ϕ is called to be a trace on $\mathcal{I}$.

Lemma 6.3.11. *Let $a \in \operatorname{Fred}_T(\mathcal{A})$. Then there is $b \in \mathcal{A}$ and $k_1, k_2 \in \mathcal{I}$ such that $ab = 1 + k_1$ and $ba = 1 + k_2$.*

Proof. $a \in \operatorname{Fred}_T(\mathcal{A})$ means that there there is $b_0 \in \mathcal{A}$ and $c_1, c_2 \in \mathcal{J}$ such that $ab_0 = 1 + c_1$ and $b_0 a = 1 + c_2$. Choose $d_1, d_2 \in \mathcal{I}$ such that $\|c_i - d_i\| < 1$, $i = 1, 2$. Put $x_1 = ab_0 - d_1$ and $x_2 = b_0 a - d_2$. Then $\|1 - x_i\| < 1$, $i = 1, 2$. Thus, $x_1, x_2 \in GL(\mathcal{A})$ and hence

$$ab_0 x_1^{-1} = 1 + d_1 x_1^{-1}, \quad x_2^{-1} b_0 a = 1 + x_2^{-1} d_2. \tag{6.3.3}$$

From (6.3.3), we get that $b_0 x_1^{-1} = x_2^{-1} b_0 + x_2^{-1}(b_0 d_1 - d_2 b_0) x_1^{-1}$. Put $b = b_0 x_1^{-1}$ and $k_1 = d_1 x_1^{-1}$, $k_2 = x_2^{-1} d_2 + x_2^{-1}(b_0 d_1 - d_2 b_0) x_1^{-1} a$. Then $k_1, k_2 \in \mathcal{I}$ and $ab = 1 + k_1$, $ba = 1 + k_2$. □

According to Lemma 6.3.11, we can define a function index: $\operatorname{Fred}_T(\mathcal{A}) \longrightarrow \mathbb{C}$ by $\operatorname{index}(a) = \phi(ab - ba)$, $\forall\, a \in \operatorname{Fred}_T(\mathcal{A})$, where $b \in \mathcal{A}$ such that $ab - 1,\ ba - 1 \in \mathcal{I}$.

index$(\cdot)$ is well–defined. In fact, if there is $b' \in \mathcal{A}$ such that $ab' - 1,\ b'a - 1 \in \mathcal{I}$, then $b' = b + (b'a - 1)b - b'(ab - 1)$ and so that

$$ab' - b'a = ab - ba + a(b'a - 1)b - (b'a - 1)ba + b'(ab - 1)a - ab'(ab - 1).$$

Noting that

$$\phi(a(b'a-1)b) = \phi((b'a-1)ba), \text{and } \phi(b'(ab-1)a) = \phi(ab'(ab-1)),$$

we have $\phi(ab - ba) = \phi(ab' - b'a)$.

The function index$(\cdot)$ satisfies following properties:

Proposition 6.3.12. *Let x, $y \in \mathrm{Fred}_T(\mathcal{A})$ and $k \in \mathcal{J}$. Then*

(1) $\mathrm{index}(xy) = \mathrm{index}(x) + \mathrm{index}(y)$.
(2) $\mathrm{index}(1 + k) = 0$.
(3) $\mathrm{index}(x + k) = \mathrm{index}(x)$.
(4) $\mathrm{index}(x) = 0$ *if* $x \in \mathrm{Fred}_T^0(\mathcal{A})$.

Proof. (1) Choose z, $z' \in \mathcal{A}$ such that $k_1 = xz - 1$, $k_2 = zx - 1$ and $k_1' = yz' - 1$, $k_2' = z'y - 1$ are all in $\mathcal{I}$ by Lemma 6.4.10. Then

$$xyz'z = x(1 + k_1')z = xz + xk_1'z = 1 + k_1 + xk_1'z$$
$$z'zxy = z'(1 + k_2)y = z'y + z'k_2y = 1 + k_2' + z'k_2y$$

and consequently,

$$\begin{aligned}\phi(xyz'z - z'zxy) &= \phi(k_1 - k_2') + \phi(xk_1'z - z'k_2y)\\ &= \phi(k_1 - k_2') + \phi(k_1'zx - k_2yz')\\ &= \phi(k_1 - k_2') + \phi(k_1'(1 + k_2) - k_2(1 + k_1'))\\ &= \mathrm{index}(x) + \mathrm{index}(y).\end{aligned}$$

(2) Pick $k' \in \mathcal{I}$ such that $\|k - k'\| < 1$. Put $c = 1 + k$. Then $\|c - k' - 1\| < 1$ and hence $d = c - k' \in GL(\mathcal{A})$. Thus, $cd^{-1} = 1 + k'd^{-1}$ and $d^{-1}c = 1 + d^{-1}k'$. So

$$\mathrm{index}(c) = \phi(cd^{-1} - d^{-1}c) = \phi(k'd^{-1} - d^{-1}k') = 0.$$

(3) Let $z \in \mathcal{A}$ and k_1, $k_2 \in \mathcal{I}$ such that $xz = 1 + k_1$, $zx = 1 + k_2$. Then

$$0 = \mathrm{index}(1 + k_1) = \mathrm{index}(xz) = \mathrm{index}(x) + \mathrm{index}(z),$$

i.e., $\mathrm{index}(z) = -\mathrm{index}(x)$ by (1) and (2).

Put $y = x + k$ and choose $k' \in \mathcal{I}$ such that $\|k - k'\| < \|z\|^{-1}$. Then $\|yz - (1 + k_1) - k'z\| < 1$ and that $w = yz - k_1 - k'z \in GL(\mathcal{A})$. Thus, $yzw^{-1} = 1 + (k_1 + k'z)w^{-1}$ and hence

$$0 = \mathrm{index}(yzw^{-1}) = \mathrm{index}(y) + \mathrm{index}(z) = \mathrm{index}(y) - \mathrm{index}(x).$$

(4) When $x \in \mathrm{Fred}_T^0(\mathcal{A})$, $x = v + k_0$ for some $v \in GL(\mathcal{A})$ and $k_0 \in \mathcal{J}$. Then $x = v(1 + v^{-1}k_0)$ and

$$\mathrm{index}(x) = \mathrm{index}(v) + \mathrm{index}(1 + v^{-1}k_0) = 0.$$

□

Corollary 6.3.13. *Let x, $y \in \mathrm{Fred}_T(\mathcal{A})$. Suppose that there exists a continuous mapping $f\colon [0,1] \to \mathrm{Fred}_T(\mathcal{A})$ such that $f(0) = x$ and $f(1) = y$. Then* $\mathrm{index}(x) = \mathrm{index}(y)$.

Proof. We have $T(x)(T(y))^{-1} \in GL_0(\mathcal{B})$. It follows from Proposition 1.4.16 (2) that there is $a \in GL(\mathcal{A})$ such that $T(x)(T(y))^{-1} = T(a)$ and hence $x = ay + k$ for some $k \in \mathcal{J}$. Therefore, $\mathrm{index}(x) = \mathrm{index}(y)$ by Proposition 6.3.12. □

Remark 6.3.14. Corollary 6.3.13 shows that $\mathrm{index}(\cdot)$ is a locally constant function on $\mathrm{Fred}_T(\mathcal{A})$. In fact, Let a, $b \in \mathrm{Fred}_T(\mathcal{A})$ such that $\|a - b\| < \dfrac{1}{\|(T(a))^{-1}\|}$ and put $f(t) = ta + (1-t)b$, $\forall\, t \in [0,1]$. Then

$$\|T(f(t)) - T(a)\| = (1-t)\|T(a) - T(b)\| < \frac{1}{\|(T(a))^{-1}\|}, \quad \forall\, t \in [0,1].$$

So $f\colon [0,1] \to \mathrm{Fred}_T(\mathcal{A})$ is continuous with $f(0) = b$ and $f(1) = a$. Hence $\mathrm{index}(a) = \mathrm{index}(b)$.

The locally constant function $\mathrm{index}(\cdot)$ is certainly continuous (for the discrete topology on $\mathbb{C}$). We call $\mathrm{index}(a)$ is the analytical index of the T–Fredholm element a.

Let $p \in \mathcal{A}\backslash\mathcal{I}$ be an idempotent element. Put $\mathcal{A}_p = \{x \in \mathcal{A} \,|\, xp - px \in \mathcal{I}\}$. Obviously, $\mathcal{A}_p$ is a linear subspace of $\mathcal{A}$ with $1 \in \mathcal{A}_p$ and $\mathcal{I} \subset \mathcal{A}_p$. Let x, $y \in \mathcal{A}_p$. Since $xyp - pxy = x(yp - py) + (xp - px)y \in \mathcal{I}$, it follows that $\mathcal{A}_p$ is also a subalgebra of $\mathcal{A}$. Moreover, we have $x^{-1} \in \mathcal{A}_p$ when $x \in \mathcal{A}_p \cap GL(\mathcal{A})$, for $x^{-1}p - px^{-1} = x^{-1}(px - xp)x^{-1} \in \mathcal{I}$.

Theorem 6.3.15. *Let p and $\mathcal{A}_p$ be as above. Define a function ω from $\mathcal{A}_p \cap GL(\mathcal{A})$ to $\mathbb{C}$ by $\omega(a) = \phi(a^{-1}(ap - pa))$, $\forall\, a \in \mathcal{A}_p \cap GL(\mathcal{A})$. Then*

(1) $\omega(ab) = \omega(a) + \omega(b)$, $\forall\, a, b \in \mathcal{A}_p \cap GL(\mathcal{A})$.
(2) $pap + 1 - p \in \mathrm{Fred}_T(\mathcal{A})$, $\forall\, a \in \overline{\mathcal{A}_p} \cap GL(\mathcal{A})$
(3) $\phi(a^{-1}(ap - pa)) = \mathrm{index}(pap + 1 - p)$, $\forall\, a \in \mathcal{A}_p \cap GL(\mathcal{A})$.

Proof. (1) We have a^{-1}, b^{-1} and $(ab)^{-1} = b^{-1}a^{-1}$ are all in $\mathcal{A}_p$. From

$$(ab)^{-1}(abp - pab) = b^{-1}(bp - pb) + b^{-1}a^{-1}(ap - pa)b,$$

we get that

$$\omega(ab) = \phi(b^{-1}(bp - pb)) + \phi(b^{-1}a^{-1}(ap - pa)b) = \omega(a) + \omega(b).$$

(2) Let $\{a_n\} \subset \mathcal{A}_p$ such that $\lim_{n\to\infty} \|a_n - a\| = 0$. Choose $n_0 \in \mathbb{N}$ such that $\|a - a_n\| < \dfrac{1}{2\|a^{-1}\|}$, $\forall\, n \geq n_0$. Then $\|1 - a^{-1}a_n\| < 1/2$ and hence $a_n \in GL(\mathcal{A})$ with $\|a_n^{-1}\| < 2\|a^{-1}\|$, $\forall\, n \geq n_0$.

Put $b = T(pap + 1 - p)$ and $b_n = T(pa_np + 1 - p)$, $\forall\, n \geq n_0$. Then $\lim_{n\to\infty} \|b_n - b\| = 0$ and $b_n^{-1} = T(a_n^{-1}p + 1 - p)$ with

$$\|b_n^{-1}\| < \|T\|(\|1 - p\| + 2\|p\|\|a^{-1}\| \triangleq M, \quad \forall\, n \geq n_0.$$

Pick $n_1 \in \mathbb{N}$ such that $\|b - b_{n_1}\| < 1/M$. Then $\|bb_{n_1}^{-1} - 1\| < 1$ and so that $b \in GL(\mathcal{B})$, i.e., $pap + 1 - p \in \mathrm{Fred}_T(\mathcal{A})$.

(3) Put $x = pap + 1 - p$ and $y = pa^{-1}p + 1 - p$. Then

$$xy - 1 = p(ap - pa)a^{-1}p \text{ and } yx - 1 = pa^{-1}(pa - ap)p$$

are all in $\mathcal{I}$, that is, $x \in \mathrm{Fred}_T(\mathcal{A})$.

Noting that $(1 - p)apa^{-1}p = (1 - p)(ap - pa)a^{-1}p \in \mathcal{I}$, we get that $\phi((1 - p)apa^{-1}p) = 0$. Similarly, $\phi((1 - p)a^{-1}pap) = 0$. Thus,

$$\begin{aligned}
\mathrm{index}(pap + 1 - p) &= \phi(xy - yx) = \phi(papa^{-1}p - pa^{-1}pap) \\
&= \phi((p - 1)apa^{-1}p + (1 - p)a^{-1}pap) \\
&\quad + \phi(apa^{-1}p - a^{-1}pap) \\
&= \phi((ap - pa)a^{-1}p + p - a^{-1}pap) \\
&= \phi(a^{-1}p(ap - pa) + p - a^{-1}pap) \\
&= \phi(a^{-1}(ap - pa)).
\end{aligned}$$

□

Remark 6.3.16. According to Remark 6.3.14 and Theorem 6.3.15 (3), $\omega(\cdot)$ is a locally constant function on $\mathcal{A}_p \cap GL(\mathcal{A})$. So we can extend it uniquely to a locally constant function : $\overline{\mathcal{A}_p} \cap GL(\mathcal{A}) \to \mathbb{C}$ by Theorem 6.3.15 (2) with $\omega(ab) = \omega(a) + \omega(b)$, $\forall\, a, b \in \overline{\mathcal{A}_p} \cap GL(\mathcal{A})$.

Definition 6.3.17. For $a \in \overline{\mathcal{A}_p} \cap GL(\mathcal{A})$, we call $\omega(a)$ is the winding number of a or the topological index of a (cf. [Murphy (2006)]).

6.4 Lifting a Moore–Penrose inverse from quotient C^*–algebras

In this section, we assume that $\mathcal{A}$ is a C^*–algebra and $\mathcal{J}$ is a closed ideal of $\mathcal{A}$. Set $\mathcal{B} = \mathcal{A}/\mathcal{J}$ and let $\pi\colon \mathcal{A} \to \mathcal{B}$ be the quotient mapping.

It is easy to verified that $\pi(G_{mp}(\mathcal{A})) \subset G_{mp}(\mathcal{B})$. It is nature to ask: When $\pi(G_{mp}(\mathcal{A})) = G_{mp}(\mathcal{B})$? To answer this question, we need to introduce the real rank of C^*–algebras.

Definition 6.4.1. Let $\mathcal{A}$ be a unital C^*–algebra. The real rank of $\mathcal{A}$ is the smallest integer, $\mathrm{RR}(A)$, such that for each n–tuple $(x_1, x_2, \cdots, x_n)$ of elements in $\mathcal{A}_{sa}$, with $n \leq \mathrm{RR}(\mathcal{A}) + 1$ and every $\epsilon > 0$, there is an n–tuple $(y_1, y_2, \cdots, y_n)$ in $\mathcal{A}_{sa}$ such that $\sum\limits_{k=1}^{n} y_k^* y_k$ is invertible in $\mathcal{A}$ and $\Big\| \sum\limits_{k=1}^{n} (x_k - y_k)^*(x_k - y_k) \Big\| < \epsilon$.

If A is non–unital, we define its real rank to be $\mathrm{RR}(\tilde{A})$.

Remark 6.4.2. Let $\mathcal{A}$ be a unital C^*–algebra. According to Definition 6.4.1, we have

(1) $\mathrm{RR}(\mathcal{A}) = 0$ iff the set of invertible self–adjoint elements in $\mathcal{A}$ is dense in $\mathcal{A}_{sa}$.
(2) $\mathrm{RR}(\mathcal{A}) \leq 1$ iff for every $a \in \mathcal{A}$ and any $\epsilon > 0$, there exists $b \in \mathcal{A}$ such that $\|a - b\| < \epsilon$ and $b^*b + bb^* \in GL(\mathcal{A})$.

The C^*–algebras with real rank zero are the most interesting. We have $\mathrm{RR}(B(H)) = 0$ and $\mathrm{RR}(\mathcal{K}(H)) = 0$ when H is a separable Hilbert space. The following results come from [Brown and Pedersen (1991)].

Theorem 6.4.3. *Let $\mathcal{A}$ be a C^*–algebra with* $\mathrm{RR}(\mathcal{A}) = 0$. *Then*

(1) *every quotient and every hereditary C^*–subalgebra of C^*–algebras are of real rank zero.*
(2) *the set of elements in $\mathcal{A}_{sa}$ with finite spectrum is dense in A_{sa}, i.e., if $x \in \mathcal{A}_{sa}$ and $\epsilon > 0$, there are mutually orthogonal projections $p_1, \cdots, p_n$ in $\mathcal{A}$ and real numbers $\lambda_1, \cdots, \lambda_n$ such that $\|x - \sum\limits_{i=1}^{n} \lambda_i p_i\| \leq \epsilon$.*

Proof. We assume that $\mathcal{A}$ is a unital C^*–algebra with $\mathrm{RR}(\mathcal{A}) = 0$.

(1) Suppose that $\mathfrak{D}$ is a hereditary C^*–subalgebra of $\mathcal{A}$. Let $b \in \mathfrak{D}_{sa}$. Given $\epsilon > 0$ and set $\delta = \dfrac{1}{4}\min\{2, \|b\|^{-1}, \|b\|^{-2}\epsilon\}$.

Pick $a_1 \in \mathcal{A}_{sa} \cap GL(\mathcal{A})$ such that $\|1 + b - a_1\| < \delta$. Put $a = a_1 - 1$. Then $a_1 = 1 + a$ and $\|b - a\| < \delta < 1$. So $1 + a - b \in GL(A) \cap \mathcal{A}_{sa}$. Put $d = 1 - (1 + a - b)^{-1} \in \mathcal{A}_{sa}$. Then

$$\|d\| \leq \|(1 + a - b)^{-1}\| \|a - b\| \leq \frac{\|a - b\|}{1 - \|a - b\|} \leq \frac{\delta}{1 - \delta} < 2\delta$$

and $\min\{\|bd\|, \|db\|\} \le \|b\|\|d\| < \frac{1}{2}$. So $1 - bd \in GL(\mathcal{A})$. Let

$$\begin{aligned} c &= (1+a-b)^{-1}(1+a)(1-db)^{-1} \\ &= (1+a-b)^{-1}(1+a-b+b)(1-db)^{-1} \\ &= (1+(1-d)b)(1-db)^{-1} = 1 + b(1-db)^{-1}. \end{aligned}$$

Then $c \in GL(\mathcal{A})$ for $1 + a = a_1 \in GL(\mathcal{A})$. From the identity

$$b(1-db)^{-1} = (1-bd)^{-1}b,$$

we get that $c^* = 1 + (1-bd)^{-1}b = c$. Moreover,

$$\begin{aligned} \|c - (1+b)\| &= \|b(1-db)^{-1} - b\| = \Big\| \sum_{n=1}^{\infty} b(db)^n \Big\| \\ &\le \frac{\|b\|^2\|d\|}{1 - \|b\|\|d\|} < 2\|b\|^2\|d\| \\ &< 4\delta\|b\|^2 < \epsilon \end{aligned}$$

and $c - 1 = b + \sum_{n=1}^{\infty} b\big[(db)^{n-1}d\big]b \in \overline{\mathfrak{D}\mathcal{A}\mathfrak{D}} \subset \mathfrak{D}$ since $\mathfrak{D}$ is hereditary and $\mathfrak{D}\mathcal{A}\mathfrak{D} \subset \mathfrak{D}$ by Proposition 1.5.29. This means that $1 + b \in \overline{\tilde{\mathfrak{D}}_{sa} \cap GL(\tilde{\mathfrak{D}})}$ by Corollary 1.5.8.

Now for every $x = \lambda + b \in \tilde{\mathfrak{D}}_{sa}$, if $\lambda \neq 0$, then $x = \lambda(1 + \lambda^{-1}b) \in \overline{\tilde{\mathfrak{D}}_{sa} \cap GL(\tilde{\mathfrak{D}})}$; if $\lambda = 0$, we set $x_\epsilon = \epsilon + b = \epsilon(1 + \epsilon^{-1}b)$. Then $\|x - x_\epsilon\| < \epsilon$ and $x_\epsilon \in \tilde{\mathfrak{D}}_{sa} \cap GL(\tilde{\mathfrak{D}})$. Therefore, $\mathrm{RR}(\mathfrak{D}) = 0$.

Let $\mathcal{I}$ be a closed ideal of $\mathcal{A}$ and $\rho\colon \mathcal{A} \to \mathcal{A}/\mathcal{I} = \mathcal{E}$ be the quotient mapping. Noting that $\rho(\mathcal{A}_{sa}) = \mathcal{E}_{sa}$ and $\rho(GL(\mathcal{A})) \subset GL(\mathcal{E})$, we get that $\mathrm{RR}(\mathcal{E}) = 0$ when $\mathrm{RR}(\mathcal{A}) = 0$.

(2) We adopt the proof of Lemma 3.2.5 (2) of [Lin (2001)].

Let $a \in \mathcal{A}_{sa}$ with $\|a\| \le 1$. Let $\epsilon \in (0, 1)$. Choose

$$-1 = t_1 < t_2 < \cdots < t_n = 1$$

such that $|t_{i+i} - t_i| < \epsilon/2$. Since $\mathrm{RR}(\mathcal{A}) = 0$, there is $x_1 \in \mathcal{A}_{sa} \cap GL(\mathcal{A})$ such that $\|(a - t_1) - x_1\| < \epsilon/4 = \epsilon_1$. Set $a_1 = x_1 + t_1$. Then $x_1 = a_1 - t_1$ is invertible, i.e., $t_1 \notin \sigma(a_1)$ and $\|a - a_1\| < \epsilon_1$. Choose $\epsilon_2 \in (0, \epsilon/8)$ so that $[t_1 - \epsilon_2, t_1 + \epsilon_2] \cap \sigma(a_1) = \emptyset$.

Repeating the procedure as above, we obtain $a_2 \in \mathcal{A}_{sa}$ such that $a_2 - t_2$ is invertible and $\|a_2 - a_1\| < \epsilon_2$. So $t_2 \notin \sigma(a_2)$. Since for each $\lambda \in \sigma(a_1)$, $|\lambda - t_1| > \epsilon_2$, we have $\|(t_1 - a_1)^{-1}\| < 1/\epsilon_2$ and hence

$$\|(a_1 - t_1)^{-1}(a_2 - t_1) - 1\| \le \|(a_1 - t_1)^{-1}\|\|a_2 - a_1\| < 1.$$

Thus, $t_1 \notin \sigma(a_2)$. By using this argument again and again, we obtain elements $a_n \in \mathcal{A}_{sa}$ such that

$$t_i, t_2, \cdots, t_n \notin \sigma(a_n) \text{ and } \|a - a_n\| < \sum_{i=1}^{n} \epsilon_i < \epsilon/2.$$

Choose $d \in (0, \epsilon/4)$ such that $(t_i - d, t_i + d) \cap \sigma(a_n) = \emptyset$, $i = 1, \cdots, n-1$. Set $F_i = [t_i + d/2, t_{i+1} - d/2]$, $i = 1, \cdots, n-1$. Then

$$\sigma(a_n) \subset \bigcup_{i=1}^{n-1} F_i, \quad F_i \cap F_j = \emptyset,\ i \neq j,\ i, j = 1, \cdots, n-1.$$

and χ_{Fi} is a continuous function on $\sigma(a_n)$, $i = 1, \cdots, n-1$. Thus $P_i = \chi_{F_i}(a_n)$ is a projection in $\mathcal{A}$. Set $b_n = \sum\limits_{i=1}^{n} t_i P_i$. Then $b_n \in \mathcal{A}_{sa}$ with $\sigma(b_n) = \{t_1, \cdots, t_n\} \cup \{0\}$ and

$$\|a - b_n\| \leq \|a - x_n\| + \|x_n - b_n\| < \epsilon/2 + \epsilon/4 + d/2 < \epsilon. \qquad \square$$

Corollary 6.4.4 ([Pedersen (1980)]). *Suppose that* $\mathrm{RR}(\mathcal{A}) = 0$. *Then for any* $a_1, \cdots, a_n \in \mathcal{A}$ *and any* $\epsilon > 0$, *there is a projection* p *in* $\mathcal{A}$ *such that* $\|a_i p - a_i\| < \epsilon$, $i = 1, \cdots, n$.

Proof. Set $b = \sum\limits_{i=1}^{n} a_i^* a_i \in \mathcal{A}_+$. Then by Theorem 6.4.3 (2), for any $\epsilon > 0$, there are mutually orthogonal projections $p_1, \cdots, p_n$ in $\mathcal{A}$ and $\lambda_1, \cdots, \lambda_n \in \mathbb{R}_+$ such that $\|b - \sum\limits_{i=1}^{n} \lambda_i p_i\| \leq \epsilon$. Put $p = \sum\limits_{i=1}^{n} p_i$. Then p is a projection in $\mathcal{A}$ and $\|bp - b\| < \epsilon$. Thus, $\|a_i p - a_i\| < \epsilon$, $i = 1, \cdots, n$. $\square$

Lemma 6.4.5. *Let* v *be a partial isometry in* $\mathcal{B}$ *and let* p *be a projection in* $\mathcal{A}$ *with* $\pi(p) = v^*v$. *Suppose that for every element* x *in* $p\mathcal{J}p$ *and any* $\epsilon > 0$, *there is always a projection* e *in* $p\mathcal{J}p$ *such that* $\|x(p-e)\| < \epsilon$. *Then there exists a partial isometry* u *in* $\mathcal{A}$ *such that* $\pi(u) = v$.

Proof. Let $x \in \mathcal{A}$ such that $\pi(x) = v$. Set $y = xp$. Then $\pi(y) = vv^*v = v$ and $\pi(y^*y) = v^*v = \pi(p)$. Thus, $p - y^*y \in p\mathcal{J}p$ and consequently, for any $\epsilon \in (0, 1)$, there is a projection $r \in p\mathcal{J}p$ such that $\|(p-r)(p-y^*y)(p-r)\| < \epsilon$ by our assumption. Therefore $(p-r)y^*y(p-r)$ is invertible in $(p-r)\mathcal{A}(p-r)$ with the inverse

$$z = (p-r)y^*y(p-r))^{-1} = \sum_{n=0}^{\infty} \left[(p-r) - (p-r)y^*y(p-r)\right]^n.$$

Put $u = yz^{1/2}$. Then $u^*u = z^{1/2}y^*yz^{1/2} = p - r$. Note that $\pi(z) = \pi(p) = v^*v$. So $\pi(u) = \pi(y)\pi(p) = v$. $\square$

The following is a special case of Lemma 6.4.5.

Corollary 6.4.6. *Let v be a unitary element or an isometry or a co–isometry in $\mathcal{B}$. Suppose that for every element x in $\mathcal{J}$ and any $\epsilon > 0$, there is always a projection e in $\mathcal{J}$ such that $\|x(1-e)\| < \epsilon$. Then there exists a partial isometry u in $\mathcal{A}$ such that $\pi(u) = v$.*

Theorem 6.4.7. *Suppose that* $\mathrm{RR}(\mathcal{J}) = 0$ *and $\mathcal{A}$ is projection–liftable (relative to $\mathcal{J}$), i.e., for any projection q in $\mathcal{B}$, there is a projection $p \in \mathcal{A}$ such that $\pi(p) = q$. Then*

(1) $\pi(G_{mp}(\mathcal{A})) = G_{mp}(\mathcal{B})$.
(2) *for any $y \in G_d(\mathcal{B})$, there is an element $x \in G_{mp}(\mathcal{A})$ and a partial isometry $u \in \mathcal{A}$ such that $\pi(x) = y$ and*

$$1 - u^*u,\ 1 - uu^* \in \mathcal{J},\ uxx^\dagger u^* - x^\dagger x \in \mathcal{J}.$$

Proof. (1) Obviously, $\pi(G_{mp}(\mathcal{A})) \subset G_{mp}(\mathcal{B})$. Now let $b \in G_{mp}(\mathcal{B})$. Then b^*b, $|b| \in G_{mp}(\mathcal{B})$ by Proposition 3.5.8. Put $v = b(|b|)^\dagger$ and $q = (|b|)^\dagger|b| = |b|(|b|)^\dagger$. Then

$$|b|q = q|b| = |b|,\ (|b|)^\dagger q = q(|b|)^\dagger = (|b|)^\dagger,\ v^*v = q.$$

So $|b|$ is invertible in $q\mathcal{B}q$ and v is a partial isometry in $\mathcal{B}$ by Proposition 1.5.5 (5).

By our assumption, there is a projection $p_0 \in \mathcal{A}$ such that $\pi(p_0) = q = v^*v$. Since $p_0\mathcal{J}p_0$ is a hereditary C^*–subalgebra of $\mathcal{J}$ and $\mathrm{RR}(\mathcal{J}) = 0$, it follows from Theorem 6.4.3 (1) that $\mathrm{RR}(p_0\mathcal{J}p_0) = 0$. Thus, by Corollary 6.4.4 and Lemma 6.4.5, we can find a partial isometry u in $\mathcal{A}$ such that $\pi(u) = v$. Put $p = u^*u$. Then $\pi(p) = q$ and so that $\pi(p\mathcal{A}p) = q\mathcal{B}q$.

Note that $0 \notin \sigma_{q\mathcal{B}q}(|b|)$. So $\delta = \min\{\lambda \mid \lambda \in \sigma_{q\mathcal{B}q}(|b|)\} > 0$ and $|b| - \delta\, q \geq 0$. Choose $c \in p\mathcal{A}p$ such that $\pi(c) = (|b| - \delta\, q)^{1/2}$. Set $d = c^*c + \delta\, p$. Then $d \in \mathcal{A}_+$ is invertible in $p\mathcal{A}p$ by Corollary 1.5.20. Put $x = ud$. Then $\pi(x) = \pi(u)\pi(d) = v|b|$ and

$$\begin{aligned} x(d+1-p)^{-1}u^*x &= ud(d+1-p)^{-1}u^*ud \\ &= ud(d+1-p)^{-1}(d+1-p)p \\ &= ud = x, \end{aligned}$$

that is, $x \in G_{mp}(\mathcal{A})$.

(2) By (1), we can choose $x \in G_{mp}(\mathcal{A})$ such that $\pi(x) = y$. Note that $yzy = y$ for some $z \in GL(\mathcal{B})$. So if we set $s = yz$ and $t = zy$, then

$zsz^{-1} = t$. Since $(yy^\dagger)s = s$, $s(yy^\dagger) = yy^\dagger$, $(y^\dagger y)t = y^\dagger y$, $t(y^\dagger y) = t$, we get that $w_1 = 1 - s + yy^\dagger$ and $w_2 = 1 - t + y^\dagger y$ are all invertible in $\mathcal{B}$ with

$$(1 - s + yy^\dagger)^{-1} = 1 + s - yy^\dagger,\ (1 - t + y^\dagger y)^{-1} = 1 + t - y^\dagger y$$

and $w_1^{-1} s w_1 = yy^\dagger$, $w_2 t w_2^{-1} = y^\dagger y$. Put $w = w_2 z w_1 \in GL(\mathcal{B})$. Then $wyy^\dagger w^{-1} = y^\dagger y$. Set $v = w|w|^{-1}$. Then $v \in U(\mathcal{B})$. Since $yy^\dagger w^* = w^* y^\dagger y$, it follows that

$$w^* w yy^\dagger = w^* y^\dagger y w = yy^\dagger w^* w,\ |w| yy^\dagger = yy^\dagger |w|.$$

Therefore, $vyy^\dagger v^* = y^\dagger y$.

By Corollary 6.4.4 and Corollary 6.4.6, there is a partial isometry $u \in \mathcal{A}$ such that $\pi(u) = v$. Then $1 - u^*u,\ 1 - uu^* \in \mathcal{J}$ and $\pi(uxx^\dagger u^*) = \pi(x^\dagger x)$, i.e., $uxx^\dagger u^* - x^\dagger x \in \mathcal{J}$. □

Remark 6.4.8. The condition that $\mathcal{A}$ is projection–liftable (relative to $\mathcal{J}$) is very important in Theorem 6.4.7. What kinds of C^*–algebras have this property ? As we know, if the C^*–algebra $\mathcal{A}$ satisfies one of the following conditions, then $\mathcal{A}$ is projection–liftable:

(1) $\mathrm{RR}(\mathcal{A}) = 0$ (cf. Theorem B.2.1).

(2) $\mathcal{J}$ is $*$–isomorphic to $\mathcal{K}(H)$ for some separable Hilbert space H.

In fact, let q be a projection in $\mathcal{B}$. Then we can find a positive element a in $\mathcal{A}$ such that $\pi(a) = q$. Thus, $a^2 - a \in \mathcal{J}$. Since every nonzero point in the spectrum of a compact operator is isolated and 0 may be the only accumulation point in its spectrum, we conclude that every point except 0 and 1 in $\sigma(a)$ is isolated and 0, 1 may be accumulation points. So we can find $\alpha_1 > \alpha_2 > \alpha_3 > 0$ such that $\sigma(a) \subset [0, \alpha_3] \cup [\alpha_2, \alpha_1]$ and $1 \in [\alpha_2, \alpha_1]$.

Let $f\colon [0, \alpha_1] \to [0, 1]$ be a continuous function such that $f\big|_{[0,\alpha_3]} \equiv 0$ and $f\big|_{[\alpha_2,\alpha_1]} \equiv 1$. Then $p = f(a) = \chi_{\sigma(a)\cap[\alpha_2,\alpha_1]}(a)$ is a projection and $\pi(p) = f(\pi(a)) = f(q) = q$.

(3) $\mathcal{J}$ is AF–algebra (cf. Theorem B.2.3).

Corollary 6.4.9 ([Rakočević (1992)]). *Let H be a separable Hilbert space. Then*

$$\begin{aligned} &G_{mp}(B(H)) \cap \overline{\Phi(H)} \\ &\quad = \Phi(H) \cup \{T \in B(H) \mid \dim \mathrm{Ker}\, T = \dim \mathrm{Ker}\, T^*,\ \mathrm{Ran}\,(T)\ \textit{closed}\}, \end{aligned}$$

where $\Phi(H) = \pi^{-1}(B(H)/\mathcal{K}(H))$.

Proof. Put $\mathcal{A} = B(H)$, $\mathcal{J} = \mathcal{K}(H)$ and $\mathcal{B} = B(H)/\mathcal{K}(H)$.

Let $A \in G_{mp}(\mathcal{A}) \cap \overline{\Phi(H)}$. Then $\operatorname{Ran}(A)$ is closed and $\pi(A) \in G_{mp}(\mathcal{B}) \cap \overline{GL(\mathcal{B})} = G_d(\mathcal{B})$ by Proposition 2.4.5. Thus, there is $A_0 \in G_{mp}(\mathcal{A})$ and a partial isometry $U \in \mathcal{A}$ with $I - U^*U,\ I - UU^* \in \mathcal{J}$ such that

$$\pi(A) = \pi(A_0) \text{ and } UA_0A_0^\dagger U^* - A_0^\dagger A_0 \in \mathcal{J}. \tag{6.4.1}$$

From (6.4.1), we get that $\pi(U)\pi(I - AA^\dagger)\pi(U)^* = \pi(I - A^\dagger A)$. So if $A \notin \Phi(H)$, then $I - AA^\dagger,\ I - A^\dagger A \notin \mathcal{J}$ and consequently, $\dim \operatorname{Ker} A = \dim \operatorname{Ker} A^* = \infty$.

On the other hand, it is clear that $\Phi(H) \subset G_{mp}(\mathcal{A}) \cap \overline{\Phi(H)}$. So we need show that if $A \in \mathcal{A}$ satisfies the condition

$$\operatorname{Ran}(A) \text{ is closed and } \dim \operatorname{Ker} A = \dim \operatorname{Ker} A^* = \infty, \tag{GW}$$

then $A \in G_{mp}(\mathcal{A}) \cap \overline{\Phi(H)}$.

Now let $A \in \mathcal{A}$ have (GW). Then

$$\dim \operatorname{Ran}(I - A^\dagger A) = \dim \operatorname{Ran}(I - AA^\dagger) = \infty.$$

and hence there is a partial isometry $V_0 \in \mathcal{A}$ such that $V_0^*V_0 = I - A^\dagger A$ and $V_0V_0^* = I - AA^\dagger$. Put $V_1 = A(|A|)^\dagger$. Then $V_1^*V_1 = A^\dagger A$ and $V_1V_1^* = AA^\dagger$. Thus $V = V_0 + V_1$ is unitary and $I - A^\dagger A = V^*(I - AA^\dagger)V$. Put $B = ((|A|)^\dagger + I - A^\dagger A)V^*$. Then B is invertible in $\mathcal{A}$ and

$$ABA = A(|A|)^\dagger V^*A = V_1V^*A = V_1V_1^*A = AA^\dagger A = A,$$

that is, $A \in G_d(\mathcal{A}) = Gi(\mathcal{A}) \cap \overline{GL(\mathcal{A})} \subset G_{mp}(\mathcal{A}) \cap \overline{\Phi(H)}$. □

Let $\{(\mathcal{A}_n, \|\cdot\|_n)\}$ be a sequence of C^*–algebras. Set

$$\prod_{n=1}^{\infty} \mathcal{A}_n = \{\{a_n\} \mid \sup_{n\geq 1} \|a_n\|_n < \infty,\ a_n \in \mathcal{A}_n,\ n \geq 1\},$$
$$\bigoplus_{n=1}^{\infty} \mathcal{A}_n = \{\{a_n\} \mid \lim_{n\to\infty} \|a_n\|_n = 0,\ a_n \in \mathcal{A}_n,\ n \geq 1\}.$$

Define the linear operation, multiplication, involution and norm on $\prod\limits_{n=1}^{\infty} \mathcal{A}_n$ and $\bigoplus\limits_{n=1}^{\infty} \mathcal{A}_n$ by

$$\lambda\{a_n\} + \mu\{b_n\} = \{\lambda a_n + \mu b_n\},\quad \{a_n\}\{b_n\} = \{a_nb_n\}$$
$$(\{a_n\})^* = \{a_n^*\}, \qquad \|\{a_n\}\| = \sup_{n\geq 1} \|a_n\|,$$

where λ, $\mu \in \mathbb{C}$ and $\{a_n\}$, $\{b_n\} \in \prod_{n=1}^{\infty} \mathcal{A}_n$ (or $\bigoplus_{n=1}^{\infty} \mathcal{A}_n$). It is easy to check that $\prod_{n=1}^{\infty} \mathcal{A}_n$ and $\bigoplus_{n=1}^{\infty} \mathcal{A}_n$ become C^*–algebras under above definitions. Furthermore, $\bigoplus_{n=1}^{\infty} \mathcal{A}_n$ is a closed ideal of $\prod_{n=1}^{\infty} \mathcal{A}_n$.

We call $\prod_{n=1}^{\infty} \mathcal{A}_n$ the direct product C^*–algebra of $\{\mathcal{A}_n\}$ and $\bigoplus_{n=1}^{\infty} \mathcal{A}_n$ the direct sum C^*–algebra of $\{\mathcal{A}_n\}$.

Lemma 6.4.10. *Let $a \in \mathcal{A}_{sa}$. If $\|a^2 - a\| < 1/4$ and $\|a\| \geq 1/2$, then there is a projection $p \in C^*(a)$ such that such that $\|a - p\| < 2\|a^2 - a\|$.*

Proof. Set $\alpha = \|a^2 - a\|$. Since $\max\limits_{\lambda \in \sigma(a)} |\lambda^2 - \lambda| = \|a^2 - a\|$, we have $\sigma(a) \subset [a_1, a_2] \cup [a_3, a_4]$ and $[a_1, a_2] \cap [a_3, a_4] = \emptyset$, where

$$a_1 = \frac{1}{2} - \Big(\frac{1}{4} + \alpha\Big)^{1/2}, \; a_2 = \frac{1}{2} - \Big(\frac{1}{4} - \alpha\Big)^{1/2},$$
$$a_3 = \frac{1}{2} + \Big(\frac{1}{4} - \alpha\Big)^{1/2}, \; a_4 = \frac{1}{2} + \Big(\frac{1}{4} + \alpha\Big)^{1/2}.$$

Define a continuous function g on $\sigma(a)$ by $g(\lambda) = 0$ when $\lambda \in [a_1, a_2]$ and $g(\lambda) = 1$ when $\lambda \in [a_3, a_4]$. Put $p = g(a) \in C^*(a)$. Then p is a projection in $\mathcal{A}$ and $\|p - a\| = \max\limits_{\lambda \in \sigma(a)} |g(\lambda) - \lambda| \leq a_2 < 2\alpha$. □

Proposition 6.4.11. *Let $\{(\mathcal{A}_n, \|\cdot\|_n)\}$ be a sequence of C^*–algebras. Then $\prod_{n=1}^{\infty} \mathcal{A}_n$ is projection–liftable with respect to $\bigoplus_{n=1}^{\infty} \mathcal{A}_n$.*

Proof. Let $\rho\colon \prod_{n=1}^{\infty} \mathcal{A}_n \to \bigoplus_{n=1}^{\infty} \mathcal{A}_n$ be the quotient mapping. Let q be a projection in $\prod_{n=1}^{\infty} \mathcal{A}_n / \bigoplus_{n=1}^{\infty} \mathcal{A}_n$. Then there is $\{a_n\} \in \Big(\prod_{n=1}^{\infty} \mathcal{A}_n\Big)_{sa}$ such that $\rho(\{a_n\}) = q$. Thus, $\sup\limits_{n \geq 1} \|a_n\|_n \geq 1$ and $\lim\limits_{n \to \infty} \|a_n^2 - a_n\|_n = 0$.

Choose $n_0 \in \mathbb{N}$ such that $\|a_n^2 - a_n\|_n < 1/4$, $\forall\, n \geq n_0$. Then we can find projections $p'_n \in \mathcal{A}_n$ such that $\|a_n - p'_n\|_n < 2\|a_n^2 - a_n\|_n$, $\forall\, n \geq n_0$ by Lemma 6.4.10. Put $p_n = 0$ when $1 \leq n < n_0$ and $p_n = p'_n$ if $n \geq n_0$. Then $p = \{p_n\}$ is a projection in $\prod_{n=1}^{\infty} \mathcal{A}_n$ and $\rho(p) = \rho(\{a_n\}) = q$. □

Definition 6.4.12. Let $\{a_n\} \subset \mathcal{A}$. We call $\{a_n\}$ is asymptotically Moore–Penrose invertible if $\sup\limits_{n \in \mathbb{N}} \|a_n\| < +\infty$ and there is a bounded sequence $\{b_n\}$

in $\mathcal{A}$ such that

$$\|a_n b_n a_n - a_n\| \to 0, \qquad \|b_n a_n b_n - b_n\| \to 0,$$
$$\|(a_n b_n)^* - a_n b_n\| \to 0, \qquad \|(b_n a_n)^* - b_n a_n\| \to 0.$$

Corollary 6.4.13. *Let $\{a_n\} \subset \mathcal{A}$ be an asymptotically Moore–Penrose invertible sequence. Suppose that $\mathrm{RR}(\mathcal{A}) = 0$. Then there is $\{x_n\} \subset G_{mp}(\mathcal{A})$ such that $\lim\limits_{n\to\infty} \|a_n - x_n\| = 0$.*

Proof. Let $\mathcal{A}_n = \mathcal{A}$. Keep the symbols in the proof Proposition 6.4.11. Then $\rho(\{a_n\}) \in G_{mp}(C)$, where $C = \prod\limits_{n=1}^{\infty} \mathcal{A}_n / \bigoplus\limits_{n=1}^{\infty} \mathcal{A}_n$.

Let $\{c_n\} \in \Big(\bigoplus\limits_{n=1}^{\infty} \mathcal{A}_n\Big)_{sa}$. Then for any $\epsilon > 0$, there is $N \in \mathbb{N}$ such that $\|c_n\| < \epsilon$, $\forall\, n > N$. Since $\mathrm{RR}(\mathcal{A}) = 0$, there exist $c'_1, \cdots, c'_N \in GL(\mathcal{A}) \cap \mathcal{A}_{sa}$ such that $\|1 + c_i - c'_i\| < \epsilon$, $i = 1, \cdots, N$. Put $d_i = c'_i$ for $i = 1, \cdots, N$ and $d_i = 1$ when $i > N$. Then $\{d_n\}$ is self–adjoint and invertible in $\bigoplus\limits_{n=1}^{\infty} \mathcal{A}_n$ with $\|\{1 + a_n\} - \{d_n\}\| < \epsilon$. This implies that $\bigoplus\limits_{n=1}^{\infty} \mathcal{A}_n$ is of real rank zero.

Finally, applying Proposition 6.4.11 and Theorem 6.4.7 (1) to $\prod\limits_{n=1}^{\infty} \mathcal{A}_n$, we can find $x = \{x_n\} \in G_{mp}\Big(\prod\limits_{n=1}^{\infty} \mathcal{A}_n\Big)$ such that $\rho(x) = \rho(\{a_n\})$, that is, $\{x_n\} \subset G_{mp}(\mathcal{A})$ and $\lim\limits_{n\to\infty} \|a_n - x_n\| = 0$. $\square$

6.5 Approximate polar decomposition in C^*–algebras

It is well–known that every bounded linear operator on a Hilbert space has the polar decomposition (Proposition 1.5.16). In a C^*–subalgebra $\mathcal{A}$ of $B(H)$ for certain Hilbert space H, $a \in \mathcal{A}\backslash\{0\}$ can be written as $a = v|a|$, where $|a| = (a^*a)^{1/2}$ and v is a partial isometry in $B(H)$ which is not in $\mathcal{A}$ generally. But if $a \in G_{mp}(\mathcal{A})$, then $|a| \in G_{mp}(\mathcal{A})$ by Proposition 3.5.8 and $v = a(|a|)^\dagger$ is a partial isometry in $\mathcal{A}$ ($v^*v = (|a|)^\dagger|a|$) and $a = v|a|$, that is, a has the polar decomposition in $\mathcal{A}$. This leads us to seek for a suitable definition so that every nonzero element in a C^*–algebra has a kind of polar decomposition. Here is an approximate version of the polar decomposition of an element as follows.

Definition 6.5.1. Let $\mathcal{A}$ be a unital C^*–algebra and $a \in \mathcal{A}\backslash\{0\}$. We call

a has the approximate polar decomposition (for short (APD)) if for any $\epsilon > 0$, there is a partial isometry v in $\mathcal{A}$ such that $\|a - v|a|\| < \epsilon$.

We will say $\mathcal{A}$ has (APD)–property if every nonzero element in $\mathcal{A}$ has the (APD). If $\mathcal{A}$ is non–unital and $\tilde{A}$ has (APD)–property, we also say $\mathcal{A}$ has (APD)–property.

Proposition 6.5.2. *Let $\mathcal{A}$ be a C^*-algebra. Then $\mathcal{A}$ has* (APD) iff $G_{mp}(\mathcal{A})$ *is dense in $\mathcal{A}$.*

Proof. Suppose that $\mathcal{A}$ has (APD). Then for any $a \in \mathcal{A}\backslash\{0\}$ and any $\epsilon > 0$, there is a partial isometry $v \in \mathcal{A}$ such that $\|a - v|a|\| < \epsilon/2$. Set $a_\epsilon = v(|a| + \epsilon)$ and $b_\epsilon = (|a| + \epsilon)^{-1}v^*$. Then a_ϵ, $b_\epsilon \in \mathcal{A}$ and $a_\epsilon b_\epsilon a_\epsilon = a_\epsilon$, that is, $a_\epsilon \in Gi(\mathcal{A}) = G_{mp}(\mathcal{A})$ by Proposition 3.5.8. So $\overline{G_{mp}(\mathcal{A})} = \mathcal{A}$.

Conversely, for each $a \in G_{mp}(\mathcal{A})$, $|a| \in G_{mp}(\mathcal{A})$ and $v = a(|a|)^\dagger \in \mathcal{A}$ is a partial isometry. Put $p = (|a|)^\dagger |a|$. Then

$$|a|(1 - p) = 0 \text{ and } (1 - p)(a^*a)(1 - p) = 0.$$

So $a(1 - p) = 0$ and hence $v|a| = a(|a|)^\dagger|a| = ap = a$. This proves the implication $\overline{G_{mp}(\mathcal{A})} = \mathcal{A} \Rightarrow \mathcal{A}$ has (APD). □

Remark 6.5.3. (1) Blackadar introduced a notation so–called the (WS)–property of a C^*–algebra $\mathcal{A}$ in his book [Blackadar (1998)], that is, the set of all well–support elements in $\mathcal{A}$ is dense in $\mathcal{A}$. But according to Proposition 3.5.8 and Proposition 6.5.2, $\mathcal{A}$ has (WS) iff $\mathcal{A}$ has (APD) and iff $\overline{G_{mp}(\mathcal{A})} = \mathcal{A}$.

(2) A unital C^*–algebra is called extremally rich if

$$\mathcal{A}_q^{-1} = \{x\,v\,y|\, v \in \text{ext}(\mathcal{A}),\ x,\, y \in GL(\mathcal{A})\}$$

is dense in $\mathcal{A}$ (cf. [Brown and Pedersen (1995)]), where

$$\text{ext}(\mathcal{A}) = \{v \in \mathcal{A}|\, (1 - v^*v)\mathcal{A}(1 - vv^*) = \{0\}\}$$

is the set of extremal points of the unit ball of $\mathcal{A}$ (cf. Proposition 1.4.7 of [Pedersen (1979)]. Since for any x, $y \in GL(\mathcal{A})$ and any partial isometry $v \in \mathcal{A}$,

$$(xvy)(y^{-1}v^*x^{-1})(xvy) = xvv^*vy = (xvy),$$

i.e., $(xvy) \in Gi(\mathcal{A}) = G_{mp}(\mathcal{A})$, we obtain that $\mathcal{A}$ has (APD) if $\mathcal{A}$ is extremally rich.

Proposition 6.5.4. *Every quotient and every hereditary C^*–subalgebra of C^*–algebras with* (APD) *has* (APD) *again.*

Proof. We assume that $\mathcal{A}$ is a unital C^*–algebra with (APD) and $\mathcal{B}$ is a hereditary C^*–subalgebra of $\mathcal{A}$. Let $b \in \mathcal{B}$. Given $\epsilon > 0$ and set $\delta = \frac{1}{4}\min\{2, \|b\|^{-1}, \|b\|^{-2}\epsilon\}$.

Since $\overline{G_{mp}(\mathcal{A})} = \mathcal{A}$, we can find $a_1 \in G_{mp}(\mathcal{A})$ such that $\|1+b-a_1\| < \delta$. Put $a = a_1 - 1$. Then $a_1 = 1+a$ and $\|b-a\| < \delta < 1$. So $1+a-b \in GL(A)$. Put $d = 1-(1+a-b)^{-1}$. Then

$$\|d\| \leq \|(1+a-b)^{-1}\|\|a-b\| \leq \frac{\|a-b\|}{1-\|a-b\|} \leq \frac{\delta}{1-\delta} < 2\delta$$

and $\|bd\| \leq \|b\|\|d\| < \frac{1}{2}$. So $1-bd \in GL(\mathcal{A})$. Let

$$\begin{aligned} c &= (1+a-b)^{-1}(1+a)(1-db)^{-1} \\ &= (1+a-b)^{-1}(1+a-b+b)(1-db)^{-1} \\ &= (1+(1-d)b)(1-db)^{-1} = 1+b(1-db)^{-1}. \end{aligned}$$

Note that $1+a \in Gi(\mathcal{A})$ means that $(1+a)z(1+a) = (1+a)$ for some $z \in \mathcal{A}$. Thus, $c(1-db)^{-1}z(1+a-b)^{-1}c = c$ and hence $c \in Gi(\mathcal{A})$. Moreover,

$$\begin{aligned} \|c-(1+b)\| = \|b(1-db)^{-1} - b\| &= \left\|\sum_{n=1}^{\infty} b(db)^n\right\| \\ &\leq \frac{\|b\|^2\|d\|}{1-\|b\|\|d\|} < 2\|b\|^2\|d\| \\ &< 4\delta\|b\|^2 < \epsilon \end{aligned}$$

and $c - 1 = b + \sum_{n=1}^{\infty} b\big[(db)^{n-1}d\big]b \in \overline{\mathcal{BAB}} \subset \mathcal{B}$ since $\mathcal{B}$ is hereditary and $\mathcal{BAB} \subset \mathcal{B}$ by Proposition 1.5.29. Finally, $c \in G_{mp}(\mathcal{B})$ by Corollary 3.5.9.

Let $\mathcal{I}$ be the closed ideal of $\mathcal{A}$ and $\pi\colon \mathcal{A} \to \mathcal{A}/\mathcal{I} = \mathcal{D}$ be the quotient mapping. From $\pi(G_{mp}(\mathcal{A})) \subset G_{mp}(\mathcal{D})$, we get that $\mathcal{D}$ has (APD). □

Remark 6.5.5. Let $\mathcal{A}$ be a unital C^*–algebra such that $M_n(\mathcal{A})$ has (APD) for some $n \geq 2$. Put $p_1 = \mathrm{diag}(1, O_{(n-1)})$, where $O_{(n-1)}$ is the zero matrix of order $n-1$. Then $\mathcal{A} \cong p_1(M_n(\mathcal{A}))p_1$ has (APD) by Proposition 6.5.4. But we do not know if $M_n(\mathcal{A})$ has (APD) for $n \geq 2$ when $\mathcal{A}$ has (APD).

Proposition 6.5.6. *Let $\mathcal{A}$ be a C^*-algebra with* (APD). Then $\mathrm{RR}(\mathcal{A}) \leq 1$.

Proof. For convenience, we assume that $\mathcal{A}$ is a unital C^*–subalgebra with (APD) in $B(H)$ for certain Hilbert space H.

To prove $\mathrm{RR}(\mathcal{A}) \leq 1$, we need only to show that for any $A \in \mathcal{A}$ and any $\epsilon > 0$, there is $B \in \mathcal{A}$ such that $\|A-B\| < \epsilon$ and $B^*B + BB^* \in GL(\mathcal{A})$

by Remark 6.4.2. Since $G_{mp}(\mathcal{A})$ is dense in $\mathcal{A}$, we will do this for every element in $G_{mp}(\mathcal{A})$.

Let $A \in G_{mp}(\mathcal{A})$ and put $B = A + \epsilon(I - AA^\dagger)$. Then A and B have the expressions $A = \begin{bmatrix} A_1 & A_2 \\ 0 & 0 \end{bmatrix}$ and $B = \begin{bmatrix} A_1 & A_2 \\ 0 & \epsilon I_2 \end{bmatrix}$ with respect to the decomposition $H = \mathrm{Ran}\,(AA^\dagger) \oplus \mathrm{Ran}\,(I - AA^\dagger)$, where $A_1 = AA^\dagger A|_{\mathrm{Ran}\,(AA^\dagger)}$, $A_2 = AA^\dagger A|_{\mathrm{Ran}\,(I-AA^\dagger)}$ and I_1 (resp. I_2) is the identity operator on $\mathrm{Ran}\,(AA^\dagger)$ (resp. $\mathrm{Ran}\,(I - AA^\dagger)$). Then $\|A - B\| = \epsilon$ and

$$B^*B + BB^* = \begin{bmatrix} A_1^*A_1 + A_1A_1^* + A_2A_2^* & A_1^*A_2 + \epsilon A_2 \\ A_2^*A_1 + \epsilon A_2^* & A_2^*A_2 + 2\epsilon^2 I_2 \end{bmatrix}$$

$$= A^*A + \begin{bmatrix} A_1A_1^* + A_2A_2^* & \epsilon A_2 \\ \epsilon A_2^* & 2\epsilon^2 I_2 \end{bmatrix}, \tag{6.5.1}$$

$$\begin{bmatrix} A_1A_1^* + A_2A_2^* & \epsilon A_2 \\ \epsilon A_2^* & 2\epsilon^2 I_2 \end{bmatrix} = C \begin{bmatrix} A_1A_1^* + \frac{1}{2}A_2A_2^* & 0 \\ 0 & 2\epsilon^2 I_2 \end{bmatrix} C^*, \tag{6.5.2}$$

where $C = \begin{bmatrix} I_1 & \frac{1}{2\epsilon}A_2 \\ 0 & I_2 \end{bmatrix} \in GL(\mathcal{A})$.

Note that $\gamma(A^*) = \gamma(A) > 0$ and

$$\|A^*x\| \geq \gamma(A^*)\|x\|, \quad \forall\, x \in (\mathrm{Ker}\, A^*)^\perp = \mathrm{Ran}\,(A) = \mathrm{Ran}\,(AA^\dagger).$$

So $(A^*x, A^*x) \geq \gamma^2(A)(x, x)$, $\forall\, x \in \mathrm{Ran}\,(AA^\dagger)$ and consequently,

$$A_1A_1^* + \frac{1}{2}A_2A_2^* \geq \frac{1}{2}(A_1A_1^* + A_2A_2^*) \geq \frac{\gamma^2(A)}{2} I_1.$$

This indicates that $\mathrm{Ker}\,(A_1A_1^* + \frac{1}{2}A_2A_2^*) = \{0\}$ and $\mathrm{Ran}\,(A_1A_1^* + \frac{1}{2}A_2A_2^*)$ is closed in $\mathrm{Ran}\,(AA^\dagger)$ and so that $A_1A_1^* + \frac{1}{2}A_2A_2^*$ is invertible in $B(\mathrm{Ran}\,(AA^\dagger))$. Combining (6.5.1) and (6.5.2) with Corollary 1.5.20, we get that $B^*B + BB^*$ is invertible in $B(H)$ and is also invertible in $\mathcal{A}$. □

Let p, q be two nonzero projections in a C^*–algebra $\mathcal{A}$. Set

$$p\mathcal{A}q = \{pxq \mid x \in \mathcal{A}\}, \;\; Gi(p\mathcal{A}q) = \{a \in p\mathcal{A}q \mid a \in G_{mp}(\mathcal{A}),\; qa^\dagger = a^\dagger p = a^\dagger\}.$$

Proposition 6.5.7. *Let p, q be two nonzero projections in a unital C^*–algebra $\mathcal{A}$ and $\mathcal{J}$ be a closed ideal of $\mathcal{A}$. Let $a \in \mathcal{J}$ with $x = paq \neq 0$. Assume that $\mathrm{RR}(\overline{|x|\mathcal{J}|x|}) = 0$. Then for any $\epsilon > 0$, there is $y \in Gi(p\mathcal{J}q)$ such that $\|x - y\| < \epsilon$.*

Proof. By Theorem 1.5.36, we assume that $\mathcal{A}$ is C^*–subalgebra of $B(H)$ for some Hilbert space H. Let $x = v|x|$ be the polar decomposition in $B(H)$ with $v^*v = P_{\overline{\mathrm{Ran}\,(x^*)}}$ and $vv^* = P_{\overline{\mathrm{Ran}\,(x)}}$ by Proposition 1.5.16. Since

$\mathrm{Ran}\,(x) \subset \mathrm{Ran}\,(p)$ and $\mathrm{Ran}\,(x^*) \subset \mathrm{Ran}\,(q)$, it follows that $qv^*v = v^*v$ and $pvv^* = vv^*$. So $pv = vq = v$. Also from $xq = x$, we have $|x|q = q|x| = |x|$.

From $v^*v = P_{\overline{\mathrm{Ran}\,(x^*)}}$, we get that $1 - v^*v = P_{\mathrm{Ker}\,x} = P_{\mathrm{Ker}\,|x|}$. So $|x|(1 - v^*v) = (1 - v^*v)|x| = 0$.

Let $b \in \overline{|x|\mathcal{J}|x|}$. Then there is $\{b_n\} \subset \mathcal{J}$ such that $\|b - |x|b_n|x|\| \to 0$ as $n \to \infty$. Then

$$(1 - v^*v)b = b(1 - v^*v) = 0, \; qb = bq = b, \; \|vb - xb_n|x|\| \to 0 \; (n \to \infty).$$

So $vb \in \mathcal{J}$. Since $\mathrm{RR}(\overline{|x|\mathcal{J}|x|}) = 0$, it follows from Theorem 6.4.3 (2) that there are $\lambda_1, \cdots, \lambda_s \in \mathbb{R}\backslash\{0\}$ and mutually orthogonal projections $q_1, \cdots, q_s$ in $\overline{|x|\mathcal{J}|x|}$ such that $\||x| - \sum\limits_{i=1}^{s} \lambda_i q_i\| < \epsilon$. Set

$$y = v\Big(\sum_{i=1}^{s} \lambda_i q_i\Big), \quad y_0 = \Big(\sum_{i=1}^{s} \lambda_i^{-1} q_i\Big)v^*.$$

Then $y, y_0 \in \mathcal{J}$, $\|x - y\| < \epsilon$. Since $q_i q = q_i$, $v^*vq_i = q_i$, $i = 1, \cdots, s$ and $pv = vq = v$, we get that $y_0 = y^\dagger$ and $py = yq = y$, $qy_0 = y_0 p = y_0$, that is, $y \in Gi(p\mathcal{J}q)$. □

From Proposition 6.5.7, we get that

Corollary 6.5.8. *Every C^*–algebra with real rank zero has* (APD).

Let $\mathcal{J}$ be a closed ideal of the C^*–algebra $\mathcal{A}$. Proposition 6.5.4 says that if $\mathcal{A}$ has (APD), then both $\mathcal{J}$ and $\mathcal{A}/\mathcal{J}$ all have (APD). Conversely, if $\mathcal{J}$ and $\mathcal{A}/\mathcal{J}$ have (APD), does $\mathcal{A}$ have (APD)? In order to solve this problem, we need following lemma:

Lemma 6.5.9. *Let $a \in \mathcal{A}_{sa}$ and let p be a non–trivial projection in $\mathcal{A}$, i.e., $p \neq 0$ or 1.*

(1) *If $\|a-p\| < 1/2$, then there is a projection $q \in C^*(a)$ such that $\|q-p\| < \|a - p\|$.*
(2) *If there is a projection $h \in \mathcal{A}$ such that $\|p - h\| < 1$, then there is a unitary element $u \in \mathcal{A}$ such that $h = upu^*$ and $\|1 - u\| < \sqrt{2}\,\|p - h\|$.*
(3) *Let $u \in U(\mathcal{A})$ with $\|1 - u\| < 1$. Then there is $a \in \mathcal{A}_{sa}$ such that $u = \mathrm{e}^{ia}$.*

Proof. (1) Set $\delta = \|a - p\|$. It is easy to check that $\sigma(p) = \{0, 1\}$ and

$$(\lambda - p)^{-1} = \frac{1}{\lambda} + \frac{1}{\lambda(\lambda - 1)}p, \quad \lambda \in \mathbb{C}\backslash\{0, 1\}.$$

Let $\lambda \in \mathbb{R}\backslash\{0,1\}$ with $\min\{|\lambda|\,|\,|1-\lambda|\} < \delta$. Then

$$\|(\lambda - p)^{-1}\| = \max\Big\{\Big|\frac{1}{\lambda} + \frac{1}{\lambda(\lambda-1)}\mu\Big|\,\Big|\,\mu \in \{0,1\}\Big\} < \delta^{-1}$$

and

$$\|(\lambda - a)(\lambda - p)^{-1} - 1\| \leq \|a - p\|\|(\lambda - p)^{-1}\| < 1.$$

Therefore, $(\lambda - a)(\lambda - p)^{-1} \in GL(\mathcal{A})$ and hence $\lambda \notin \sigma(a)$. This means that $\sigma(a) \subset [-\delta, \delta] \cup [1-\delta, 1+\delta]$. Take a real–valued continuous function f on $[-\delta, 1+\delta]$ such that $f(\lambda) = 0$ when $\lambda \in [-\delta, \delta]$ and $f(\lambda) = 1$ if $\lambda \in [1-\delta, 1+\delta]$. Since $[-\delta,\delta]\cap[1-\delta,1+\delta] = \emptyset$, we have $f\big|_{\sigma(a)} = \chi_{\sigma(a)\cap[1-\delta,1+\delta]}$. Set $q = f(a) \in C^*(a)$. Then q is a projection and

$$\|q - a\| = \max_{\lambda\in\sigma(a)} |f(\lambda) - \lambda| \leq \delta.$$

(2) Set $v = 1 - p - h + 2hp$. Then simple computation shows that

$$v^*v = 1 - (p-h)^2 = vv^*, \ vp = hp = hv, \ pv^*v = v^*hv = v^*vp.$$

Since $\|1 - v^*v\| = \|(p-h)^2\| = \|p-h\|^2 < 1$, it follows that $v^*v = vv^* \in GL(\mathcal{A})$ and consequently, $v \in GL(\mathcal{A})$. Put $u = v|v|^{-1}$. Then $u \in U(\mathcal{A})$. From $pv^*v = v^*vp$ and $vv^*v = v^*vv$, we get that $p|v| = |v|p$ and $v|v| = |v|v$. Thus, $upu^* = v|v|^{-1}p|v|^{-1}v^* = h$.

Note that $u^* + u = |v|^{-1}(v^* + v) = 2|v|^{-1}(1-(p-h)^2) = 2|v|$. So

$$\|1-u\|^2 = \|(1-u)^*(1-u)\| = \|2 - u^* - u\| = 2\|1 - |v|\|.$$

Since $v^*v \leq 1$, we have $|v| \leq 1$ and

$$v^*v = (v^*v)^{1/4}|v|(v^*v)^{1/4} \leq (v^*v)^{1/4}1(v^*v)^{1/4} = |v|$$

by Proposition 1.5.19 (2). Thus, $\|1 - |v|\| \leq \|1 - |v|^2\| = \|(p-h)^2\|$ and hence $\|1-u\| \leq \sqrt{2}\|p-h\|$.

(3) Since $u - 1$ is a normal element and $\|u-1\| < 1$, we have $|\lambda - 1| < 1$, $\forall\,\lambda \in \sigma(u)$. Note that $f(\lambda) = \operatorname{In}\lambda$ is holomorphic in $|\lambda - 1| < 1$. Thus, $b = f(u) \in \mathcal{A}$ and $\mathrm{e}^b = u$ for $\mathrm{e}^{f(\lambda)} = \lambda$.

By Proposition 1.4.15 (3) and Proposition 1.5.5 (2), $\{\mathrm{e}^\lambda|\,\lambda \in \sigma(b)\} \subset \mathbf{S}^1$. This means that $\operatorname{Re}\lambda = 0$, $\forall\,\lambda \in \sigma(b)$. So $b = ia$ for some $a \in \mathcal{A}_{sa}$. □

Using Lemma 6.5.9, we can prove the following theorem, which answers the question mentioned above.

Theorem 6.5.10. *Let $\mathcal{A}$ and $\mathcal{J}$ be as above. Assume that $\mathcal{J}$ and $\mathcal{B} = \mathcal{A}/\mathcal{J}$ have* (APD). *Then $\mathcal{A}$ has* (APD) *if and only if following two conditions are satisfied:*

(1) *For each* $v \in \mathrm{Ais}(\mathcal{B})$ *there is* $w \in \mathrm{Pis}(\mathcal{A})$ *such that* $\pi(w)v^*v = v$ *and*
(2) $\mathrm{Pis}(\mathcal{A}) + \mathcal{J} \subset \overline{Gi(\mathcal{A})}$,

where $\mathrm{Pis}(\mathcal{A})$ *is the set of all partial isometries in* $\mathcal{A}$, $\mathrm{Ais}(\mathcal{B}) = \{b(|b|)^\dagger \,|\, b \in Gi_0(\mathcal{B})\} \subset \mathrm{Pis}(\mathcal{A})$ *and* $Gi_0(\mathcal{B}) \subset Gi(\mathcal{B})$ *such that* $Gi_0(\mathcal{B})$ *is dense in* $\mathcal{B}$.

Proof. We assume that $\mathcal{A}$ is unital.

($\Leftarrow$) Let $x \in \mathcal{A}$. Then for any $\epsilon > 0$, there is $y \in Gi_0(\mathcal{B})$ such that $\|\pi(x) - y\| < \frac{\epsilon}{3}$. Set $v = y|y|^\dagger \in \mathrm{Ais}(\mathcal{B})$. By hypotheses, we have a partial isometry $w \in \mathcal{A}$ such that $\pi(w)v^*v = v$. Pick $a \in \mathcal{A}_+$ with $\pi(a) = |y|$. Noting that $v^*v|y| = |y|$, we have $\pi(wa) = \pi(w)v^*v|y| = v|y| = y$. Put $x_0 = w\big(a + \frac{\epsilon}{3}\big)$. Then $\|\pi(x - x_0)\| < \frac{2\epsilon}{3}$ and $x_0 \in Gi(\mathcal{A})$. Thus, we can find an element $k \in \mathcal{J}$ such that $\|x - x_0 - k\| < \frac{2\epsilon}{3}$. Applying Condition (2) to $w + k\big(a + \frac{\epsilon}{3}\big)^{-1}$, we obtain $z_\epsilon \in Gi(\mathcal{A})$ such that

$$\|w + k\big(a + \frac{\epsilon}{3}\big)^{-1} - z_\epsilon\| < \frac{\epsilon}{3\|a + \frac{\epsilon}{3}\|}.$$

Put $x_\epsilon = z_\epsilon\big(a + \frac{\epsilon}{3}\big)$. Then $x_\epsilon \in Gi(\mathcal{A})$ and $\|x - x_\epsilon\| < \epsilon$.

($\Rightarrow$) Let v be in $\mathrm{Ais}(\mathcal{B})$. Then there is $x \in Gi(\mathcal{A})$ such that $\|\pi(x) - v\| < \frac{1}{100}$. Let $x = u|x|$ be the polar decomposition of x, i.e., $u = x|x|^\dagger$ and set $p' = \pi(u^*u)$, $q' = \pi(uu^*)$, $p = v^*v$, $q = vv^*$, $b = \pi(|x|)$. Simple computation shows that

$$\|p - b^2\| < \frac{3}{100}, \quad \|(1 - p')p\| < \frac{3}{100}, \quad \|q - \pi(u)p\pi(u^*)\| < \frac{6}{100}. \tag{6.5.3}$$

Note that $\|b\| < 2$. Thus, from (6.5.3), we get that

$$\|pb - bp\| < \frac{6\|b\|}{100} < \frac{12}{100}, \quad \|(pbp)^2 - p\| < \frac{27}{100}.$$

Let $\lambda \in \sigma(pbp)$. Then $\lambda \geq 0$ and $|\lambda^2 - 1| < \frac{27}{100}$. Thus $|\lambda - 1| < \frac{27}{100}$ and so that $\|pbp - p\| < \frac{27}{100}$. Furthermore, $\|bp - p\| < \frac{39}{100}$ and

$$\|v - \pi(u)p\| = \|(v - \pi(u)b)p - \pi(u)(p - bp)\| < \frac{40}{100}.$$

Now by (6.5.3), $\|p - p'pp'\| < \frac{6}{100}$. Thus, by Lemma 6.5.9, there is a projection $p_0' \in p'\mathcal{B}p'$ and a unitary element $v_0 \in \mathcal{B}$ such that

$$\|p - p_0'\| < 2\,\|p - p'pp'\| < \frac{12}{100}, \quad \|1 - v_0\| < \frac{12\sqrt{2}}{100} \text{ and } p = v_0^*p_0'v_0,$$

Since $v_0 = e^{ia}$ for some $a \in \mathcal{B}$ by Lemma 6.5.9 (3), we can pick $x \in \mathcal{A}_{sa}$ such that $\pi(x) = a$. Put $u_0 = e^{ix}$. Then $u_0 \in U(\mathcal{A})$ and $\pi(u_0) = v_0$. Put $w_0 = uu_0$. Then $w_0 \in \mathrm{Pis}(\mathcal{A})$ and $p \leq v_0^* p' v_0 = \pi(w_0^* w_0)$. From (6.5.3), we get that

$$\begin{aligned}\|q - \pi(w_0)\, p\, \pi(w_0^*)\| &= \|q - \pi(u)\, p_0' \pi(u^*)\| \\ &\leq \|q - \pi(u)\, p\, \pi(u^*)\| + \|p - p_0'\| \\ &< \frac{18}{100}.\end{aligned}$$

Since $\pi(w_0)\, p\, \pi(w_0^*)$ is a projection, it follows from Lemma 6.5.9 that there is $v_1 \in U(\mathcal{B})$ with $\|1 - v_1\| < \dfrac{18\sqrt{2}}{100}$ such that $q = v_1 \pi(w_0)\, p\, \pi(w_0^*) v_1^*$. Pick $u_1 \in U(\mathcal{A})$ with $\pi(u_1) = v_1$ and set $w_1 = u_1 w_0$. Then $w_1 \in \mathrm{Pis}(\mathcal{A})$ and $q = \pi(w_1)\, p\, \pi(w_1^*)$. Therefore, $q\, \pi(w_1) = \pi(w_1)\, p$ and $v^* \pi(w_1)$ is unitary in $p\mathcal{B}p$. Put $z = v^* \pi(w_1) p + 1 - p \in U(\mathcal{B})$. Since

$$\begin{aligned}\|p - v^* \pi(w_1) p\| &\leq \|v - q\, \pi(w_1)\, p\| = \|v - \pi(w_1)\, p\| \\ &\leq \|v - \pi(u) p\| + \|(v_1 - 1)\pi(u) v_0 p\| + \|\pi(u)(v_0 - 1) p\| \\ &< \frac{40}{100} + \frac{12\sqrt{2}}{100} + \frac{18\sqrt{2}}{100} < 1,\end{aligned}$$

we have $\|1 - z\| < 1$. Thus, there is $u_2 \in U(\mathcal{A})$ such that $\pi(u_2) = z$ and hence $p\, \pi(u_2) = \pi(u_2)\, p$. Put $w = w_1 u_2^*$. Then $w \in \mathrm{Pis}(\mathcal{A})$ and $v = \pi(w)\, p$. This proves Condition (1).

The Condition (2) is obvious since $\overline{Gi(\mathcal{A})} = \mathcal{A}$. □

Let $a, b \in \mathcal{A}$ and let p, q be two nonzero projections in $\mathcal{A}$. Write

$$a_{11} = qap,\ a_{12} = qa(1-p),\ a_{21} = (1-q)ap,\ a_{22} = (1-q)a(1-p)$$
$$b_{11} = paq,\ b_{12} = pb(1-q),\ b_{21} = (1-p)bq,\ b_{22} = (1-p)b(1-q).$$

Then $a = a_{11} + a_{12} + a_{21} + a_{22}$, $b = b_{11} + b_{12} + b_{21} + b_{22}$. For convenience, we write a (resp. b) as a matrix $a = \begin{bmatrix} a_{11} & a_{12} \\ a_{21} & a_{22} \end{bmatrix}$ (resp. $b = \begin{bmatrix} b_{11} & b_{12} \\ b_{21} & b_{22} \end{bmatrix}$). It is easy to check that $ba = \begin{bmatrix} b_{11}a_{11} + b_{12}a_{21} & b_{11}a_{12} + b_{12}a_{22} \\ b_{21}a_{11} + b_{22}a_{21} & b_{21}a_{12} + b_{22}a_{22} \end{bmatrix}$.

The Condition (2) of Theorem 6.5.10 is rather complex. The following statement will give a simple sufficient condition.

Proposition 6.5.11. *Let $\mathcal{J}$ be a closed ideal of the C^*-algebra $\mathcal{A}$. Suppose that for any non-zero projections p_1 and p_2 in $\mathcal{A}$ with $p_1 \mathcal{J} p_2 \neq \{0\}$, $Gi(p_1 \mathcal{J} p_2)$ is dense in $p_1 \mathcal{J} p_2$ and $Gi(p_1 \tilde{\mathcal{J}} p_1)$ is dense in $p_1 \tilde{\mathcal{J}} p_1$. Then* $\mathrm{Pis}(\mathcal{A}) + \mathcal{J} \subset \overline{Gi(\mathcal{A})}$.

In particular, when $\mathrm{RR}(\mathcal{J}) = 0$, $\mathrm{Pis}(\mathcal{A}) + \mathcal{J} \subset \overline{Gi(\mathcal{A})}$.

Proof. Let $k \in \mathcal{J}$ and let $v \in \mathrm{Pis}(\mathcal{A})$. Put $p = v^*v$, $q = vv^*$ and

$$\begin{aligned} b_{11} &= q(v+k)p = qv(p+v^*k)p, \\ b_{12} &= q(v+k)(1-p) = qk(1-p), \\ b_{21} &= (1-q)(v+k)p = (1-q)kp, \\ b_{22} &= (1-q)(v+k)(1-p) = (1-q)k(1-p). \end{aligned}$$

Since $Gi(p\tilde{\mathcal{J}}p)$ is dense in $p\tilde{\mathcal{J}}p$, we can pick $k_{11} \in Gi(p\tilde{\mathcal{J}}p)$ such that $\|p+v^*k-k_{11}\| < \frac{\epsilon}{4}$. Put $a_{11} = vk_{11}$. Then $a_{11} \in Gi(q\mathcal{A}p)$ for $a_{11}^\dagger = k_{11}^\dagger v^*$ satisfying $pa_{11}^\dagger = a_{11}^\dagger q = a_{11}^\dagger$. If we write

$$v+k = \begin{bmatrix} b_{11} & b_{12} \\ b_{21} & b_{22} \end{bmatrix} \qquad x_0 = \begin{bmatrix} a_{11} & b_{12} \\ b_{21} & b_{22} \end{bmatrix},$$

then $\|v+k-x_0\| < \frac{\epsilon}{4}$. Note that x_0 can be decomposed as

$$x_0 = \begin{bmatrix} q & 0 \\ b_{21}a_{11}^\dagger & 1-q \end{bmatrix} \begin{bmatrix} a_{11} & (q-a_{11}a_{11}^\dagger)b_{12} \\ b_{21}(p-a_{11}^\dagger a_{11}) & b_{22}-b_{21}a_{11}^\dagger b_{12} \end{bmatrix} \begin{bmatrix} p & a_{11}^\dagger b_{12} \\ 0 & 1-p \end{bmatrix}. \tag{6.5.4}$$

Now put

$$b'_{12} = (q-a_{11}a_{11}^\dagger)b_{12},\ b'_{21} = b_{21}(p-a_{11}^\dagger a_{11}),\ b'_{22} = b_{22}-b_{21}a_{11}^\dagger b_{12}.$$

Then there exist

$$a_{12} \in Gi((q-a_{11}a_{11}^\dagger)\mathcal{J}(1-p)),\ a_{21} \in Gi((1-q)\mathcal{J}(p-a_{11}^\dagger a_{11}))$$

such that

$$\max\{\|b'_{12}-a_{12}\|,\ \|b'_{21}-a_{21}\|\} \le \frac{\epsilon}{4\|1+b_{21}a_{11}^\dagger\|\|1+a_{11}^\dagger b_{12}\|}.$$

Put $x_1 = \begin{bmatrix} a_{11} & a_{12} \\ a_{21} & b'_{22} \end{bmatrix}$. Then by (6.5.4),

$$\left\| x_0 - \begin{bmatrix} q & 0 \\ b_{21}a_{11}^\dagger & 1-q \end{bmatrix} x_1 \begin{bmatrix} p & a_{11}^\dagger b_{12} \\ 0 & 1-p \end{bmatrix} \right\| < \frac{\epsilon}{2}. \tag{6.5.5}$$

Decompose x_1 as

$$x_1 = \begin{bmatrix} q & 0 \\ b'_{22}a_{12}^\dagger & 1-q \end{bmatrix} \begin{bmatrix} a_{11} & a_{12} \\ a_{21} & d \end{bmatrix} \begin{bmatrix} p & a_{21}^\dagger b'_{22}(1-p-a_{12}^\dagger a_{12}) \\ 0 & 1-p \end{bmatrix},$$

where $d = (1-q-a_{21}a_{21}^\dagger)b'_{22}(1-p-a_{12}^\dagger a_{12})$. Pick

$$a_{22} \in Gi((1-q-a_{21}a_{21}^\dagger)\mathcal{J}(1-p-a_{12}^\dagger a_{12}))$$

such that $\|d - a_{22}\| < \eta^{-1}\epsilon$, where

$$\eta = 4\,\|1 + b_{21}a_{11}^{\dagger}\|\,\|1 + a_{11}^{\dagger}b_{12}\|\,\|1 + b'_{22}a_{12}^{\dagger}\|\,\|1 + a_{21}^{\dagger}b'_{22}(1-p-a_{12}a_{12}^{\dagger})\|.$$

Set $x_2 = \begin{bmatrix} a_{11} & a_{12} \\ a_{21} & a_{22} \end{bmatrix}$ and

$$x = \begin{bmatrix} q & 0 \\ b_{21}a_{11}^{\dagger} & 1-q \end{bmatrix} \begin{bmatrix} q & 0 \\ b'_{22}a_{12}^{\dagger} & 1-q \end{bmatrix} x_2 \begin{bmatrix} p & a_{21}^{\dagger}b'_{22}(1-p-a_{12}^{\dagger}a_{12}) \\ 0 & 1-p \end{bmatrix} \begin{bmatrix} p & a_{11}^{\dagger}b_{12} \\ 0 & 1-p \end{bmatrix}.$$

Then $\|v+k-x\| < \epsilon$ by (6.5.5). Noting that $a_{11}^*a_{12} = a_{11}^*(q-a_{11}a_{11}^{\dagger})a_{12} = 0$ and $a_{21}^*a_{22} = 0$, we get that

$$x_2^*x_2 = \begin{bmatrix} a_{11}^*a_{11} + a_{21}^*a_{21} & 0 \\ 0 & a_{12}^*a_{12} + a_{22}^*a_{22} \end{bmatrix}.$$

But we also have $(a_{11}^*a_{11})(a_{21}^*a_{21}) = 0$, $(a_{12}^*a_{12})(a_{22}^*a_{22}) = 0$. So

$$\sigma(x_2^*x_2) = \sigma(a_{11}^*a_{11}) \cup \sigma(a_{12}^*a_{12}) \cup \sigma(a_{21}^*a_{21}) \cup \sigma(a_{22}^*a_{22}).$$

This means that $0 \notin \sigma(x_2^*x_2)$ or $0 \in \sigma(x_2^*x_2)$ is an isolated point by Proposition 3.5.8. Therefore, $x \in Gi(\mathcal{A})$ by using by Proposition 3.5.8 again.

When $\mathrm{RR}(\mathcal{J}) = 0$, we obtain that the hypotheses in Proposition 6.5.11 are satisfied by Proposition 6.5.7. □

Corollary 6.5.12. *Let $\mathcal{J}$, $\mathcal{A}$ and $\mathcal{B}$ be as in Theorem* 6.5.10. Suppose that $\mathcal{A}$ is unital and $\mathrm{RR}(\mathcal{J}) = 0$. If $\mathcal{B}$ satisfies one of the following conditions, then $\mathcal{A}$ has (APD).

(1) $\mathcal{B}$ has (APD) and $\mathcal{A}$ is projection–liftable with respect to $\mathcal{J}$.
(2) The set of all one–sided invertible elements in $\mathcal{B}$ is dense in $\mathcal{B}$, especially, $\mathrm{tsr}(\mathcal{B}) = 1$.

Proof. (1) Let $v \in \mathrm{Ais}(\mathcal{B})$. Then by Corollary 6.4.4 and Lemma 6.4.5, there is $w \in \mathrm{Pis}(\mathcal{A})$ such that $\pi(w) = v$. Therefore $\mathcal{A}$ has (APD) by Theorem 6.5.10 and Proposition 6.5.11.

(2) Since the isometry (or co–isometry or unitary) in $\mathcal{B}$ can be lifted to a partial isometry in $\mathcal{A}$ by Corollary 6.4.4 and Corollary 6.4.6 and $\mathrm{Ais}(\mathcal{B})$ consists of isometries or co–isometries in $\mathcal{B}$ by the assumption, it follows from Theorem 6.5.10 and Proposition 6.5.11 that $\mathcal{A}$ has (APD).

If $\mathrm{tsr}(\mathcal{B}) = 1$, then $\mathrm{Ais}(\mathcal{B}) = U(\mathcal{B})$. Thus, $\mathcal{A}$ has (APD). □

Remark 6.5.13. (1) Rødam showed in [Rørdam (1991)] that if the C^*–algebra $\mathcal{B}$ in Corollary 6.5.12 is a unital simple purely infinite C^*–algebra (cf. Definition B.1.4), then the set of all one–sided invertible elements in $\mathcal{B}$

is dense in $\mathcal{B}$ (cf. Theorem 4.5 and Theorem 3.3 of [Rørdam (1991)]. So $\mathcal{A}$ has (APD) by Corollary 6.5.12 (2) when $\mathrm{RR}(\mathcal{J}) = 0$.

(2) Proposition 2.6 of [Brown and Pedersen (1995)] says that for every partial isometry $v \in \mathcal{B}$, we can find a $v' \in \mathrm{ext}(\mathcal{B})$ such that $v'v^*v = v$ when $\overline{\mathcal{B}_q^{-1}} = \mathcal{B}$. So if we assume that $\mathrm{ext}(\mathcal{B}) \subset \pi(\mathrm{Pis}(\mathcal{A}))$, then $\mathcal{A}$ has (APD) by Theorem 6.5.10 and Proposition 6.5.11.

Corollary 6.5.12 shows that if $\mathrm{RR}(\mathcal{J}) = 0$ and $\mathrm{tsr}(\mathcal{B}) = 1$, then $\mathcal{A}$ has (APD). A natural problem is when $\mathcal{A}$ is extremally rich. We will use K–theory data to describe the extremal richness of $\mathcal{A}$ in Proposition 6.5.14.

Let τ be a non–zero quasitrace on $\mathcal{J}$ (cf. Definition B.3.2). Then τ induces a homomorphism $K_0(\mathcal{A}) \to \mathbb{R}$ defined by $\tau_*(x) = \tau(p) - \tau(q)$, for $x = [p] - [q] \in K_0(\mathcal{J})$, where p (resp. q) is a projection in $M_n(\mathcal{J})$ (resp. $M_m(\mathcal{J})$).

Proposition 6.5.14. *Let $\mathcal{A}$ be a unital C^*–algebra and $\mathcal{J}$ be a closed ideal of $\mathcal{A}$. Put $\mathcal{B} = \mathcal{A}/\mathcal{J}$ and $\pi\colon \mathcal{A} \to \mathcal{B}$ be the quotient mapping. Suppose that* $\mathrm{tsr}(\mathcal{B}) = 1$ *and $\mathcal{J}$ is a σ–unital and simple C^*–algebra with* $\mathrm{RR}(\mathcal{J}) = 0$ *and* $\mathrm{tsr}(\mathcal{J}) = 1$. *Let τ be a quasitrace on $\mathcal{J}$ such that for any projections p_1, p_2 in $\mathcal{J}$, if $\tau(p_1) < \tau(p_2)$, then p_1 is equivalent to a projection $q_2 \le p_2$ in $\mathcal{J}$. Then $\mathcal{A}$ is extremally rich iff* $\mathrm{Ker}\,\tau_* \cap \mathrm{Ran}\,(\partial_1) = \{0\}$, *where ∂_1 is given in §B.3 of Appendix B.*

Proof. By Corollary 6.5.12 (2), $\mathcal{A}$ has (APD).

Now suppose that $\mathrm{Ker}\,\tau_* \cap \mathrm{Ran}\,(\partial_1) = \{0\}$. Let u be in $U(\mathcal{B})$. Then there is $v \in \mathrm{Pis}(\mathcal{A})$ such that $\pi(v) = u$ by Corollary 6.4.4, Corollary 6.4.6 and $\partial([u]) = [1 - v^*v] - [1 - vv^*]$ by Proposition B.3.1. So if $\tau(1 - v^*v) = \tau(1 - vv^*)$, then $[1 - v^*v] = [1 - vv^*]$. Since $\mathrm{tsr}(\mathcal{J}) = 1$, we can find $z_0 \in \mathcal{J}$ such that $1 - v^*v = z_0^*z_0$, $1 - vv^* = z_0z_0^*$. Put $w = u + z_0$. Then $w \in U(\mathcal{A})$ and $\pi(w) = u$. If $\tau(1 - v^*v) > \tau(1 - vv^*)$, there is $z_1 \in \mathcal{J}$ such that $z_1z_1^* = 1 - vv^*$, $z_1^*z_1 \le 1 - v^*v$ by the assumption of τ. Put $w_1 = v + z_1$. Then w_1 is a co–isometry in $\mathcal{A}$ and $\pi(w_1) = u$; Similarly, if $\tau(1 - v^*v) < \tau(1 - vv^*)$, we can find an isometry w_2 in $\mathcal{A}$ such that $\pi(w_2) = v$.

Let $x \in \mathcal{A}$. Then for any $\epsilon \in (0, 1)$, there is $y \in GL(\mathcal{B})$ such that $\|\pi(x) - y\| < \epsilon$. Set $v = y|y|^{-1} \in U(\mathcal{B})$. Choose $a \in \mathcal{A}_{sa} \cap GL(\mathcal{A})$ and an isometry or a co–isometry s in $\mathcal{A}$ such that $\pi(a) = |y|$ and $\pi(s) = v$, respectively. Pick $k \in \mathcal{J}$ such that $\|x - sa - k\| < \epsilon$. Since $1 - ka^{-1}s^*$ and $1 - s^*ka^{-1}$ are all in $\tilde{\mathcal{J}}$, it follows that there is $a_0, a_1 \in GL(\tilde{\mathcal{J}})$ such that

$$\|1 - ka^{-1}s^* - a_0\| < \frac{\epsilon}{\|a^{-1}\|}, \quad \|1 - s^*ka^{-1} - a_1\| < \frac{\epsilon}{\|a^{-1}\|}.$$

Thus, $\|x - a_0 sa\| < 2\epsilon$ when $s^*s = 1$ or $\|x - sa_1a\| < 2\epsilon$ when $ss^* = 1$. Therefore, $\mathcal{A}$ is extremally rich.

Conversely, if $\operatorname{Ker}\tau_* \cap \operatorname{Ran}(\partial_1) \neq \{0\}$, then there is $y \in K_1(\mathcal{B})$ such that $\partial_1(y) \neq 0$ and $\tau_*(\partial(y)) = 0$. $\operatorname{tsr}(\mathcal{B}) = 1$ implies that there is $u \in U(\mathcal{B})$ such that $x = \partial(y) = \partial([u])$. Let $w \in \operatorname{Pis}(\mathcal{A})$ such that $\pi(w) = u$. Then $x = [1 - w^*w] - [1 - ww^*]$ in $K_0(\mathcal{J})$ and $\tau(1 - w^*w) = \tau(1 - ww^*)$. Consequently, $w^*w \neq 1$ and $ww^* \neq 1$. Since $\mathcal{J}$ is simple, the closed ideal $I(1 - w^*w) = \overline{span}\{c(1 - w^*w)d \,|\, c, d \in \mathcal{J}\}$ generated by $1 - w^*w$ is $\mathcal{J}$. If $w \in \operatorname{ext}(\mathcal{A})$, then $(1 - ww^*)(1 - ww^*) = 0$, a contradiction.

According to Corollary 6.3 of [Brown and Pedersen (1995)], $\mathcal{A}$ is extremally rich iff every unitary element in $U(\mathcal{B})$ can be lifted to an extremal point of $D(\mathcal{A})$. Since u could not be lifted to an extremal point, $\mathcal{A}$ is not extremally rich. □

6.6 The reduced minimum modulus relative to a quotient C^*–algebra

Throughout the section, $\mathcal{A}$ is a unital C^*–algebra and $\mathcal{J}$ is a closed ideal of $\mathcal{A}$. Put $\mathcal{B} = \mathcal{A}/\mathcal{J}$ and let $\pi\colon \mathcal{A} \to \mathcal{B}$ be the quotient mapping.

We have defined the reduced minimum modulus $\gamma_{\mathcal{A}}(a)$ of a non–zero element $a \in \mathcal{A}$ by

$$\gamma_{\mathcal{A}}(a) = \inf\{\|b - a\| \mid K_r(a) \subsetneqq K_r(b), b \in \mathcal{A}\}$$

in Definition 3.5.10 and showed that $\gamma_{\mathcal{A}}(a) = \inf\{\lambda \,|\, \lambda \in \sigma(|a|)\backslash\{0\}\}$ in Theorem 3.5.12. In this section, we will discuss the relation between $\gamma_{\mathcal{A}}(a)$ and $\gamma_{\mathcal{B}}(\pi(a))$.

Definition 6.6.1. An element a in $\mathcal{A}$ is called a left (resp. right) topological zero divisor relative to $\mathcal{J}$ if there exists a normalized sequence $\{\pi(b_n)\}$ in $\mathcal{B}$ such that $\lim\limits_{n\to\infty} \|\pi(ab_n)\| = 0$ (resp. $\lim\limits_{n\to\infty} \|\pi(b_na)\| = 0$).

We denote the set of these elements $Z_l(\mathcal{A}, \mathcal{J})$ (resp. $Z_r(\mathcal{A}, \mathcal{J})$) and put $Z(\mathcal{A}, \mathcal{J}) = Z_l(\mathcal{A}, \mathcal{J}) \cap Z_r(\mathcal{A}, \mathcal{J})$.

Let $\Phi_l(\mathcal{A})$ (resp. $\Phi_r(\mathcal{A})$) denote the set of all elements a in $\mathcal{A}$ such that $\pi(a)$ is left (resp. right) invertible in $\mathcal{B}$. Put $\Phi(\mathcal{A}) = \Phi_l(\mathcal{A}) \cap \Phi_r(\mathcal{A})$. Set $\Phi_l^c(\mathcal{A}) = \mathcal{A}\backslash\Phi_l(\mathcal{A})$, $\Phi_r^c(\mathcal{A}) = \mathcal{A}\backslash\Phi_r(\mathcal{A})$ and $\Phi^c(\mathcal{A}) = \mathcal{A}\backslash\Phi(\mathcal{A})$.

Proposition 6.6.2. $Z_l(\mathcal{A}, \mathcal{J}) = \Phi_l^c(\mathcal{A})$, $\quad Z_r(\mathcal{A}, \mathcal{J}) = \Phi_r^c(\mathcal{A})$.

Proof. Let $a \in Z_l(\mathcal{A}, \mathcal{J})$. Then there is a sequence $\{\pi(b_n)\}$ in $\mathcal{B}$ such that $\|\pi(b_n)\| = 1$, $\forall n \geq 1$ and $\|\pi(ab_n)\| \to 0$ as $n \to \infty$. Thus $\pi(a)$ is not left invertible in $\mathcal{B}$, i.e., $a \in \Phi_l^c(\mathcal{A})$ and $Z_l(\mathcal{A}, \mathcal{J}) \subset \Phi_l^c(\mathcal{A})$.

Now let $a \in \Phi_l^c(\mathcal{A})$. Then $\pi(a^*a)$ is not invertible in $\mathcal{B}$. Define a sequence of continuous functions $\{f_n\}$ on $[0, \|\pi(a^*a)\|]$ by

$$f_n(t) = \begin{cases} 1 & 0 \leq t \leq \frac{1}{2n} \\ 2n(\frac{1}{n} - t) & \frac{1}{2n} \leq t \leq \frac{1}{n} \\ 0 & \frac{1}{n} \leq t \leq \|\pi(a^*a)\| \end{cases}$$

for n large enough. Set $b_n = f_n(\pi(a^*a)) \geq 0$. Then $b_n \in \mathcal{B}$, $\pi(a^*a)b_n = b_n\pi(a^*a)$, $\|b_n\| = 1$ and

$$\|\pi(a)b_n^{\frac{1}{2}}\|^2 = \|b_n^{\frac{1}{2}}\pi(a^*a)b_n^{\frac{1}{2}}\| = \|\pi(a^*a)b_n\| \leq \frac{1}{2n}.$$

Therefore, $a \in Z_l(\mathcal{A}, \mathcal{J})$.

Noting that $a \in Z_l(\mathcal{A}, \mathcal{J})$ iff $a^* \in Z_r(\mathcal{A}, \mathcal{J})$ and $a \in \Phi_l^c(\mathcal{A})$ iff $a^* \in \Phi_r^c(\mathcal{A})$, we get that $Z_r(\mathcal{A}, \mathcal{J}) = \Phi_r^c(\mathcal{A})$. □

Set

$$m(a) = \inf\{\lambda | \lambda \in \sigma(|a|)\}, \quad m_{\mathcal{J}}(a) = \inf\{\lambda | \lambda \in \sigma(\pi(|a|))\}.$$

Proposition 6.6.3. *Let a be a nonzero element in $\mathcal{A}$. Then*

$$\begin{aligned} &\operatorname{dist}(a, \Phi_l^c(\mathcal{A})) = m_{\mathcal{J}}(a), \ \operatorname{dist}(a, \Phi_r^c(\mathcal{A})) = m_{\mathcal{J}}(a^*) \\ &\operatorname{dist}(a, \Phi^c(\mathcal{A})) = \min\{m_{\mathcal{J}}(a), \ m_{\mathcal{J}}(a^*)\} \end{aligned}$$

Proof. We prove first equality. If $a \in \Phi_l^c(\mathcal{A})$, then $\pi(|a|)$ is not invertible in $\mathcal{B}$. So in this case, the statement holds.

Now suppose that $\pi(a)$ is left invertible. By Theorem 3.5.12,

$$\gamma_{\mathcal{B}}(\pi(a)) = \min\{\lambda \,|\, \lambda \in \sigma(\pi(|a|))\} = m_{\mathcal{J}}(a).$$

By the definition of $\gamma_{\mathcal{B}}(\pi(a))$, we can find $b_\epsilon \in \mathcal{B}$ for any $\epsilon > 0$, such that

$$\gamma_{\mathcal{B}}(\pi(a)) > \|\pi(a) - b_\epsilon\| - \epsilon, \qquad K_r(\pi(a)) \subsetneqq K_r(b_\epsilon).$$

Since $K_r(\pi(a)) = 0$, b_ϵ is not left invertible in $\mathcal{B}$. Choose $c_\epsilon \in \mathcal{A}$ such that $\pi(c_\epsilon) = b_\epsilon$ and $k_\epsilon \in \mathcal{J}$ such that $\|\pi(a) - b_\epsilon\| > \|a - c_\epsilon - k_\epsilon\| - \epsilon$. Then $c_\epsilon + k_\epsilon \in \Phi_l^c(\mathcal{A})$. Thus

$$\gamma_{\mathcal{B}}(\pi(a)) > \|\pi(a) - \pi(c_\epsilon)\| - \epsilon > \|a - c_\epsilon - k_\epsilon\| - 2\epsilon \geq \operatorname{dist}(a, \Phi_l^c(\mathcal{A})) - 2\epsilon$$

and so that $m_{\mathcal{J}}(a) = \gamma_{\mathcal{B}}(\pi(a)) \geq \operatorname{dist}(a, \Phi_l^c(\mathcal{A}))$.

Since $\pi(a)$ is left invertible, it follows that $\pi(a^*a)$ is invertible in $\mathcal{B}$. Put $b = (\pi(a^*a))^{-1}\pi(a^*)$. It is easy to check that $b = (\pi(a))^\dagger$ and

$$\|b\|^2 = \|bb^*\| = \|(\pi(a^*a))^{-1}\| = \|(\pi(|a|))^{-1}\|^2,$$

so that $\|b\| = \|(\pi(|a|))^{-1}\| = \max\{\lambda^{-1}|\lambda \in \sigma(\pi(|a|))\}$.

We now show that $\|a - z\| \geq m_{\mathcal{J}}(\mathcal{A})$, $\forall z \in \Phi_l^c(\mathcal{A})$. If there exists $c_0 \in \Phi_l^c(\mathcal{A})$, such that $\|a - c_0\| < m_{\mathcal{J}}(a)$, then

$$\|1 - b\pi(c_0)\| \leq \|b\|\|\pi(a - c_0)\| < \|(\pi(|a|))^{-1}\| \, m_{\mathcal{J}}(a) < 1.$$

Thus $b\pi(c_0)$ is invertible in $\mathcal{B}$ and hence $c_0 \in \Phi_l(\mathcal{A})$, a contradiction. So, $m_{\mathcal{J}}(a) \leq \operatorname{dist}(a, \Phi_l^c(\mathcal{A}))$.

Since $a \in \Phi_r(\mathcal{A})$ iff $a^* \in \Phi_l(\mathcal{A})$, we get that $\operatorname{dist}(a, \Phi_r^c(\mathcal{A})) = m_{\mathcal{J}}(a^*)$ by above argument. Similarly, we can obtain third equality. □

From Proposition 6.6.2 and Proposition 6.6.3, we have

Corollary 6.6.4. *Let $a \in \mathcal{A}\backslash\{0\}$. Then $a \in Z_l(\mathcal{A}, \mathcal{J})$ iff $m_{\mathcal{J}}(a) = 0$ and $a \in Z(\mathcal{A}, \mathcal{J})$ iff $m_{\mathcal{J}}(a) = m_{\mathcal{J}}(a^*) = 0$.*

Remark 6.6.5. Let H be a separable Hilbert space. Take $\mathcal{A} = B(H)$ and $\mathcal{J} = \mathcal{K}(H)$ in Proposition 6.6.3. Then $\mathcal{B}$ is the Calkin algebra. In this situation, $m_{\mathcal{K}(H)}(T) = m_e(T)$ for $T \in B(H)$. Proposition 6.6.3 shows that $\operatorname{dist}(T, \Phi_l^c(B(H))) = m_e(T)$, $\forall T \in B(H)$. This result was proved by Zemánek in [Zemáneck (1982)]. Corollary 6.6.4 also generalizes Theorem 4.2 and Corollary 4.3 of [Gopalraj and Stroh (2004)].

In Hilbert spaces, there are two important properties about reduced minimum modulus of operators. These are:

(1) $A \in \Phi_l(B(H)) \cup \Phi_r(B(H))$ implies that $\gamma(A + K) > 0$, $\forall K \in \mathcal{K}(H)$.
(2) For any $A \in B(H)\backslash\mathcal{K}(H)$, there exists $K \in \mathcal{K}(H)$ such that $\gamma(\pi(A)) = \gamma(A + K)$.

Property (2) was generalized to the set of von Neumann algebras in [Gopalraj and Stroh (2004)]. We will extend above two properties to the set of C^*–algebras as follows.

Proposition 6.6.6. *The following conditions are equivalent:*

(1) $\gamma_{\mathcal{A}}(a) > 0$, *for any* $a \in \Phi_l(\mathcal{A}) \cup \Phi_r(\mathcal{A})$.
(2) $\gamma_{\mathcal{A}}(1 + c) > 0$, *for any* $c \in \mathcal{J}_{sa}$.
(3) $\sigma(c)$ *is countable and* 0 *is the only accumulation point of* $\sigma(c)$ *for any* $c \in \mathcal{J}_{sa}$ *if* $\sigma(c)$ *is infinite.*

(4) *$\mathcal{J}$ is $*$-isomorphic to a C^*-subalgebra of $\mathcal{K}(H)$.*

Proof. (1)$\Rightarrow$(2) is Obvious.

(2)$\Rightarrow$(3) Let c be a self-adjoint element in $\mathcal{J}$ and let $\lambda \in \sigma(c)\backslash\{0\}$. By our assumption we have $\gamma_{\mathcal{A}}(1-\lambda^{-1}c) > 0$. Thus, 0 is an isolated point of $\sigma(|1-\lambda^{-1}c|)$ by Theorem 3.5.12 and hence 1 is an isolated point of $\sigma(\lambda^{-1}c)$. So λ is an isolated point of $\sigma(c)\backslash\{0\}$. This indicates that $\sigma(c)\backslash\{0\}$ is countable and so is the $\sigma(c)$.

The equivalence of (3) and (4) was shown by Proposition B.1.2.

(3)$\Rightarrow$(1) We assume that $a \in \Phi_l(\mathcal{A})$. Then $\pi(|a|)$ is invertible and there are $b \in \mathcal{A}$ and $k \in \mathcal{J}$ such that $b|a| = 1+k$. Set $h = k^*k + kk^*$. If $0 \in \sigma(h)$ is an isolated point, we obtain a projection $q \in \mathcal{J}$ such that $h(1-q) = 0$. If $0 \in \sigma(h)$ is an accumulation point, then $\sigma(h) = \{0\} \cup \{\lambda_n |\, n \geq 1\}$ and $\lim\limits_{n\to\infty} \lambda_n = 0$, according to the assumption. Let p_i be the spectral projection of h corresponding to $\{\lambda_i\}$ in $\mathcal{J}$ (i.e., $p_i = \chi_{\{\lambda_i\}}(h)$), $i = 1, 2, \cdots$. Then $h = \sum\limits_{i=1}^{\infty} \lambda_i p_i$. Put $e_n = \sum\limits_{i=1}^{n} p_i \in \mathcal{J}$. Then

$$\|he_n - h\| \leq \max_{i \geq n+1} |\lambda_i| \to 0 \text{ as } n \to \infty.$$

Now choose n_0 such that $\|k(1-e_{n_0})\| < 1$ (when 0 is an isolated point of $\sigma(h)$, take $e_{n_0} = q$) so that

$$\|(1-e_{n_0})b|a|(1-e_{n_0}) - (1-e_{n_0})\| < 1$$

and hence there is $x \in (1-e_{n_0})\mathcal{A}(1-e_{n_0})$ such that $xb|a|(1-e_{n_0}) = 1-e_{n_0}$, that is, $a_{11} = (1-e_{n_0})(a^*a)(1-e_{n_0})$ is invertible in $(1-e_{n_0})\mathcal{A}(1-e_{n_0})$. From now on a_{11}^{-1} denotes the inverse of a_{11} in $(1-e_{n_0})\mathcal{A}(1-e_{n_0})$.

Put $a_{12} = (1-e_{n_0})(a^*a)e_{n_0}$, $a_{21} = e_{n_0}(a^*a)(1-e_{n_0})$ and $a_{22} = e_{n_0}(a^*a)e_{n_0}$. Write a^*a as $a^*a = \begin{bmatrix} a_{11} & a_{12} \\ a_{21} & a_{22} \end{bmatrix}$. Then we have

$$a^*a = \begin{bmatrix} a_{11} & 0 \\ a_{21} & e_{n_0} \end{bmatrix} \begin{bmatrix} 1-e_{n_0} & 0 \\ 0 & a_{22} - a_{21}a_{11}^{-1}a_{12} \end{bmatrix} \begin{bmatrix} 1-e_{n_0} & a_{11}^{-1}a_{12} \\ 0 & e_{n_0} \end{bmatrix}.$$

Put $m = a_{22} - a_{21}a_{11}^{-1}a_{12}$, $z_0 = e_{n_0} - m \in \mathcal{J}$ and $z = 1 - z_0$. Sine $z_0^* + z_0 - z_0^*z_0 = 1 - z^*z \in \mathcal{J}$, we have $1 \notin \sigma(1-z^*z)$ or $1 \in \sigma(1-z^*z)$ is an isolated point. Thus $z^\dagger$ exists. Put

$$y = \begin{bmatrix} 1-e_{n_0} & -a_{11}^{-1}a_{12} \\ 0 & e_{n_0} \end{bmatrix} z^\dagger \begin{bmatrix} a_{11}^{-1} & 0 \\ -a_{21}a_{11}^{-1} & e_{n_0} \end{bmatrix}.$$

Then $a^*aya^*a = a^*a$ and hence $\gamma_{\mathcal{A}}(a) > 0$ by Proposition 3.5.8 and Proposition 3.5.11.

If $a \in \Phi_r(\mathcal{A})$, then $a^* \in \Phi_l(\mathcal{A})$. So $\gamma_{\mathcal{A}}(a^*) = \gamma_{\mathcal{A}}(a) > 0$ by Corollary 3.5.14. □

Theorem 6.6.7. *Let $a \in \mathcal{A}\backslash\mathcal{J}$ and suppose that $\mathcal{A}$ is of real rank zero. Then $\gamma_{\mathcal{B}}(\pi(a)) = \sup\{\gamma_{\mathcal{A}}(a+k) \,|\, k \in \mathcal{J}\}$.*

In addition, if $\mathcal{J}$ is a σ-unital essential ideal of $\mathcal{A}$ with $\mathrm{RR}(M(\mathcal{J})) = 0$, then there is $k \in \mathcal{J}$ such that $\gamma_{\mathcal{B}}(\pi(a)) = \gamma_{\mathcal{A}}(a+k)$.

Proof. Since $\sigma(\pi(|a|) \subset \sigma(|a+k|)$, $\forall\, k \in \mathcal{J}$, it follows from Theorem 3.5.12 that $\gamma_{\mathcal{A}}(a+k) \leq \gamma_{\mathcal{B}}(\pi(a))$, $\forall\, k \in \mathcal{J}$. Thus

$$\gamma_{\mathcal{B}}(\pi(a)) \geq \sup\{\gamma_{\mathcal{A}}(a+k) \,|\, k \in \mathcal{J}\}.$$

Put $b = \pi(a)$. If $0 \in \sigma(|b|)$ is not an isolated point, then $\gamma_{\mathcal{B}}(b) = 0$ and $\gamma_{\mathcal{A}}(a+k) = 0\ \forall\, k \in \mathcal{J}$ by Lemma 3.5.13. In the following, we assume that $0 \notin \sigma(|b|)$ or $0 \in \sigma(|b|)$ is an isolated point.

Let $b = u|b|$ be the polar decomposition in $\mathcal{B}$. Set $q = u^*u$. Then $q|b| = |b|$ and $|b| \in GL(q\mathcal{B}q)$. Let $\epsilon \in \big(0, \dfrac{1}{3}\big)$. Since $\mathrm{RR}(\mathcal{A}) = 0$, it follows from Theorem 6.4.3 (1) that $\mathrm{RR}(\mathcal{J}) = \mathrm{RR}(\mathcal{B}) = 0$. Thus, we can find real numbers $\mu_1, \cdots, \mu_m$ and mutually orthogonal projections $q_1, \cdots, q_m$ in $q\mathcal{B}q$ such that

$$\sum_{i=1}^{m} q_i = q, \quad \Big\||b| - \sum_{i=1}^{m} \mu_i q_i\Big\| < \frac{\epsilon}{\||b|^\dagger\|}, \tag{6.6.1}$$

where $|b|^\dagger$ is the inverse of $|b|$ in $q\mathcal{B}q$. By Corollary 6.4.4, Lemma 6.4.11 and Corollary B.2.2, there is a partial isometry w in $\mathcal{A}$ with $\pi(w) = u$ and mutually orthogonal projections $s_1, \cdots, s_m$ in $p\mathcal{A}p$, where $p = w^*w$, such that $\sum\limits_{i=1}^{m} s_i = p$ and $\pi(s_i) = q_i$, $i = 1, \cdots, m$.

Choose a positive invertible element $d \in p\mathcal{A}p$ such that $\pi(d) = |b|$ and set $a_0 = wd$. Then $\pi(a_0) = \pi(a) = u|b|$. From (6.6.1), we have $\Big\|\pi(d - \sum\limits_{i=1}^{m} \mu_i s_i)\Big\| < \dfrac{\epsilon}{\||b|^+\|}$. Take $k_0 \in p\mathcal{J}p$ such that

$$\Big\|d + k_0 - \sum_{i=1}^{m} \mu_i s_i\Big\| < \Big\|\pi(d - \sum_{i=1}^{m} \mu_i s_i)\Big\| + \frac{\epsilon}{\||b|^+\|} < \frac{2\epsilon}{\||b|^+\|}. \tag{6.6.2}$$

By (6.6.1), $\Big\||b|^+\Big(\sum\limits_{i=1}^{m} \mu_i q_i\Big) - q\Big\| < \epsilon$. Thus $\sum\limits_{i=1}^{m} \mu_i q_i$ is invertible in $q\mathcal{B}q$ and

$$\max_{1 \leq i \leq m} |\mu_i^{-1}| = \Big\|\sum_{i=1}^{m} \mu_i^{-1} q_i\Big\| < \frac{\||b|^+\|}{1-\epsilon}.$$

It follows from (6.6.2) that

$$\Big\|(d + k_0)\big(\sum_{i=1}^{m} \mu_i s_i\big)^{-1} - p\Big\| < \frac{2\epsilon}{\||b|^+\|}\Big\|\big(\sum_{i=1}^{m} \mu_i s_i\big)^{-1}\Big\| < \frac{2\epsilon}{1-\epsilon} < 1.$$

Thus, $d + k_0$ is invertible in $p\mathcal{A}p$ and

$$\|(d+k_0)^{-1}\| < \frac{1-\epsilon}{1-3\epsilon}\Big\|\sum_{i=1}^{m}\mu_i^{-1}s_i\Big\| < \frac{\||b|^+\|}{1-3\epsilon}.$$

Now put $x = w(d+k_0)$, $y = (d+k_0)^{-1}w^*$, $k_\epsilon = k_1 + wk_0$, where $k_1 = a_0 - a \in \mathcal{J}$. Then $xyx = x$, $a + k_\epsilon = a_0 + wk_0 = x$. Therefore, by Proposition 3.5.11,

$$\gamma_{\mathcal{A}}(a+k_\epsilon) \geq \frac{1}{\|y\|} \geq \frac{1}{\|(d+k_0)^{-1}\|} > \frac{1-3\epsilon}{\||b|^+\|}.$$

On the other hand, by Lemma 3.5.13 and Theorem 3.5.12,

$$\gamma_{\mathcal{B}}(\pi(a)) = \gamma_{\mathcal{B}}(|b|) = \frac{1}{\||b|^+\|} = \min\{\lambda |\, \lambda \in \sigma(|b|)\backslash\{0\}\}.$$

So $\gamma_{\mathcal{A}}(a+k_\epsilon) > (1-3\epsilon)\gamma_{\mathcal{B}}(\pi(a))$. This proves that

$$\gamma_{\mathcal{B}}(\pi(a)) = \sup\{\gamma_{\mathcal{A}}(a+k) \,|\, k \in \mathcal{J}\}.$$

Now we assume that $\mathrm{RR}(M(\mathcal{J})) = 0$. Let a, b, u, q be as above. By Lemma 6.4.5, there is a partial isometry $w \in \mathcal{A}$ such that $\pi(w) = u$. Let $p = w^*w$. Set $\mathcal{B}_0 = C^*(q, |b|)$ (the C^*–subalgebra generated by $|b|$ and q in $q\mathcal{B}q$) and $\mathcal{A}_0 = \pi^{-1}(B_0) \subset p\mathcal{A}p$. Consider the commutative diagram of two short exact sequences

$$\begin{array}{ccccccccc} 0 & \longrightarrow & p\mathcal{J}p & \longrightarrow & \mathcal{A}_0 & \xrightarrow{\pi|_{\mathcal{A}_0}} & \mathcal{B}_0 & \longrightarrow & 0 \\ & & \| & & \downarrow{\scriptstyle i} & & \downarrow{\scriptstyle \tau} & & . \\ 0 & \longrightarrow & p\mathcal{J}p & \longrightarrow & M(p\mathcal{J}p) & \xrightarrow{\hat{\pi}} & M(p\mathcal{J}p)/p\mathcal{J}p & \longrightarrow & 0 \end{array}$$

Since $\mathcal{J}$ is essential in $\mathcal{A}$, we get that $p\mathcal{J}p$ is essential in $p\mathcal{A}p$ so that $\mathcal{A}_0$ can be viewed as a C^*–subalgebra of $M(p\mathcal{J}p)$ (cf. Proposition 5.1.5 of [Lin (2001)]). Note that $M(p\mathcal{J}p) = pM(\mathcal{J})p$. So $\mathrm{RR}(M(p\mathcal{J}p)) = 0$. Since $\sigma(|b|)$ is a compact subset in real line, we have by Corollary 1.10, Proposition 1.5 in [Lin (1995a)], there is an approximate unit $\{e_n\}$ of $p\mathcal{J}p$ consisting of increasing projections and a dense sequence $\{\xi_n\}$ in $\sigma(|b|)\backslash\{0\}$ with isolated points repeated infinitely often such that

$$\tau(y) = \hat{\pi}\big[\sum_{n=1}^{\infty}\phi_n(y)(e_n - e_{n-1})\big], \ (e_0 = 0) \quad \forall\, y \in \mathcal{B}_0,$$

here ϕ_n is the character on $\mathcal{B}_0$ with $\phi_n(|b|) = \xi_n$, $n \geq 1$. Put

$$\hat{a}_0 = \sum_{n=1}^{\infty}\xi_n(e_n - e_{n-1}) \in M(p\mathcal{J}p).$$

Then

$$\hat{\pi}(\hat{a}_0) = \tau(|b|) = \tau \circ \pi|_{\mathcal{A}_0}(p|a|p) = \hat{\pi}(p|a|p) \quad \text{and} \tag{6.6.3}$$

$$\sigma(\hat{a}_0) = \{0\} \cup \overline{\{\xi_n \,|\, n \geq 1\}} = \{0\} \cup \sigma(|b|)\,. \tag{6.6.4}$$

By (6.6.3), $\hat{a}_0 - p|a|p = \hat{k} \in p\mathcal{J}p$ and hence $\hat{a}_0 \in p\mathcal{A}p$. Put $\hat{a} = w\hat{a}_0$. Then $|\hat{a}| = (\hat{a}_0 w^* w \hat{a}_0)^{\frac{1}{2}} = \hat{a}_0$ and $\pi(\hat{a}) = \pi(a) = b$ for $q|b| = |b|q$. Set $k = \hat{a} - a \in \mathcal{J}$. We have by Theorem 3.5.12 and (6.6.4),

$$\begin{aligned}\gamma_{\mathcal{A}}(a+k) &= \gamma_{\mathcal{A}}(\hat{a}) = \gamma_{\mathcal{A}}(|\hat{a}|) = \gamma_{\mathcal{A}}(\hat{a}_0)\\ &= \min\{\lambda \,|\, \lambda \in \sigma(\hat{a}_0)\backslash\{0\}\} = \min\{\lambda \,|\, \lambda \in \sigma(|b|)\backslash\{0\}\}\\ &= \gamma_{\mathcal{B}}(\pi(b))\,.\end{aligned}$$

The proof is completed. □

6.7 Notes

Section 6.1. The reduce minimum modulus of Banach spaces is closely related to the spectral theory. The interesting readers can refer to [Badea and Mbekhta (1997)], [Badea and Mbekhta (1999)] and [Mbekhta (1993)].

Section 6.2. Some results about perturbation analysis for operator equations on Hilbert spaces can be found in [Stănică (2010)].

Section 6.3. Proposition 6.3.12 and Corollary 6.3.13 come from [Grobler and Raubenheimer (2008)]. Theorem 6.3.15 generalizes Theorem 3.1 of [Murphy (2006)]. Applications of $\omega(\cdot)$ and index$(\cdot)$ will be published elsewhere.

Section 6.4. Some results about the lifting of generalized inverses in certain Banach algebras can found in [Djordjević (1994)]. Lemma 6.4.10 comes from [Lin (2001)]. The conception of "asymptotically Moore–Penrose invertible sequence" comes from [Hagen *et al.* (2001)]. [Roch and Silbermann (1999)] characterized the "asymptotically Group invertible sequence".

Section 6.5. Proposition 6.5.4 and Proposition 6.5.6 come from [Jeong and Osaka (1998)]. Proposition 6.5.7 slightly generalizes Proposition 6.5.4 of [Blackadar (1998)]. Lemma 6.5.9 comes from [Lin (2001)]. Theorem 6.5.10 and Proposition 6.5.11 come from [Xue (2007a)].

Appendix A

Some Results From Topology And Analysis

A.1 Some topological results

Definition A.1.1. Let M be a locally compact Hausdorff space. A complex function f on M is said to vanish at infinity if for every $\epsilon > 0$ there is a compact subset K in M such that $|f(x)| < \epsilon$, $\forall\, x \in M \backslash K$.

Proposition A.1.2. *Let f be a continuous function on the locally compact Hausdorff space M. If f vanishes at infinity, then $f(M)$ is bounded in $\mathbb{C}$ and f is uniformly continuous on M.*

For the locally compact Hausdorff space M, let $C_0(M)$ denote the set of all continuous functions on M, vanishing at infinity. Then $C_0(M)$ is a Banach space under the linear operations of functions and the norm

$$\|f\| = \sup\{|f(x)|\,|\, x \in M\}.$$

Let X be a finite–dimensional Banach space. Then $S(X)$ is a compact Hausdorff space. We have Borsuk antipodal theorem as follows.

Theorem A.1.3. *Let X and Y be finite–dimensional Banach spaces with* $\dim X > \dim Y$. *Let $f\colon S(X) \to Y$ be a continuous mapping satisfying $f(-x) = -f(x)$, $\forall\, x \in S(X)$. Then there is $x \in S(X)$ such that $f(x) = 0$.*

A.2 Holomorphic vector–valued functions

Let $\mathbb{Z}_+$ denote the set of all nonnegative integers. For $\alpha = (\alpha_1, \cdots, \alpha_n) \in \mathbb{Z}_+^n$ and $z = (z_1, \cdots, z_n) \in \mathbb{C}^n$, we write

$$|\alpha| = \sum_{i=1}^{n} \alpha_i, \quad \alpha! = \alpha_1! \cdots \alpha_n!, \quad z^\alpha = z_1^{\alpha_1} \cdots z_n^{\alpha_n}.$$

Theorem A.2.1. *Let G be an open subset of $\mathbb{C}^n$ and let f be a function from G to a Banach space X. Then following properties are equivalent:*

(1) *the function $z \mapsto \phi(f(z))$ is holomorphic for each $\phi \in X^*$.*
(2) *for every $w \in G$, there exists a neighborhood U of w and elements $x_\alpha \in X$ $(\alpha \in \mathbb{Z}_+^n)$ such that $f(z) = \sum\limits_{\alpha \in \mathbb{Z}_+^n} x_\alpha (z-w)^\alpha, \quad \forall\, z \in U$.*
(3) *f is holomorphic in each variable separately.*

A function satisfying the conditions of Theorem A.2.1 is called to be holomorphic on G.

Theorem A.2.2. *Let G be an open subset of $\mathbb{C}^n$ and X be a Banach space. Let $f\colon G \to X$ be holomorphic.*

(1) *If there is an open subset $U \subset G$ such that $f\big|_U \equiv 0$, then $f \equiv 0$.*
(2) *For any $w \in G$, there is $U(w,r) = \{z \in \mathbb{C}^n \,|\, |z_j - w_j| \le r, 1 \le j \le n\} \subset G$, such that $f(z) = \sum\limits_{\alpha \in \mathbb{Z}_+^n} x_\alpha (z-w)^\alpha, \quad \forall\, z \in U(w,r)$, where*

$$x_\alpha = \frac{1}{\alpha!} D^\alpha f(w) = \frac{1}{(2\pi i)^n} \oint_\Gamma \frac{f(\zeta)}{(\zeta - w)^{\alpha+1}} \mathrm{d}\,\zeta,$$

and $\Gamma = \{z \in G \,|\, |z_j - w_j| = r,\ 1 \le j \le n\}$.

A.3 Inverse Function Theorem

Let X and Y be Banach spaces. Let U be an open subset of X and let $f\colon U \to Y$ be a C^1–mapping. Then we have the Inverse Function Theorem of f as follows.

Theorem A.3.1. *Let $x_0 \in U$. If $f'(x_0)$ is a linear homeomorphism of X onto Y, then f is a local homeomorphism of a neighborhood $U(x_0)$ of x_0 in U to a neighborhood of $f(x_0)$. Furthermore, if $\|y - f(x_0)\|$ is sufficiently small, the sequence $x_{n+1} = x_n + [f(x_0)]^{-1}(y - f(x_n))$ converges to the unique solution of $f(x) = y$ in $U(x_0)$.*

References for Appendix A: [Rudin (1973)], [Muller (2007)], [Fritzsche and Grauert (2002)], [Berger (1977)].

Appendix B

More On C^*–Algebras

B.1 Some basic results in C^*–algebras

Let $\mathcal{A}$ be a C^*–algebra. For every subset M of $\mathcal{A}$, we denote by $R(M)$ (resp. $L(M)$) the set of $x \in \mathcal{A}$ such that $Mx = \{0\}$ (resp. $xM = \{0\}$).

Definition B.1.1 ([Kaplansky (1948)]). *A C^*–algebra $\mathcal{A}$ is called to be dual if for every closed right (left) ideal $\mathcal{I}$, we have $R(L(\mathcal{I})) = \mathcal{I}$ and $L(R(\mathcal{I})) = \mathcal{I}$ respectively.*

The following proposition is quoted from Addenda 4.7.20 of [Dixmier (1977)].

Proposition B.1.2. *Let $\mathcal{A}$ be a C^*–algebra. Then the following conditions are equivalent:*

(1) *$\mathcal{A}$ is a dual C^*–algebra.*
(2) *The sum of the minimal left (resp. right) ideals of $\mathcal{A}$ is dense in $\mathcal{A}$.*
(3) *$\mathcal{A}$ is $*$–isomorphic to a C^*–subalgebra of $\mathcal{K}(H)$ for some Hilbert space.*
(4) *The spectrum of every maximal commutative C^*–algebra of $\mathcal{A}$ is discrete.*
(5) *For each $x \in \mathcal{A}_{sa}$, every non-zero point of $\sigma(a)$ is isolated in $\sigma(a)$.*

Definition B.1.3. Let $\mathcal{A}$ be a separable C^*–algebra. We call $\mathcal{A}$ is an AF–algebra if for any $\epsilon > 0$ and $a_1, \cdots, a_n \in A$, there exists a finite dimensional C^*–subalgebra $\mathcal{B} \subset A$ and $b_1, \cdots, b_n \in \mathcal{B}$ such that $\|a_i - b_i\| < \epsilon$, $i = l, \cdots, n$.

Definition B.1.4. A projection p in a C^*–algebra $\mathcal{A}$ is called infinite if there is a projection q in $\mathcal{A}$ and a partial isometry $s \in \mathcal{A}$ such that $p = s^*s$

and $ss^* \leq q$, $ss^* \neq q$. According to [Cuntz (1981)], we say that $\mathcal{A}$ is purely infinite if for any $a \in \mathcal{A}_+\backslash\{0\}$, $\overline{a\mathcal{A}a}$ contains an infinite projection.

Definition B.1.5. Let $\mathcal{A}$ be a non–unital C^*–algebra. $\mathcal{A}$ is called to be σ–unital if $\mathcal{A}$ has a sequential approximate unit.

Definition B.1.6. Let $\mathcal{A}$ be a non–unital C^*–algebra and (π_u, H_u) be the universal representation of $\mathcal{A}$. An element $x \in B(H_u)$ is called a (two–sided) multiplier for $\mathcal{A}$ if $x\pi_u(\mathcal{A}) \subset \pi_u(\mathcal{A})$ and $\pi_u(\mathcal{A})x \subset \pi_u(\mathcal{A})$. The set of all these is called the multiplier algebra for $\mathcal{A}$, i.e.,

$$M(\mathcal{A}) = \{x \in B(H_u) |\ x\pi_u(a),\ \pi_u(a)x \in \pi_u(\mathcal{A}), \forall\, a \in \mathcal{A}\}.$$

Clearly, $M(\mathcal{A})$ is a C^*–subalgebra of $B(H_u)$. We may regard $\mathcal{A}$ as a C^*–subalgebra of $M(\mathcal{A})$ by identifying a with $\pi_u(a)$, $\forall\, a \in \mathcal{A}$ and moreover, $\mathcal{A}$ is an essential closed ideal of $M(\mathcal{A})$, that is, for any $x \in M(\mathcal{A})$ with $x\mathcal{A} = \{0\}$, we have $x = 0$.

From [Lin (1995b)] and [Zhang (1992)], we have

Proposition B.1.7. *Let $\mathcal{A}$ be a non–unital and σ–unital C^*–algebra. If $\mathcal{A}$ satisfies one of following conditions, then* $\mathrm{RR}(M(\mathcal{A})) = 0$:

(1) $\mathrm{RR}(\mathcal{A}) = 0$, $\mathrm{tsr}(\mathcal{A}) = 1$ *and* $\{e^{ia} | a \in \tilde{\mathcal{A}}\}$ *is dense in* $U(\tilde{\mathcal{A}})$;
(2) $\mathcal{A}$ *is purely infinite simple with* $K_1(\mathcal{A}) = \{0\}$;

where $K_1(\mathcal{A})$ is the K_1-group of $\mathcal{A}$, defined in §B.3.

B.2 Lifting problems about projections

Throughout the section, $\mathcal{A}$ is a unital C^*–algebra and $\mathcal{J}$ is a closed ideal of $\mathcal{A}$. Let $\mathcal{B} = \mathcal{A}/\mathcal{J}$ be the quotient C^*–algebra and let $\pi\colon \mathcal{A} \to \mathcal{B}$ be the quotient mapping.

Theorem B.2.1 ([Brown and Pedersen (1991)], [Zhang (1992)]). *Suppose that* $\mathrm{RR}(\mathcal{J}) = \mathrm{RR}(\mathcal{B}) = 0$. *Then* $\mathrm{RR}(\mathcal{A}) = 0$ *iff for every projection $q \in \mathcal{B}$, there is a projection $p \in \mathcal{A}$ such that $\pi(p) = q$.*

Corollary B.2.2. *Suppose that* $\mathrm{RR}(\mathcal{A}) = 0$. *Let q be a projection in $\mathcal{B}$ and p be a projection in $\mathcal{A}$ with $\pi(p) = q$. Assume that there are mutually orthogonal projections $q_1, \cdots, q_n$ in $\mathcal{B}$ such that $q = \sum\limits_{i=1}^{n} q_i$. Then there are* *mutually orthogonal projections* $p_1, \cdots, p_n$ *in* $\mathcal{A}$ *such that* $p = \sum_{i=1}^{n} p_i$ *and* $\pi(p_i) = q_i$, $i = 1, \cdots, n$.

Proof. Since $\pi(p\mathcal{A}p) = q\mathcal{B}q$ and $p\mathcal{A}p/p\mathcal{J}p = q\mathcal{B}q$, it follows from Theorem 6.4.3 (1) that $\mathrm{RR}(p\mathcal{J}p)$, $\mathrm{RR}(\mathcal{B})$ and $\mathrm{RR}(q\mathcal{B}q)$ are all zero. Then by Theorem B.2.1, there is a projection $p_1 \in p\mathcal{A}p$ such that $\pi(p_1) = q_1 \in q\mathcal{B}q$. Similarly, there exists a projection $p_2 \in (p - p_1)\mathcal{A}(p - p_1)$ such that $\pi(p_2) = q_2 \in (q - q_1)\mathcal{B}(q - q_1)$.

In this way, we can find mutually orthogonal projections $p_1, \cdots, p_{n-1}, p'_n \in p\mathcal{A}p$ such that $\pi(p_i) = q_i$, $i = 1, \cdots, n-1$ and $\pi(p'_n) = q_n$. Thus, $\pi(p_1 + \cdots + p_{n-1} + p'_n) = \pi(p)$. Put $p_0 = p - p_1 - \cdots - p_{n-1} - p'_n$. Then p_0 is a projection in $p\mathcal{J}p$. Take $p_n = p_0 + p'_n$. Then $p_1, \cdots, p_n$ are mutually orthogonal and $\sum_{i=1}^{n} p_i = p$. □

Theorem B.2.3 ([Choi (1983)]). *Let* $\mathcal{J}$ *be an AF–algebra. Then for any projection* q *in* $\mathcal{B}$*, there is a projection* p *in* $\mathcal{A}$ *such that* $\pi(p) = q$.

B.3 K–theory for C^*–algebras

This section is based on [Blackadar (1998)], [Lin (2001)] and [Rørdam *et al.* (2000)]. Throughout the section, $\mathcal{A}$ is always a C^*–algebra.

Let 0_n denote the $n \times n$ matrix all of whose entries are 0. If $\mathcal{A}$ is unital, we denote by 1_n the unit of $M_n(\mathcal{A})$. For any $n \in \mathbb{N}$, we have $M_n(\mathcal{A}) \subset M_{n+1}(\mathcal{A})$ in the sense that $a \mapsto \mathrm{diag}(a, 0)$, $\forall\, a \in M_n(\mathcal{A})$. Put $M_\infty(\mathcal{A}) = \bigcup_{n=1}^{\infty} M_n(\mathcal{A})$ and $\mathcal{P}(\mathcal{A}) = \{p \in M_\infty(\mathcal{A}) |\, p \text{ is a projection}\}$.

Let p, $q \in \mathcal{P}(\mathcal{A})$. Then there is $m \in \mathbb{N}$ such that p, $q \in M_m(\mathcal{A})$. Set $p \oplus q = \mathrm{diag}(p, q) \in M_{2m}(\mathcal{A})$. We say p and q are equivalent, and write $p \sim q$, if there is v in $M_m(\mathcal{A})$ such that $p = v^*v$ and $q = vv^*$ (see Definition 2.4.13).

Let p, q, $p', q' \in \mathcal{P}(\mathcal{A})$. Then

(1) If $p \sim p'$ and $q \sim q'$, then $p \oplus q \sim p' \oplus q'$.
(2) $p \oplus q \sim q \oplus p$.

Suppose that $\mathcal{A}$ has unit 1. Let p, $q \in \mathcal{P}(\mathcal{A})$. p and q are called to be stably equivalent, denoted as $p \approx q$, if there are n, $m \in \mathbb{N}$ such that $p \oplus 1_n$, $q \oplus 1_n \in M_m(\mathcal{A})$ and $p \oplus 1_n \sim q \oplus 1_n$ in $M_m(\mathcal{A})$. We let $[p]$ denote the equivalence class of p with respect to "$\approx$" in $\mathcal{P}(\mathcal{A})$ and set

$V(\mathcal{A}) = \{[p] \mid p \in \mathcal{P}(\mathcal{A})\}$. We can define an addition "+" on $V(\mathcal{A})$ by

$$[p] + [q] = [p \oplus q], \quad \forall p, q \in \mathcal{P}(\mathcal{A}).$$

Define an equivalence relation "$\asymp$" on $V(\mathcal{A}) \times V(A)$ by $(x_1, y_2) \asymp (x_2, y_2)$ if there exists $z \in V(\mathcal{A})$ such that $x_1 + y_2 + z = x_2 + x_1 + z$. Write $K_0(\mathcal{A})$ for the quotient $V(\mathcal{A}) \times V(\mathcal{A})/ \backsim$ and let $< x, y >$ denote the equivalence of $(x, y) \in V(\mathcal{A}) \times V(A)$ with respect to "$\asymp$". The operation

$$< x_1, y_1 > + < x_2, y_2 >=< x_1 + x_2, y_1 + y_2 >, \ \forall (x_1, y_1), (x_2, y_2) \in V(\mathcal{A})$$

is well-defined and turns $(K_0(\mathcal{A}), +)$ into an Abelian group with

$$- < x, y >=< y, x > \quad \text{and} \quad < x, x >= 0, \ \forall x, y \in V(\mathcal{A}).$$

Since $x \mapsto < x, 0 >$ of $V(\mathcal{A})$ to $K_0(\mathcal{A})$ is a homomorphism and is injective, we write $K_0(\mathcal{A})$ as $K_0(\mathcal{A}) = \{[p] - [q] \mid p, q \in \mathcal{P}(\mathcal{A})\}$.

$K_0(\mathcal{A})$ is called the K_0–group of $\mathcal{A}$.

If $\mathcal{A}$ is non–unital, we let $\pi\colon \tilde{\mathcal{A}} \to \mathbb{C}$ be the canonical homomorphism. Let π_n be the induced homomorphism of π on $M_n(\tilde{\mathcal{A}})$. Then we have a homomorphism $\pi_*\colon K_0(\tilde{\mathcal{A}}) \to K_0(\mathbb{C})$ given by

$$\pi_*([p] - [q]) = [\pi_n(p)] - [\pi_m(q)], \forall n, m \in \mathbb{N}, \ p \in M_n(\mathcal{A}), \ q \in M_m(\mathcal{A}).$$

In this case, the K_0–group $K_0(\mathcal{A})$ of $\mathcal{A}$ is defined by $K_0(\mathcal{A}) = \operatorname{Ker} \pi_*$. We can express $K_0(\mathcal{A})$ as the set $\{[p_k] - [p] \in K_0(\tilde{\mathcal{A}}) | \, p - p_k \in M_n(\mathcal{A})\}$, where $p_k = \operatorname{diag}(1_k, 0_{n-k})$.

Let $\mathcal{A}$ be a unital C^*–algebra. Put $U_n(\mathcal{A}) = U(M_n(\mathcal{A}))$ and $U_n^0(\mathcal{A}) = U_0(M_n(\mathcal{A}))$. Then we have $U_n(\mathcal{A}) \subset U_{n+1}(\mathcal{A})$ and $U_n^0(\mathcal{A}) \subset U_{n+1}^0(\mathcal{A})$ in the sense that $x \mapsto \operatorname{diag}(x, 1)$. Put

$$U_\infty(\mathcal{A}) = \bigcup_{n=1}^{\infty} U_n(\mathcal{A}), \quad U_\infty^0(\mathcal{A}) = \bigcup_{n=1}^{\infty} U_n^0(\mathcal{A}).$$

Let $u \in U_n(\mathcal{A})$ and $v \in U_m(\mathcal{A})$. We say u and v are equivalent in $U_\infty(\mathcal{A})$, if there exist $k_1, k_2 \in \mathbb{N}$ such that $n + k_1 = m + k_2$ and $\operatorname{diag}(u, 1_{k_1}) \operatorname{diag}(v^*, 1_{k_2}) \in U_{n+k_1}^0(\mathcal{A})$. Let $[u]$ denote the equivalence class of u with respect to this equivalence relation. Set $K_1(\mathcal{A}) = \{[u] \mid u \in U_\infty(A)\}$. Define an addition on $K_1(\mathcal{A})$ by

$$[u] + [v] = [\operatorname{diag}(u, v)], \quad \forall u, v \in U_\infty(\mathcal{A}).$$

This addition is well–defined because $\operatorname{diag}(u, v)$ is equivalent to $\operatorname{diag}(v, u)$ and $\operatorname{diag}(\operatorname{diag}(u, v), 1_s)$ is equivalent to $\operatorname{diag}(\operatorname{diag}(u, 1_k)\operatorname{diag}(v, 1_p), 1_m)$ for some $s, k, p, m \in \mathbb{N}$ by Whitehead Lemma.

Therefore, we call $K_1(\mathcal{A})$ is the K_1–group of $\mathcal{A}$. If $\mathcal{A}$ is non–unital, we set $K_1(\mathcal{A}) = K_1(\tilde{\mathcal{A}})$.

Now we state some important properties of the K_0–groups and K_1–groups of C^*–algebras as follows.

(i) **Functoriality**. Let $\mathcal{B}$ be a C^*-algebra and let $\phi\colon \mathcal{A} \to \mathcal{B}$ be a $*$-homomorphism. Then we have homomorphisms $\phi_n\colon M_n(\mathcal{A}) \to M_n(\mathcal{B})$ given by $\phi_n((a_{ij})_{n\times n}) = (\phi(a_{ij})_{n\times n})$, $\forall\ (a_{ij})_{n\times n} \in M_n(\mathcal{A})$, $n \in \mathbb{N}$. $\{\phi_n\}$ induces homomorphisms $K_i(\phi)\colon K_i(\mathcal{A}) \to K_i(\mathcal{B})$, $i = 0, 1$.

(ii) **Homotopy invariance**. Let $\mathcal{B}$ be a C^*-algebra and ϕ, $\psi\colon \mathcal{A} \to \mathcal{B}$ be a $*$-homomorphisms. Suppose that ϕ is homotopic to ψ, that is, there is a family of $*$-homomorphisms $\{\rho_t | t \in [0,1]\}$ from $\mathcal{A}$ to $\mathcal{B}$ such that $\rho_0 = \phi$, $\rho_1 = \psi$ and $t \mapsto \rho_t(a)$ is a norm continuous mapping from $[0,1]$ to $\mathcal{B}$ for every fixed a in $\mathcal{A}$. Then $K_i(\phi) = K_i(\psi)$, $i = 0, 1$.

(iii) **Periodicity**. Set

$$S\mathcal{A} = C_0(\mathbb{R}, \mathcal{A}) = \{f\colon \mathbb{R} \to \mathcal{A} \text{ continuous}, \lim_{x\to\infty} \|f(x)\| = 0\}.$$

$S\mathcal{A}$ is a C^*-algebra under the norm $\|f\| = \sup_{x\in\mathbb{R}} \|f(x)\|$, $f \in S\mathcal{A}$. We call $S\mathcal{A}$ the suspension of $\mathcal{A}$. Then we have

$$K_0(S\mathcal{A}) \cong K_1(\mathcal{A}) \text{ and } K_1(S\mathcal{A}) \cong K_0(S\mathcal{A}).$$

(iv) **The cyclic six term exact sequence**. Let $\mathcal{J}$ be a closed two side ideal of $\mathcal{A}$ and put $\mathcal{B} = \mathcal{A}/\mathcal{J}$. Let $\pi\colon \mathcal{A} \to \mathcal{B}$ be the quotient mapping. Then we have following exact sequence:

$$\begin{array}{ccccc} K_0(\mathcal{J}) & \xrightarrow{K_0(i)} & K_0(\mathcal{A}) & \xrightarrow{K_0(\pi)} & K_0(\mathcal{B}) \\ {\scriptstyle \partial_1}\uparrow & & & & \downarrow{\scriptstyle \partial_0} \\ K_1(\mathcal{B}) & \xleftarrow[K_1(\pi)]{} & K_1(\mathcal{A}) & \xleftarrow[K_1(\pi)]{} & K_1(\mathcal{J}) \end{array}$$

where is the exponential map and ∂_1 is the index map, which are defined as follows.

Let p be a projection in $M_n(\tilde{\mathcal{B}})$ with $p - p_k \in M_n(\mathcal{B})$ for some $k \in \mathbb{N}$. Choose $a \in \mathcal{A}_{sa}$ such that $\pi_n(a) = p$. Then $\mathrm{e}^{ia} \in U_n(\tilde{\mathcal{J}})$. So we define ∂_0 by $\partial_0([p] - [p_k]) = [\mathrm{e}^a]$.

Let $v \in U_n(\tilde{\mathcal{B}})$. Pick $u \in U_{2n}^0(\mathcal{A})$ such that $\pi_{2n}(u) = \operatorname{diag}(v, v^*)$. Then $up_nu^* - p_n \in M_{2n}(\mathcal{J})$. Thus we define by $\partial_1([v]) = [up_nu^{-1}] - [p_n]$.

By the definition of ∂_1, we have

Proposition B.3.1. *Let $\mathcal{J}$, $\mathcal{A}$, $\mathcal{B}$ and π be as above. Suppose that $\mathcal{A}$ is unital and $\mathcal{J}$ admits an approximate unit consisting of projections in $\mathcal{J}$. Then for any $v \in U_n(\mathcal{B})$, there exists a partial isometry $u \in M_n(\mathcal{A})$ such that $\pi_n(u) = v$ and $\partial_1([u]) = [1_n - v^*v] - [1_n - vv^*]$.*

Definition B.3.2. A quasitrace on a $\mathcal{A}$ is a function $\tau\colon M_\infty(\mathcal{A}) \to \mathbb{C}$ such that:

(1) $0 \leq \tau(x^*x) = \tau(xx^*)$ for all x in $M_\infty(\mathcal{A})$.
(2) τ is linear on commutative $*$–subalgebras of $M_\infty(\mathcal{A})$.
(3) If $x = a + ib$ with $a, b \in (M_\infty(\mathcal{A}))_{sa}$, then $\tau(x) = \tau(a) + i\tau(b)$.

Bibliography

Apostol, C. (1976). Inner derivations with closed range, *Rev. Roum. Math. Pures et Appl.* **21**, 3, pp. 249–265.

Apostol, C. (1985). The reduced minimum modulus, *Mich. Math. J.* **32**, 3, pp. 279–294.

Arghiriade, E. (1968). Sur linverse generalisee dun operateur lineaire dans les espaces de Hilbert, *Atti Accad. Naz. Lincei Rend. Cl. Sci. Fis. Mat. Natur.* **45**, 8, pp. 471–477.

Aubin, J. P. and Cellina, A. (1984). *Differential Inclusions* (Springer–Verlag, Berlin).

Badea, C. and Mbekhta, M. (1997). Generalized inverses and the maximal radius of regularity of a Fredholm operator, *Integr. Equat. Oper. Th.* **28**, pp. 133–146.

Badea, C. and Mbekhta, M. (1999). Compressions of resolvents and maximal radius of regularity, *Trans. Amer. Math. Soc.* **351**, pp. 2949–2960.

Bartels, S. G. (1991). *Two Topics in Matrix Analysis: Structured Sensitivity for Vandermonde-like Systems and a Subgradient Method for Matrix Norm Estimation*, Ph.D. thesis, Univ. of Dundee.

Beauzamy, B. (1985). *Introduction to Banach Spaces and their Geometry*, 2nd edn., Math. Studies 68 (North–Holland Publishing Company, Amsterdam New York Oxford).

Ben-Israel, A. and Greville, T. N. E. (2003). *Generalized Inverse: Theory and Applications*, 2nd edn. (Springer–Verlag, New York).

Berger, M. (1977). *Non-linearity and Functional Analysis* (Academic press, New York).

Blackadar, B. (1998). *K-Theory for Operator Algebras*, 2nd edn. (CUP, London).

Bott, R. and Duffin, R. J. (1953). On the algebra of networks, *Trans. Amer. Math. Soc.* **74**, pp. 99–109.

Brown, L. G. and Pedersen, G. K. (1991). C^*–alegbras of real rank zero, *J. Funct. Anal.* **98**, pp. 131–149.

Brown, L. G. and Pedersen, G. K. (1995). On the geometry of the unit ball of a C^*–algebra, *J. Reine Angew. Math.* **469**, pp. 113–147.

Campbell, S. L. and Meyer, C. D. (1979). *Generalized Inverses of Linear Transformations* (Pitman, London, New York).

Caradus, S. R. (1978). *Generalized Inverses and Operator Theory*, no. 50 in Queen's paper in pure and applied mathematics (Quenn's University, Kingston, Ontario).

Castro-González, N., Dopazo, M. and Martinez-Serrano, M. (2009). On the Drazin inverse of the sum of two operators and its application to operator matrices, *J. Math. Anal. Appl.* **350**, pp. 207–215.

Castro-González, N. and Koliha, J. J. (2004). New additive results for the g–Drazin inverse, *Proc. Royal Soc. Edinburgh Sect. A* **134**, pp. 1085–1097.

Castro-González, N., Koliha, J. J. and Rakočević, V. (2002a). Continuity and general perturbation of the Drazin inverse for closed linear operators, *Abstract and Applied Analysis* **76**, pp. 335–347.

Castro-González, N., Koliha, J. J. and Wei, Y. (2002b). Error bounds for perturbation of the Drazin inverse of closed operators with equal spectral projections, *Applicable Analysis* **81**, 4, pp. 915–928.

Castro-González, N. and Martínez-Serrano, M. F. (2010). Expressions for the g–Drazin inverse of additive perturbed elements in a Banach algebra, *Linear Algebra Appl.* **432**, pp. 1885–1895.

Castro-González, N. and Vélez-Cerrada, J. (2008). On the perturbation of the group generalized inverse for a class of bounded operators in Banach spaces, *J. Math. Anal. Appl.* **341**, pp. 1213–1223.

Chen, G., Liu, G. and Xue, Y. (2002). Perturbation theory for the generalized Bott–Duffin inverse and its applications, *Applied Math. Comput.* **129**, pp. 145–155.

Chen, G., Liu, G. and Xue, Y. (2003). Perturbation analysis of the generalized Bott–Duffin inverse of L–zero matrices, *Linear Multilinear Algebra* **51**, 1, pp. 11–20.

Chen, G., Wei, M. and Xue, Y. (1996). Perturbation analysis of the least square solution in hilbert spaces, *Linear Algebra Appl.* **244**, pp. 69–80.

Chen, G., Wei, Y. and Xue, Y. (2004). The generalized conditi,on numbers of bounded linear operators in Banach spaces, *J. Aust. Math. Soc.* **76**, pp. 281–290.

Chen, G. and Xue, Y. (1997). Perturbation analysis for the operator equation $Tx = b$ in Banach spaces, *J. Math. Anal. Appl.* **212**.

Chen, G. and Xue, Y. (1998). The expression of generalized inverse of the perturbed operators under type I perturbation in Hilbert spaces, *Linear Algebra Appl.* **285**, pp. 1–6.

Chen, Y. (1990). The generalized Bott–Duffin inverse and its applications, *Linear Algebra Appl.* **134**, pp. 71–91.

Chen, Y., Hu, X. and Xu, Q. (2009). The Moore–Penrose inverse of $A - XY^*$, *J. Shanghai Normal Univ.* **38**, pp. 15–19.

Choi, M. (1983). Lifting pojections from quotient C^*–algebras, *J. Operator Theory* **10**, pp. 21–30.

Christensen, O. (1999). Operators with closed range, pseudo–inverses, and perturbation of frames for a subspace, *Canad. Math. Bull.* **42**, 1, pp. 37–45.

Christensen, O. (2003). *An Introduction to Frames and Riesz Bases* (Birkahuser, Boston).

Cline, R. E. (1965). Representations for the generalized inverses of sums of matrices, *SIAM J. Numer. Anal.(Ser. B)* **2**, pp. 99–114.

Conway, J. B. (2000). *A Course in Operator Theory, Graduate Studies in Mathematics*, Vol. 21 (Amer. Math. Soc.).

Cowen, M. J. and Douglas, R. G. (1978). Complex geometry and operator theory, *Acta Math.* **141**, pp. 187–261.

Cuntz, J. (1981). K–Theory for certain C^*–algebras, *Ann. of Math.* **113**, pp. 181–197.

Demmel, J. W. (1987). On condition numbers and the distance to the nearest ill–posed problem, *Numer. Math.* **51**, pp. 251–289.

Deng, C. (2009). On the invertibility of the operator $A - XB$, *Numer. Linear Algebra Appl.* **16**, pp. 817–831.

Deng, C. and Du, H. (2009). Representations of the Moore–Penrose inverse of 2×2 block operator valued matrices, *J. Korean Math. Soc.* **46**, 6, pp. 1139–1150.

Deng, C. and Du, H. (2010). Representations of the Moore–Penrose inverse for a class of 2–by–2 block operator valued partial matrices, *Linear and Multilinear Algebra* **58**, 1, pp. 15–26.

Deng, C. and Wei, H. (2011). A note on additive results for the generalized Drazin inverse, *Linear and Multilinear Algebra* , pp. 1–11.

Ding, J. (2001). Lower and upper bounds in the perturbation of general linear algebraic equations, *Applied Math. Letters* **14**, pp. 49–52.

Ding, J. and Huang, L. (1996). Perturbation of generalized inverses of linear operators in Hilbert spaces, *J. Math. Anal. Appl.* **198**, pp. 505–516.

Ding, J. and Huang, L. (1999). A generalization of a classic theorem in the perturbation theory for linear operators, *J. Math. Anal. Appl.* **239**, pp. 118–123.

Ding, J. and Wei, Y. (2002). Relative errors versus residuals of approximate solutions of weighted least squares problems in Hilbert spaces, *Compters Math. Appl.* **44**, pp. 407–411.

Dixmier, J. (1977). *C^*-algebras* (North–Holland, Amsterdam).

Djordjević, D. (1994). Regular and T–Fredholm elements in Banach algebras, *Publ. Math. Inst. (Belgrade)* **56**, 70, pp. 90–94.

Djordjević, D. (2001). Unified approach to the reverse order rule for generalized inverses, *Acta Sci. Math. (Szeged)* **167**, pp. 761–776.

Djordjević, D. (2007). Further results on the reverse order law for generalized inverses, *SIAM J. Matrix Anal. Appl.* **29**, 4, pp. 1242–1246.

Djordjević, D. and Dinčić, N. (2010). Reverse order law for the Moore–Penrose inverse, *J. Math. Anal. Appl.* **361**, pp. 252–261.

Djordjević, D. and Stanimirović (2001). Splitting of operators and generalized inverses, *Publ. Math. Debrecen* **59**, pp. 147–159.

Djordjević, D., Stanimirović and Wei, Y. (2004). The representation and approximations of outer generalized inverses, *Acta Math. Hungar.* **104**, 1–2, pp. 1–26.

Djordjević, D. and Wei, Y. (2004). Operators with equal projections related to their generalized inverses, *Applied Math. Comput.* **155**, 3, pp. 655–664.

Drazin, M. (1958). Pseudo–inverse in associative rings and semigroups, *Am. Math. Mon.* **65**, pp. 506–514.

Du, F. and Xue, Y. (2009). Expression of the bounded linear operator $A^{(2)}_{T,S}$ in Banach spaces, *Numer. Math. J. Chinese Univ.* **31**, 3, pp. 240–245.

Du, F. and Xue, Y. (2010). Expression of the Moore-Penrose inverse of $A - XY^*$, *J. East China Norm. Univ. (Natur. Sci.)* , 5, pp. 33–37.

Du, F. and Xue, Y. (2011). perturbation analysis of the outer generalized inverse with with prescribed range and kernel, Submitted.

Fredholm, I. (1903). sur une classe d'équations fonctionnelles, *Acta Math.* **27**, pp. 365–390.

Fritzsche, K. and Grauert, H. (2002). *From Holomorphic Functions to Complex Manifolds, Graduate Texts in Mathematics*, Vol. 213 (Spinger–Verlag, New York, Inc.).

Goldberg, S. (1985). *Unbounded Linear Operators* (Dover, New York).

Gopalraj, P. and Stroh, A. (2004). On the essential bound of elements in von Neumann algebras, *Integr. Equ. Oper. Th.* **49**, pp. 379–386.

Grobler, J. J. and Raubenheimer, H. (2008). The index for Fredholm elements in a Banach algebra via a trace, *Studia Math.* **187**, 3, pp. 281–297.

Hagen, R., Roch, S. and Silbermann, B. (2001). *C^*-Algebras and Numberical Analysis*, no. 236 in Pure and Applied mathematics (Marcel Dekker, Inc., New York Basel).

Harte, R. (1982). Fredholm theory relative to a Banach algebra homomorphism, *Math. Z.* **179**, pp. 431–436.

Harte, R. (1987). Regular boundary elements, *Proc. Amer. Math. Soc.* **99**, pp. 328–330.

Harte, R. and Mbekhta, M. (1992). Generalized inverses inverses in C^*-algebras, *Studia Math.* **103**, pp. 71–77.

Harte, R. and Mbekhta, M. (1993). Generalized inverses inverses in C^*-algebras (II), *Studia Math.* **106**, pp. 129–138.

Hartwig, R., Wang, G. and Wei, Y. (2001). Some additive results on Drazin inverse, *Linear Algebra Appl.* **322**, pp. 207–217.

Hestenes, M. R. (1961). Relative self-adjoint operators in Hilbert space, *Pacific J. Math.* **11**, pp. 1315–1357.

Higham, D. and Higham, N. (1992). Componentwise perturbation theory for linear systems with multiple right–hand sides, *Linear Algebra Appl.* **174**, pp. 111–129.

Huang, D. (1992). Generalized inverses over Banaeh algebras, *Integr. Equat. Oper. Th.* **15**, pp. 454–469.

Huang, D. (1993). Group inverses and Drazin inverses over Banach algebras, *Integr. Equat. Oper. Th.* **17**, pp. 54–67.

Huang, Q. (2011). On perturbations for oblique projection generalized inverses of closed linear operators in Banach spaces, *Linear Algebra Appl.* **434**, pp. 2468–2474.

Huang, Q. and Ma, J. (2005). Continuity of generalized inverses of linear operators in Banach spaces and its applications, *Appl. Math. Mech.* **26**, pp. 1657–1663.

Huang, Q. and Zhai, W. (2011). Perturbations and expressions for generalized inverses in Banach spaces and Moore–Penrose inverses in Hilbert spaces of closed linear operators, *Linear Algebra Appl.* **435**, pp. 117–127.

Izumino, S. (1982). The product of operators with closed range and an extension of the reverse order law, *Tôhoku Math. J.* **34**, pp. 43–52.

Izumino, S. (1983). Convergence of generalized inverses and spline projectors, *J. Approx. Theory* **38**, pp. 269–278.

Jeong, J. and Osaka, H. (1998). Extremally rich C^*–crossed products and the cancellation property, *J. Aust. Math. Soc.* **64**, pp. 285–301.

Jiang, C. and Wang, Z. (1998). *Strongly Irreducible Operators on Hiblert Spaces, Pitman Research Notes in Mathematics Series*, Vol. 389 (Addison–Wesley–Longman Company).

Kaniuth, E. (2009). *A Course in Commutative Banach Algebras, Graduate Texts in Mathematics*, Vol. 246 (Springer, New York).

Kaplansky, I. (1948). Dual rings, *Ann. of Math.* **49**, pp. 689–701.

Kato, T. (1984). *Perturbation Theory for Linear Operator*, 2nd edn. (Springer–Verlag, Berlin Heidelberg, New York, Tokyo).

Koliha, J. J. (1996). A generalized drazin inverse, *Glasgow Math. J.* **38**, pp. 367–381.

Koliha, J. J. and Rakočević, V. (1998). Continuity of the drazin inverse II, *Studia Math.* **131**, pp. 167–177.

Koliha, J. J. and Tran, T. D. (2001). The Drazin inverse for closed linear operators and the asymptotic convergence of C_0–semigroups, *J. Operator Theory* **46**, pp. 323–336.

Kurt, S. and Riedel, A. (1991). A Shermen–Morrison–Woodbury identity for rank augmenting matrices with application to centering, *SIAM J. Math. Anal.* **12**, 1, pp. 80–95.

Li, Y. (2008). The Moore–Penrose inverses of products and differences of projections in a C^*–algebra, *Linear Algebra Appl.* **428**, pp. 1169–1177.

Lin, H. (1995a). C^*–algebra extensions of $C(X)$, *Mem. Amer. Math. Soc.* **115**, 550.

Lin, H. (1995b). Generalized Weyl–von Neumann theorems II, *Math. Scand.* **77**, pp. 129–147.

Lin, H. (2001). *An Introduction to the Classification of Amenable C^*–Algebras* (World Scientific, Singapore).

Ma, J. (1999). $(1,2)$ inverses of operators between Banach spaces and local conjugacy theorem, *Chin. Ann. of Math.* **20B**, 1, pp. 57–62.

Ma, J. (2000). Local conjugacy theorem, rank theorems in advanced calculus and a generalized principle for constructing banach manifolds, *Sci. China Ser. A* **43**, 12, pp. 1233–1237.

Ma, J. (2001). A generalized preimage theorem in global analysis, *Sci. China Ser. A* **44**, 3, pp. 299–303.

Ma, J. (2008). A generalized transversality in global analysis, *Pacific J. Math.* **236**, 2, pp. 357–371.

Markus, A. (1959). On some properties of linear operators connected with the notion of the gap (Russian), *UChen. Zap. Kishinev. Gos. Univ.* **39**, pp. 265–272.

Mbekhta, M. (1993). Generalized spectrum and a problem of Apostol, *Proc. Amer. Math. Soc.* **118**, pp. 857–859.

Moore, E. H. (1920). On the reciprocal of the general algebraic matrix, *Bull. Amer. Math. Soc.* **26**, pp. 394–395.

Muller, V. (2007). *Spectral Theory of Linear Operators and Spectral Systems in Banach Algebras* (Birkhauser, Basel Boston Berlin London).

Murphy, G. J. (2006). Topological and analytical indices in C^*–algebras, *J. Funct. Anal.* **234**, pp. 261–276.

Nashed, M. Z. and Zhao, Y. (1992). The Drazin inverse for singular evolution equations and partial differential operators, *World Sci. Ser. Appl. Anal.* **1**, pp. 441–456.

Nashed, N. Z. (1976a). *Perturbations and approximations for generalized inverses and linear operator equations* (Academic Press, New York, San Franciso, London), pp. 325–396.

Nashed, N. Z. (1976b). *A Unified operator theory of generalized Inverses* (Academic Press, New York, San Franciso, London), pp. 1–110.

Patrício, P. and Costa, A. D. (2009). On the Drazin index of regular elements, *Central Euro. J. Math.* **7**, 2, pp. 200–205.

Pedersen, G. K. (1979). *C^*-Algebras and their Automorphism Groups* (Academic Press, London New York San Francisco).

Pedersen, G. K. (1980). The linear span of projections in simple C^*–algebras, *J. Operator Theory* **4**, pp. 289–296.

Penrose, R. (1955). A generalized inverse for matrices, *Proc. Cambridge Philos. Soc.* **51**, pp. 406–413.

Petryshyn, W. V. (1967). On generalized inverses and on the uniform convergence of $(I - \beta K)^n$ with application to iterative methods, *J. Math. Anal. Appl.* **18**, pp. 417–439.

Rakočević, V. (1992). A note on regular elements in Calkin algebras, *Collect. Math.* **43**, 1, pp. 37–42.

Rakočević, V. (1999). On the continuity of the Drazin inverse, *J. Operator Theory* **41**, pp. 55–68.

Rakočević, V. and Wei, Y. (2001). The perturbation theory for the Drazin inversean its applications, II, *J. Asutral. Math. Soc.* **70**, pp. 189–197.

Rickart, C. E. (1960). *General Theory of Banach Algebras* (Robert E. Krieger Publishing CO., INC., New York).

Rieffel, M. A. (1983). Dimension and stable rank in the K–theory of C^*–algebras, *Proc. London Math. Soc.* **46**, pp. 301–333.

Roch, S. and Silbermann, B. (1999). Continuity of generalized inverses in Banach algebras, *Studia Math.* **136**, 3, pp. 197–227.

Rørdam, M. (1991). On the structure of simple C^*–algebras tensored with a UHF–algebra, *J. Funct. Anal.* **100**, pp. 1–17.

Rørdam, M., Larsen, F. and Laustsen, N. (2000). *An Introduction to K–Theory for C^*–Algebras, London Math. Soc. Student Texts*, Vol. 49 (Cambridge University Press).

Rudin, W. (1973). *Functional Analysis* (McGraw-Hill Book Company, New York).

Stewart, G. W. (1977). On the perturbation of pseudo–inverses, projections, and linear least squares problems, *SIAM Rev.* **19**, pp. 634–662.

Stănică, D. (2010). The stability of some operator equations in Hilbert spaces, *Annals of the University of Bucharest (math. ser.)* **1**, pp. 155–164.

Taylor, A. E. (1958). *Introduction to Functional Analysis* (John Wiley & Sons, Inc., New York).

Tseng, Y.-Y. (1949). Generalized inverses of unbounded operators between two unitary spaces, *Doklady Akad. Nauk SSSR (N.S.)* **67**, pp. 431–434.

Wang, J., Li, Z. and Xue, Y. (2009). Perturbation analysis for the minimal norm solution of a consistent operator equation in Banach spaces, *J. East China Norm. Univ. (Natur. Sci.)* , 1, pp. 48–52.

Wang, Y. and Zhang, H. (2007). Perturbation analysis for oblique projection generalized inverses of closed linear operators in Banach spaces, *Linear Algebra Appl.* **426**, pp. 1–11.

Wei, Y. and Ding, J. (2001). Representations for Moore–Penrose inverses in Hilbert spaces, *Appl. Math. Letters* **14**, pp. 599–604.

Xu, Q. and Hu, X. (2008). Particular formulae for the Moore–Penrose inverses of the partitioned bounded linear operators, *Linear Algebra Appl.* **428**, pp. 2941–2946.

Xu, Q., Song, C. and Wei, Y. (2010a). The stable perturbtation of the Drazin inverse of the square matrices, *SIAM J. Matrrix Anal. Appl.* **31**, 3, pp. 1507–1520.

Xu, Q., Wei, Y. and Gu, Y. (2010b). Sharp norm-estimations for Moore–Penrose inverses of stable perturbations of Hilbert C^*–module operators, *SIAM J. Numer. Anal.* **47**, 6, pp. 4735–4758.

Xue, Y. (1989). Some peoperties of the C^*–algebras generated by $B_n(\Omega)$ class operators, *Chin. Ann. of Math.* **10A**, 4, pp. 441–446.

Xue, Y. (2007a). (APD)–property of C^*–algebras by extensions of C^*–algebras with (APD), *Proc. Amer. Math. Soc.* **135**, 3, pp. 705–711.

Xue, Y. (2007b). The reduced minimum modulus in C^*–algebras, *Integr. Equat. Oper. Th.* **59**, 2, pp. 269–280.

Xue, Y. (2007c). Stable perturbation in Banach algebras, *J. Aust. Math. Soc.* **83**, pp. 1–14.

Xue, Y. (2008a). A new characterization of the reduced minimum modulus of an operator on Banach spaces, *Publ. Math. Debrecen* **72**, 1–2, pp. 155–166.

Xue, Y. (2008b). A note about a theorem of R. Harte, *Filomat* **22**, 2, pp. 95–98.

Xue, Y. and Chen, G. (2002). The expression of the generalized Bott–Duffin inverse and its perturbation theory, *Applied Math. Comput.* **132**, pp. 437–444.

Xue, Y. and Chen, G. (2004). Some equivalent conditions of stable perturbation of operators in Hilbert spaces, *Applied Math. Comput.* **147**, pp. 765–772.

Xue, Y. and Chen, G. (2007). Perturbation analysis for the Drazin inverse under stable perturbation in Banach space, *Missouri J. Math. Sci.* **19**, 2, pp. 106–120.

Xue, Y. and Wang, Z. (1999). A C^*–algebra approach to the irreducibility of Cowen–Douglas operators, *Chin. Ann. of Math.* **20B**, 3, pp. 1–4.

Yosida, K. (1980). *Functional Analysis* (Springer–Verlag, Berlin Heidelberg New York).

Zemáneck, J. (1982). *Geometric interpretation of the essential minimum modulus* (Birkhauser Verlag, Basel).

Zhang, S. (1992). Certain C^*–algebras with real rank zero and their corona and multiplier algebras, part I, *Pacific J. Math.* **155**, 1, pp. 169–197.

Zhang, X., Chen, G. and Xue, Y. (2005). Perturbation analysis of the generalized Bott–Duffin inverse of L–zero matrices (II), *J. East China Norm. Univ. (Natur. Sci.)* , 5–6, pp. 72–77.

Zhu, C., Cai, J. and Chen, G. (2003). Perturbation analysis for the reduced minimum modulus of bounded linear operator in Banach spaces, *Appl. Math. Comput.* **145**, pp. 13–21.

Index